AF477527

EVOLUTION OF MASSIVE STARS

EVOLUTION OF MASSIVE STARS

A Confrontation between Theory and Observation

Edited by

D. Vanbeveren

W. van Rensbergen

and

C. de Loore

Vrije Universiteit Brussel

Reprinted from *Space Science Reviews*, Vol. 66, Nos 1–4, 1994

KLUWER ACADEMIC PUBLISHERS

DORDRECHT / BOSTON / LONDON

A C.I.P. Catalogue record for this book is available from the Library of Congress

ISBN 0-7923-2799-3

Published by Kluwer Academic Publishers,
P.O. Box 17, 3300 AA Dordrecht, The Netherlands.

Kluwer Academic Publishers incorporates
the publishing programmes of
D. Reidel, Martinus Nijhoff, Dr W. Junk and MTP Press.

Sold and distributed in the U.S.A. and Canada
by Kluwer Academic Publishers,
101 Philip Drive, Norwell, MA 02061, U.S.A.

In all other countries, sold and distributed
by Kluwer Academic Publishers Group,
P.O. Box 322, 3300 AH Dordrecht, The Netherlands.

Printed on acid-free paper

All Rights Reserved
© 1994 Kluwer Academic Publishers
No part of the material protected by this copyright notice may be reproduced or
utilized in any form or by any means, electronic or mechanical,
including photocopying, recording or by any information storage and
retrieval system, without written permission from the copyright owner.

Printed in the Netherlands

TABLE OF CONTENTS.

III. STRUCTURE AND EVOLUTION OF MASSIVE STARS.

PREFACE

Massive stars occupy an exceptional place in general astrophysics. They trigger many if not all of the important processes in galactic evolution whereas due to their intrinsic brightness, they offer the (only until now) possibility to study the stellar content and stellar behaviour in distant galaxies.

The last, say, 25 years, massive stars have been the subject of numerous meetings discussing the influence of massive stars on population synthesis, the number distribution of different types of massive stars, the LBV phenomenon, WR stars, X-ray binaries, stellar winds in massive stars, chemical pecularities in massive stars, supernova explosions of massive stars and the important SN1987A event, the influence of massive stars and chemical evolution of galaxies.

It is clear that without a theory of stellar evolution, the study of these topics loses a lot of its significance. Massive star evolution therefore got a chance in these meetings, but rarely as a prime subject. The state of the art, the physical processes and the uncertainties in stellar evolution were barely touched. Even more, the influence of close binaries in all these massive star meetings slowly disappeared the last, say, 13 years without any scientific justification, although a significant fraction of stars occurs in close binaries with periods small enough so that both components will interact during their evolution. Denying the binaries or not discussing their influence on results and conclusions, makes the latter very uncertain or even completely unreliable.

Massive stars end their live as a supernova, returning nuclearly processed matter back to the interstellar medium, in this way significantly influencing the evolution of galaxies. What we learn from galaxies and from detailed observations of supernova explosions is fundamental for the evolution of massive stars.

It was the aim of the meeting in Brussels (1993) to bring together and mutually inform the communities concerned respectively with the actual state of the art of stellar evolution, single stars and close binaries and to discuss these computations in connection to extended observations of individual stars and stellar groups, supernova explosions and chemistry of galaxies. The influence of stellar wind on evolution is generally studied by including some semi–empirical formalism into the computer codes. This mass loss process is of fundamental importance for stellar evolution and it was therefore logical to dedicate a significant part of the conference to the massive star atmospheres, theory and observations.

The meeting was held at Brussels University (VUB) from 3 to 6 August in a less exotic place than many of previous massive star meetings, but animated by the inspiring breath of a culture which lives since thousand years.

The Editors.

Space Science Reviews **66**: 1–2, 1994.
© 1994 *Kluwer Academic Publishers. Printed in Belgium.*

Members of the SOC:

M. Arnould (Belgium), E.P.J. van den Heuvel (The Netherlands), C. de Jager
(The Netherlands), C. de Loore (Belgium), R. Kudritzki (Germany), A. Maeder
(Switserland), D. Vanbeveren (Belgium, Chairman), C. Wheeler (USA), A. Willis
(UK), J.P. Zahn (France).

Members of the LOC:

W. Hummel, C. de Loore, N. Rons, M. Runacres, D. Vanbeveren, W. Van
Rensbergen.

PARTICIPANTS LIST

NAME	E-MAIL
ANNUK K.	annuk@mars.aai.ee
ASIDA S.	shimon@bilbo.fis.huji.ac.il
BAGNUOLO W.	bagnuolo@chara.gsu.edu
BEAULIEU J.P.	span17649 earn@friap51
BEECH M.	a2670@nve.uwo.ca
BIANCHI L.	bianchi@stcsi.edu
BLOMME R.	ronny@atmos.oma.be
BOHANNAN B.	bruce@noao.edu
BOMANS D.	bomans@babsy.mpifr-bonn.mpg.de
BRANDSE F.	frankb@sron.ruu.nl
BRAUN H.	hartl@ibm-2.MPA-Garching.MPG.DE
BRESSAN A.	bressan@astrpd.astro.it
BROWN A.	brown@rulhsw.leidenuniv.nl
BROWNSBERGER K.	brownsbe@ucsc.colorado.edu
BRUZUAL G.	p75@vm.urz.uni-heidelberg.de
CANANZI K.	kgafa05@cc2.kuleuven.ac.be
CHIOSI C.	chiosi@astrpd.astro.it
CONTI P.	pconti%jila@colorado.bitnet
CROWTHER P.	pac@starlink.ucl.ac.uk
DAMINELI A.	daminelli%iagusp@brfapesp.bitnet
DE GREVE J.P.	jpdgreve@vnet3.vub.ac.be
DE GROOT M.	mdg@starlink.armagh.queens-belfast.ac.uk
DE JAGER C.	keesj@sron.ruu.nl
DE LOORE C.	wvanrens@vnet3.vub.ac.be
DENG L.	licai@tsmi19.sissa.it
DENISSENKOV P.	dpa@cvxastro.mpa-garching.mpg.de
DREIZLER S.	dreizler@sternwarte.uni-erlangen.d400.de
EL EID M.	meid@ibm.gwdg.de
ESTEBAN C.	cfr. Herrero
FAGOTTO F.	cfr. Chiosi
FELDMEIER A.	cfr. d. Lennon
FIGUEIREDO J.	jmfiguei@ciup1.ncc.up.pt
FREEMAN K.	kcf@mso.anu.edu.au
GABLER R.	cfr. D. Lennon
GANG T.	
GIES D.	gies@chara.gsu.edu
GLATZEL W.	wglatze@ibm.gwdg.de
GOCHERMANN J.	astrorub@ruba.rz.ruhr-uni-bochum.dbp.de
GOLDSTEIN U.	raphael@bgumail.bgu.ac.il
GREBEL E.	grebel@stsci.edu
GRISON P.	span17649 earn@friap51
GUMMERSBACH C.	dc1@vm.urz.uni-muenchen.de
HAFNER M.	c81@mvs.urz.uni-heidelberg.de
HAMANN W.R.	pas81@rz.uni-kiel.dbp.de
HAMMERSCHLAG G.	godelieve@astro.uva.nl

Space Science Reviews **66**: 3–5, 1994.

HASER S.	haser@usm.uni-mienchen.de
HEAP S.	hrsheap@stars.gsfc.nasa.gov
HERRERO A.	ahd@iac.es
HERZIG K.	kherzi@ibm.gwdg.de
HILDITCH R.	rwh@star.st-andrews.ac.uk
HUMMEL W.	whummel@vub.ac.be
JORDI C.	carme@facjn0.ub.es
KEMNADE E.	
KINGSBURGH R.	robin@bufadora.astrosen.unam.mx
KIRIAKIDIS M.	mkiriak@ibm.gwdg.de
KOVETZ A.	attay@etoile.tau.ac.il
KUDRITZKI R.	kudritzki@usm.uni-muenchen.de
LANDSMAN W.	landsman@uit.dnet.nasa.gov
LANGER N.	ntl@ibma.ipp-garching.mpg.de
LEITHERER C.	leitherer@stsci.bitnet
LEMKE M.	michael@io.as.utexas.edu
LENNON D.	djl@usm.uni-muenchen.de
LOBEL A.	alexl@ruurcl.sron.ruu.nl
LUCY L.	
LUDKE E.	
MAEDER A.	maeder@cgeuge11.bitnet
MATIAS J.	matias@srvdec.obs-mip.fr
MAZZALI P.	mazzali@vxoat.astro.it
MEYNET G.	meynet@scsun.unige.ch
MOFFAT A.	moffat@astro.umontreal.ca
MORRIS P.	cfr. Brownsberger
MOWLAVI N.	nmowlavi@astro.ulb.ac.be
NASI E.	cfr. Chiosi
NETZER N.	netzer@techunix.technion.ac.il
NIEMELA V.	virpi@iafe.edu.ar
NIEUWENHUYZEN H.	hansn@sron.ruu.nl
NOTA A.	nota@stsci.edu
OBERLACK U.	oberlack@cesrvb.dnet.nasa.gov
PARKER J.	joel@hrssun.gsfc.nasa.gov
PAULDRACH A.	cfr. d. Lennon
PENNY L.	cfr. Gies
PODSIADLOWSKI P.	ppodsi@starlink.astronomy.cambridge.ac.uk
PRIALNIK D.	
RAMINELLI A.	
RITOSSA C.	ritossa@alma02.cineca.it
ROED OEDEGAARD K.	knutjo@astro.uio.no
ROLLESTON R.	wrjr@starlink.physics.queens-belfast.ac.uk
RONS N.	nrons@vnet3.vub.ac.be
ROXBURGH I.	i.w.roxburg@qmw.ac.uk
RUELAS A.	rarm@astroscu.unam.mx
RUNACRES M.	mrunacre@vub.ac.be
SANTOLAYA REY E.	esr@iac.es
SCHAERER D.	schaerer@scsun.unige.ch

SCHMUTZ W.	schmutz@astro.phys.ethz.ch
SCHULTE-LADBECK R.	rsl@vms.cis.pitt.edu
SHARA M.	shara@stsci.edu
SIMON K.	simon@usm.uni-muenchen.de
SMITH L.	ljs@starlink.ucl.ac.uk
SMITH K.	kcs@starlink.ucl.ac.uk
SMARTT S.	cfr. Rolleston
SMOLINSKI J.	jansmol@pltumk11
SREENIVASAN R.	srs@acs.ucalgary.ca
STECHER T.	stecher@uit.dnet.nasa.gov
TALON S.	talon@iap.fr
UNDERHILL A.	undhil@geop.ubc.ca
VALLENARI A.	vallenari@astrpd.astro.it
VAN DEN HEUVEL E.P.J.	edvdh@astro.uva.nl
VAN DER HUCHT K.	karelh@sron.ruu.nl
VAN GENT J.	
VAN RENSBERGEN W.	wvanrens@vnet3.vub.ac.be
VANBEVEREN D.	wvanrens@vnet3.vub.ac.be
VENN K.	kim@zon.spa.umn.edu
VERSCHUEREN W.	vershue@rulhsw.leidenuniv.nl
VOLOSHINA I.	sai@crao.crimea.ua
VOORS R.	robertv@sron.ruu.nl
VRANCKEN M.	myriam@tena2.vub.ac.be
VREUX J.M.	
WERNER K.	pas75@rz.uni-kiel.dbp.de
WESSOLOWSKI U.	pas77@rz.uni-kiel.dbp.de
WHEELER C.	wheel@astro.as.utexas.edu
WHITELOCK P.	paw@mv.saao.ac.za
WILLIS A.	ajw@starlink.ucl.ac.uk
YAARI A.	atara@bilbo.fiz.huji.ac.il
ZAHN J.P.	zahn@iap.fr
ZAVAGNO A.	zavagno@obmara.cnrs-mrs.fr

MASSIVE STARS: SETTING THE STAGE

C. DE JAGER

Laboratory for Space Research, Sorbonnelaan 2, 3584 CA Utrecht, The Netherlands

Abstract. The paper gives a summary of the situation mid-1993 of theory and observations regarding massive stars. I describe: stellar mass loss and its implications, pre-main-sequence evolution, the main sequence, problems of atmospheric instability, Luminous Blue Supergiants, Yellow Hypergiants, Wolf–Rayet stars and supernovae.

Key words: mass-loss – Luminous Blue Variables – Hypergiants – supernovae

1. Mass-loss and non-spherical flow

Massive stars have a high rate of mass loss, and in the most massive stars mass loss has a decisive influence on their evolution. Five years ago we (De Jager, Nieuwenhuijzen and Van der Hucht, 1988) published a comprehensive table of all mass loss rates, $\dot{M}$, known at that time. The list contains 271 objects, but we guess that at present the number of stars with known rates of mass loss may well be twice as large, while some of the earlier determinations have been repeated and/or improved. A revision of our earlier work is therefore useful because it will lead to a better knowledge of the variation of mass loss over the Hertzsprung-Russell diagram and to clearer views on the significance of mass loss for evolution. Such a revision has just been started.

Many authors, including us, have tried to parametrize the dependence of $\dot{M}$ on luminosity L, effective temperature T_{eff} and other primary stellar parameters. This was either done for the whole complex of stars or for parts thereof. It appears invariably that $\dot{M}$ depends primarily on L. The dependence on other variables is secondary but certainly not negligible. The strong L-dependence is e.g. shown for early–type stars in a recent parametrization by Lamers and Leitherer (1993), who found a linear dependence of $\dot{M}$ on L. However, K and M stars deviate by about 3 dex from the average hot star dependence, while the $\dot{M}(L)$ line has, even for early–type stars, still a width of about 1.5 dex which is larger than the intrinsic error of the individual mass loss rate determinations. Hence there must be other parameters than L that influence mass loss.

It appeared from our 1988 paper that the standard error of one $\dot{M}$-value, if derived by application of one method only (a "unit weight determination"), is about 0.37 dex. Since for many stars $\dot{M}$ has been determined in various ways, e.g. from U.V. resonance lines, subordinate lines in the visible, I.R. or radio data, molecular data etc., many of the values found in literature have larger weights than unity and consequently are more precise. The "best" determinations have standard errors of 0.1 to 0.15 dex, which corresponds to "weights" of 15 to 6.

For hot stars, the winds are driven by radiation. Lamers and Leitherer (1993) give a diagram, here partly reproduced as Figure 1, which shows that the observed

Space Science Reviews **66**: 7–19, 1994.
© 1994 *Kluwer Academic Publishers. Printed in Belgium.*

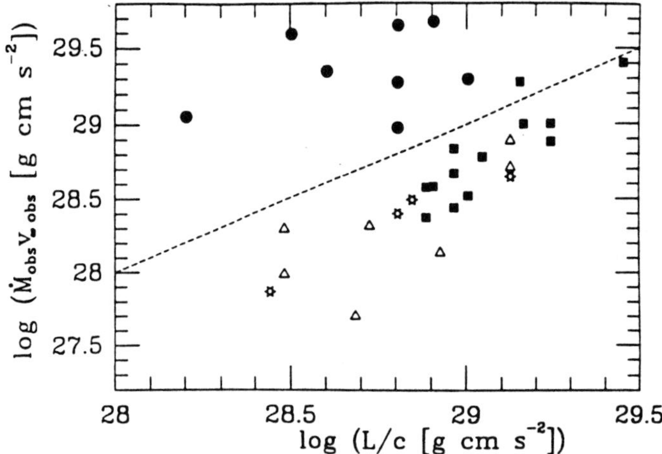

Fig. 1. Observed momentum losses of stellar winds, compared with photon momentum losses(abscissa). Filled dots refer to WNL stars. From Lamers & Leitherer, (1993). Ap.J. publication

momentum losses $\dot{M}v$ of the stellar winds are smaller then the radiative momentum losses L/c, but not for the WNL stars (shown as solid dots in the diagram). This observation must be related to the fact that the winds of the WNL stars are optically thick so that multiple photon scattering occurs. Indeed, Abbot and Conti (1987) presented evidence that this is true: an extreme WR star such as V444 Cyg is situated in the region of optically thick winds in the HR diagram.

For the most luminous hot stars, as well as for several yellow hypergiants, the winds have already large velocities at photospheric levels. Massa et al. (1992) reproduced an interesting set of B-type stellar spectra of different luminosity, showing luminosity dependent violet displacements of Fraunhofer lines. The upper-photospheric velocities are above 100 km/s for the Ia$^+$ type star Zeta-1 Sco. For such extreme stars a photosphere is a dynamic structure and its theoretical treatment should be dealt with accordingly.

The 1991 Baltimore Symposium on non-spherical flow from stars (Drissen et al., 1992) brought abundant evidence, observational as well as theoretical, that for most if not all stars the wind flow is not spherically symmetric. Due to rotation or to magnetic fields or both, gas that is leaving the star has a strong tendency to concentrate towards the equatorial plane, thus producing a disk rather than a spherical wind structure. This effect has seldom been taken into account in the interpretation of observational data relevant to atmospheres or winds, obviously because it introduces another parameter, difficult to handle. A treatment of stellar atmospheres and winds on the basis of the concept of non-spherical flow is important.

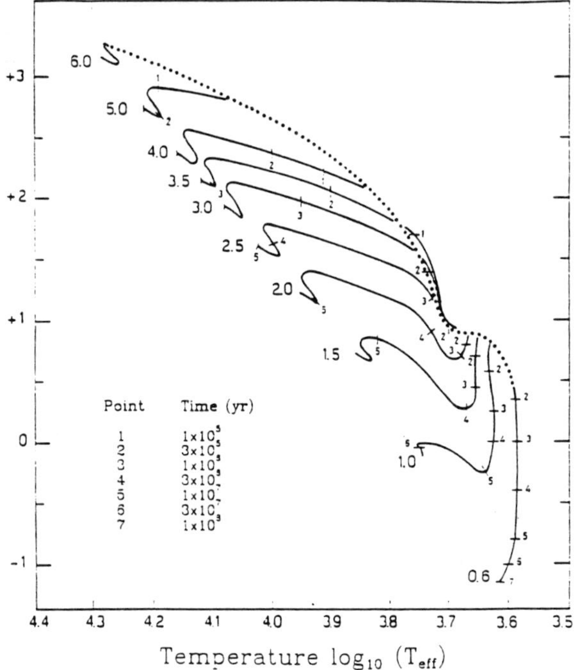

Fig. 2. Contraction lines in the HR diagram and the birth line (dotted) of stars up to 6 solar masses. The ordinate gives log L in solar units. From Palla & Stahler (1993). Ap.J. publication

2. Pre–main sequence evolution

Conventionally, star formation takes place in three steps:
- phase of protostellar accretion;
- phase of contraction; not necessarily homologous or fully convective;
- nuclear burning phase.

Palla and Stahler (1993) presented numerical calculations of contraction lines in the HR diagram for stars of masses up to 15 solar values. They found that the line connecting the end of the contraction phase in the HR diagram for stars of different masses (the "birth line") moves towards the ZAMS for increasing stellar mass. It joins the ZAMS at about 6 $M_\odot$ (Figure 2). For the present symposium, which deals with stars of larger masses, this result means that we may safely assume that massive stars contract right onto the ZAMS: there are only two phases: accretion, directly followed by the nuclear burning phase, without a contraction phase in between.

The first two phases (those of accretion and contraction) are certainly not as smooth as theory wants us to believe. There is an increasing body of evidence that the process of star formation may be powered by mass flow and/or radiation from nearby massive stars. Marti et al. (1993) presented impressive observations of the HH80-81 complex, a number of aligned pre-MS objects with in the middle a massive young star with L= 1.4 10^4 L$_\odot$. O'Dell et al. (1993) described HST observations of pre-MS stars near the Orion Trapezium. The objects show non-spherical shapes like 'cometary' tails, suggesting that they are powered by interstellar shocks, most probably related to nearby supernovae or to gas flow from massive stars.

It remains to be investigated whether powering by massive stars is the rule or an exception in the process of star formation. More observations are essential to that end.

3. The HR diagram for massive stars; the main sequence

Figure 3 (upper part) gives the HR diagram for 8 well-studied OB associations (Garmany and Stencel, 1992), while the lower part shows it for the LMC (Chiosi, Bertelli and Bressan, 1992b). Prominent aspects of the diagram are
- the main sequence, defined as the area of core hydrogen burning;
- the Humphreys–Davidson limit, being the line connecting the most luminous stars above which no stars are assumed to exist;
- the "ledge", an apparent discontinuity in the star density at about T_{eff}= 10 kK and $\log L/L_\odot$ = 5;
- the "yellow void", a nearly empty area near T_{eff} = 5 kK and $\log L/L_\odot$ = 5.6.
- the "red turning line" or "red turning area", where stars turn blueward or upward after their initial redward evolution.

Each of these presents its own characteristic problems.

We discuss the main sequence. Although the physical process of hydrogen burning seems to be well understood, it is remarkable that calculations of the width of the main sequence do not yet present good agreement with observations, as appears from the discrepancy between the main sequence as defined by the most densely occupied area in the HR diagram and the positions of the first evolutionary turning points (which mark the end of core hydrogen burning). Conspicuous is also the lack of O-stars close to the ZAMS. The remedy to these problems is not yet clear. It may be found in a more precise treatment of convective overshooting, a problem that so far has only been handled by the introduction of ad hoc overshooting parameters. The answer may either be found in the treatment of mass loss: there are indications that early–type B supergiant atmospheres are enriched in N (are all early B's enriched?) There is also some hope to obtain better agreement between theory and observations by using the new OPAL opacities (Iglesias et al., 1993) which are considered an improvement as compared to the LOAL (Los Alamos) opacity tables.

The other aspects listed above will be discussed in next chapters.

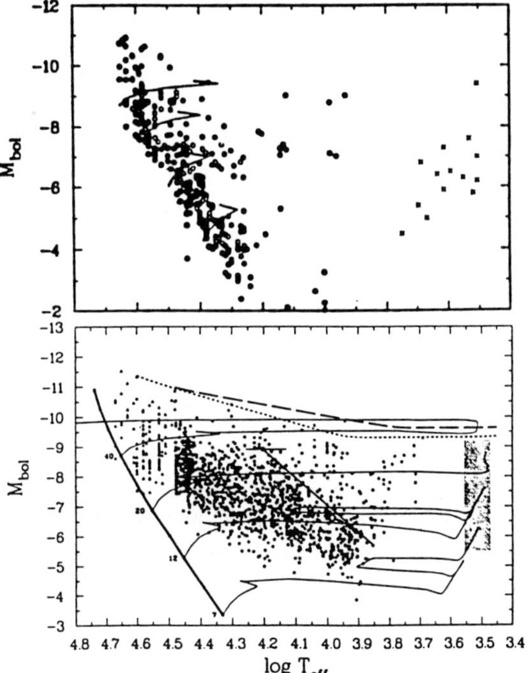

Fig. 3.   The Hertzsprung-Russell diagram for massive stars. Top figure: for 8 well studied galactic associations. Garmany and Stencel (1992). A&A. publication. Lower figure: For the LMC. From Chiosi, Bertelli & Bressan (1992b). North Holland Publ. Corp. publication.

4. Instability of non-streaming atmospheres

Winds, even at photospheric level, are an important property of massive stars. For some hot Ia–type atmospheres the winds are supersonic already at photospheric levels, cf. Section 1. A "non-streaming" atmosphere is therefore a simplification that is not always allowed. Current photospheric models, among them most prominently the impressive set of Kurucz models, are all based on the assumption of hydrostatic equilibrium. They are an important tool for present–day analyses of stellar spectra. Moreover, some considerations on the (in)stability of stellar atmospheres are also based on such models.

The Eddington limit, either in its classical form (with opacity defined by electron scattering) or in its "generalized" form (with the flux-mean opacity, lines included) has been used in attempts to explain the absence of stars above the Humphreys–Davidson limit. Lamers and Fitzpatrick (1988), who used the generalized Eddington expression, found by extrapolating Kurucz models down to gravity zero, a fair agreement for hot stars between the limit thus derived and the Humphreys–Davidson limit. The weak point in this approach is that a star with gravity zero does not exist (having infinite extension) and one may expect that a real star, in which gravity slowly decreases towards zero during its evolution, will expand and thus strongly change the atmospheric properties, which makes straight extrapolation a dangerous procedure. Particularly the reduced atmospheric density will increase ionisation and thus the opacity will approach the classical electron scattering value.

In the range of effective temperatures below about 9000 K the opacity decreases with decreasing effective temperature and the Eddington limit rises up to large luminosity values. Therefore, cool atmospheres would be stable even for extremely luminous stars. The fact that such stars do not exist above about logL = 5 is ascribed to the fact that stars can not reach that area because they meet the sloping branch of the Humphreys–Davidson limit already earlier in their redward evolution, whereupon the atmosphere becomes unstable.

In cool stars with effective temperatures between about 6000 and 8000 K, a density inversion occurs in photospheric or subphotospheric layers, depending on the value of T_{eff}. This inversion is due to the fact that for temperatures near 8000 K the opacity increases steeply with temperature, hence with atmospheric optical depth. Since in a hydrostatic atmosphere the increase of pressure with depth (defined by the pressure scale height) goes more slowly than that, the ratio P/T, which is proportional to the density, may locally decrease with height. Thus a density inversion is produced.

Gustafsson and Plez (1992) have studied Kurucz models in order to find in which stars such density inversions occur. The lower limit of the atmospheric gravity values of stars without such an inversion at any place in the photosphere is shown in Figure 4, along with a number of stars situated in that region. These stars are chiefly yellow super- and hypergiants, objects known for their extreme near-instability. The question arises if this inversion may cause truly instable atmospheres. The answer is negative, first of all for the observational reason that stars appear to exist beyond the limit. (But we have to remark that these stars have a large rate of mass loss and show mass ejections).

From the theoretical point of view, the answer has been given already as early as 1958 (Bohm-Vitense, 1958) and it is also negative: as long as the pressure keeps increasing with depth, the effective acceleration vector remains directed downward and therefore a density inversion gives at most rise to some kind of shell formation but not to atmospheric disruption. The process of shell formation has been elaborated by Maeder (1992) in his "geyser model", but in a rather preliminary way.

5. Dynamic atmospheric instability

Observations of stellar spectra show the existence of motions on a variety of time scales. As a rule, the velocity amplitude of the motions increases with luminosity, and it surpasses the sonic value for hypergiants (stars of the Ia[+] luminosity class).

In a non-magnetic atmosphere two kinds of waves are expected to exist: gravity and pressure waves. The regions of occurrence of such waves in the

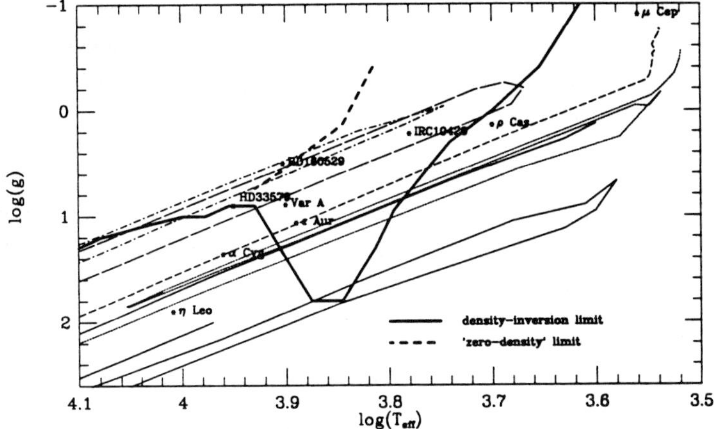

Fig. 4.   The density inversion limit in the (log g, log T_{eff}) plane. The thin lines are evolutionary tracks from Maeder and Meynet (1991). A few stars are indicated. From Gustafsson and Plez (1992). North Holland Publishing Corp. publication

P,L-plane (where P is the wave period and L the wavelength), can be studied by a diagnostic diagram (De Jager et al., 1991), like the one shown in Figure 5 for Alpha Per. Pressure waves, which transform into shock waves within a time shorter than one wave period, occur only on the line labelled "Shock". Gravity waves are possible in the area above the curve labeled "Grav" if they are not dampened by radiation, hence to the right of the curve labeled "Tdamp". For the red supergiant Alpha Ori their predicted "wavelengths" are about 0.1 stellar radius (De Jager, 1993). Such waves were confirmed observationally through interferometric observations of the surface of Alpha Ori by Wilson et al. (1992, with reference to earlier work).

Most supergiants show evidence of such large-scale, long-wavelength gravity waves.

Outward running shock waves occur in all supergiants with a subphotospheric convection layer, hence in stars with effective temperatures below about 8000 K. "Microturbulence" is also observed in atmospheres of hotter stars as evident from spectral observations, but their cause in these stars is unclear. The significance of shock waves in massive stars is twofold: they strongly change the atmospheric model as compared to the smooth Kurucz models, and they cause an outward mass flux which can lead to mass loss in cool stars (Cuntz, 1987; Bowen, 1988). The former studied shocks of small scale and wavelength related to the subphotospheric convection; the latter investigated larger scale waves that may be related to non-radial stellar pulsations. In both cases mass loss may result.

The influence of small–scale motions on the stability of massive stars is being

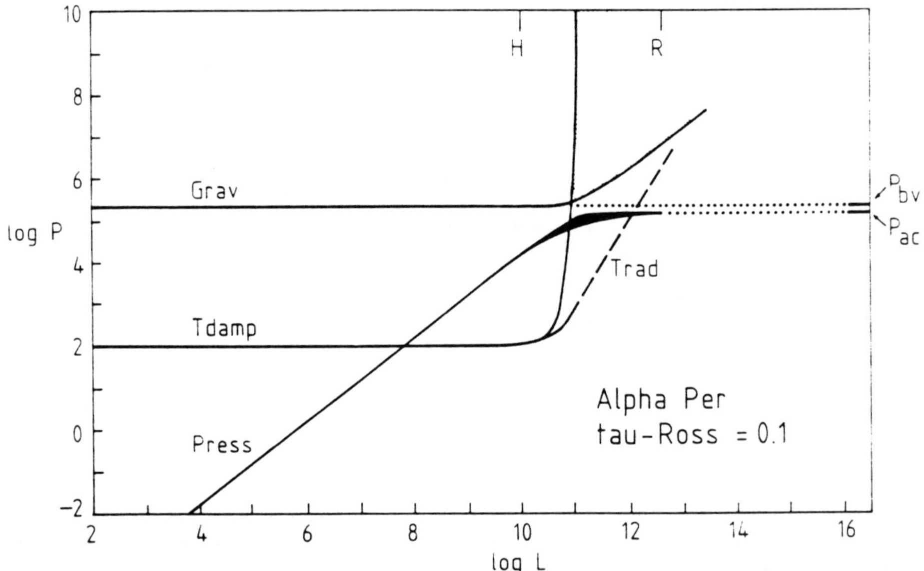

Fig. 5.   Diagnostic P (period in s) - L (wavelength in cm) diagram for motions in the atmosphere of Alpha Per, showing the regions where gravity and pressure (shock) waves can exist. P_{bv} and P_{ac} are the Brunt–Vaisala and acoustic cut-off periods. From De Jager et al. (1991). A&A publication

studied by Nieuwenhuijzen and De Jager (1992) who calculated the variation of the effective acceleration in stellar atmospheres over the upper part of the HR diagram. The calculations are based on five fundamental observables: the luminosity L, the effective temperature T_{eff}, the mass M, the rate of mass loss $M_{\odot}$ and the average gravity wave velocity amplitude, all quantities of which the variation over the HR diagram is (assumed to be) known. The output shows the variation of the resulting vector of atmospheric acceleration g_{eff} over the HR diagram (Figure 6). For blueward evolving stars, the line where g_{eff} passes the zero line agrees fairly well with the high-temperature part of the Humphreys–Davidson limit, but for redward evolution (hence: very evolved stars) there is an area of negative g_{eff}-values. This area coincided fairly well with the "yellow evolutionary void" (cf. Section 3) and also with the region where most yellow hypergiants are located.

6. Observational evidence of massive star evolution; LBV's and Yellow Hypergiants

Figure 7 from to Chiosi et al. (1992b) gives a summary of the upper part of the theoretical HR diagram, with its main characteristics and specific properties relevant to evolution. Here, EAGB stands for Early Asymptotic Giant Branch, the RGB is the Red Giant Branch, and the letters TP denote the thermally pulsing

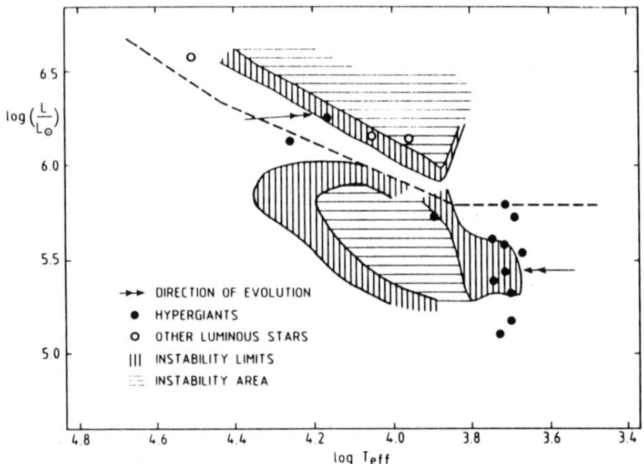

Fig. 6. Limiting lines (the vertically hatched areas, the widths of which are a measure of the accuracy) of atmospheric stability based on a dynamic treatment of stellar atmospheres. Upper line: for redward evolving stars. Lower line: for blueward evolution. From Nieuwenhuijzen & De Jager (1992). North Holland Publishing Corp. publication

regime. ZAHB is the zero-age horizontal branch which contains core He burning stars of low mass. D-up stands for dredge-up and PN for planetary nebula.

Evolution goes rapidly for the most massive stars and it is of interest to collect observational data on noticeable evolution in historic times. Thus far, two cases are known to me. There are more cases known of stars that underwent important changes in temperature and/or spectrum, often in a few tens of years only (HR 5171A; Van Genderen, 1992; Var-A; Humphreys, 1988 and other examples), but in these cases the variation reflects temporal rather than secular evolutionary changes.

P Cygni: Lamers and De Groot (1992) collected historic observations of this prototype of the Luminous Blue Variables for the period from between 1700 to the present, and concluded from the observed gradual change in visual brightness that the star's effective temperature must have decreased by 7000 K in 300 years. This is ascribed to redward evolution of this star.

AG Car is another LBV. The star is surrounded by a nebular detached shell. A study by Robberto et al. (1993) yielded the following past evolutionary history. The gas contained in the shell is due to a slow wind produced when the star was a red supergiant. When it evolved blueward, the wind velocity as well as the wind flux increased, sweeping up the earlier material, thus producing the detached shell. The authors published a diagram from which one concludes that the shell was produced when the star was a yellow hypergiant (T_{eff} about 4000 K) while its present T_{eff} is about 25000 K.

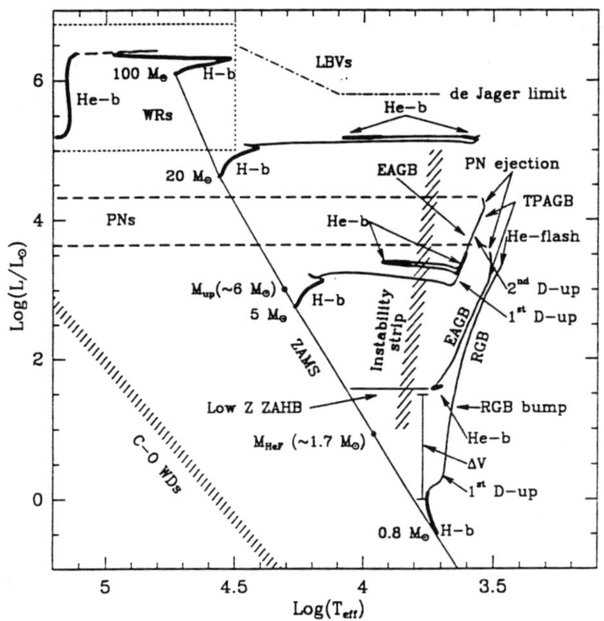

Fig. 7.   HR diagram showing main phases of stellar evolution. From Chiosi, Bertelli & Bressan (1992). Ann. Rev. Astron. Astrophys. publication

Hence we are meeting two LBV's with contrary evolutionary evidences: P Cyg moving redward and AG Car moving blueward. Can both evidences be correct? Or is there more than one type of LBV? Abundance determinations of LBV's show atmospheric nuclear burning products, which may suggest that the LBV's are on the blueward track.

The class of LBV stars forms an ill-determined group. Apart from the rare giant outbursts, like the one shown by Eta Car (in fact, the only well-recorded giant outburst), the common feature is a brightness eruption of one or a few magnitudes coupled to a redward displacement, apparently caused by an increase of the stellar radius at constant bolometric luminosity (Figure 7). The initial suggestion (Davidson, 1987), that the increase of radius during these intermediate-type outbursts is due to an outward displacement of the photosphere caused by an increased rate of mass loss during the eruption, does not appear to be correct, as was shown by De Koter and Lamers (1993): the enhancement of mass loss is insufficient to that end. They suggest that the real cause of the phenomenon must be seated deeper inside the stars.

Next to the intermediate outbursts, the LBV's also show variability of smaller amplitude, a property they have in common with the "Alpha Cygni variables", supergiants with (as a rule: fairly small) semi-regular variations of brightness. The variation of the observed quasi-periods of their light variations over the HR diagram obeys to the same regularities as the Alpha Cygni's and even of all other supergiants (Van Genderen et al. 1992). As compared to the main body of

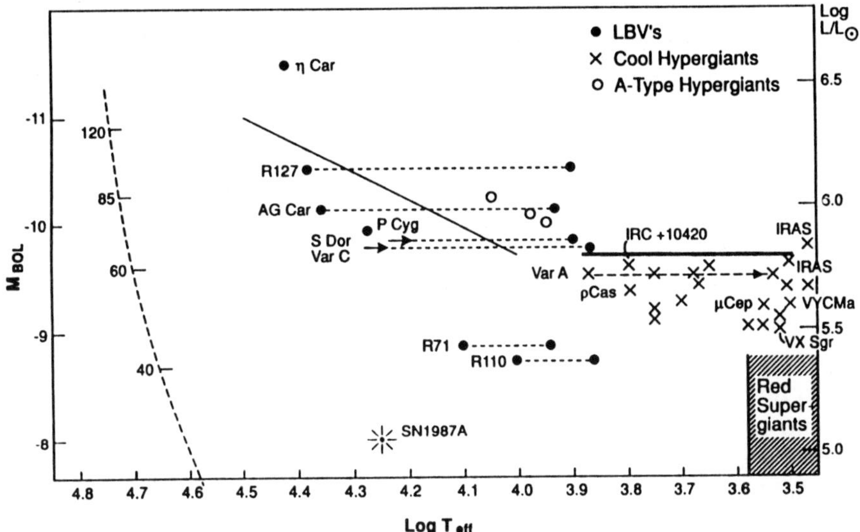

Fig. 8.   LBV's and yellow hypergiants in the HR diagram. From Humphreys (1993).

the Alpha Cyg variables the LBV's have the largest brightness amplitudes (Van Genderen et al., 1992).

Humphreys (1993) questioned the homogeneity of the group. There is indeed a large difference, also in luminosity, between an impressive eruptive star like Eta Car and fairly quiet LBV's like R71 and R110. The more pronounced ones are situated close to the hot side of the sloping part of the HD limit as is obvious from Figure 7. The suggestion that the mechanism of their instability is different from that of less luminous LBV's finds support in stability studies of massive stars by Glatzel & Kiriakidis (1993) who discovered that violent mode-coupling operating on dynamic time scales may be the reason for the observed blowing-up of stars with ZAMS masses above about 60 solar values.

The yellow hypergiants, with prototypes Rho Cas, HR 8752, HR 5171, IRC 10420 ... show features that resemble the LBV phenomenon (cf. Figure 8): redward displacements at short time scales (typically years to some tens of years), and it may be that this phenomenon is related to that of the moderate LBV's of the R71 class. The yellow hypergiants have also a strong rate of mass loss, a feature that may be related to the fact that they are situated close to the Yellow Evolutionary Void (Figure 6).

The Wolf–Rayet stars with their extremely high mass fluxes are considered objects at the end of the blueward evolutionary phase, as is supported by abundance studies, which show dominance of CN cycle nuclear fusion products, and reduced H abundance. Their observed variability seems to be seated fully in the spectral lines (Van Genderen, private comm.), and must therefore be a wind feature. Underhill and Yang (1991) advanced arguments that the winds may be mainly flowing in the equatorial plane, forming a detached ring. Davidson et al. (1993) describe He3-519, a star that resembles the LBV AG Car, and that may be in a post-LBV phase, on the way to the WN phase. If confirmed, this would link LBV's with WR stars and that would identify both types as very evolved blueward moving stars.

7. Supernovae

An elementary subdivision divides the supernovae in two classes, according to their lightcurves. Spectral observations show that the Type I's have little or no hydrogen in their mantle, and have binary progenitors. The Type II's do show hydrogen lines and have single star progenitors. But there are many SN's that do not fit into either of these two types.

Studies, mainly based on the frequency of occurrence in comparison with the galactic Initial Mass Function, yield that the type II's must have masses above about 8 solar. Since the number of stars on the ZAMS decreases rapidly with mass, most Type II progenitors must therefore have masses in the range of 8 to 16 solar (hence ZAMS masses between 9 and 19), but more massive SN progenitors, although rare, will certainly also occur.

Until 1987, the classical scenario for a type II SN was that of an about 12 $M_\odot$ red supergiant that exploded due to core implosion at the end of the nuclear burning phase. A typical example of such a possible evolution is Alpha Ori (Betelgeuze), a star that is at present very close to the end phase of the nuclear burning part of its evolution.

It was therefore somewhat of a surprise when the first progenitor that could be observed, that of SN 1987A, appeared to be a blue star. This can, however, be understood realizing that the end of the nuclear burning phase of stars more massive than about 20 solar values does not occur in the red supergiant phase, since these stars return blueward after the red turning point. In order to fit the progenitor of SN 1987A to an evolutionary track the rate of mass loss had to be increased in one of the proposed solutions by about a factor 3 above standard values (Arnett et al., 1989), which seems offhand very large (cf. section 1). We may remark, though, that it helps to realize that the progenitor passed through the instability region of the Yellow Evolutionary Void during the last part of its evolution, which involves a period of enhanced mass loss.

It is therefore very interesting that on March 26 of this year another supernova exploded with an observed progenitor: 1993J. The progenitor seems reddish (K-type?). But, as in so many cases: the lightcurve shows that this again is not a standard supernova. It has been described as a Type II with Type I characteristics. Spectral observations showed H–lines for about one month, which demonstrates that the progenitor had only a thin hydrogen shell. In addition, the lightcurve could indeed be fitted by assuming a Hshell of less than one solar mass. This must be related to the indications that the star was part of a binary.

8. Acknowledgements

Many thanks are due to Dany Vanbeveren for useful suggestions.

References

Abbot, D. and Conti, P.: 1987, *Ann. Rev. Astron. Astrophys.* **25**, 113.

Arnett, W., Bahcall, J., Kirshner, R. and Woosley, S.: , *1989* **Ann. Rev. Astron. Astrophys.,** 27 629.

Bohm-Vitense, E.: 1958, *Z. f. Astrophysik* **46**, 108.

Bowen, G.: 1988, *Astrophys. J.* **329**, 299.

Chiosi, C., Bertelli, G, and Bressan, A.: 1992, *Ann. Rev. Astron. Astrophys.* **30**, 235.

Chiosi, C., Bertelli, G, and Bressan, A.,1992, in C. de Jager and H. Nieuwenhuijzen (eds.): Instabilities in Evolved Super- and Hypergiants, North Holland Publ. Corp., Amsterdam, 145.

Cuntz, M.: 1987, *A&A* **188**, L5.

Davidson, K.: 1987, *Astrophys. J.* **317**, 760.

Davidson, K., Humphreys, R., Hajian, A. and Terzian, Y.: 1993, *Astrophys. J.* **preprint,**

De Jager, C.: 1993, *The Observatory* **113**, 43.

De Jager, C., Nieuwenhuijzen H. and Van der Hucht, K.: 1988, *A&A Suppl.* **72**, 259.

De Jager, C., De Koter, A., Carpay, J. and Nieuwenhuijzen, H.: 1991, *A&A* **244**, 131.

De Koter, A. and Lamers, H.: 1993, *A&A* **preprint,**

Drissen, L., Leitherer, C. and Nota, A.: 1992, *Nonisotropic and variable outflow from stars*, ASP Conference Series, 22

Garmany, C. and Stencel, R.: 1992, *A&A Suppl.* **94**, 211.

Glatzel, W. and Kiriakidis, M.: 1993, *MNRAS* **263**, 375.

Gustafsson, B. and Plez, B,1992, in C. de Jager and H. Nieuwenhuijzen (eds.): Instabilities in Evolved Super- and Hypergiants, North Holland Publ. Corp., Amsterdam, 86.

Humphreys, R., Jones, R. and Gehrtz, R.: , *Astron. J.* **94**, 315.

Humphreys,R.: 1993, *preprint* ,

Iglesias, C., Rogers, F. and Wilson, B.: 1993, *preprint* ,

Lamers, H. and Leitherer, C.: 1993, *Astrophys. J.* **412**, 771.

Lamers, H. and Fitzpatrick, E.: 1988, *Astrophys. J.* **324**, 279.

Maeder, A. ,1992, in C. de Jager and H. Nieuwenhuijzen (eds.): Instabilities in Evolved Super- and Hypergiants, North Holland Publ. Corp., Amsterdam, 138.

Maeder, A. and Meynet, G.: 1991, *A&A Suppl.* **89**, 451.

Marti, J., Rodriguez, L. and Reipurth, B.: 1993, *Astrophys. J.* **preprint,**

Massa, D., Shore, S. and Wynne, D.: 1992, *A&A* **264**, 169.

Nieuwenhuijzen, H. and De Jager, C. ,1992, in C. de Jager and H. Nieuwenhuijzen (eds.): Instabilities in Evolved Super- and Hypergiants, North Holland Publ. Corp., Amsterdam, 171.

O'Dell, C., Wen, Z. and Hu, X.: 1993, *Astrophys. J.* **410**, 696.

Palla, F. and Stahler, S.: 1993, *Astrophys. J.* **preprint,**

Robberto, M., Ferrari, A., Nota, A. and Paresce, F.: 1993, *A&A* **269**, 330.

Underhill, A.B. and Yang, S.: 1991, *Astrophys. J.* **368**, 588.

Van Genderen, A. ,1992, in C. de Jager and H. Nieuwenhuijzen (eds.): Instabilities in Evolved Super- and Hypergiants, North Holland Publ. Corp., Amsterdam, 37.

Van Genderen, A., Van de Bosch, F., Dessing, F., Fehmers, G., Van Grunsven, J., Van der Heiden, R., Janssens, A., Kalter, R., Van der Meer, R., Van Ojik, R., Smit, J. and Zijderveld, M.: 1992, *A&A* **264**, 88.

Wilson, R., Baldwin, J., Buscher, D. and Warner, P.: 1992, *MNRAS* **257**, 369.

THE MASSIVE STAR DISTRIBUTION IN THE GALAXY AND THE MAGELLANIC CLOUDS

KAREL A. VAN DER HUCHT

Space Research Organization Netherlands, Sorbonnelaan 2, 3584 CA Utrecht

Abstract. The numbers and distribution of Population I O-type stars and Wolf-Rayet stars are reviewed. The numbers of known WR stars in the Galaxy, the LMC and the SMC are 185, 114, and 9, respectively. Distances and galactic distributions determined by various authors are compared. The single star and binary distributions are discussed in the light of evolutionary studies.

Key words: stars: Wolf-Rayet – stars: individual – stars: distribution – stars: binaries – stars: evolution of

1. Introduction

This paper is restricted to O-type stars and Wolf-Rayet (WR) stars, representing the main-sequence and evolved phase, respectively, of massive stars. O-type stars have hot stellar winds with velocities in the range $v_\infty = 1000\text{-}3500\ kms^{-1}$ and mass loss rates of $\dot{M} = [3 \times 10^{-5} - 6 \times 10^{-9}]M_\odot yr^{-1}$ (Garmany 1988). WR stars have hot dense stellar winds with velocities in the range $v_\infty = 1000\text{-}2500\ km.s^{-1}$ and mass loss rates of $\dot{M} = [2 - 10] \times 10^{-5} M_\odot yr^{-1}$. WR stellar winds are responsible for the strong broad emission lines in WR spectra, which define the WR phenomenon, and for their *free-free* μm to cm radiation.

In recent years a fair number of reviews and conference proceedings on WR stars and related hot massive stars have appeared, of which we note since 1985 (without claiming completeness) reviews written by Massey (1985), Chiosi & Maeder (1986), Abbott & Conti (1987), Willis & Garmany (1987), and van der Hucht (1992); and proceedings containing numerous reviews edited by de Loore, Willis & Laskarides (1986), Lamers & de Loore (1987), Coyne *et al.* (1988), Nugis & Pustyl'nik (1988), Davidson, Moffat & Lamers (1989), Garmany (1990a), Willson & Stalio (1990), Leitherer, Walborn, Heckman & Norman (1991), van der Hucht & Hidayat (1991), Heber & Jeffery (1992), and Drissen, Leitherer & Nota (1992). A monograph on O-type and WR stars has been composed by Conti & Underhill (1988).

In the past decade a consensus has been reached that WR stars are evolved hot massive stars, close to the end of their nuclear burning phase. Evidence for this has come from atmospheric studies, evolutionary studies and studies of their ambient circumstellar *c.q.* interstellar environment. Massive stars are responsible for the nucleosynthesis of the majority of the heavy elements, and thus they dominate the chemical enrichment in at least the early phases of galactic evolution.

Space Science Reviews **66**: 21–35, 1994.

© 1994 *Kluwer Academic Publishers. Printed in Belgium.*

2. Inventory

The galactic WR census of van der Hucht *et al.* (1981, 1988) has been comple-
mented by the discovery of 28 new WR stars:
- Acker & Stenholm (1990) reclassified the alleged planetary nebula Th3-28 (The
1964) as a new WN2.5-3 star (WR93a);
- Cohen *et al.* (1991) classified IRAS 17380-3031 as a new WC8-9 star (WR98a);
- Shara *et al.* (1991) discovered in a dedicated survey 13 new WR stars (11 WN
and 2 WC);
- Crawford & Barlow (1991) classified the emission line star We21 (Weaver
1974) as a new WN8 star (WR47a);
- van Kerkwijk *et al.* (1992) discovered that Cygnus X-3 has in the IR a variable
WN4-7 spectrum (WR145a);
- Mereghetti *et al.* (1993) identified the X-ray source 1E 1024.0-5732 with the
emission line star Th35-42 (The 1966) and classified it as WN6 (WR20c);
- Shara *et al.* (1993, these proceedings) continued their dedicated survey to dis-
cover 10 new WR stars (7 WN and 3 WC).
This brings the number of known galactic WR stars to 185, of which 101 are
WN types, 7 are WN/WC types (Conti & Massey 1989, Conti *et al.* 1990), 75
are WC types and 2 are WO types (Barlow & Hummer 1982).

For the LMC, Lortet (1991) refers to 27 new WR stars since the census
by Breysacher (1981). Excluding the 12 Of/WN stars she lists, this brings the
number of LMC WR stars up to 114. This number can readily increase where
high spatial resolution imaging has shown some LMC and SMC WR stars to
be actually double or multiple (*e.g.*, Heydari-Malayeri 1991; Schild and Testor,
1992. An alleged possible new LMC WC9-type (Heydari-Malayeri *et al.* 1990),
a subtype otherwise unknown in the LMC, appears to be rather a unique Of?p
like object (Moffat 1991; Heydari-Malayeri & Melnick 1992).

Since the WR census in the SMC by Azzopardi & Breysacher (1979), one
new WN2.5+O5: system has been discovered by Morgan *et al.* (1991), bringing
the total number to 9 WR stars, among which 8 WN types and 1 WO type.

3. Basic parameters and galactic distribution

WR stars are highly luminous and therefore visible at great distances. Because of
their concentration to the galactic plane, many WR stars suffer significant extinction, sometimes exacerbated by additional local extinction either by material left
over from their formation or circumstellar dust made by themselves. Knowledge
of their luminosities and distances is required for many purposes, *e.g.*, location
in the HR diagram for comparison with stellar models, and location in the Milky
Way to study the striking dependence of WR subtype with galactocentric distance. The first is required to help resolve the question of how evolved WR stars
are and the second is believed to result from the influence of metallicity on the

efficiency of radiation pressure in driving the stellar wind.

3.1. BASIC PARAMETERS

Intrinsic parameters of Wolf-Rayet stars, like intrinsic colours, colour excesses and absolute visual magnitudes, have been derived and discussed by Hidayat *et al.* (1982; 1984), Lundstrom & Stenholm (1984), Massey (1984), van der Hucht *et al.* (1988), Torres-Dodgen & Massey (1988) and Vacca & Torres-Dodgen (1990). New empirical methods, based on observed emission line strengths and ratios, have been developed by Conti & Massey (1989), Conti & Morris (1990) and Smith *et al.* (1990) to determine absolute visual magnitudes of WN stars, interstellar reddening of WN stars and distances of WC stars, respectively. Schmutz & Vacca (1991) developed a model dependent method to determine the interstellar reddening from *ubv* photometry.

Hidayat *et al.* (1982, 1984) and van der Hucht *et al.* (1988) used available narrow-band filter photometry in the system of Smith (1968a,b) and Lundstrom & Stenholm (1979, 1984). They derived colour excesses, intrinsic colours and absolute visual magnitudes for galactic WR stars only, adopting distances and colour excesses found in studies of WR stars in galactic open clusters and associations by Lundstrom & Stenholm (1984) as the basis of their calibration.

In spite of the limited number of WR stars in galactic clusters and associations (42, distributed over 18 WR subtypes), van der Hucht *et al.* (1988) warn against averaging of galactic and LMC basic parameters. Their concern is shared by Hamann (1991) and Koesterke *et al.* (1991), who found from quantitative spectroscopy that LMC WN stars have on the average lower luminosities than their galactic counterparts. In the meantime, verification of cluster and association membership of WR and OB stars is receiving attention (Garmany 1990b, 1991).

Massey (1984) and Torres-Dodgen & Massey (1988) were able to derive synthetic line-free photometry from spectrophotometry. In addition, Vacca & Torres-Dodgen (1990) derived colour excesses from the strength of the 2200A absorption feature in low resolution *IUE* spectra of galactic and LMC WR stars. They averaged the parameters found per subclass for the Milky Way with those found for the LMC.

A new approach to reddening and distance determinations restricts itself to the infrared wavelength region, which is less affected by extinction than the optical wavelength region. IR photometry of WR stellar winds can be used to deriving photometric distances by determining IR absolute luminosities from observations of WR stars whose distances are known independently from their membership in stellar clusters and associations. The IR flux from WR stars not having thermal emission by circumstellar dust is a combination of photospheric and stellar wind *free-free* emission, with the latter dominating at longer wavelength. Since the classification of WR stars is based on emission lines formed in strong stellar winds, the stellar wind continua may be better correlated with spectral subtype than the photospheric continua (which show considerable dispersion, *e.g.*, van der Hucht et al. 1988). Also, because the stellar wind has a flatter spectrum ($\propto \nu^{0.8}$) than the photosphere ($\propto \nu^2$), it will come do dominate the continuum at some wavelength in the IR, reducing its sensitivity to the possible presence of a non-WR companion, even if the latter were brighter than the WR star in the optical. Finally, the IR is much less susceptible to reddening than the optical. This method was used in the IR photometric distance determinations of WR147 (AS 431, WN8) and WR125 (IC14-36, WC7+O9III) by Churchwell *et al.* (1992) and Williams *et al.* (1992), respectively, showing the importance of these improved distances determinations for assessment of X-ray and radio luminosities and mass loss rates. This method is being explored further by van der Hucht *et al.* (1993).

3.2. GALACTIC DISTRIBUTION

Studies of the galactic distribution of WR stars as a diagnostic for the determination of their initial masses and (subtype) evolution have been attempted by Maeder *et al.* (1980), Hidayat *et al.* (1982), Meylan & Maeder (1983) and Hidayat *et al.* (1984) on the basis of narrow-band galactic data compiled by van der Hucht *et al.* (1981); by Conti *et al.* (1983) on the basis of broad-band LMC data of Prevot-Burnichon (1981); by van der Hucht *et al.* (1988) using improved narrow-band galactic data only; by Conti & Vacca (1990) mixing galactic and LMC data; and by Maeder (1991a,b) comparing his evolutionary model predictions with the results of both van der Hucht *et al.* (1988) and Conti & Vacca (1990). The latter, using WR absolute visual magnitudes determined by Vacca & Torres (1990) by averaging galactic data (from cluster and association distances) and LMC data, re-calculated the distances of the over 40 galactic WR stars initially considered by them to be members of open clusters and associations, which constitute about 50% of the number of WR stars within 2.5 *kpc* from the Sun. This inconsistency causes the main discrepancy between their WR galactic distribution and the one determined by van der Hucht *et al.* (1988), the latter showing good agreement in subtype ratios *vs.* galactocentric distance with evolutionary calculations of Maeder (1991a,b). Comparison of the two observational studies also indicates that photometric distances of WR *field* stars have an accuracy of about 50%. In all mentioned studies the overall galactic WR distributions found agree surprisingly well with each other, in spite of the different data sets and methods used. Notably the concentration of WR stars in galactic spiral arms, their absence toward the galactic anti-center (Orion spur) (Roberts 1962), and the concentration of late-WC stars toward the galactic center (Smith 1968c), are confirmed by the recent studies. For the probably complete sample of WR stars with $d < 2.5$ *kpc* van der Hucht *et al.* (1988) find an observed galactocentric distribution with $N_{WN}/N_{WC} \approx 1$ for $R > 7.5$ *kpc* and $N_{WN}/N_{WC} \approx 0.4$ for $R < 7.5$ *kpc*. For the same sample they find that perpendicular to the galactic plane $|z| = 46$ *pc*, almost exactly like the $|z|$-distribution of O-type stars (Garmany *et al.* 1982). Repercussions of the distributions found are discussed further in section 5.

4. Binaries

After the numerous studies of WR binaries in the early 80-s by Massey (1981) and co-workers, the main breakthrough in WR binary research has been the determination of orbital inclinations from linear polarization modulation measurements by Moffat and co-workers, as reviewed by Moffat & Robert (1991). The physical basis of this method is that the same free electrons in a WR wind that cause a dip in the light-curve of a WR+O binary when the O star is behind, can scatter O star light into the line-of-sight. This scattered light will be polarized, depending on the scattering angle, or rather on the vectorial sum of all individual scatterings. Polarimetric and spectropolarimetric studies in this sense have been performed by Luna (1982, 1985), Drissen *et al.* (1986a,b, 1987), St.-Louis *et al.* (1987, 1988), Bastien (1988), Moffat (1988), Piirola & Linnaluoto (1988), Schmidt (1988), Drissen *et al.* (1989), Moffat & Seggewiss (1989), Robert & Moffat (1989), Robert *et al.* (1989), Schulte-Ladbeck & van der Hucht (1989), Moffat *et al.* (1990), Robert *et al.* (1990) and Drissen *et al.* (1992b). Resulting WR masses listed by those authors range from 6 to 40 $M_\odot$ for WN stars and from 5 to 20 $M_\odot$ for WC stars in binaries.

Those and other studies of established binaries, *i.e.*, those with radial velocity solutions, have been summarized, *e.g.*, by Smith & Maeder (1989) for the Milky Way and by Moffat *et al.* (1990) for the LMC and SMC. Additional new radial velocity studies have been performed by, *e.g.*, Annuk (1990, 1991) for the long-period binaries WR137 (HD 192641), WR138 (HD 193077) and WR140 (HD 193693); by Stickland & Lloyd (1990) and Stickland (1991) for WR11 (γ^2 Vel); by Grandchamp & Moffat (1991) for WR141 (HD 193928), WR148 (HD 197406) and WR155 (CQ Cep); by Lewis *et al.* (1991, 1992) for WR151 (CX Cep); and by Niemela (1991) for WR8 (HD 62910), WR98 (HDE 318016) and the LMC WR object Sk−71°34. Duplicity of the WC4/WN6 object WR8, however, has been questioned by Willis & Stickland (1990) who found that both the WN and WC emission lines arise in the *same* stellar wind, as happens in the WN/WC objects WR145 (MR111) and WR153 (GP Cep) (Massey & Grove 1989). Vrancken *et al.* (1991) calculated from evolutionary models that the expected number of WR+WR binaries in the solar neighbourhood is less than one. That one may have been found by Panov & Seggewiss (1990) who argue on the basis of a light-curve analysis that the quadruple system WR153 (GP Cep) harbours two WR+O systems.

From a light-curve analysis Gosset *et al.* (1991) discovered that WR22 (HD 92740), a bright WN7 binary with a period of 80.35 d but with a hitherto unseen companion (Conti *et al.* 1979), is eclipsing.

WR stars which may be accompanied by a compact companion left after a supernova explosion of the primary, *cf.* the evolutionary scheme of van den Heuvel & de Loore (1973; see also van den Heuvel 1976), were listed by Hidayat *et al.* (1984). Their estimated number has dwindled lately, where the attention has drifted from compact companions to intrinsic variability (*e.g.*, Vreux 1985), although some groups continue the search (*e.g.*, Antokhin & Cherepashchuk 1989; Moffat 1992). Among the dozen or so left, the two most likely candidates are still WR6 (EZ CMa, WN5 + neutron star?, $P = 3.77$ *d*) (Firmani *et al.* 1980; Drissen *et al.* 1989; van der Hucht *et al.* 1990; Moffat 1992; Koenigsberger & Auer 1992; Robert *et al.* 1992 (; although questioned by Willis *et al.* 1989; Schulte-Ladbeck 1990, 1991; St-Louis *et al.* 1993) and WR148 (WN7 + black hole?, $P = 4.32$ *d*) (Drissen *et al.* 1986a; Kundt & Fischer 1989; Moffat 1992).

A recent discovery, which may constitute the missing link in the scenario of van den Heuvel & de Loore (1973) is the WN classification of the X-ray binary Cygnus X-3 by van Kerkwijk *et al.* (1992), commented on by Conti (1992). Although the short orbital period (4.79 *hr*) precludes WN radii of a few or more solar radii, this discovery should lead to renewed modelling of both WR stars and of Cyg X-3 itself. If it can be confirmed that the companion of the neutron star in Cyg X-3 is indeed a regular Population I WR star, this case will also provide an additional convincing argument for the evolved nature of WR stars (van Kerkwijk 1993, Terasawa, N., Nakamura, H. 1993).

The binary frequency in the solar neighbourhood was found to be 37% (van der Hucht *et al.* 1988). The most complete recent listing of galactic WR+OB binaries has been presented by Smith & Maeder (1989), who note that the difference in distribution of period (and, by implication, separation) between WN and WC binaries, found first by Smith (1973b) but challenged by Massey (1981), persists. Eight (50%) of the WN binaries have a period of less than seven days, corresponding to a separation of 60 $R_\odot$, but not one of the galactic WC binaries has a period this short.

New short-period binaries ($P < 1$ *d*) may be discovered from high time-resolution photometric observations as demonstrated by, *e.g.*, van Genderen *et al.* (1990, 1991), although in most cases some degree of intrinsic variability cannot be excluded. The light-curves and radial velocity curves of the WN3pec star WR46 (HD 104994) yield $P = 6.8$ *hr*, the shortest orbital period known among the WR binaries, with the exception of Cyg X-3. The radial velocity curve of the *HeII*λ6560 emission line yields a mass function $f(m) = 0.002 M_\odot$. The variations superimposed on the light-curves may be caused by a variable geometry of the continuum and line emitting regions. The equivalent widths of the most prominent lines show phase dependent variations of 10-25%, implying that the line emitting region suffers more distortion than the continuum emitting region (van Genderen *et al.* 1991).

Studies of colliding winds have been performed for the WN5+O6 binary WR139 (V444 Cyg, $P = 4.21$ *d*) by Shore & Brown (1988), who explained the variability of its UV P-Cygni profiles in terms of shock-dominated wind-wind interactions, and by Luo *et al.* (1990) and Usov (1990), who addressed its X-ray emission modulation.

A rewarding albeit time-consuming way of detecting long-period (P of the order of 10 yr) and highly eccentric WC binaries is being explored by Williams *et al.* (1985, 1987a,b, 1990a,b, 1991a,b, 1992; see also Churchwell *et al.* 1992 and van der Hucht *et al.* 1992) by monitoring excess X-ray fluxes, episodic/periodic dust formation and non-thermal radio emission. WR binaries with orbit sizes of the order of magnitude of their WR radio photosphere sizes or larger reveal their wind collision interaction regions, which are otherwise hidden in short-period binaries, because of extinction in the WR stellar wind. Where single star models (*e.g.*, White 1985; Chen & White 1991), trying to explain the non-thermal radio emission and excess X-ray emission of WR stars, failed, the binary model of Williams *et al.* (1990a) has been successful in explaining all these phenomena as well as their variability as being associated with colliding winds in binary systems, in particular for the best studied case of WR140 (HD 193793, WC7+O4-5). Theoretical support for this model has been provided by Pollock *et al.* (1991), Usov (1991, 1992) and Stevens *et al.* (1992), explaining the observed X-rays from WR140 and showing that dust can form in cooling regions between hot colliding winds. Van der Hucht *et al.* (1992) suggest that all non-thermal WR stars, as well as all (variable) non-thermal radio OB stars (*e.g.*, θ^1 Ori A, Felli *et al.* 1991), are actually long-period binaries with their observed non-thermal phenomena originating in colliding winds. The five new long-period binaries within 2.5 kpc from the Sun (van der Hucht *et al.* 1992), if all confirmed by radial velocity studies, would bring the binary frequency in that volume (*cf.* van der Hucht *et al.* 1988) to 48%, thus significantly increasing the galactic WR binary frequency as well as the relative importance of the binary channel for WR star formation and evolution (Maeder 1982, 1991a).

5. Distribution and evolution

Since the review of Chiosi & Maeder (1986) on the evolution of single massive stars and the productions of WR stars, who argued for the evolutionary sequences

- O $\rightarrow$ Of $\rightarrow$ BSG and LBV $\rightarrow$ WR $\rightarrow$ SN   for $M_i > 60\ M_\odot$
- O $\rightarrow$ BSG $\rightarrow$ RSG $\rightarrow$ WR $\rightarrow$ SN   for $25\ M_\odot < M_i < 60\ M_\odot$

and since the comparative study of de Loore (1988), new major evolutionary studies have appeared, like those of Langer (1989a,b, 1990), Maeder (1990,1991a,b,c,d), Schaller *et al.* (1992), Chin & Stothers (1990, 1991), Stothers (1991) and Stothers & Chin (1991a,b). Langer and Maeder, both emphasizing the effects of metallicity, differ in that the former applied mass loss and semi-convection (the effect of molecular weight gradients on convection) as the mixing process to bring nuclear burning products to the surface, while the latter relied on mass loss and an increased convective core size due to overshooting (*cf.* Maeder & Meynet 1989). Both investigators adopted a $\dot{M} \propto M^\alpha$ dependence, with $\alpha = 2.5$ for WNE and WC stars and $\alpha = 0$ for WNL stars. When constant mass loss rates are applied, the evolutionary models lead to larger WR luminosities than have been observed (Schmutz *et al.* 1989).

Maeder *et al.* (1980) proposed an explanation of the observed frequency of WR stars as a function of galactocentric distance, by a connection between the local metallicity Z and single WR mass loss rates. (In case of binaries the picture may change, *cf.*, *e.g.*, Vanbeveren 1991; Vanbeveren & de Loore 1993; Vanbeveren 1994, these proceedings.) At large Z (*e.g.*, inner galactic regions) the gas opacities are larger and consequently more momentum is transferred to the stellar wind by radiation pressure. Therefore in large Z environments mass loss by stellar winds in massive O stars is larger and thus more WR stars are formed (Maeder 1991b). To quantify this, Maeder (1990, 1991a) calculated grids of evolutionary models for various values of the metallicity Z ($Z = 0.002$-0.040), and adopted $\dot{M} \propto Z^{0.5}$, as indicated by stellar wind models of Kudritzky *et al.* (1987) and confirmed by Leitherer and Langer (1991). Maeder found WR lifetimes as a function of Z of the order of 5×10^5 yr and initial masses M_i as low as 20 $M_\odot$, which compares well with the observed minimum value of 25 $M_\odot$ found by van der Hucht *et al.* (1988), and the minimum value of 23 $M_\odot$ found by Vanbeveren (1991) in comparing binary observations and binary evolution models. For $Z \geq 0.02$ all stars with $M_i \geq 25 M_\odot$ finish their evolution with masses in the range 5-10 $M_\odot$ (Maeder 1991d), while the length of the WR phase increases strongly with increasing metallicity and mass loss rate. WR/O, WC/WR and WC/WN number ratios following from Maeder's models compare well with ratios observed in the Milky Way (van der Hucht *et al.* 1988), in the LMC (Azzopardi & Breysacher 1985), and in M31 and M33 (Smith 1988). The observed WC/WN ratio in the LMC requires $\dot{M}$-rates which are about 50 % smaller than those in the solar neighbourhood. The work of Maeder also indicates that, if many WR stars are observed in regions with low Z, at least a fraction of those WR stars have to originate through an alternative 'channel' of WR formation, *e.g.*, from massive binary evolution.

Maeder's model calculations confirm an overall general evolutionary sequence WNL $\rightarrow$ WNE $\rightarrow$ WCL $\rightarrow$ WCE $\rightarrow$ WO. For binaries, studies by Moffat et al. (1990 and references therein) also point toward a continuous WCL $\rightarrow$ WCE/WO subtype evolution. Of the allowed subtype evolution paths, based on the observed WR galactic distribution (van der Hucht *et al.* 1988), *i.e.*,

- at galactocentric radius $R < 8.5\ kpc$: WNL $\rightarrow$ WCL
- at $R > 6.5\ kpc$: WNL $\rightarrow$ WCE $\rightarrow$ WO
- and in general: WNE $\rightarrow$ no WC stars,

the first two paths agree well with models with high M_i and low Z values, respectively. The third path corresponds only to lower M_i at low Z, but may be somewhat relaxed since WNE stars and WCE stars have some overlap in galactocentric distances (Maeder 1991a). The galactic distribution of the WC subtypes as determined by van der Hucht *et al.* (1988) has been explained quantitatively in terms of the galactic metallicity gradient by Smith & Hummer (1988) and Smith & Maeder (1991): high Z_i and M_i lead to WCL subtypes, while low Z_i and M_i lead to WCE subtypes.

In spite of the different physics, the results of the models by the Geneva group, notably the recent ones of Schaller *et al.* (1992), agree generally well with those of Langer, both predicting similar abundances in the WR stellar winds and the ~ 3 % fraction of WN/WC 'transition' WR stars classified as such by Conti & Massey (1989).

6. WR stars as supernova progenitors

No clear picture exists as yet as to the exact type of supernovae to which WR stars will eventually evolve. Suggestions for a evolutionary link between WR stars and Type Ib/Ic supernovae (no hydrogen) by, *e.g.*, Wheeler & Levrault (1985), Begelman & Sarazin (1986), Filippenko & Sargent (1986) and Schaeffer *et al.* (1987) have been evaluated by Ensman & Woosley (1988), who concluded that no definite answer could yet be given.

Nomoto (1991) found that the maximum brightness and fast decline of typical Type Ib/Ic supernova light-curves can be well accounted for by helium star models with masses of 3-5 $M_\odot$ (or 7 $M_\odot$, Panagia & Laidler 1991), which form from stars with $M_i \approx 12 - 18 M_\odot$ in binary systems (Shigeyama *et al.* 1990). Thus only a WR star with a mass reduced to 3-5 $M_\odot$ could be a Type Ib/Ic progenitor. However, such a star would not undergo extensive mixing and without mixing and the associated formation of clumps the resulting model light-curve is too broad to explain the observed Type Ib/Ic supernovae. Furthermore, the birth rate of massive stars is too small to explain the observed frequency of Type Ib/Ic supernovae (Nomoto 1991; Filippenko 1991; Matteucci 1991).

Observational tests to distinguish between the various evolutionary models leading to Type Ib/Ic supernovae could be provided by X-ray and γ-ray observations of such events (The *et al.* 1991).

Filippenko (1991) noted that the most massive stars may end their lives as blackholes not preceded by supernova events. Black holes may form more likely in low Z regions, where the theoretical final WR masses are relatively large (Maeder 1991d).

Van den Bergh (1992), assuming that WR stars are the progenitors of massive core-collapse supernovae, argues on the basis of the observed magnitude and reddening distribution of galactic WR stars that the galactic frequency of massive ($M \geq 35 M_\odot$) core-collapse supernovae with $V_{max} < 0$ is ~ 0.2 per millennium.

7. Isotopes from WR winds

Products of nuclear processing which should appear in WR winds according to evolutionary models have been discussed, *e.g.*, by Maeder (1983a,b, 1991a) and Maeder & Meylan (1993) who argue for ^{12}C, ^{16}O, and ^{22}Ne abundance increments of more than two orders of magnitude at the transition from the WN to the WC phase, and abundances of ^{26}Mg and ^{25}Mg equal to that of ^{24}Mg in

WC stars.

Anomalous abundances found in cosmic rays and meteorites (*e.g.*, Eberhardt *et al.* 1981; Mewaldt 1981; Casse & Paul 1982; Webber & Soutoul 1989) could indicate a WR origin. However, the *Ne* abundance derived for the wind of WR star in WR11 (γ^2 Vel, WC8+O9I) from IR emission lines by Barlow *et al.* (1988) is only slightly larger than cosmic. IR spectroscopy with *ISO* will provide *Ne* abundance determinations for many WR stars in the near future.

WR stars should also be copious producers and emitters of ^{26}Al as argued from interior studies by, *e.g.*, Dearborne & Blake (1988). By postulating large numbers of WR or related massive stars in the galactic center, it is possible to predict a concentration of ^{26}Al adequate to produce the ^{26}Al 1.809 MeV γ-ray emission observed in that direction (Blake & Dearborne 1989). Recent IR observations show indeed the signature of evolved hot massive stars in the galactic center (Krabbe *et al.* 1991). Massive stars, supernovae and novae are likely competing contributors to the ^{26}Al concentration there (Signore & Vedrenne 1988; Signore & Dupraz 1990; von Ballmoos 1991). Recent *Compton Observatory* 1.809 MeV observations have been discussed in terms of massive stars being probably the main contributors of radiactive ^{26}Al in the Galaxy (Prantzos 1993a,b; Signore & Dupraz 1993; Clayton *et al.* 1993).

References

Abbott, D.C. 1982, Astrophysical Journal**259**, 282

Abbott, D.C., Bieging, J.H., Churchwell, E. 1984, Astrophysical Journal**280**, 671

Abbott, D.C., Bieging, J.H., Churchwell, E., Torres, A.V. 1986, Astrophysical Journal**303**, 239

Abbott, D.C., Conti, P.S. 1987, Annual Review of Astronomy and Astrophysics**25**, 113

Acker, A., Stenholm B. 1990, Astronomy and Astrophysics, Supplement Series**86**, 219

Annuk, K. 1990, *Acta Astron.* **40**, 267

Annuk, K. 1991, in: K.A. van der Hucht & B. Hidayat (eds.), Wolf-Rayet Stars and Interrelations with Other Massive Stars in Galaxies, *Proc. IAU Symp. No. 143* (Dordrecht: Kluwer), p. 245

Antokhin, I.I., Cherepashchuk, A.M. 1989, *Pis'ma Astron. Zh.* **15**, 701 (= *Sov. Astron. Letters* **15**, 303)

Azzopardi, M., Breysacher, J. 1979, Astronomy and Astrophysics**75**, 120

Azzopardi, M., Breysacher, J. 1985, Astronomy and Astrophysics**149**, 213

Azzopardi, M., Lequeux, J., Maeder, A. 1988, Astronomy and Astrophysics**189**, 34

von Ballmoos, P. 1991, Astrophysical Journal**380**, 98

Barlow, M.J., Hummer, D.G. 1982, in: C. de Loore & A.J. Willis (eds.), Wolf-Rayet Stars: Observations, Physics, Evolution, *Proc. IAU Symp. No. 99* (Dordrecht: Reidel), p. 387

Barlow, M.J., Roche, P.F., Aitken, D.K. 1988, Monthly Notices of the RAS**232**, 821

Bastien, P. 1988, in: G.V. Coyne, A.M. Magalhaes, A.F.J. Moffat, R.E. Schulte-Ladbeck, S. Tapia, D.T., Wickramasinghe (eds.), Polarized Radiation of Circumstellar Origin (Vatican: Vatican Observatory), p. 595

Begelman, M.C., Sarazin, C.L. 1986, Astrophysical Journal, Letters to the Editor**302**, L59

van den Bergh, S. 1992, Astrophysical Journal**390**, 133

Blake, J.B., Dearborn, D.S.P. 1989, Astrophysical Journal, Letters to the Editor**338**, L17

Breysacher, J. 1981, Astronomy and Astrophysics, Supplement Series**43**, 209

Casse, M., Paul, J.A. 1982, Astrophysical Journal**258**, 860

Chen, W., White, R.L. 1991, Astrophysical Journal**366**, 512

Chin, C.-W., Stothers, R.B. 1990, Astrophysical Journal, Supplement Series**73**, 821

Chin, C.-W., Stothers, R.B. 1991, Astrophysical Journal, Supplement Series**77**, 299

Chiosi, C., Maeder, A. 1986, Annual Review of Astronomy and Astrophysics**24**, 329

Churchwell, E., Bieging, J.H., van der Hucht, K.A., Williams, P.M., Spoelstra, T.A.Th., Abbott, D.C. 1992, Astrophysical Journal**393**, 329

Clayton, D.D., Hartmann, D.H., Leising, M.D. 1993, Astrophysical Journal, Letters to the Editor**415**, L25

Cohen, M., van der Hucht, K.A., Williams, P.M., The, P.S. 1991, Astrophysical Journal**378**, 302

Conti, P.S., Niemela, V.S., Walborn, N.R. 1979, Astrophysical Journal**228**, 206

Conti, P.S., Garmany, C.D., de Loore, C., Vanbeveren, D. 1983, Astrophysical Journal**274**, 302

Conti, P.S, Underhill A.B. (eds.) 1988, O Stars and Wolf-Rayet Stars, *NASA SP-497*

Conti, P.S., Massey, P. 1989, Astrophysical Journal**337**, 251

Conti, P.S., Morris, P.W. 1990, Astronomical Journal**99**, 898

Conti, P.S., Vacca, W.D. 1990, Astronomical Journal**100**, 431

Conti, P.S., Massey, P., Vreux, J.-M. 1990, Astrophysical Journal**354**, 359

Conti, P.S. 1992, *Nature* **355**, 680

Coyne, G.V., Magalhaes, A.M., Moffat, A.F.J., Schulte-Ladbeck, R.E., Tapia, S. & Wickramasinghe, D.T. (eds.) 1988, Polarized Radiation of Circumstellar Origin (Vatican: Vatican Observatory)

Crawford, I.A., Barlow, M.J. 1991, *A&A* (Letters) **252**, L39

Dearborn, D.S.P., Blake, T.B. 1988, Astrophysical Journal**332**, 305

Drissen, L., Lamontagne, R., Moffat, A.F.J., Bastien, P., Seguin, M. 1986a, Astrophysical Journal**304**, 188

Drissen, L., Moffat, A.F.J., Bastien, P., Lamontagne, R. 1986b, Astrophysical Journal**306**, 215

Drissen, L., St-Louis, H., Moffat, A.F.J., Bastien, P. 1987, Astrophysical Journal**322**, 888

Drissen, L., Robert, C., Lamontagne, R., Moffat, A.F.J., St-Louis, N., van Weeren, N., van Genderen, A.M. 1989, Astrophysical Journal**343**, 426

Drissen, L., Leitherer, C. & Nota, A. (eds.) 1992a, Non-Isotropic and Variable Outflows from Stars, *ASP Conf. Series* **22**

Drissen, L., Robert, C., Moffat, A.F.J. 1992b, Astrophysical Journal**386**, 288

Eberhardt, P., Junck, M.H.A., Meier, F.O., Niederer, F.R. 1981 *Geochimica et Cosmochimica Acta* **45**, 1515

Ensman, L.M., Woosley, S.E. 1988, Astrophysical Journal**333**, 754

Felli, M., Massi, M., Catarzi, M. 1991, Astronomy and Astrophysics**248**, 45

Filippenko, A.V., Sargent, W.L.W. 1986, Astronomical Journal**91**, 691

Filippenko, A.V. 1991, in: K.A. van der Hucht & B. Hidayat (eds.), Wolf-Rayet Stars and Interrelations with Other Massive Stars in Galaxies, *Proc. IAU Symp. No. 143* (Dordrecht: Kluwer), p. 529

Firmani, C., Koenigsberger, G., Bisiacchi, G.F., Moffat, A. F.J., Isserstedt, J. 1980, Astrophysical Journal**239**, 607

Garmany, C.D., Conti, P.S., Chiosi, C. 1982, Astrophysical Journal**263**, 777

Garmany, C.D. 1988, in: Conti, P.S, Underhill A.B. (eds.) 1988, O Stars and Wolf-Rayet Stars, *NASA SP-497*, p. 157

Garmany, C.D (ed.) 1990a, Intrinsic Properties of Hot Luminous Stars, Proc. Boulder-Munich Workshop, *ASP Conf. Series* **7**

Garmany, C.D 1990b, in: C.D. Garmany (ed.), Intrinsic Properties of Hot Luminous Stars, *ASP Conf. Series* **7**, p. 16

Garmany, C.D. 1991, in: K. Janes (ed.), The Formation and Evolution of Star Clusters, *ASP Conf. Series* **13**, p. 23

van Genderen, A.M., van der Hucht, K.A., Larsen, I. 1990, Astronomy and Astrophysics**229**, 123

van Genderen, A.M., Verheijen, M.A.W., van der Hucht, K.A., de Loore, C.W.H., Schwarz, H.E., van Esch, B.P.M., Greidanus, H., van der Heiden, R., van Kampen, E., Kuulkers, E., le Poole, R.S., Reijns, R.A., Robijn, F.H.A., Spijkstra. L. 1991, in: K.A. van der Hucht & B. Hidayat (eds.), Wolf-Rayet Stars and Interrelations with Other Massive Stars in Galaxies, *Proc. IAU Symp. No. 143* (Dordrecht: Kluwer), p. 129

Gosset, E., Remy, M., Manfroid, J., Vreux, J.-M., Balona, L.A. 1991, *Inf. Bull. Var. Stars* No. 3571

Grandchamps, A., Moffat, A.F.J. 1991, in: K.A. van der Hucht & B. Hidayat (eds.), Wolf-Rayet

Stars and Interrelations with Other Massive Stars in Galaxies, *Proc. IAU Symp. No. 143* (Dordrecht: Kluwer), p. 258

Hamann, W.-R. 1991, in: K.A. van der Hucht & B. Hidayat (eds.), Wolf-Rayet Stars and Interrelations with Other Massive Stars in Galaxies, *Proc. IAU Symp. No. 143* (Dordrecht: Kluwer), p. 81

Heber U. & Jeffery, S. (eds.) 1992, Atmospheres of Early Type Stars, *Lecture Notes in Physics* No. 401 (Berlin: Springer)

van den Heuvel, E.P.J., de Loore, C.W.H. 1973, Astronomy and Astrophysics**25**, 387

van den Heuvel, E.P.J. 1976, in: P. Eggleton, S. Mitton & J. Whelan (eds.), Structure and Evolution of Close Binary Systems, *Proc. IAU Symp. No. 73* (Dordrecht: Reidel), p. 35

Heydari-Malayeri, M, Melnick, J., Van Drom, E. 1990, *A&A* (Letters) **236**, L21

Heydari-Malayeri, M.: in: K.A. van der Hucht & B. Hidayat (eds.), Wolf-Rayet Stars and Interrelations with Other Massive Stars in Galaxies, *Proc. IAU Symp. No. 143* (Dordrecht: Kluwer), p. 647

Heydari-Malayeri, M., Melnick, J. 1992, *A&A* (Letters) **258**, L13

Hidayat, B., Supelli, K., van der Hucht, K.A. 1982, in: C. de Loore & A.J. Willis (eds.), Wolf-Rayet Stars: Observations, Physics, Evolution, *Proc. IAU Symp. No. 99* (Dordrecht: Reidel), p. 27

Hidayat, B., Admiranto, A.G., van der Hucht, K.A. 1984, *Ap. Space Sci.* **99**, 175

van der Hucht, K.A., Conti, P.S., Lunstrom, I., Stenholm, B. 1981, Space Science Reviews**28**, 227

van der Hucht, K.A., Hidayat, B., Admiranto, A.G., Supelli, K.R., Doom, C. 1988, Astronomy and Astrophysics**199**, 217

van der Hucht, K.A., van Genderen, A.M., Bakker, P.R. 1990, Astronomy and Astrophysics**228**, 108

van der Hucht, K.A. 1991, *Transactions IAU* **XXIA**, Reports on Astronomy, p. 324

van der Hucht, K.A. & Hidayat, B. (eds.) 1991, Wolf-Rayet Stars and Interrelations with Other Massive Stars in Galaxies, *Proc. IAU Symp. No. 143* (Dordrecht: Kluwer)

van der Hucht, K.A., Williams, P.M., Spoelstra, T.A.Th., de Bruyn, A.G. 1992, in: L. Drissen, C. Leitherer, A. Nota (eds.), Non-isotropic and Variable Outflows from Stars, *ASP Conf. Series* **22**, p. 253

van der Hucht, K.A., Williams, P.M., Setia Gunawan, D.Y.A. 1993, in preparation

van Kerkwijk, M.H., Charles, P.A., Geballe, T.R., King, D.L., Miley, G.K., Molnar, L.A., van den Heuvel, E.P.J., van der Klis, M., van Paradijs, J. 1992, *Nature* **355**, 703

van Kerkwijk, M.H. 1993, *A&A* (Letters) **276**, L9

Koenigsberger, G., Auer, L.H. 1992, in: L. Drissen, C. Leitherer, A. Nota (eds.), Non-isotropic and Variable Outflows from Stars, *ASP Conf. Series* **22**, 239

Koesterke, L., Hamann, W.-R., Schmutz, W., Wessolowski, U. 1991, Astronomy and Astrophysics**248**, 166

Krabbe, A., Genzel, R., Drapatz, S., Rotaciuc, V. 1991, Astrophysical Journal, Letters to the Editor**382**, L19

Kudritzki, R.P., Pauldrach, A., Puls, J. 1987, Astronomy and Astrophysics**173**, 293

Kundt, W., Fischer, D. 1989, *J. Astrophys. Astron.* **10**, 119

Lamers, H. & de Loore, C. (eds.) 1987, Instabilities in Luminous Early Type Stars (Dordrecht: Reidel)

Langer, N. 1989a, Astronomy and Astrophysics**210**, 93

Langer, N. 1989b, Astronomy and Astrophysics**220**, 135

Langer, N. 1990, in: C.D. Garmany (ed.), Intrinsic Properties of Hot Luminous Stars, *ASP Conf. Series* **7**, p. 328

Leitherer, C., Walborn, N.R., Heckman, T.M., Norman, C.A. (eds.) 1991, Massive Stars in Starbursts, *StScI Symposium Series No. 5* (Cambridge: CUP)

Leitherer, C., Langer, N. 1991, in: R.F. Haynes, D.K. Milne (eds.), The Magellanic Clouds, *Proc. IAU Symp. No. 148* (Dordrecht: KLuwer), p. 480

Lewis, D., Moffat, A.F.J., Robert, C. 1991, in: K.A. van der Hucht & B. Hidayat (eds.), Wolf-Rayet Stars and Interrelations with Other Massive Stars in Galaxies, *Proc. IAU Symp. No. 143* (Dordrecht: Kluwer), p. 256

de Loore, C.W.H, Willis & A.J., Laskarides, P. (eds.) 1986, Luminous Stars and Associations in

Galaxies, *Proc. IAU Symp. No. 116* (Dordrecht: Reidel)

de Loore, C. 1988, Astronomy and Astrophysics**203**, 71

Lortet, M.-C. 1991, in: K.A. van der Hucht & B. Hidayat (eds.), Wolf-Rayet Stars and Interrelations with Other Massive Stars in Galaxies, *Proc. IAU Symp. No. 143* (Dordrecht: Kluwer), p. 513

Luna, H.G. 1982, Publications of the ASP**94**, 695

Luna, H.G. 1985, *Rev. Mexicana Astron. Astrof.* **10**, 267

Lundstrom, I., Stenholm, B. 1979, Astronomy and Astrophysics, Supplement Series**35**, 303

Lundstrom, I., Stenholm, B. 1984, Astronomy and Astrophysics, Supplement Series**58**, 163

Luo, D., McCray, R., Mac Low, M.-M. 1990 Astrophysical Journal**362**, 267

Maeder, A., Lequeux, J., Azzopardi, M. 1980, *A&A* (Letters) **90**, L17

Maeder, A. 1983a, Astronomy and Astrophysics**120**, 113

Maeder, A. 1983b, Astronomy and Astrophysics**120**, 130

Maeder, A., Meynet, G. 1989, Astronomy and Astrophysics**210**, 155

Maeder, A. 1990, Astronomy and Astrophysics, Supplement Series**84**, 139

Maeder, A. 1991a, Astronomy and Astrophysics**242**, 93

Maeder, A. 1991b, in: C. Leitherer, N.R. Walborn, T.M. Heckman & C.A. Norman (eds.), Massive Stars in Starbursts, *StScI Symposium Series No. 5* (Cambridge: CUP), p. 97

Maeder, A. 1991c, in: K.A. van der Hucht & B. Hidayat (eds.), Wolf-Rayet Stars and Interrelations with Other Massive Stars in Galaxies, *Proc. IAU Symp. No. 143* (Dordrecht: Kluwer), p. 445

Maeder, A. 1991d, Quarterly Journal of the RAS**32**, 217

Maeder, A., Meynet, G. 1993 **278**, 406

Massey, P. 1981, Astrophysical Journal**246**, 153

Massey, P. 1984, Astrophysical Journal**281**, 789

Massey, P. 1985, Publications of the ASP**97**, 5

Massey, P., Grove, K. 1989, Astrophysical Journal**344**, 870

Matteucci, F. 1991, in: K.A. van der Hucht & B. Hidayat (eds.), Wolf-Rayet Stars and Interrelations with Other Massive Stars in Galaxies, *Proc. IAU Symp. No. 143* (Dordrecht: Kluwer), p. 625

Mereghetti, S., Belloni, T., Shara, M., Drissen, L. 1993, Astrophysical Journal, in press.

Mewaldt, R.A. 1981, *Proc. 17th Int. Cosmic Ray Conf.* Vol 13, p. 49

Meylan, G., Maeder, A. 1983, Astronomy and Astrophysics**124**, 84

Moffat, A.F.J. 1988, in: G.V. Coyne, A.M. Magalhaes, A.F.J. Moffat, R.E. Schulte-Ladbeck, S. Tapia, D.T., Wickramasinghe (eds.), Polarized Radiation of Circumstellar Origin (Vatican: Vatican Observatory), p. 607

Moffat, A.F.J., Niemela, V.S., Marraco, H.G. 1990, Astrophysical Journal**348**, 232

Moffat, A.F.J. 1991, *A&A* (Letters) **244**, L9

Moffat, A.F.J., Robert, C. 1991, in: K.A. van der Hucht & B. Hidayat (eds.), Wolf-Rayet Stars and Interrelations with Other Massive Stars in Galaxies, *Proc. IAU Symp. No. 143* (Dordrecht: Kluwer), p. 109

Moffat, A.F.J. 1992, Astronomy and Astrophysics**253**, 425

Morgan, D.H., Vassiliadis, E., Dopita, M.A. 1991, Monthly Notices of the RAS**251**, 51P

Niemela, V.S. 1991, in: K.A. van der Hucht & B. Hidayat (eds.), Wolf-Rayet Stars and Interrelations with Other Massive Stars in Galaxies, *Proc. IAU Symp. No. 143* (Dordrecht: Kluwer), p. 201

Nomoto, K. 1991, in: K.A. van der Hucht & B. Hidayat (eds.), Wolf-Rayet Stars and Interrelations with Other Massive Stars in Galaxies, *Proc. IAU Symp. No. 143* (Dordrecht: Kluwer), p. 515

Nugis, T. & Pustyl'nik I. (eds.), 1988, Wolf-Rayet Stars and Related Objects, *Tartu Astrofuus. Obs. Teated* No. 89

Panagia, N., Laidler, V.G. 1991, in: K.A. van der Hucht & B. Hidayat (eds.), Wolf-Rayet Stars and Interrelations with Other Massive Stars in Galaxies, *Proc. IAU Symp. No. 143* (Dordrecht: Kluwer), p. 567

Panov, K.P., Seggewiss, W. 1990, Astronomy and Astrophysics**227**, 117

Piirola, V., Linnaluoto, S. 1988, in: G.V. Coyne, A.M. Magalhaes, A.F.J. Moffat, R.E. Schulte-Ladbeck, S. Tapia, D.T., Wickramasinghe (eds.), Polarized Radiation of Circumstellar Origin (Vatican: Vatican Observatory), p. 655

Pollock, A.M.T., Blondin, J., Stevens, I. 1991, in: K.A. van der Hucht & B. Hidayat (eds.), Wolf-Rayet Stars and Interrelations with Other Massive Stars in Galaxies, *Proc. IAU Symp. No. 143*

(Dordrecht: Kluwer), p. 253

Prantzos, N. 1993a, Astrophysical Journal, Letters to the Editor**405**, L55

Prantzos, N. 1993b, Astronomy and Astrophysics, Supplement Series**97**, 119

Prevot-Burnichon, M.L., Prevot, L., Rebeirot, E., Rousseau, J., Martin, N. 1981, Astronomy and Astrophysics**103**, 83

Robert, C., Moffat, A.F.J., Bastien, P., St-Louis, N., Drissen, L. 1990, Astrophysical Journal**359**, 211

Robert, C., Moffat, A.F.J., Drissen, L., Lamontagne, R., Seggewiss, W., Niemela, V.S., Cerruti, M.A., Barrett, P., Bailey, J., Garcia., J., Tapia, S. 1992, Astrophysical Journal**397**, 277

Roberts, M.S. 1962, Astronomical Journal**67**, 79

Schaller, G., Schaerer, D., Meynet, G., Maeder, A. 1992, Astronomy and Astrophysics, Supplement Series**96**, 269

Schild, H., Testor, G. 1992, Astronomy and Astrophysics, Supplement Series**92**, 729

Schmidt, G.D. 1988, in: G.V. Coyne, A.M. Magalhaes, A.F.J. Moffat, R.E. Schulte-Ladbeck, S. Tapia, D.T., Wickramasinghe (eds.), Polarized Radiation of Circumstellar Origin (Vatican: Vatican Observatory), p. 641

Schmutz, W., Vacca, W.D. 1991, Astronomy and Astrophysics, Supplement Series**89**, 259

Schulte-Ladbeck, R.E., van der Hucht, K.A. 1989, Astrophysical Journal**337**, 872

Schulte-Ladbeck, R.E., Nordsieck, K.H., Nook, M.A., Magalhaes, A.M., Taylor, M., Bjorkman, K.S., Anderson, C.M. 1990, Astrophysical Journal, Letters to the Editor**365**, L19

Schulte-Ladbeck, R.E., Nordsieck, K.H., Taylor, M., Nook, M.A., Bjorkman, K.S., Magalhaes, A.M., Anderson, C.M. 1991, Astrophysical Journal**382**, 301

Shara, M.M., Moffat, A.F.J., Smith, L.F., Potter, M. 1991, Astronomical Journal**102**, 716

Shara, M.M., Moffat, A.F.J., Smith, L.F., Potter, M., Niemela, V.S. 1993, these proceedings.

Shigeyama, T., Nomoto, K., Tsujimoto, T. 1990, Astrophysical Journal, Letters to the Editor**361**, L23

Shore, S.N., Brown, D.N. 1988, Astrophysical Journal**334**, 1021

Signore, M., Vedrenne, G. 1988, Astronomy and Astrophysics**201**, 379

Signore, M., Dupraz, C. 1990, *A&A* (Letters) **234**, L15

Signore, M., Dupraz, C. 1993, Astronomy and Astrophysics, Supplement Series**97**, 141

Smith, L.F. 1968a, Monthly Notices of the RAS**138**, 109

Smith, L.F. 1968b, Monthly Notices of the RAS**140**, 409

Smith, L.F. 1968c, Monthly Notices of the RAS**141**, 317

Smith, L.F. 1973, in: M.K.V. Bappu & J. Sahade (eds.), Wolf-Rayet and High-Temperature Stars, *Proc. IAU Symp. No. 49* (Dordrecht: Reidel), p. 228

Smith, L.F. 1988, Astrophysical Journal**327**, 128

Smith, L.F., Hummer, D.G. 1988, Monthly Notices of the RAS**230**, 511

Smith, L.F., Maeder, A. 1989, Astronomy and Astrophysics**211**, 71

Smith, L.F., Shara, M.M., Moffat, A.F.J. 1990, Astrophysical Journal**358**, 229

Smith, L.F., Maeder, A. 1991, Astronomy and Astrophysics**241**, 77

Stevens, I.R., Blondin, J.M., Pollock, A.M.T. 1992, Astrophysical Journal**386**, 265

Stickland, D.J., Lloyd, C. 1990, *The Observatory* **110**, 1

Stickland, D.J. 1991, in: K.A. van der Hucht & B. Hidayat (eds.), Wolf-Rayet Stars and Interrelations with Other Massive Stars in Galaxies, *Proc. IAU Symp. No. 143* (Dordrecht: Kluwer), p. 237

St-Louis, H., Drissen, L., Moffat, A.F.J., Bastien, P. 1987, Astrophysical Journal**322**, 870

St-Louis, N., Moffat, A.F.J., Drissen, L., Bastien, P., Robert, C. 1988, Astrophysical Journal**330**, 286

St-Louis, N., Howarth, I.D., Willis, A.J., Stickland, D.J., Smith, L.J., Conti, P.S., Garmany, C.D. 1993, Astronomy and Astrophysics**267**, 447

Stothers, R.B. 1991, Astrophysical Journal**383**, 820

Stothers, R.B., Chin, C.-W. 1991a, Astrophysical Journal**374**, 288

Stothers, R.B., Chin, C.-W. 1991b, Astrophysical Journal, Letters to the Editor**381**, L67

Stothers, R.B. 1992, Astrophysical Journal**392**, 706

The, L.-S., Clayton, D.D., Burrows, A. 1991, in: K.A. van der Hucht & B. Hidayat (eds.), Wolf-

Rayet Stars and Interrelations with Other Massive Stars in Galaxies, *Proc. IAU Symp. No. 143* (Dordrecht: Kluwer), p. 537

The, P.S. 1964, *Contr. Bosscha Obs.* No. 26

The, P.S. 1966, *Contr. Bosscha Obs.* No. 35

Terasawa, N., Nakamura, H. 1993, Monthly Notices of the RAS**265**, L1

Torres-Dodgen, A.V., Massey, P. 1988, Astronomical Journal**96**, 1076

Usov, V.V. 1990, *Ap. Space Sci.* **167**, 297

Usov, V.V. 1991, Monthly Notices of the RAS**252**, 49

Usov, V.V. 1992, Astrophysical Journal**389**, 635

Vacca, W.D., Torres-Dodgen, A.V. 1990, Astrophysical Journal, Supplement Series**73**, 685

Vanbeveren, D. 1991 Astronomy and Astrophysics**252**, 159

Vanbeveren, D. 1991 Space Science Reviews**56**, 249

Vanbeveren, D., de Loore, C.W.H. 1993, in: J.P. Cassinelli *et al.* (eds), *ASP Conf. Series* **35**, 257

Vrancken, M., De Greve, J.P., Yungelson, L. Tutukov, A. 1991, Astronomy and Astrophysics**249**, 411

Vreux, J.-M. 1985, Publications of the ASP**97**, 274

Weaver, W.B. 1974, Astrophysical Journal**189**, 263

Webber, W.R., Soutoul, A. 1989, Astronomy and Astrophysics**215**, 128

Wheeler, J.G., Levreault, R. 1985, Astrophysical Journal, Letters to the Editor**294**, L17

White, R.L. 1985, Astrophysical Journal**289**, 698

Williams, P.M., Longmore, A.J., van der Hucht, K.A., Talevera, A., Wamsteker, W.M., Abbott, D.C., Telesco, C.M. 1985, Monthly Notices of the RAS**215**, 23P

Williams, P.M., van der Hucht, K.A., The, P.S. 1987a, Astronomy and Astrophysics**182**, 91

Williams, P.M., van der Hucht, K.A., van der Woerd, H., Wamsteker, W.M., Geballe, T.R., Garmany, C.D., Pollock, A.M.T. 1987b, in: H. Lamers & C.W.H. de Loore (eds.), Instabilities in Luminous Early Type Stars (Dordrecht: Reidel), p. 221

Williams, P.M., Eenens, P.R.J. 1989, Monthly Notices of the RAS**240**, 445

Williams P.M., van der Hucht, K.A., Pollock, A.M.T., Florkowski, D.R., van der Woerd, H., Wamsteker, W.M. 1990a, Monthly Notices of the RAS**243**, 662

Williams P.M., van der Hucht, K.A., The, P.S., Bouchet, P. 1990b, Monthly Notices of the RAS**247**, 18P

Williams, P.M., van der Hucht, K.A., The, P.S., Bouchet, P., Roberts, G. 1991a, in: K.A. van der Hucht & B. Hidayat (eds.), Wolf-Rayet Stars and Interrelations with other Massive Stars in Galaxies, *Proc. IAU Symp. No. 143* (Dordrecht: Kluwer), p. 417

Williams, P.M., van der Hucht, K.A., Tapia, M., Caldwell, J.J., Anthony, D.M., Fitzsimmons, A., Conlon, E.S. 1991b, in: C. Jaschek & Y. Andrillat (eds.), The Infrared Spectral Region of Stars (Cambridge: CUP), p. 351

Williams, P.M., van der Hucht, K.A. 1992, in: L. Drissen, C. Leitherer & A. Nota (eds.), Non-Isotropic and Variable Outflows from Stars, *ASP Conf. Series* **22**, p. 269.

Williams, P.M., van der Hucht, K.A., Bouchet, P., Spoelstra, T.A.Th., Eenens, P.R.J., Geballe, T., Kidger, M.R., Churchwell, E.B. 1992, Monthly Notices of the RAS**258**, 461

Willis, A.J., Garmany, C.D. 1987, in Y. Kondo (ed.), Exploring the Universe with the *IUE* satellite (Dordrecht: Reidel), p. 157

Willis, A.J., Howarth, I.D., Smith, L.J., Garmany, C.D., Conti, P.S. 1989, Astronomy and Astrophysics, Supplement Series**77**, 269

Willis, A.J., Stickland, D.J. 1990, Astronomy and Astrophysics**232**, 89

Willis, A.J. 1991, in: K.A. van der Hucht & B. Hidayat (eds.), Wolf-Rayet Stars and Interrelations with Other Massive Stars in Galaxies, *Proc. IAU Symp. No. 143* (Dordrecht: Kluwer), p. 265

Willson, L.A., Stalio, R. (eds) 1990, Angular Momentum and Mass Loss for Hot Stars, *NATO ASI Series* **C316** (Dordrecht: Kluwer)

MASSIVE STAR DISTRIBUTION IN EXTERNAL GALAXIES AND STARBURST REGIONS

PETER S. CONTI

Joint Institute for Laboratory Astrophysics, Campus Box 440
University of Colorado, Boulder, CO 80309 USA

Abstract. Counts of hot and luminous stars in a number of associations in the Galaxy and Magellanic Clouds enable one to directly investigate the numbers and types of massive stars. There seems to be little, if any, dependence of the slope of the Intial Mass Function, or the $\mathcal{M}_{upper}$ on the initial composition of the stars. Indirect estimates of numbers of massive stars in various more distant environments are reviewed and discussed within a framework of a *calibration* of the methods using the stellar census of 30 Doradus. Very young starbursts, containing large numbers of massive stars, seem to be composed of smaller sub-units similar or somewhat larger than that object. These units might be newly born globular clusters.

Key words: Stars, O-type – Galaxies, Local Group – Galaxies, Starburst

1. Introduction

The term "Massive Stars" will be taken here to be those whose masses are sufficiently large that they will end their lives as supernovae; this typically means $\gtrsim 8\mathcal{M}_{\odot}$, corresponding to O to mid-B main sequence spectral subtypes. When I speak of "Distribution", I think of the following: "Where are they located?"; "How many and which kinds are there?" In this paper, I shall primarily be concerned with the values of the slope of the *Initial Mass Function* (IMF) and *Star Formation Rate* (SFR), or *Star Formation History* (SFH).

Massive stars are, for the most part, born in giant molecular clouds (GMCs) in stellar associations; a powerful enough birth event would be called a "starburst". Initially these stars will be surrounded by the dense molecular gas cloud and shrouded by the commonly associated dust. These ensembles will radiate strongly in the IR and radio regions due to the dust heating and gas excitation but might be completely hidden optically (*e.g.* W49, W51 in our Galaxy). After some time, the molecular clouds are dissociated and the dust is dissipated by the radiation and stellar winds from the O stars within, and the region becomes visible as an optical H II or giant H II (GH II) region (*e.g.* 30 Doradus). The appearance of the spiral arms of galaxies in the visible is primarily determined by the distributions of the H II and GH II regions within them. For the nearer galaxies, the individual stars of these ensembles may be investigated, but for more distant ones, only the integrated properties of the association as it affects the excited gas (and dust) can be studied. There is a great body of literature on the distribution of massive (hot) stars in our Galaxy and others of the Local Group and the location and numbers of GH II regions in more distant objects.

Space Science Reviews **66**: 37–53, 1994.

© 1994 *Kluwer Academic Publishers. Printed in Belgium.*

Actual counts of massive stars in associations can be used *directly* to give estimates of the slope of the IMF, along with the related quantities $\mathcal{M}_{upper}$ and $\mathcal{M}_{lower}$. I will discuss these parameters for a group of associations in our Galaxy and the Magellanic Clouds in §2. I will then turn to *indirect* methods used to confront the questions of "How many and what kinds?" of massive stars are present in more distant GH II regions and starburst galaxies where the individual stars cannot be identified. In §3 I will consider indirect methods that make use of spectra; in §4 I shall discuss two global properties of galaxies, their far-IR (FIR) luminosities and optical and UV imaging. For *indirect* methods, I will examine the use of 30 Doradus as a fundamental *calibrator*.

2. Direct Star Counts - Census

Pioneering efforts to elucidate the numbers and types of massive stars in various environments have been made primarily by Massey and associates (references to follow). I shall briefly review the procedure and then summarize the results. This body of literature is a homogeneous approach to the determination of the IMF for various associations of the Galaxy and Magellanic Clouds. While it is possible that systematic errors remain, the similarity of analyses indicates that the comparisons among various stellar groupings are quantitatively correct.

2.1. PROCEDURE

One begins with the acquisition of deep CCD UBV frames of the relevant stellar associations. Accurate photometry (to 0.02 mag) must be accomplished, and color-color plots used to estimate the extinction and identify the bluest stars. As Massey (1985) has shown, even the unreddened UBV colors for the hottest stars are degenerate, that is, one cannot distinguish between the hottest and coolest O type stars on the basis of their photometry alone, as is commonly done with "luminosity functions". The UBV colors *can* be used to determine the brightness of the stars but spectra suitable for classification are needed to determine the T_{eff} of all stars earlier in type than B1V or so. Obtaining spectra is a time consuming effort requiring large telescopes; the photometry can be done on modest ones.

The next step is to derive a distance for the association by "classic" spectroscopic parallax for the Galactic clusters. For the Magellanic Clouds, one adopts the standard distances. Next, one converts the parameters of M_v and spectral type (or unreddened color) to M_{bol} and T_{eff}, using a calibration procedure. Then one plots these "observed" parameters for the association stars on a "theoretical" HR diagram with evolutionary tracks, typically those of Maeder and associates. Finally, one counts the numbers of stars in each mass interval along the track, and plots the values as a function of mass. The *slope* of this relationship is referred to as Γ, defined in the equation

$$f(\mathcal{M}) = A\mathcal{M}^{\Gamma-1} \tag{1}$$

where f($\mathcal{M}$) is the fractional number of stars per unit mass interval $\mathcal{M}$, A is a scaling constant, and the *Salpeter* value for Γ is -1.35.

By the above procedure, one is essentially measuring the slope of the *Present Day Mass Function* (PDMF). An important assumption is that this number is identical to the slope of the IMF; in other words, there are no stellar deaths. This is reasonable for the very youngest O associations, and one can, if necessary, account for the already highly evolved W-R stars that are present in some of the regions studied. Another assumption is that the formation time is of the order of, or less than, the evolution time, which seems reasonable for O associations. Finally, one ignores the binary membership. Unless the binary fraction, typically considered to be 40% for hot stars (Garmany *et al.* 1982), is different from place to place, this assumption is also not unreasonable in the context of seeking similarities and differences in the numbers and types of massive stars in various environments.

2.2. RESULTS

In Table I, I have summarized the results for ten associations in the Galaxy and Magellanic Clouds, along with that for the solar vicinity. I have grouped the values by galaxy, which can be thought of as sampling the composition of the environment out of which these stars formed. The two entries for the solar vicinity were obtained by a method similar to the others but the models used were older and the numbers not quite comparable; the values are just noted for completeness. The authors of the various papers cited suggest that Γ has been determined to an accuracy of ± 0.2 in each association. If there is a dependence on "metal abundance", Z, which is by no means clear, then Γ gets *shallower* as the metal abundance *increases*. This is completely at odds with the prevailing view (*e.g.* Shields and Tinsley 1976) where the Γ is suggested to get *steeper* with *increasing* abundance.

In Table I, I also give a quantitative indication of $\mathcal{M}_{upper}$ of the most massive objects by listing the actual numbers of stars with masses inferred from the M_{bol} and T_{eff} to be larger than $60 \mathcal{M}_{\odot}$. This is obtained by inspection of the HR diagrams plotted in the various papers. The values, which are more or less proportional to the total numbers of stars in each region, range further upwards in mass to somewhere between 80 to $100 \mathcal{M}_{\odot}$; this is a reasonable estimate of $\mathcal{M}_{upper}$ for starburst modeling purposes. There is NO dependence of either the numbers of massive stars or the $\mathcal{M}_{upper}$ on the galaxy environment, or on Z.

It has been suggested from *indirect* arguments that the $\mathcal{M}_{lower}$ limit in some starburst regions might be significantly larger than the canonical 0.1 $\mathcal{M}_{\odot}$ found near the sun, and more like a few $\mathcal{M}_{\odot}$ (*e.g.*, M 82 - McLeod *et al.* 1993). Is there *any* direct evidence for this? Among well studied energetic GH II regions, 30 Dor would be the place to look. However, Parker's (1991) survey was complete only to an apparent magnitude corresponding to a few $\mathcal{M}_{\odot}$, *i.e.*, just where it begins to get interesting. Despite the (crowding) difficulties, a deeper CCD photometric

TABLE I

Massive Star Parameters for Various Regions

Association	$-\Gamma$	$\# > 60\mathcal{M}_\odot$	Galaxy	References
Field $< R_\odot$	1.3		Milky Way	Garmany *et al.* 1982
Field $> R_\odot$	2.1			Garmany *et al.* 1982
Cyg OB2	1.0	7		Massey and Thompson 1991
Car OB1	1.3	7		Massey and Johnson 1993
Ser OB1	1.1	1		Hillenbrand *et al.* 1993
LH9	1.6	0	LMC	Parker *et al.* 1992
LH10	1.1	4		Parker *et al.* 1992
LH58	1.7	0		Garmany *et al.* 1993
LH117	1.8	2		Massey *et al.* 1989a
LH118	1.8	0		Massey *et al.* 1989a
30 Dor	1.4	21		Parker 1991
NGC 346	1.8	3	SMC	Massey *et al.* 1989b

survey of 30 Dor needs to be made to investigate its $\mathcal{M}_{lower}$ limit and this issue in general.

In their study of NGC 6611 = Ser OB1, Hillenbrand *et al.* (1993) have gone sufficiently deep in their CCD survey to be able to say something about the stellar population at 3 - 7 $\mathcal{M}_\odot$. Remarkably, they find that these stars are *above* the main sequence, in the pre-main sequence phase. Furthermore, the ages of these still contracting stars are a few $\times 10^5$ years, appreciably *less than* the turn-off time of the upper main sequence, which is a few $\times 10^6$ years! Thus in this association, the presence of (at least) thirteen O stars has *not* inhibited further star formation of lower mass stars. Whether ten or one hundred times that many O stars would inhibit subsequent lower mass star formation remains problematic.

Hillenbrand *et al.* (1993) also call attention to several luminous stars which sit well to the right of the main body of massive stars in NGC 6611. They argue that these stars are indeed cluster members, which must have formed *before* most of the rest. Eye examination of the rest of the associations referenced above also invariably reveals a few stars in similar, advanced, evolutionary stages. I would propose to call these stars "blue leaders" (by analogy with the better known "blue stragglers" - which seem to be *less advanced* in their evolutionary state). Hillenbrand *et al.* suggest star formation might proceed much like "popcorn" when it is heated; a few kernels "pop" before the main body, and a few more lag behind. The "blue leaders", which seem almost ubiquitous in stellar associations, might be understood in a "popcorn" model of star formation.

2.3. Luminosity Functions

Massey (1985) noted that using only photometry, it is difficult to distinguish among the most massive stars employing luminosity functions alone. In particular, in Massey *et al.* (1989b) they show that had they taken only their UBV photometry for the analysis of NGC 346, they would have found Γ to be -2.5, instead of the value of -1.8 listed in Table I above. Massey *et al.* (1986) have done CCD photometry of several associations in M31. Using plots and evolution tracks similar to those discussed above, they obtained the curious result (shown in their Figs. 31 and 32) that there are *no* stars in the luminous associations OB78 and OB48 more massive than 40 $\mathcal{M}_\odot$, although both have W-R stars present. I suspect that this inference is due purely to the absence of spectroscopic data. The use of luminosity functions for massive stars must be thus treated with caution, as both the derived Γ and the $\mathcal{M}_{upper}$ might be suspect. For luminous stars of M31 and other Local Group galaxies, spectra are difficult but not impossible to obtain on the largest telescopes, especially with multiple slit instrumentation.

2.4. Wolf-Rayet Stars

These highly evolved objects are descended from the most massive O-type single stars and close binaries. W-R star spectra are made up predominantly of emission lines from their strong stellar winds. While this complicates our understanding of the physics of the outflow, it does allow us to identify and classify these objects at fainter apparent magnitudes than is practicable for absorption line stars. Surveys of all of the Local Group and even more distant galaxies for W-R stars has been carried out by several groups. The presence of W-R stars is a nice indication of the presence of massive stars which have been born only a few million years ago. A summary of the distribution of individual W-R stars in our Galaxy and others is given by Conti and Vacca (1990) and Massey and Armandroff (1991; see also Maeder 1991a).

3. Spectra of Galaxies

The spectra of galaxies are primarily those of the underlying stellar population (absorption lines) with the addition of nebular emission lines from the GH II and young starburst regions which are present if recent star formation has occurred. I will first consider how one might infer the distribution of massive stars from the nebular line analyses, and then discuss spectral synthesis of the stellar features.

3.1. Nebular Line Analyses

3.1.1. H II and GH II Regions

It is well known that nebular emission lines may be used to infer abundances of certain elements, in particular oxygen (*e.g.* Osterbrock 1989). From extensive studies of H II and GH II regions in our Galaxy and others (*e.g.* Shields 1990) has come the concept that there exist, in at least some spiral galaxies, "chemical gradients": typically a *decrease* of abundance with *increasing* galacto-centric distance (e.g., Shaver *et al.* 1983; McCall *et al.* 1985). In very simple terms, what is observed is an *increase* in the [O III] / $H\beta$ ratio with galacto-centric distance. While initially attributed to the T_{eff} of the exciting stars (Aller 1942), it was later realized to be primarily an abundance effect in the nebulae (Searle 1971). This is due to the fact that as the oxygen abundance *decreases*, the nebular cooling due to the "heavy" elements also *decreases*, the electron temperature - T_e *increases* and the [O III] (and other) lines *increase* in strength, relative to the Balmer series. I need to stress that this reasoning is relatively *independent* of the nebular or stellar models. If the abundance of oxygen in spiral galaxies typically decreases with increasing galacto-centric distance, then, since this element is produced primarily by massive stars, their SFH has been more energetic towards the centers of such objects. Thus we would expect that there would have been more massive stars towards the centers of galaxies rather than outwards.

The nebular T_e arguments above have been taken another step, *i.e.* to infer the T_{eff} of the exciting stars (*e.g.* Vilchez and Pagel 1988 - but see McGaugh 1991). This is carried even further to say something about their masses (*e.g.* Vilchez *et al.* 1988). In M33 these authors found that, based upon the nebular analyses, the T_{eff} of the exciting stars got hotter with increasing galacto-centric distance, and with decreasing abundance. If this were true generally, one would then conclude either that the $\mathcal{M}_{upper}$ becomes larger with these two parameters or the IMF slope Γ *shallower*, or both. We see that this result is opposite that of Table I. There I suggested that neither Γ nor the $\mathcal{M}_{upper}$ had much dependence on Z (if anything, Γ gets shallower with increasing abundance). It is also opposite what is implied by the oxygen gradient itself and the presumed SFH.

Inferences about the T_{eff} from nebular line analyses are *indirect* and very highly dependent on the nebular and stellar models. In the former case, it is assumed that dust and clumping play no role, and that the nebula is spherically symmetric. In the latter case, the emergent energy distribution is taken from the Kurucz (1979) blanketed LTE models, in which no allowance is made for the effect of the overlying wind.

Is there any way to check the appropriatness of deriving a stellar T_{eff} from nebular line analyses? For example, do nearby well-studied Galactic H II regions show the same T_{eff} as that inferred from the spectral types of their exciting star(s)? Curiously, this has only been confirmed for the Orion nebula, in which the inferred T_{eff} of 39000 K is similar to that expected from the central star, type O7V. In a few other cases for which I have been able to extract the numbers from the literature, the nebular and spectral type T_{eff} do not always agree. For example, Mathis and Rosa (1991) have provided predictions of the (O^+/O^{++}) and (S^+/S^{++}) abundances as a function of the T_{eff} from *models*. Taking *observed* values of (O^+/O^{++}) and (S^+/S^{++}) in some well-studied H II regions from, say, Dennefeld and Stasinska (1983), one can compare their T_{eff} with that inferred from the spectral types of the exciting stars. In η Car, the nebular T_{eff} is 44000

K; but there are several O3 stars in this association, with T_{eff} closer to 50000 K. By contrast, in the H II region NGC 346, which also contains O3 stars, T_{eff} (nebular) is inferred to be 75000 K. What's going on here? A concerted effort to investigate the consistency of nebular line analyses in a number of Galactic H II regions is now under way by my PhD student, Victor Robledo. Perhaps in a year or two we will be better able to model the nebular T_{eff} procedure.

While I do not believe that T_{eff} of the exciting stars, or their *masses* may be inferred from nebular line analyses of H II or GH II regions, the *numbers* of exciting stars can readily be found (*e.g.* Kennicutt 1984; Shields 1990). One observes the spatially integrated Hα flux (or the Hβ or radio recombination measure, *etc.*), accounts for the extinction (if necessary), and from simple recombination theory infers the total number of Lyman continuum photons ($NLyc$) being emitted. This last step contains the nominal assumptions for the nebula of "Case B", no dust, and ionization bounded. The $NLyc$ is a direct measure of the number of exciting stars. The total number released is a product of the numbers of hot stars, the slope of the IMF, and the $NLyc$ at each spectral subtype.

Vacca (1991; 1993) has recently thoroughly reviewed and quantified this procedure. He introduces the parameter η_0 as defined in the equation

$$NO^* = NO7V/\eta_0 \tag{2}$$

where $NO7V$ is the number of "equivalent" O7V type stars, and NO^* the total number of O stars. The $NO7V$ is simply the *observed* $NLyc$ divided by the number of Lyman continuum photons emitted by an O7V star (1×10^{49} s^{-1}). The η_0 can be calculated as a function of Γ, T_{eff} and $\log g$ using the Kurucz stellar atmosphere models. Its value, tabulated by Vacca, depends also on $\mathcal{M}_{upper}$, and the $\mathcal{M}_{lower}$ for Lyman photon production, roughly the OB star boundary. This latter parameter depends on the metal abundance Z via the stellar structure models of Maeder and associates. This procedure makes the assumption that all O stars in H II and GH II regions are main sequence; one can separately allow for massive star evolution.

How accurately does this method work? It has been calibrated using 30 Dor in the LMC. Parker (1991) has made a detailed census of the hot star population in an $7' \times 7'$ area centered on R136. He finds about 400 O stars. Vacca (1991, 1993) has analyzed the nebular spectrum of a trailed, 15 minute spectrophotometric exposure of an $8' \times 8'$ area scan centered on R136 (taken by Mark Phillips). Using the Z value for the LMC, Parker's value for Γ (-1.4), he finds η_0 to be 0.44 and estimates about 330 O stars are present, within 30% of the actual count (allowing for the slight difference in areas)! This gives us confidence in the procedure.

3.1.2. Emission Line Galaxies - Starbursts

These are galaxies with emission line spectra like those of H II regions, first noted by Sargent and Searle (1970). Given the often substantial numbers of O type stars found within these galaxies, they can be understood to be examples of very young "starbursts". Using slit spectroscopy, one may obtain the numbers of exciting stars by a procedure similar to that outlined above for 30 Dor. For galaxies, however, we might have additional complications if the slit width does not include the entire region of interest, if the nebulosity is density bounded, or if dust is present in sufficient quantities to absorb a considerable fraction of the Lyman photons.

3.1.3. Wolf-Rayet Galaxies

These are a subset of emission line galaxies in which, in addition to the nebular line spectrum, one observes broad emission at $\lambda 4686$ A due to the presence of Wolf-Rayet *stars* (*e.g.* Conti 1991 and references therein). Since evolutionary models indicate that the W-R phase lasts only a small fraction of the main sequence lifetime of massive O stars (Maeder 1991b), W-R features in the spectra of emission-line galaxies also provide direct evidence of a recent starburst episode. The starburst phenomena illustrated by W-R galaxies represent an extreme "burst" of star formation, in which hundreds to thousands (or more) of massive stars have been born.

There are currently about 50 examples of such systems, most of which have been discovered serendipitously. Many are Markarian or Zwicky galaxies and exhibit disturbed morphologies which may be the result of interactions ormergers. Examples of W-R galaxies may be found among Blue Compact Dwarf Galaxies (*e.g.* Sargent 1970), isolated extragalactic H II regions (*e.g.* Sargent and Searle 1970), dwarf irregulars (*e.g.* Dinerstein and Shields 1986), "amorphous" galaxies (*e.g.* Walsh and Roy 1987), spiral galaxies containing knots or GH II regions (*e.g.* Keel 1982), recent galaxy mergers (*e.g.* Rubin *et al.* 1990), or powerful IRAS galaxies (Armus *et al* 1988). As examples of substantial massive star formation, W-R galaxies may be paradigms for much more distant faint blue galaxies.

The broad emission feature(s) are seen in contrast to the galaxy stellar population continua. The dilution is such that the few hundred A equivalent width of an emission line of typical single W-R stars is usually only a few A in W-R galaxies. Examples of the spectra are given in Vacca and Conti (1992) and Conti (1993). Many of these galaxies are "metal weak" but this may be a selection effect given that starburst galaxies with more normal composition are likely to have a brighter underlying stellar population, which could "drown out" the W-R stars even if present.

Vacca and Conti (1992) have recently analyzed optical spectra of ten Wolf-Rayet and four other emission line galaxies. The nebular line ratios indicate the excitation is caused by stars (following Veilleux and Osterbrock 1987). The strength of the $\lambda 4686$ A may be used to infer the numbers of W-R stars present; this uses a calibration of the line flux for single W-R stars in the LMC, dividing that number into the measured line flux in the galaxy. Typically, tens to hundreds of W-R stars are inferred to be present in the starbursts. This procedure has been

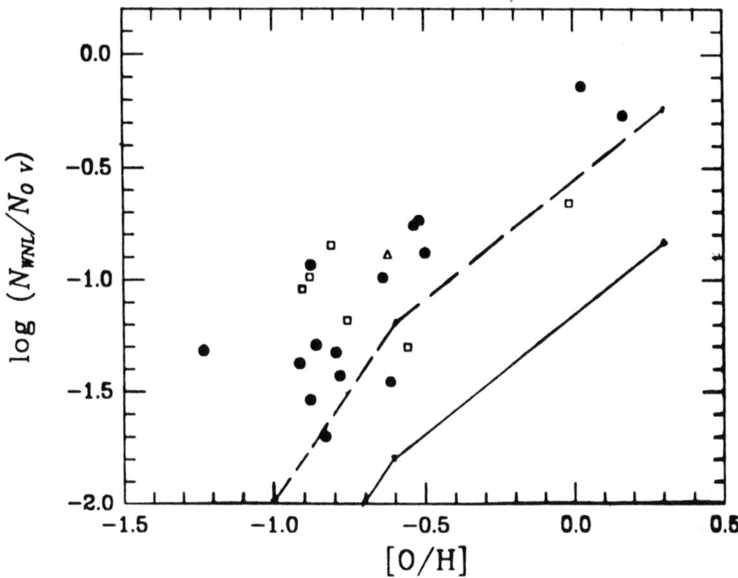

Fig. 1. WNL/OV ratio in W-R galaxies, as a function of the oxygen abundance [O/H], adapted from Conti (1993). Filled circles, galaxies of Vacca and Conti (1992); open boxes, galaxies of Vacca (private communication); open triangle, 30 Dor (from Vacca 1991). The solid line denotes the predictions of Maeder (1991b) for "continuous" star formation; the dashed line, the values estimated for a "burst" (see text).

quantitatively checked by using the area spectrophotometry of 30 Dor in which a broad $\lambda 4686$ He II line is measured; the inferred numbers of W-R stars (20) are similar to the census of Moffat *et al.* (1987). The contribution of the W-R stars to the observed $NLyc$ is subtracted from the observed value for the galaxy, and the remainder is treated, as above, to determine the number of O-type stars present. As it is *not* possible to derive the slope of an IMF from this procedure, a Γ equivalent to that of 30 Dor was adopted for each galaxy; similarly the $\mathcal{M}_{upper}$ was assumed to be the same ($100\mathcal{M}_{\odot}$). With the usual assumptions that low mass stars have formed along with the high mass ones, typical SFR range from 2×10^{-2} to 3 $\mathcal{M}_{\odot}$ per year (the former is the value for 30 Dor).

Highly evolved W-R stars are (mostly) in the helium burning stages of stellar evolution; their anomalous surface compositions are primarily due to mass loss (and mixing) in earlier phases. This process depends strongly on the initial composition (Z) of the (single) star: larger abundances will result in the production of more W-R stars, partly because the lower mass limit for their production decreases and partly because the helium burning phase will be longer. Thus one predicts a dependence of the W-R/O star ratio on Z (e.g. Maeder 1991b). Star formation of massive stars in starbursts might go on for a time long with respect to the evolution time; such a mode would be called "continuous" ($\tau_{for} \gtrsim 10^7$ years). Alternatively, it might be shorter ($\tau_{for} \lesssim 10^6$ years) and labeled as a "burst".

Maeder (1991b) has given predictions of the W-R/O star ratio as a function of Z in terms of [O/H] for the case of "continuous star formation. Following Arnault *et al.* (1989), Vacca and Conti (1992) estimated that the "burst" values would be about four times larger. In Fig. 1, I show the predicted and observed values for the W-R/O star ratio in a selected sample of W-R galaxies. The observed values are, strictly speaking, the number of WN stars derived from the strength of $\lambda4686$ He II, but in most of these W-R galaxies, there is little or no evidence of WC stars. The largest source of uncertainty in the observed values is the potential for a mismatch between the slit width (1.5$''$) and the starburst (typically 2 to 3$''$). This underestimates the number of O stars. Arbitrarily correcting for this would *lower* the ratio by a factor 2 (see a more thorough discussion in Conti 1993).

From Fig. 1 one can conclude that for most W-R galaxies, the W-R/O ratios are well above the predictions for "continuous" star formation, but a better match to the "burst" models. From this we can infer that the starbursts we are observing in W-R galaxies are typically going on for only a relatively "brief" interval, typically $\tau_{for} \lesssim 10^6$ years. The energetics are similar to 30 Doradus at the faint end to more than $100\times$ larger. Are these nearby starbursts paradigms for the first phases of massive star evolution in very young galaxies?

3.2. SPECTRAL SYNTHESIS

Optical spectroscopy of most galaxies shows absorption line spectra from an old evolved population of late type giant stars. In the starbursts of the W-R galaxies studied by Vacca and Conti (1992) the optical absorption spectra are those of late B and A type stars (narrow K line, upper Balmer series in absorption, no other features). In these objects, even though there are large numbers of very hot and very young stars, the optical spectra are dominated by a somewhat older population. However, in the UV the O (and W-R) stars become the dominant contributors to the continua. In emission line galaxies containing OB stars, P Cygni spectral features at $\lambda1400$ Si IV and $\lambda1550$ C IV are seen; if W-R stars are present, an emission line at $\lambda1640$ He II is found. As examples of such spectra, in Fig. 2 I show an IUE spectrum of a $3' \times 3'$ area centered on 30 Doradus (analogous to the optical spectra discussed above) and one of the W-R galaxy NGC 1741. Their spectral features look remarkably similar!

The spectra shown in Fig. 2 can be used directly to estimate the numbers of hot stars present by spectral synthesis techniques (Leitherer *et al.* 1992; Robert *et al.* 1993). We can also ascertain the age of the starburst and tell whether or not the formation of the massive stars has been "continuous" or a "burst". In Fig. 3, I show a time series of models of a "burst" of star formation matching the line profiles observed in the 30 Dor ($3' \times 3'$ area) IUE spectrum. There is a reasonable fit for the Si IV, C IV and He II profiles for ages between 1.5 and 2 million years. For younger ages, insufficient W-R stars have been produced to match the $\lambda 1640$ emission line; for older ages, the Si IV and C IV profiles begin to diverge from the observations. Using analogous plots as in Fig. 3 but for the "continuous" star formation models, the agreement is not as good at any age.

It is also possible to estimate the number of O stars from the extinction corrected IUE continuum of 30 Dor, as the models predict this quantity. The agreement with the census from the Parker (1991) thesis is good. My use of 30 Dor as a calibrator for UV spectra of galaxies comes from work in progress by Vacca *et al.* (1993). Bill Vacca and I also have efforts under way to do analogous spectral synthesis of IUE spectra of several W-R galaxies. Unfortunately, galaxies such as these are at the limit of IUE sensitivity. The $10'' \times 20''$ slit is typically larger than the starburst, so dilution of the spectral signatures can be a problem. For the number counts, but not for the age estimate, the UV extinction is critical and remains a serious problem for more general application of this method.

4. Global Properties

4.1. FAR INFRARED LUMINOSITIES

Radiation from dusty H II, GH II regions and galaxies in the far infrared (FIR) appears to come from heating of the dust by their stellar populations. The FIR luminosities have been argued to relate directly to the content of hot stars by Devereux and Young (1991). They have derived the FIR luminosity of 124 spiral galaxies from the IRAS archives, which also have $H\alpha$ measurements. From the individual IRAS colors, they find that this dust is at a temperature of 30 to 40 K, corresponding to heating by *hot* stars; they suggest that ambient heating by stars of all types would be more like 15 - 20 K. In their Fig. 5 they show a correlation between the FIR and the $H\alpha$ emission extending over two orders of magnitude. Part of the dispersion in their relation could come from the differences in T_{eff} of the exciting stars. Although Devereux and Young claim that one can infer the actual T_{eff} values, I don't believe that the modeling used is adequate.

4.2. IMAGING OF EMISSION LINE AND W-R GALAXIES

Optical CCD photometry of galaxies is a growth industry: spectral syntheses of the continua can give us valuable information on stellar populations of all *but* the hottest and most massive stars. As I have already noted, even in W-R galaxies, demonstrably the youngest examples of starburst phenomena, the optical light comes primarily from stars of type A. Thus in this wavelength region we are sampling stars with lifetimes (not necessarily ages) of 50 million years or more.

As PI on a Cycle 2 Hubble project, I have recently been able to acquire UV photometry of a number of W-R galaxies. We (Filippenko, Leitherer, Robert, Sargent, Vacca) have used the FOC camera and the intermediate $\lambda 2200$ A filter to obtain images. At this wavelength one is primarily sampling the OB stars, which have lifetimes of a few $\times$ 10^6 to 10^7 years. Given the presence of W-R stars in these galaxies, we know the *ages* are closer to the lower of those numbers. Also,

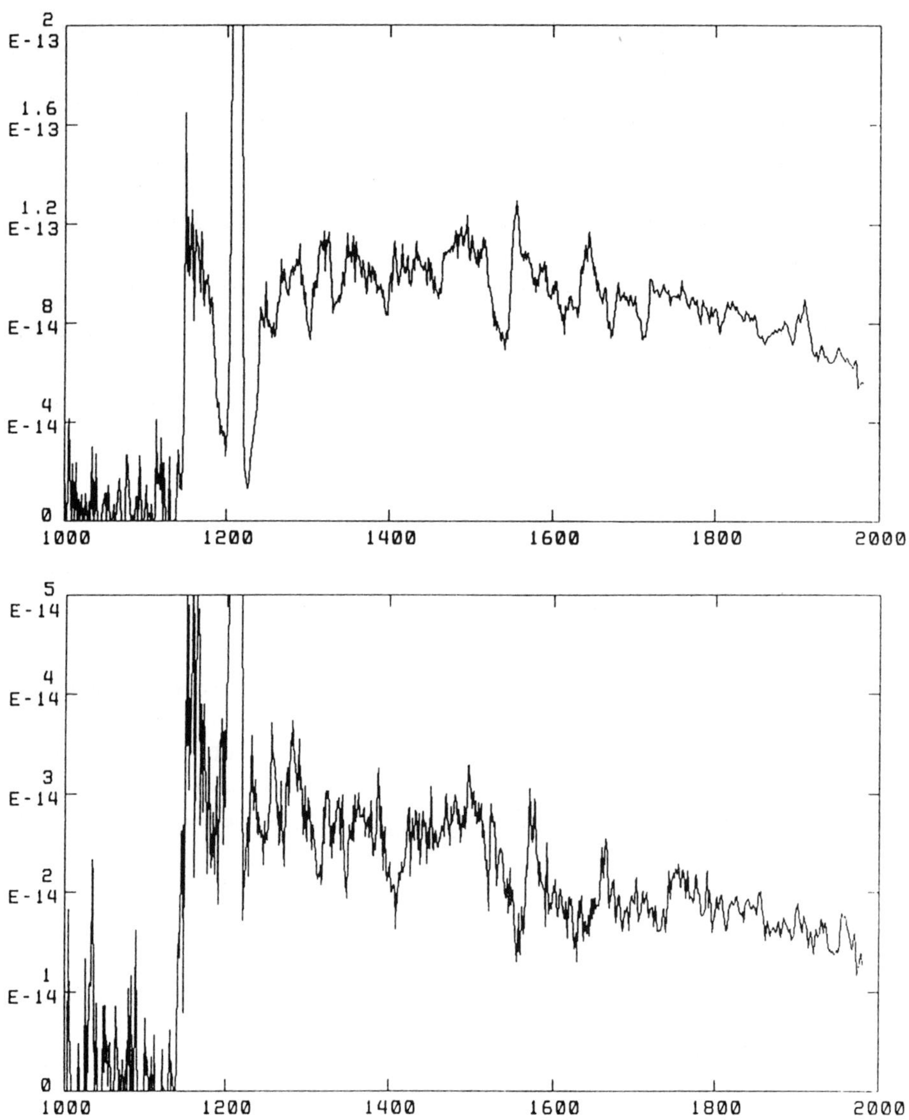

Fig. 2. Observed IUE SWP spectra of 30 Dor (top) and NGC 1741, a W-R galaxy (bottom). Stellar line features at λ1400 Si IV, λ1550 C IV, λ1640 He II and λ1718 N IV are seen in both objects. The lines shortward of Si IV are primarily interstellar.

With some caution, one can perhaps use the FIR luminosities of star forming regions to infer the total numbers of hot stars present, but, absent other information, one *cannot* infer an IMF slope (Γ) or an $\mathcal{M}_{upper}$. The FIR luminosities for normal spiral galaxies range from 10^9 to 10^{11} $L_\odot$ and, according to Devereux and Young (1991), show no dependence on the Hubble type. It would be nice to confirm that the

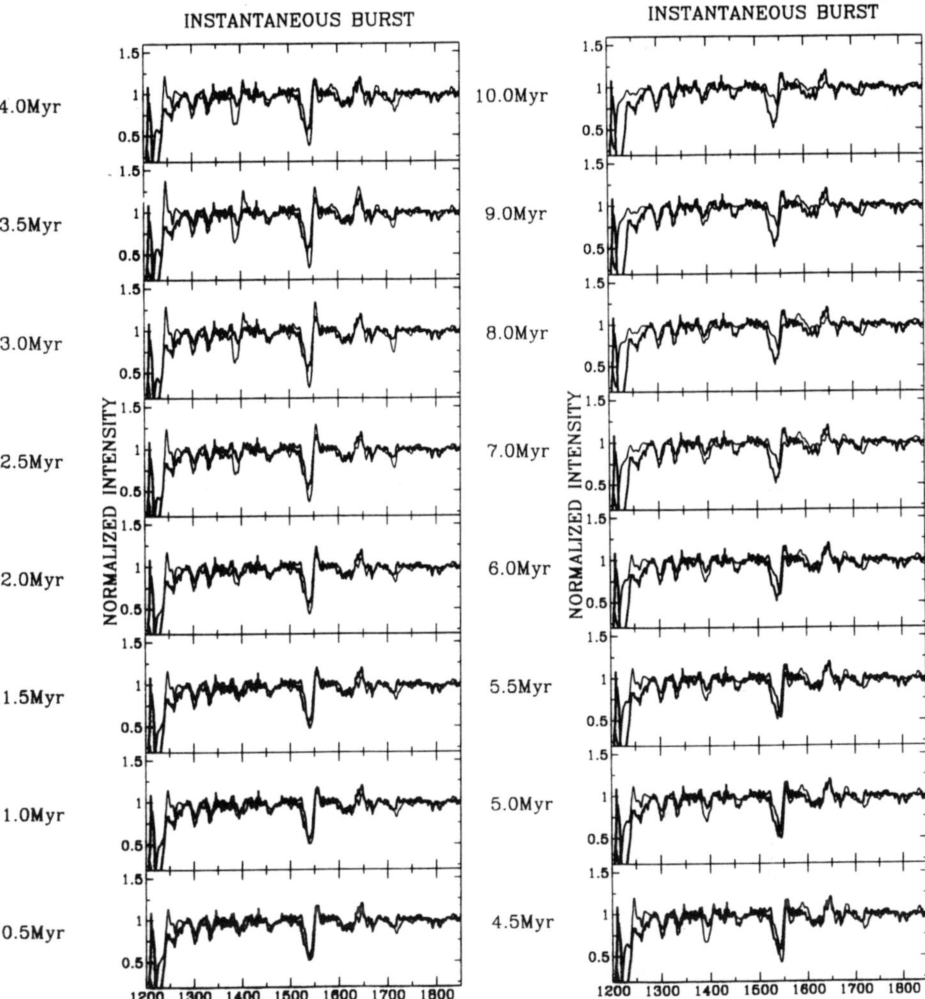

Fig. 3.   Spectral synthesis of IUE spectrum of 30 Dor. The dark line is the observations; the light line, the various models. The age of the burst is given in the LH margins. There is a reasonable fit for ages of 1.5 to 2 million years.

relationship between FIR luminosity and numbers of hot exciting stars is as orderly as claimed. This might be accomplished with identifications and counts of stars in otherwise obscured GH II regions of our Galaxy, such as W51, through K band spectroscopy (Conti *et al.* 1993).

the spatial resolution of HST, with image restoration, is $\approx 0.1''$.

In *all* of the W-R galaxies for which we have obtained FOC UV imaging, now some 15 in number, we consistently see starbursts that are unresolved in the optical breakup into smaller sub-units in the UV high spatial resolution FOC environment. Several of our HST team are currently engaged in evaluating the UV fluxes of these starburst knots. I can report on a preliminary reduction of similar objects found in the blue compact dwarf galaxy He2-10 (Conti aand Vacca 1993). This W-R galaxy has recently been found by Corbin *et al.* (1993), using deep CCD photometry, to be at the center of a faint elliptical galaxy! There are three starburst centers, the strongest (containing W-R stars) at the center, another one $10''$ east, and a fainter one W (and slightly N). The interested reader should peruse Figs. 1 and 2 of Corbin *et al.* These starbursts are resolved optically but the image sizes are still limited by seeing ($1.2''$).

Figure 4 is a reconstructed HST image of the central starburst in He2-10. We see ten individual knots of activity, which I will call "globs" for a reason to become apparent presently. These seem to lie along a curved line, with linear extent about 100 pc. Careful investigation of the physical sizes of these "globs"in the HST image reveals them to be about 10 pc in diameter, with separations of about twice this number. In obtaining their optical spectra of He2-10, Vacca and Conti (1992) had used a long slit, $1.5''$ in width, lying EW (so as to include both the nucleus and the E knot). Although the existence of individual sub-units of star formation was unknown at the time, their aperture encompassed all the "globs" shown in Fig. 4.

Absent additional color or spectral information of the individual "globs", we assume each of them is at a similar stage of evolution, as represented by the integrated optical spectra and bright UV continuum. We believe each contains O stars but whether or not W-R stars are also present depends on the precise ages. An interesting question is whether or not these objects have formed "simultaneously" or "sequentially"; HST UV spectra might allow us to address this issue. A potentially analogous object is NGC 4214, a face-on starburst galaxy with four sub-units of activity. Sargent and Filippenko (1991), using optical spectroscopy, have found W-R stars in two of these; they suggest an age effect might be responsible.

It is appropriate to measure the $\lambda 2200A$ intensities of each of the He2-10 objects, resulting in numbers of a few $\times 10^{-16}$ ergs/s/cm^s/A each. Allowing for the extinction through the FOC 2200A filter, by assuming an (intrinsic) power law index for the galaxy in this wavelength region of -3 (due to hot stars), we find $A_{2200A} = 7.86 \times E(B - V)$. Given the latter value from Vacca and Conti, correcting for the distance of He2-10 (9 Mpc), $\lambda 2200A$ fluxes of a few $\times 10^{38}$ ergs/s/A are found for each "glob" (with a scatter of a factor of a few). These numbers can now be compared to the $\lambda 2200A$ corrected fluxes of the 30 Dor $3' \times 3'$ area from our IUE spectra: 4×10^{37} ergs/s/A. We find that *each* of the He2-10 "globs" is from slightly more to ten times more luminous than 30 Dor at $\lambda 2200A$. (These numbers are preliminary; more secure values will be discussed in Conti and Vacca 1993.)

What is the nature of the starburst "globs" in He2-10? *If* each "glob" is similar

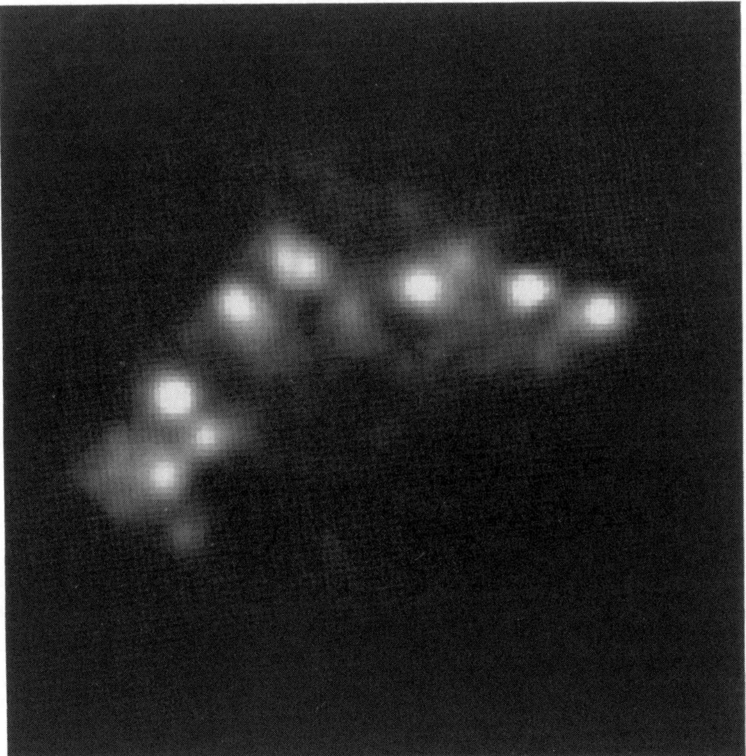

Fig. 4.   Reconstructed UV image of the central starburst in He2-10. Nine individual starburst knots are readily seen. The spatial scale across the figure is $3''$, corresponding to 125 pc; the individual units have linear dimensions of about 10 pc. North is up and East is left.

to 30 Dor in age and mass distribution parameters, and lower mass stars have also formed in addition to the high mass ones we know are there, then the total masses of each are a few $10^6 \mathcal{M}_\odot$. *This is very similar to that for globular clusters!* Could these "globs" be newly born globular clusters? Such a concept is not without precedent: Whitmore *et al.* (1993) have recently identified about 40 blue objects in the galaxy merger remnant NGC 7252 as being globular clusters of age about 500 million years, on the basis of their optical colors and absolute magnitudes. What would the NGC 7252 objects have looked like at 5 million years of age? The "globs" we are observing in the core of the starburst of He2-10 contain many massive stars. Unless they have made *no* low mass ones, their connection to star groups later known as globular clusters seems quite plausible.

5. Conclusions

Using star counts of massive stars in nearby associations, it is possible to evaluate the slope of the IMF, Γ, and the $\mathcal{M}_{upper}$ directly. While there may be distinct differences in the value of Γ in the various regions studied, the slope does *not* have a simple dependence on Z; similarly, there is no evidence for a systematic change in $\mathcal{M}_{upper}$ with composition. There is an indication that at least some *low mass* stars have formed *after* the O type stars in the association NGC 6611.

Indirect methods to evaluate the total numbers of massive stars in GH II and starburst regions include nebular emission line analyses, spectral synthesis, continuum fitting, and FIR radiation. None of these methods can be safely utilized to estimate Γ or $\mathcal{M}_{upper}$; however, if these parameters are specified, the numbers of hot stars can be estimated. Preliminary calibrations of three of the techniques, using a census of 30 Dor, and evaluating its spectroscopic properties, are presented. The agreement with the actual counts is good to about 30%.

Newly acquired HST FOC UV imaging of W-R galaxies at $\lambda2200A$ is presented. At this wavelength, one is examining directly the OB star populations. In all cases so far studied, some 15 in number, the starbursts appear to resolve into individual sub-units of activity. In the BCDG galaxy He2-10, the units have sizes of some 10 pc or so with separations of several times this number. Their UV luminosities range from a few to some ten times larger than the value for 30 Dor. Under a reasonable assumption that these units are morphologically similar to that of the GH II region, we derive masses of millions of solar masses. These numbers are like those of Globular Clusters. Perhaps in this very young and active starburst we are witnessing the birth of such clusters.

Acknowledgements

I am indebted to Dr. William Vacca for our continuing fruitful collaborations. Drs. Leitherer and Robert have also aided in some of the work reported upon here. I appreciate receipt of preprints from Drs. Garmany and Massey. Enthusiastic discussions with Drs. Heap and Whitmore should not go unnoticed. I appreciate continuing support by the NSF under Grant AST90-15240 and NASA (STScI) for HST observations.

References

Aller, L.H.: 1942, *ApJ* **95**, 52
Armus, L., Heckman, T., & Miley, G.: 1988, *ApJL* **326**, 45
Arnault, Ph., Kunth, D. & Schild, H.: 1989, *AA* **224**, 73
Conti, P.S. & Vacca, W.D.: 1990, *AJ* **100**, 431
Conti, P.S.: 1991, *ApJ* **377**, 115
Conti, P.S.: 1993, '' in J.P. Cassinelli & E.B. Churchwell, ed(s)., *Massive Stars: Their Lives in the Interstellar Medium*, ASP Conf. Ser.35:Provo, 449
Conti, P.S., Block, D.L., Geballe, T.R. & Hanson, M.M.: 1993, *ApJL* **406**, L21

Conti, P.S. & Vacca, W.D.: 1993, *ApJL* , to be submitted
Corbin, M.R., Korista, K.T., & Vacca, W.D.: 1993, *AJ* **105**, 1313
Dennefeld, M. & Stasinska, G.: 1983, *AA* **118**, 234
Devereux, N.A. & Young, J.S.: 1991, *ApJ* **371**, 515
Dinnerstein, H.L. & Shields, G.A.: 1986, *ApJ* **311**, 45
Garmany, C.D., Conti, P.S., & Chiosi, C.: 1982, *ApJ* **263**, 777
Garmany, C.D., Massey, P., & Parker, J.W.: 1993, *AJ* , in press
Hillenbrand, L.A., Massey, P., Strom, S.E., & Merrill, K.M.: 1993, *AJ* , in press
Keel, W.C.: 1982, *PASP* **94**, 765
Kennicutt, R.C.,Jr.: 1984, *ApJ* **287**, 116
Kurucz,: , ,
Leitherer, C., Robert, C. & Drissen, L.: 1992, *ApJ* **401**, 596
Maeder, A.: 1991a, '' in K.A. van der Hucht and B. Hidiyat, ed(s)., *Wolf-Rayet Stars and Interre-lations with Other Massive Stars in Galaxies*, Kluwer:Dordrecht, 445
Maeder, A.: 1991b, *AA* **242**, 93
Massey, P.: 1985, *PASP* **97**, 5
Massey, P., Armandroff, T.E., & Conti, P.S.: 1986, *AJ* **92**, 1303
Massey, P., Garmany, C.D., Silkey, M., & DeGioa-Eastwood, K.: 1989a, *AJ* **97**, 107
Massey, P., Parker, J.W., & Garmany, C.D.: 1989b, *AJ* **98**, 1305
Massey, P. & Thompson, A.B.: 1991, *AJ* **101**, 1408
Massey, P. & Armandroff, T.E.: 1991, '' in K.A. van der Hucht and B. Hidiyat, ed(s)., *Wolf-Rayet Stars and Interrelations with Other Massive Stars in Galaxies*, Kluwer:Dordrecht, 575
Massey, P. & Johnson, J.: 1993, *AJ* **105**, 980
Mathis, J.S. & Rosa, M.R.: 1991, *AA* **245**, 625
McCall, M.L., Rybski, P.M., & Shields, G.A.: 1985, *ApJS* **57**, 1
McGaugh, S.S.: 1991, *ApJ* **380**, 140
McLeod, K.K., Rieke, G.H., Rieke, M.J., & Kelley, D.M.: 1993, *ApJ* **412**, 99
Moffat, A.F.J., Niemela, V.S., Phillips, M.M., Chu, Y.-H., & Seggewiss W.: 1987, *ApJ* **312**, 612
Osterbrock D.E.: 1989, *Astrophysics of Gaseous Nebulae and Active Galactic Nuclei*, Univ. Sci. Books; Mill Valley, Cal, 408
Parker, J.W.: 1991, Ph.D Thesis, University of Colorado, Boulder
Parker, J.W., Garmany, C.D., Massey, P., & Walborn, N.R.: 1992, *AJ* **103**, 1205
Robert, C. Leitherer, C. & Heckman, T.M.: 1993, *ApJ* , in press
Rubin, V.C., Hunter, D.A., & Ford, W.K., Jr.: 1990, *ApJ* **365**, 86
Sargent, W.L.W.: 1970, *ApJ* **160**, 405
Sargent, W.L.W. & Searle, L.: 1970, *ApJL* **162**, 155
Sargent, W.L.W. & Filippenko. A.V.: 1991, *AJ* **102**, 105
Searle, L.: 1971, *ApJ* **168**, 327
Shaver, P.A., McGee, R.X., Newton, L. M., Danks, A.C., & Pottash,S.R.: 1983, *MNRAS* **204**, 53
Shields, G.A.: 1990, *ARAA* **28**, 525
Shields, G.A. & Tinsley, B.M.: 1976, *ApJ* **203**, 66
Vacca, W.D.: 1991, Ph.D Thesis, University of Colorado, Boulder
Vacca, W.D.: 1993, *ApJ* , in press
Vacca, W.D. & Conti, P.S.: 1992, *ApJ* **401**, 533
Vacca, W.D., Conti, P.S., Leitherer, C. & Robert, C.: 1993, *ApJ* , to be submitted
Veilleux, S. & Osterbrock, D.E.: 1987, *ApJS* **63**, 295
Vilchez, J.M. & Pagel, B.E.J.: 1988, *MNRAS* **231**, 257
Vilchez, J.M., Pagel, B.E.J., Diaz, A.I., Terlevich, E., & Edmonds,M.G.: 1988, *MNRAS* **235**, 633
Walsh, J.R. & Roy, J.-R.: 1987, *ApJL* **319**, 57
Whitmore, B.C., Schweizer, F., Leitherer, C., Borne, K. & Robert, C.: 1993, *AJ* **106**, 1354

A DISSECTION OF 30 DORADUS AND A DISCUSSION OF OTHER MAGELLANIC CLOUD OB ASSOCIATIONS

JOEL WM. PARKER
National Research Council Fellow
NASA/Goddard Space Flight Center
Greenbelt, MD 20771 USA

Abstract. I Present the results of ground-based and Hubble Space Telescope photometry and spectroscopy of the stars in the central region (roughly 7×7 arcmin) of 30 Doradus in the Large Magellanic Cloud (LMC). Using photometric data for over 2400 stars (complete to $V \approx 18$ mag), and spectroscopic observations of over 150 stars in the region, the best estimate of the initial mass function (IMF) yields a slope of $\Gamma = -1.5 \pm 0.2$ for masses $> 12\mathcal{M}_\odot$, where the Salpeter slope is $\Gamma = -1.35$. I compare these results to other measurements of the IMF for OB associations in the Magellanic Clouds.

Key words: 30 Doradus – Magellanic Clouds – Massive Stars – Mass Function

1. The Stellar Content of 30 Doradus

The work referenced here is based on the thesis work of Parker (1992). Figure 1 shows the region of 30 Doradus covered by this study. Magnitudes have been determined for 2350 stars in the central region (1438 have UBV, and 912 have BV magnitudes), and for 45 stars in the detached northern region. 54 new spectroscopic observations have resulted in new classifications of 43 OB stars. The current inventory of stars with spectroscopic classifications in 30 Doradus is presented in Table I. The details of the data reduction, final photometry, spectroscopic observations and classification, H-R diagrams, and other analyses can be found in Parker (1992, 1993) and Parker & Garmany (1993). This work has also led to the discovery (Parker *et al.* 1993) that R 143 is a Luminous Blue Variable (LBV), the first and perhaps the lone LBV in the central cluster of 30 Doradus, and only the sixth known LMC LBV.

2. The Initial Mass Function of 30 Doradus

Figure 2 shows the IMF for 30 Doradus, taken from Figures 9 and 10 in Parker & Garmany (1993). The IMF seems to turnover at the high mass end, which could be due to either incompleteness or evolutionary effects. Since these results are from ground-based data that do not resolve the core cluster of R 136, it is possible that some massive stars have been missed. In fact, the results from Malumuth & Heap (1993) indicate that the core cluster may have proportionately more massive stars and the IMF is flatter. Other incompleteness effects may be due to the number of O stars that still do not have spectroscopic classifications (and, therefore, the

Space Science Reviews **66**: 55–59, 1994.
© 1994 *Kluwer Academic Publishers. Printed in Belgium.*

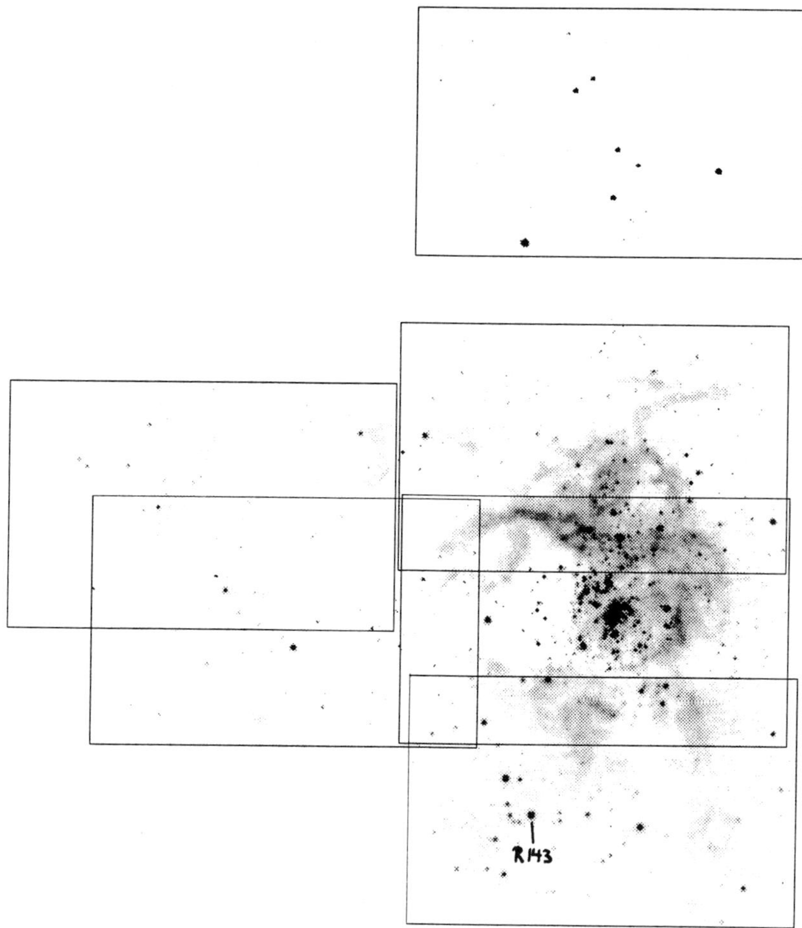

Fig. 1. This is a mosaic of the V filter CCD images of 30 Doradus used in this study. The outlines indicate the CCD fields, each of which are 2.6 × 4.1 arcsec. North is up and East is to the left.

temperatures and masses have probably been underestimated). Analysis of the Lyman luminosity of the stars (Parker 1992) compared to the Balmer flux of the 30 Doradus H II region (Kennicutt & Hodge 1986) supports that possibility. For comparison, Parker & Garmany (1993) calculated the IMF for the region to the southeast of the crowded R 136 cluster, and found that the IMF is well-fitted by a power law slope of $\Gamma = -1.7$ down to at least $9\mathcal{M}_{\odot}$.

TABLE I

30 Doradus Stars with Spectroscopic Classifications

(Adapted from Table 3 in Walborn 1991)

Spectral Types	Melnick, Walborn	Walborn & Blades	Parker	other	TOTAL
Wolf-Rayet				15	15
LBV			1		1
O3-6 V	14	17	7		38
O3-6 III	3	3			6
O3 If*	5				5
O6-9 V	4	11	13		28
O6-9 III	5	10	4		19
O6-8 Iaf	1	1			2
O9 I	2	1			3
B0-1 V		1	11		12
B0-1 III		7	1		8
B0-1 I	11	3	2		16
B1-3 III		2	1		3
B1-3 I	1	2	2		5
B5-A I	4	3	1		8
TOTAL	50	61	43	15	169

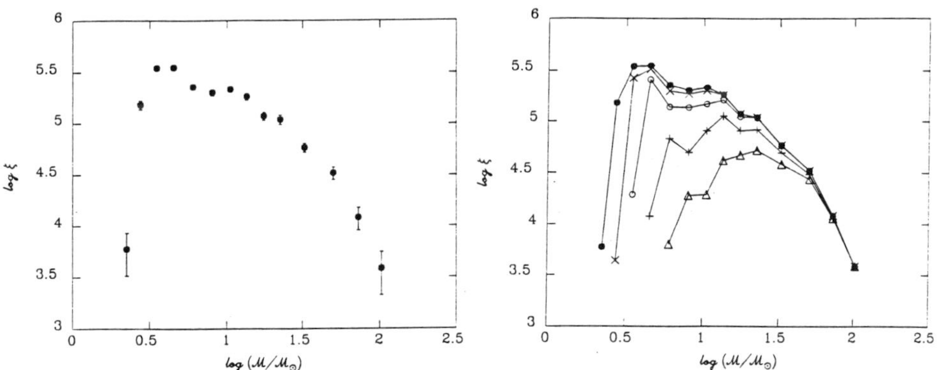

Fig. 2. Plot (a) shows the IMF and the $\sqrt{N}$ error bars. Plot (b) shows how the observed IMF changes as a function of the observed V magnitude limit. The symbols indicate all stars ("•"), and stars brighter than $V = 19$ mag ("×"), 18 mag ("o"), 17 mag ("+"), and 16 mag ("△"). The catalog is complete (except for the crowded central core of R 136) to $V = 18$ mag, indicating a completeness to at least $12\mathcal{M}_\odot$.

TABLE II

Comparisons of 30 Doradus Initial Mass Functions

Region	IMF slope (Γ)	source
7 × 7 arcmin	-1.5 ± 0.2	Parker & Garmany (1993)
southeast section	-1.7 ± 0.1	Parker & Garmany (1993)
outer 3.3–35 arcsec	-1.8	Malumuth & Heap (1993)
inner 3.3 arcsec	-0.9	Malumuth & Heap (1993)
2 × 3 arcmin	-0.9 ± 0.4	Meylan (1993)
2.5 × 4 arcmin	-0.8 ± 0.4	Lapierre & Moffatt (1991)

TABLE III

Comparisons of Initial Mass Functions for Other Magellanic Cloud OB Associations Analyzed in Related Studies

Association	IMF slope (Γ)	source
LH 117 & 118	-1.8 ± 0.1	Massey *et al.* (1989a)
NGC 346 (SMC)	-1.8 ± 0.2	Massey *et al.* (1989b)
LH 9	-1.6 ± 0.1	Parker *et al.* (1992)
LH 10	-1.1 ± 0.1	Parker *et al.* (1992)
LH 58	-1.7 ± 0.2	Garmany *et al.* (1993)
LH 2	-1.3 ± 0.2	Parker (1994)

3. Comparison of The Initial Mass Functions of 30 Doradus And Other OB Associations In The Magellanic Clouds

The IMF slopes for 30 Doradus from different studies given in Table II. It is evident that studies of 30 Doradus have found a wide range of IMF slopes. A number of these 30 Doradus studies have relied on photometry alone; but as discussed by Massey (1985), the UBV colors of OB stars are degenerate. Temperatures (and the resulting masses) cannot be determined accurately from UBV photometry alone.

The IMF slopes for some other Magellanic Cloud OB associations are given in Table III. These other associations have been observed and analyzed in a manner similar to the 30 Doradus study presented here. Although there seems to be a tendency for the slopes to be on the order of $\Gamma \approx -1.8$, there is a significant range of slopes. Most notable are those of LH 9 and 10. They are two neighboring associations, and it would be expected that their star-formation environments would be similar. The differences may be due to the effects of sequential star formation, which are discussed in more detail by Parker *et al.* (1992) and in comparison to 30 Doradus by Walborn & Parker (1992).

Acknowledgements

This research has made use of the Simbad database, operated at CDS, Strasbourg, France. Support for this study has come from the National Research Council, NASA/Goddard Space Flight Center, NASA Graduate Student Research Fellowship NGT-50483, a Graduate Fellowship from the University of Colorado, and a Fellowship from the Boettcher Foundation.

References

Garmany, C. D., Massey, P., & Parker, J. Wm. 1993, *AJ*, submitted
Kennicutt, R. C., & Hodge, P. W.: 1986, *ApJ* **306**, 130
Lapierre, N., & Moffat, A. F. J: 1991, '' in K. Janes, ed(s)., *The Formation and Evolution of Star Clusters*, ASP Conference Series, v. 13, 155
Malumuth, E. M. & Heap, S. R., 1993, *ApJ*, submitted
Massey, P.: 1985, *PASP* **97**, 5
Massey, P., Garmany, C. D., Silkey, M., & DeGioia-Eastwood, K.: 1989a, *AJ* **97**, 107
Massey, P., Parker, J. Wm., & Garmany, C. D.: 1989b, *AJ* **98**, 1305
Melnick, J.: 1985, *A&A* **153**, 235
Meylan, G., 1993, preprint
Parker, J. Wm 1992, Ph. D. thesis, University of Colorado
Parker, J. Wm.: 1993, *AJ* **106**, 560
Parker, J. Wm., Clayton, G, C., Winge, C., & Conti, P. S.: 1993, *ApJ* **409**, 770
Parker, J. Wm., & Garmany, C. D.: 1993, *AJ* **106**, 1471
Parker, J. Wm., Garmany, C. D., Massey, P., & Walborn, N. R.: 1992, *AJ* **103**, 1205
Walborn, N. R.: 1986, '' in C. W. H. de Loore, A. J. Willis, and P. Laskarides, ed(s)., *Luminous Stars and Associations in Galaxies, IAU Symposium No. 116*, Dordrecht Reidel, 185
Walborn, N. R.: 1991, '' in R. Haynes and D. Milne, ed(s)., *The Magellanic Clouds, IAU Symposium No. 148*, Kluwer Academic Publishers: The Netherlands, 145
Walborn, N. R., & Parker, J. Wm.: 1992, *ApJ* **399**, L87

BLUE AND RED SUPERGIANTS AND THE AGE STRUCTURE OF THE NGC 330 REGION *

DOMINIK J. BOMANS
Sternwarte Univ. Bonn, Auf dem Hugel 71, D-53121 Bonn, Germany

and

EVA K. GREBEL
European Southern Observatory, Casilla 19001, Santiago 19, Chile

Abstract. We analysed new UBVRI CCD photometry of the massive, blue SMC cluster NGC 330 and its surrounding field. The age structure and a new reddening value for the stellar population in this region of the SMC are derived and the implications for star formation in this part of the SMC and for stellar evolution are discussed.

Key words: NGC 330 – star formation – stellar evolutionary models – supergiants

1. The blue populous cluster NGC 330 in the SMC

NGC 330 is the brightest blue populous cluster in the SMC. Since the analysis of Arp (1959) it is known that this cluster is very young and hosts a surprising number of red *and* blue supergiants. The metallicity of NGC 330 (e.g., Spite et al. 1991) is not well settled but clearly quite low. Grebel & Richtler (1992) analysed CCD Stromgren photometry of the cluster and the surrounding field and found NGC 330 more metal-poor than the field stars. The 'field' was chosen as all stars having a radial distance of more than $100''$ from the cluster center, a definition which we also adopt in this paper.

The determination of the metallicity for the stars in this region is tightly coupled with the determination of the reddening (Bessell et al. 1992). The values advocated are between $E_{B-V} = 0.03$ and 0.12. Recently some hot stars in NGC 330 were analysed using IUE low dispersion spectra (Caloi et al. 1993). One of the main results was an independent determination of the reddening, $E_{B-V} = 0.09$.

Carney et al. (1985), who used photographic and photoelectric photometry and some low dispersion spectra for spectral classification, determined an age of 12 Myr for NGC 330 and of 18 Myr for the young surrounding field population. They found an age spread of 2.5 Myr for the cluster.

Grebel et al. (1992) showed that NGC 330 is extremely rich in Be stars. Be stars are located somewhat redward of the main sequence, a fact further complicating the interpretation of the color-magnitude diagram (CMD).

In their study Carney et al. (1985) pointed out the importance of NGC 330 for stellar evolutionary theories because of its large number of supergiants, a

---

* Based on observations obtained at ESO, La Silla

Space Science Reviews **66**: 61–64, 1994.

© 1994 *Kluwer Academic Publishers. Printed in Belgium.*

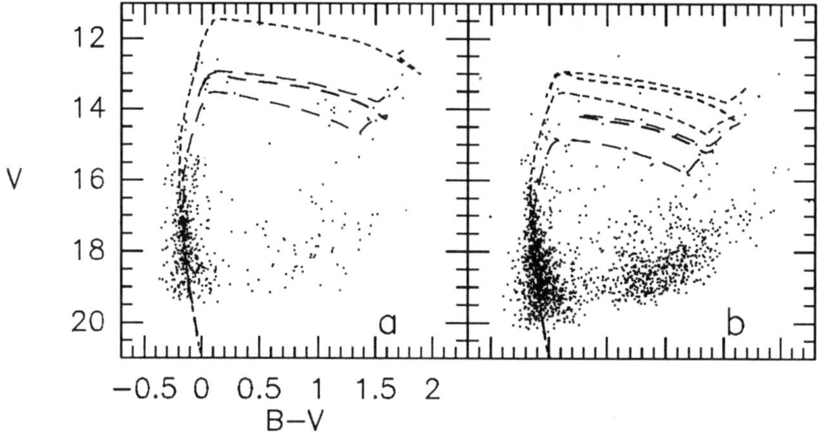

Fig. 1.   (B–V), V CMDs of NCG 330 cluster (a), and field (b). Overplotted are Z=0.004 isochrones of 12 and 25 Myr in Fig. 1a, 25 and 50 Myr in Fig. 1b.

statement emphasized by Stothers & Chin (1992). Recently Brocato & Castellani (1993) proposed stellar models using the Ledoux criterion for convection to explain the features in the CMD of NGC 330. It is really surprising that up to now only one study of NGC 330 using CCD photometry has been published (Cayrel et al. 1988).

2. New photometry and new Isochrones

We present here an analysis of new UBVRI CCD photometry of NGC 330 and a neighbouring field. The data were taken with the ESO-MPI 2.2m telescope at La Silla in December 1991 during a run of the Key Program on 'Coordinated Investigations of Selected Regions in the Magellanic Clouds' (de Boer et al. 1991). After the basic reductions crowded field photometry was performed with DAOPHOT II. For the photometric transformation we used standard stars close to NGC 330 (Alcaino & Alvarado 1988). The analysed field has a size of about 10′ by 5.8′.

For the interpretation of the data we used isochrones calculated from the latest evolutionary tracks of the Geneva group, the latest Kurucz flux tables, and standard filter passbands (Roberts & Grebel, this conference). For our isochrone fits, we used their isochrones with Z=0.004 and Z=0.002.

3. Reddening

For a reliable determination of the reddening based on the (U–B),(B–V) two-colour diagram, Be stars should be taken out because of their additional intrinsic reddening (Bomans et al. 1993, in prep.). From the location of the main sequence

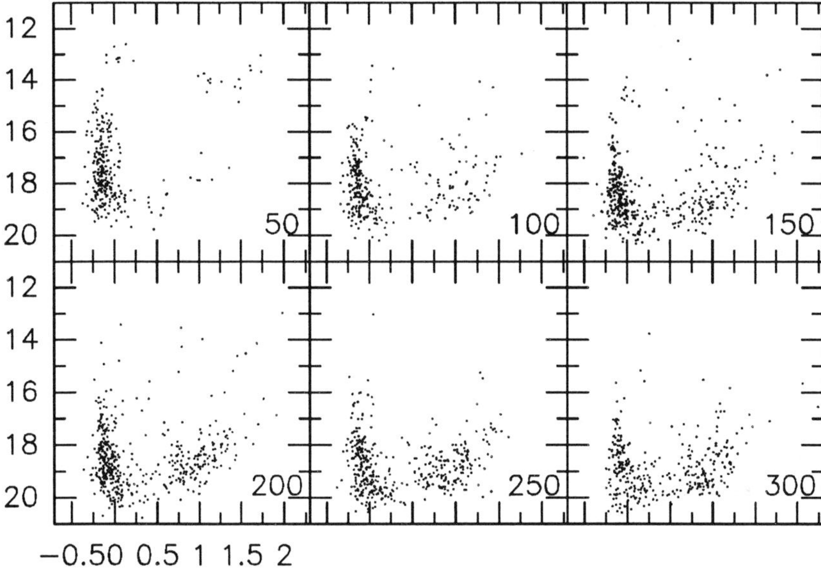

Fig. 2.   (B–V), V diagrams of concentric annuli of radius $50''$ each, all centered on NGC 330. Note that the area of each annulus is different.

in the (B–V), V and (V–I), I CMDs and the isochrones we derived $E_{B-V} = 0.07$ as the optimum reddening value in course of our analysis of the age structure of the NGC 330 region.

In this paper we adopt an SMC distance modulus of 18.9 mag.

4. Age structure of the NGC 330 region

Problems of the isochrones to fit all features of the 'cluster' region are obvious (Fig. 1). Note especially the problem of fitting the very blue stars between V=14 and 15.5 mag and simultaneously the lower main sequence. Nevertheless the structure of the CMD with clumps of red and blue supergiants is reproduced with these models that use Schwarzschild's criterion.

It is important to keep in mind that the colour offset of the Be stars with respect to the main sequence is intrinsic to this kind of stars.

A 25 Myr isochrone fits well the turn-off and the location of both red and blue supergiants (see Fig. 1). Some non-supergiant stars can be seen above the turn-off of the 25 Myr isochrone. These stars can be fitted by an 12 Myr isochrone, which would imply an age spread of about 10 Myr. However, the influence of crowding on the individual measurements still needs to be assessed by a photometry under subarcsecond seeing conditions.

The isochrones indicate a metallicity for NGC 330 between -0.7 and -1.0 dex.

The 'field' CMD (Fig. 1b) shows an underlying stellar population somewhat older than 50 Myr plus several blue stars above the main sequence turnoff. The existence of these stars and the distribution of the red supergiants populating the Hayashi line over 2 mag in V implies that this episode of star formation was more continuous than a sudden burst would have been. When looking at CMDs of stars in concentric annuli around NGC 330 (Fig. 2) one can see that NGC 330 stands out with its luminous supergiants and hot main sequence stars that are highly concentrated to the cluster center. The lower luminosity red supergiants and the numerous blue stars above V = 16.5 mag are unlikely to be cluster members.

No obvious difference in the (V–I), I location of the Hayashi line between the red supergiants in NGC 330 and the field due to metallicity can be seen. An exception is the group of intermediate red supergiants strongly confined to the cluster area. These stars are in a different evolutionary phase and are somewhat CN enhanced as indicated by the Stromgren photometry of Grebel & Richtler (1992) and Hilker et al. (1993, in prep.)

All this implies that before NGC 330 was formed, a lower level of massive star formation was already present in this region, maybe linked to the outer regions of the same giant molecular cloud which formed NGC 330.

The location and number ratios of blue to red supergiants in NGC 330 can be perfect tools to finetune stellar model parameters. With the current dataset the number of red supergiants predicted by the models is too low, as already pointed out by Meynet (1993). He also stressed the importance of mass loss as an additional parameter concerning supergiant number ratios.

More tests of models with different input physics and an analysis of the number distribution of the supergiant and main sequence stars, using the synthetic CMD method, are in preparation (Bomans et al. 1993).

References

Alcaino, G., Alvarado, F.: 1988, *AJ* **95**, 1724
Arp, H.: 1959, *AJ* **64**, 254
Bessell, M.S.: 1991, *A&A* **242**, L17
Brocato, E., Castellani, V.: 1993, *ApJ* **410**, 99
Caloi, V., Cassatella, A., Castellani, V., Walker, A.: 1993, *A&A* **271**, 109
Carney, B.W., Janes, K.A., Flower, P.J.: 1985, *AJ* **90**, 1196
Cayrel, R., Tarrab, I., Richtler, T.: 1988, *ESO Messenger* **54**, 29
de Boer, K.S., et 31 alii: 1991, *ESO Messenger* **66**, 14
Grebel, E.K., Richtler, T.: 1992, *A&A* **253**, 359
Grebel, E.K., Richtler, T., de Boer, K.S.: 1992, *A&A* **254**, L5
Meynet, G.: 1993, in: 'The feedback of chemical evolution on the stellar content of galaxies', Eds.:
    D. Alloin, G. Stasinska, Observatoire de Paris, p. 40
Spite, F., Richtler, T., Spite, M.: 1991, *A&A* **252**, 557
Stothers, R.B., Chin, C.-W.: 1992, *ApJ* **390**, 136

BE STARS IN YOUNG CLUSTERS IN THE MAGELLANIC CLOUDS *

EVA K. GREBEL

European Southern Observatory, Casilla 19001, Santiago 19, Chile

WM JAMES ROBERTS

Space Telescope Science Institute, CSC, 3700 San Martin Dr., Baltimore, MD 21218, USA

and

JEAN–MARIE WILL and KLAAS S. DE BOER

Sternwarte der Universitat Bonn, Auf dem Hugel 71, D-53121 Bonn, FRG

Abstract.

A substantial fraction (typically 10%) of Galactic B stars consists of Be stars. While Galactic Be stars have been fairly well investigated, very little is known about the Be star content of the Magellanic Clouds (MCs). We present a refined method of Be star identification by CCD photometry and apply it to four young clusters and associations in the MCs. We find NGC 330 in the SMC to be exceptionally rich in Be stars, while the fraction is clearly lower in the similarly aged LMC clusters NGC 2004 and NGC 1818. NGC 2044, a very young region in the LMC, contains almost no Be stars. Among very early–type B stars in the investigated MC clusters we find the largest number of Be stars, while in the Milky Way this is shifted to somewhat later types. In the LMC, there may be a second frequency peak around B6.

Key words: Stars: Be – Clusters: NGC 330, NGC 2004, NGC 1818, NGC 2044 – Magellanic Clouds

1. Photometric Identification

Be stars are non-supergiant B stars that show (or once have shown) Balmer emission lines. Therefore, currently active Be stars can be detected by using (Grebel et al., 1992)

- an Hα filter to detect Hα emission,

- an R–like filter for the continuum,

- and another broadband filter to use the information about colour and temperature of a colour–magnitude diagram (CMD) to distinguish between non-supergiant B stars and others.

CCD imaging with two broadband filters and an Hα filter is an efficient tool to identify Be stars and makes crowded regions, such as clusters (Grebel et al., 1992), accessible. From the definition of a Be star it is clear that *all current detections establish only a lower limit to the true number of Be stars.*

* Based on observations obtained at the 2.2m MPIA telescope at ESO, La Silla, Chile, partly on time granted by the MPIA, Heidelberg.

© 1994 *Kluwer Academic Publishers. Printed in Belgium.*

2. Calibration of the Hα luminosity

The Bessell R filter is an adequate continuum filter for a measurement of the Hα flux to 30% accuracy or better. The energy flux per unit wavelength outside the atmosphere at R=0 is $\log(f_R) = -7.76$ in erg cm^{-2} nm^{-1} (Allen 1972). In the following, f denotes flux, L luminosity, D distance in kpc, and P power. Since the ratio of our r and $h\alpha$ instrumental passbands is 40 (=4 mag), the Hα luminosity of any star is given by

$$\log(L_{H\alpha}/L_\odot) = 2\log(D/\text{kpc}) + 10.5 + \log(P_{H\alpha}), \text{ where}$$
$$P_{H\alpha} = (3.5\text{nm})f_{H\alpha} = 3.5 \times 10^{(-0.4R)}10^{0.4(r-h\alpha+4.0)}, \text{ so that}$$
$$\log(L_{H\alpha}/L_\odot) = 2\log(D/\text{kpc}) + 3.4 + 0.4([r-h\alpha]-R) \text{ with}$$
$$[r-h\alpha] \equiv (r-h\alpha+4.0).$$

3. Results for our program clusters

Fig. 1a shows (B–R) vs. our passband-corrected Hα index for NGC 330, a young blue cluster in the SMC. The clump at (B–R) = −0.2, $[r-h\alpha] \leq 0$ comprises blue supergiants and blue MS stars without Hα emission. The corresponding area redwards of (B–R) $\geq$ 0.5 consists of red supergiants and giants. Above the blue clump, a band of blue stars bright in Hα can be seen. As a criterion for Be stars, we chose $[r-h\alpha] \geq 0.2$. That this is a reasonable distinction was confirmed for NGC 330 using spectroscopic classifications by Feast (1972) and an unpublished grism survey by M. Azzopardi (see Meyssonnier & Azzopardi 1991).

In the dereddened CMD of NGC 330 (Fig. 1b) Be stars are marked by filled triangles. Most of them lie redwards of the upper main-sequence (MS) and they clearly contribute to the apparent width of the MS. The colour index (B–R) makes the effect more dramatic than (B–V) or (V–I). Using our synthetic isochrones based on tracks of the Geneva group, Kurucz model atmospheres, and standard filter passbands (Roberts & Grebel, this conference), we obtain the best fit for 20 Myr and [Fe/H] = −0.7 dex. The derived age automatically constrains the evolutionary status of the Be stars in NGC 330 and indicates that they are MS stars.

The young LMC cluster NGC 2004 (age 20 Myr, [Fe/H] $\approx$ −0.4 dex) contains many fewer Be stars (Fig. 1c) and shows weaker Hα emission. In the CMD of NGC 2004 (Fig. 1d), a first group of Be stars is concentrated around 15 mag, while a second group can be found at 18 mag.

For the LMC cluster NGC 1818 (age about 25 Myr, [Fe/H]= −0.8 (Richtler et al., 1989)), again we find only relatively few Be stars, and again we see two apparent concentrations.

In Fig. 2, we compare Be star fractions by spectral type with the Galactic values. The numbers of Galactic Be stars were taken from Jaschek & Jaschek (1983). In the MC clusters, the highest percentages of Be stars can be found among B0–B1, while in the Milky Way the frequency peak is in the range B1–

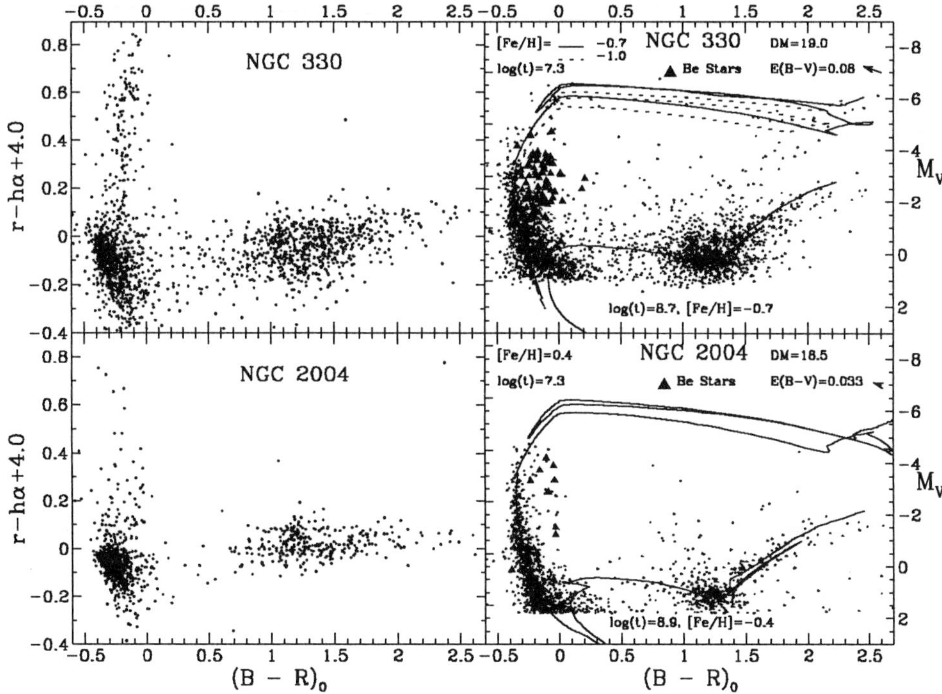

Fig. 1. (a–d) Two-colour identification diagram for Be stars in NGC 330 (Fig. 1a) and NGC 2004 (Fig. 1c) and colour-magnitude diagrams with Be stars (filled triangles) in NGC 330 (Fig. 1b) and NGC 2004 (Fig. 1d). See sect. 3 for more explanation.

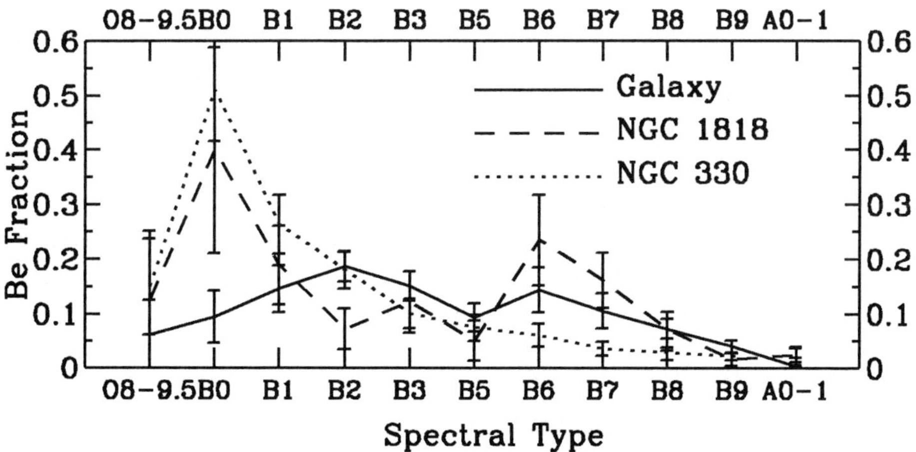

Fig. 2. Comparison between Be star fractions by spectral type in the Milky Way, NGC 330, and NGC 1818. Note the different frequency peaks.

B3. In addition, two of the LMC clusters show indications of a second frequency peak at later B types (B6-B7). In the Milky Way, such a peak may be present, too. The distributions are significantly different.

The LMC OB association NGC 2044 contains four very young clusters (ages $\leq$ 10 Myr, Grebel & Melnick, in prep.). Though very rich in B stars, we find only up to 3 Be stars per cluster. In comparison with the other clusters this may indicate a lower age limit for Be stars, possibly associated with an evolutionary spin-up of young B stars leading to the Be phenomenon (Bessell & Wood 1993).

4. Discussion

With its large number of Be stars, NGC 330 is an exception among the MC clusters studied thus far. Balona (1992) found that a number of the Be stars in this cluster are λ Eri variables. Of the three suspected λ Eri candidates he discovered in NGC 2004 (1993), which we found to be not very rich in Be stars, we confirm only one as a Be star. Are λ Eri variables a clue to the Be star phenomenon in the MCs? – In Galactic open clusters, the presence of λ Eri variables seems to exclude the presence of β Ceps and vice versa. Does this hold also in young MC clusters?

As can be seen in Fig. 1b, Be stars can broaden the MS considerably which must be taken into account for isochrone fitting. A possible reason for such red excess could be free-free emission from circumstellar material.

More precise spectral classifications and a determination of the rotational velocities would disclose more about the nature of the MC Be stars. How can theory account for the different frequency peaks in the MCs? Are high rotational velocities a general property of young MC clusters rich in Be stars? In NGC 330 in particular, high $v \sin i$ values might be expected.

Obviously, we are only beginning to understand the Be stars in the MCs.

References

Allen, C. W. 1972, in "Astrophysical Quantities", Athlone Press, London, p. 202
Balona, L. 1992, MNRAS, 256, 425
Balona, L. 1993, MNRAS, 260, 795
Bessell, M. S., & Wood, P. R. 1993, in "New Aspects of Magellanic Clouds Research", Eds. B. Baschek, G. Klare, J. Lequeux, Springer Lecture Notes in Physics 416, p. 271
Feast, M. W. 1972, MNRAS, 159, 113
Grebel, E.K., Richtler, T., de Boer, K.S. 1992, A&A, 254, L 5
Jaschek, C., Jaschek, M. 1983, A&A, 117, 357
Meyssonnier, M., Azzopardi, M. 1991, IAU Symp. 148, Eds. R. Haynes & D. Milne, Kluwer, Dordrecht, p. 196
Richtler, T., Spite, M., Spite, F. 1989, A&A, 225, 351

THE POPULATION OF MASSIVE STARS IN R136 FROM HST/FOC UV OBSERVATIONS *

ANTONELLA NOTA,** GUIDO DE MARCHI*** and CLAUS LEITHERER

Space Telescope Science Institute 3700 San Martin Drive, Baltimore, MD, USA

and

ROBERTO RAGAZZONI[‡] and CESARE BARBIERI[‡‡]

Abstract. New ultraviolet ($\lambda \simeq 1300$ A, $\lambda \simeq 3400$ A), *HST FOC* observations have been used to derive the UV color–magnitude diagram (CMD) of R136, with the main scientific goal of studying the upper end of the stellar mass function at ultraviolet wavelengths where the color degeneracy encountered in visual CMDs is less severe. The CMD has been compared to a set of theoretical isochrones, which have been computed using the latest generation of evolutionary models and model atmospheres for early type stars. Wolf-Rayet stars are included. Comparison of the *theoretical* and *observed* CMD suggests that there are no stars brighter than $M_{130} \simeq -11$. We use the observed main sequence turn-off and the known spectroscopic properties of the stellar population to derive constraints on the most probable age of R136. The presence of WNL stars and the lack of red supergiants suggests a most likely age of 3 ± 1 Myr. A theoretical isochrone of 3 ± 1 Myr is consistent with the observed stellar content of R136 if the most massive stars have initial masses around $\simeq 50$ M$_\odot$.

Key words: massive stars - IMF - clusters

1. Introduction

R136 is the brightest cluster in the giant HII region 30 Doradus in the Large Magellanic Cloud. R136a is its bright core, which remained unresolved for many years. The interpretation of the nature and the stellar content of R136a has been controversial. Once it was believed to be a single superluminous supermassive star $\simeq 2100$ M$_\odot$ (Feitzinger et al. 1980; Cassinelli et al. 1981; Savage et al. 1983) or a small group of supermassive stars formed by ordinary stellar collapse in a region containing peculiar dust or by the coalescence of stars in a dense region. Speckle interferometry observations (Weigelt & Baier 1985; Neri & Grewing 1988) later confirmed that R136a is a cluster. However, only several years later the Hubble Space Telescope (*HST*) unambiguously resolved and measured the flux of the brightest components, both with the Faint Object Camera (*FOC*) (Weigelt et al. 1991) and with the Wide Field Planetary Camera (Campbell et al. 1992, Malumuth et al. 1992, Heap et al. 1993).

* Based on Observations with the NASA/ESA Hubble Space Telescope, obtained at the STScI, which is operated by AURA, Inc., under NASA contract NAS5-26555.

** Affiliated to the Astrophysics Division, Space Science Department, ESA

*** on leave from Dipartimento di Astronomia, Universita di Firenze, Italy

[‡] Osservatorio Astronomico di Padova, Italy.

[‡‡] Dipartimento di Astronomia, Universita' di Padova, Italy.

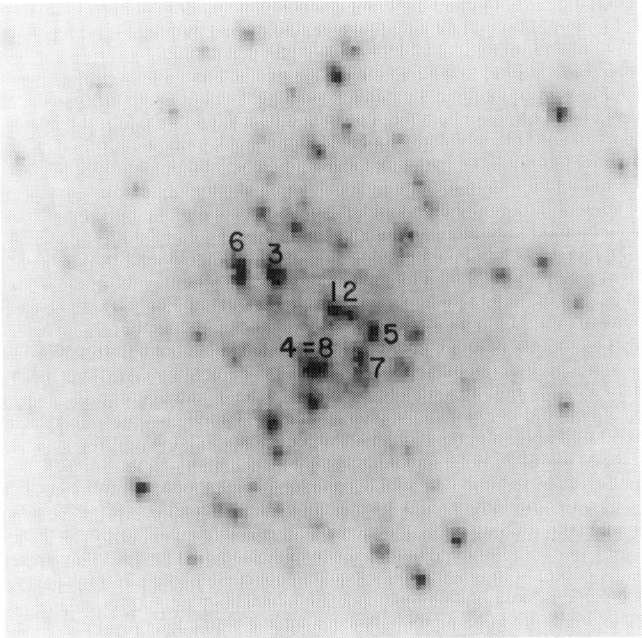

Fig. 1.   Enlargement of the 4.44" × 4.44" region around the R136 cluster center, taken with the FOC F130M filter. North is up and East to the left and Weigelt's components a_1 - a_8 have been labelled.

R136a itself is resolved into at least 12 components (Campbell et al. 1992), three of which are found to be Wolf-Rayet (W-R) stars from the emission in the HeII $\lambda4686$ line. The brightest stars within R136 are found to have luminosities and colors of *normal* massive stars still on the main sequence or already evolved into the supergiant or Wolf-Rayet phases, in overall agreement with the claim of Moffat et al. (1985) who had already suspected that R136 *does not* contain extraordinarily bright stars.

2. Observations

Observations of R136 were performed on 22 August 1990, 4 January 1991, and 2 February 1991 with the *HST-FOC* in its F/96 mode. The 22 August 1990 observation was obtained with the filter combination F346M + F8ND, and an exposure time of 597 seconds. A few months later, 4 additional images of the same region were taken with the filter F130M and an exposure time of 900s each. The two filters used, their characteristics, and the instrument set-up are described in detail in De Marchi et al. (1993).   The F130M images, largely overlapping, have been used to create a single, high signal-to-noise image of the central area of the cluster. The final summed F130M image is shown in Figure 1, with a plate scale of 0.022" pix^{-1}. The F346M image has been aligned to the F130M sum. The resulting overlap area of the two filters is $\simeq$ 113 arcsecs^2, but the two images differ considerably in quality: using the brightest stars we can estimate that the S/N ratio in the F130M summed image is $\simeq$ 28 while it is $\simeq$ 7 in the F346M frame.

3. The Color Magnitude Diagram

We located 221 stars in the F130M summed image, which were also found in the F346M frame with, at least, a SNR of $\simeq$ 2.5 in their peak. The photometric reduction followed the "core aperture photometry" technique, and was calibrated through the comparison with standard stars taken with the same filters. While internal photometric errors are better than 0.10 mag at worst, the quite large uncertainty in the UV calibration of the FOC causes the overall accuracy of our fluxes to be about 20%. Magnitudes measured this way are used to derive the CMD shown in Figure 2. To make easier the comparison with theoretical isochrones, absolute dereddened magnitudes were plotted. We assumed a distance modulus of 18.55 $\pm$ 0.13 (Panagia et al. 1991) and estimated the total intervening absorption as due to three different components: the Milky Way and the LMC (Fitzpatrick 1985), and the 30 Doradus nebular dust (Fitzpatrick & Savage 1984). Our best estimate for the intervening absorption is E(B-V) = 0.41, A_{130} = 4.31 mag, and A_{346} = 2.21 mag. The eight components (actually only seven, because a4 = a8) of R136a originally resolved by Weigelt & Baier (1985) are labelled by number in our CMD. Interestingly, they are the brightest stars in the cluster and also among the reddest, and are clearly concentrated within the innermost 1" radius.

4. The Isochrones

Theoretical isochrones have been derived using evolutionary models by Maeder (1990) and Maeder & Maynet (1988) to cover masses up to 120 $M_\odot$. LTE model atmospheres with line blanketing (Kurucz 1992) are adopted for stars which are not in their W-R phase. Emergent fluxes of W-R stars are calculated with the theoretical continuum energy distribution of Schmutz et al. (1992, 1993). Model atmospheres have been renormalized to derive the expected fluxes in the FOC bandpasses, and finally converted into HST instrumental magnitudes by using the FOC-specific FOCSIM simulator. We compared the observed CMD to theoretical isochrones to estimate the age and the mass spectrum of the cluster. The presence (already known from the ground) of W-R stars in R136 (Melnick 1985) sets a lower limit for the age of this cluster of at least 2 Myr. Conversely, the presence of WNL stars, along with the detected lack of red supergiants, suggests an age no larger than $\simeq$ 4 Myr. The isochrones we overlay on our CMD in Figure 2 correspond to 3 and 4 Myr, for Z = 0.25 $Z_\odot$. The fit between observed

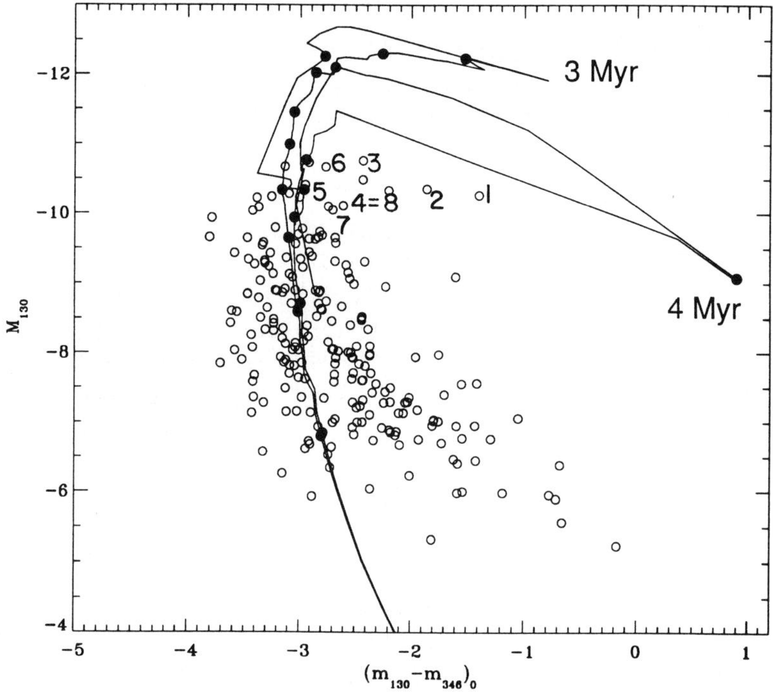

Fig. 2.   Isochrones corresponding to 3 and 4 Myrs and $Z = 0.25\ Z_\odot$ are superimposed on our observed CMD. Filled circles along the isochrones represent mass steps starting from 10 $M_\odot$ (at $M_{130} \simeq$ -6.8), with a sampling of 10 $M_\odot$. The dashed line indicates the 4σ detection limit.

and theoretical distribution is clearly better for main sequence stars than fainter than $M_{130} \simeq$ -10 than for brighter stars. In other words, we do not observe the theoretically expected population of very bright and very massive stars which have recently left the main sequence towards the supergiant and W-R.

Among the possible interpretations of this finding, we favor the idea that the most massive stars in R136 have ZAMS masses less than about 50 $M_\odot$. To support this hypothesis, we observe that, adopting a classical IMF with a slope of -2.35, one would expect 11.2 for the ratio of the number of stars with 10 $M_\odot < M <$ 50 $M_\odot$ over those with 50 $M_\odot < M <$ 120 $M_\odot$. Applied to Figure 2, this would predict about 20 stars with $M_{130} < -11$. These stars are not observed. This is consistent with earlier results by Moffat et al. (1985), Campbell et al (1992), and Parker (1992), who conclude that R136 does not contain extraordinary massive and/or bright stars. In addition, the colors of stars R136a$_1$ through r136a$_8$ can be understood if they are either evolved blue supergiant or W-R stars. It is difficult to estimate their initial masses purely form their positions in our CMD (because all W-R stars pass through this part of the diagram), but if their progenitors had masses above $\sim$ 80 $M_\odot$, then we would expect a significant MS population in the range 50 $M_\odot < M <$ 80 $M_\odot$, as the stellar lifetime decreases with mass. Since this is not observed, it is very likely that they originate form stars of $\sim$ 50

$M_\odot$.

We conclude that the lack of extremely luminous stars in R136 is due to an intrinsic deficit of stars with ZAMS masses above $\simeq 50\ M_\odot$, if a standard Salpeter IMF is extrapolated upward from the mass interval $10\ M_\odot < M < 50\ M_\odot$. The individual components of R136a are the most luminous — and presumably amomg the most massive — stars in the R136 region. Their ZAMS masses may be as high as 80 $M_\odot$ but values around 50 $M_\odot$ are more likely.

References

Campbell, B., Hunter, D.A., Holtzman, J.A., Lauer, T.R., Shaya, E.J., Code, A., Faber, S.M., Groth, E.J., Light, R.M., Lynds, R., O'Neil, E.J., Westphal, J.A. 1992, A. J., 104, 1721.

Cassinelli, J.P., Mathis, J.S., and Savage, B.D. 1981, Science, 212, 1497.

De Marchi, G., Nota, A., Leitherer, C., Ragazzoni, R., and Barbieri, C. 1993, Ap.J., Dec 20 issue.

Feitzinger, J. V., Schlosser, W., Schmidt-Kaler, Th., and Winkler, C. 1980, A.&A., 84, 50.

Fitzpatrick, E.L., Savage, B.D. 1984, Ap.J., 279, 578.

Heap, S. 1993, these conference proceedings.

Kurucz 1992, IAU Symp. 149, in *The Stellar Populations of Galaxies*, Ed. B. Barbuy, A. Renzini, (Dordrecht: Kluwer), p.225.

Maeder, A. 1990, A.&A., 84, 139.

Maeder, A., Meynet, G. 1988, A.&A.Suppl., 76, 411.

Malumuth, E., Heap, S. 1992, in *Science with the Hubble Space Telescope*, P. Benvenuti and E. Schreier edrs. p.297.

Melnick, J. 1985, A.&A. 153, 235.

Moffat, A.F.J., Seggewiss, W., and Shara, M.M. 1985, Ap.J., 295, 109.

Neri, R., and Grewing, M. 1988, A.&A., 196, 338.

Panagia, N., Gilmozzi, R., Macchetto, F., Adorf, H.-M., and Kirshner, R.P. 1991, Ap.J., 380, L23.

Parker, J. 1993, these conference proceedings.

Savage,B.D., Fitzpatrick, E.L., Cassinelli, J P., and Ebbetts, D.C. 1983, Ap.J., 273, 597.

Schmutz, W., Leitherer, C., Gruenwald, R. 1992, P.A.S.P., 104, 1164.

Schmutz, W., Vogel, M., Hamann, W.-R., and Wessolowski, U. 1993, A.&A.Suppl., in prep.

Weigelt, G., and Baier, G. 1985, A.&A., 150, L18.

Weigelt, G., et al. 1991, Ap.J., 378, L21.

SPECTRAL SYNTHESIS OF SPECTRAL POPULATIONS USING BALMER LINES

K.CANANZI[1], R.AUGARDE[1],J.LEQUEUX[2]

[1]:*Observatoire de Marseille, Place Leverrier, F-13248 Marseille Cedex 04;* [2]:*Observatoire de Meudon, Place Jules Janssen, F-92195 Meudon Principal Cedex*

1. Introduction

The goal of spectral synthesis is to determine the characteristics of a stellarpopulation (age and Initial Mass function) by comparing its observed spectrum to synthetic spectra built by adding contributions of stars (or star clusters) of various ages. When the population is young, its visible spectrum is dominated by Balmer lines. These lines have rarely been used for spectral synthesis because Balmer lines in emission are superimposed to the corresponding stellar absorption lines if there is gas ionized by the hottest young stars, a very frequent case. Still it is here that the absorption Balmer lines are the most useful. An example of a problem that can be attacked only by observing the Balmer lines is that of *the possible existence of a low-mass cutoff* in the IMF of an extragalactic starburst. This is a very important problem as it may affect our knowledge of star formation in general and also our ideas about evolution of galaxies. The colors of a young starburst as well as most of its spectral features are relatively unaffected by stars of a few solar masses which contribute little to the integrated luminosity and to the ionization and excitation properties. However, as the Balmer lines are much stronger for those stars than for the more massive, hotter ones, one may hope that the strength of the Balmer lines in the integrated spectrum of the starburst can give information on the presence or absence of those hot stars. It has been suggested by Augarde and Lequeux (1985) that the starburst galaxy Mkr 171 = Arp 299 = NGC 3690/IC 694 contains relatively few stars of a few solar masses as the Balmer lines in absorption seem relatively weak. Separation of the absorption from the emission component of those lines requires a good spectral resolution with a high dynamical range. Modern detectors make this possible. We have thus undertaken a program to study the Balmer lines in absorption in extragalactic starburst galaxies. This program has necessitated a new, systematic calibration of the Balmer lines equivalent widths of 70 well-classified galactic stars from spectral types O3 to G0 and luminosity classes I, III, V obtained with the same instrument and in the same conditions as the extragalactic observations (Cananzi et al.,1993,paper I). The study is limited to the $H\delta$ and $H\gamma$ lines as they are easily accessible and less contaminated by emission lines than $H\alpha$ and $H\beta$. We present here the model we have computed showing the variation of the equivalent widths of a star cluster as a function of the slope x, the higher mass M_{up}, the lower

Space Science Reviews **66**: 75–79, 1994.

© 1994 *Kluwer Academic Publishers. Printed in Belgium.*

mass M_{inf} of the IMF for 2 histories of star formation: an instantaneous burst in which all stars are formed at the same time on the ZAMS and subsequently evolve in the HR diagram until 10^7 years and a constant star formation scenario also lasting up to 10^7 years. The time step used is 10^6 years. Because of the non-unicity of the solution inherent to spectral synthesis model, constraints like L_{Bol}, L_V, L_B, L_U, the number of Lyman continuum α photons and the Hβ luminosity (Osterbrock, 1989) have been synthesized as well to compare with what we observe in galaxies (Cananzi et al.,1993: paper II).

2. Description of the model

In this model, all the parameters are functions of the stellar masses and luminosity classes. The star sample is in the mass range 1 to 120 $M_\odot$ with a bin of 0.25 $M_\odot$ to make it as continuous as possible. Hence, the values of the parameters have been interpolated linearly when missing. Then, the evolution with time of the different stellar parameters is given by the variation of their values when a star goes from a luminosity class to another, i.e., when a star evolves from the main sequence up to the supergiant stage if it is massive enough. In particular, we have derived a mean value of the luminosities L_{Bol}, L_V, L_B and L_U within a given class. The time step of 10^6 years minimizes the error made by not following more accurately the star evolution. Anyway, the comparison of our modelled parameters like L_B or $L_{H\beta}$ with those from other models (Leitherer, 1990: Mas–Hesse and Kunth, 1991) has shown discrepancies not larger than 5%. We have defined the IMF as $\Phi(M) = \frac{dN}{dM} = kM^x$ with x=-2.7, -2.35, -3, -2, i.e., the value derived by Scalo and Salpeter for the solar neighbourhood, an upper and a lower limit, respectively. If N(M) represents the number of stars of mass M, the equivalent width of Hδ for stars in the range $[M_{inf},M_{up}]$ is:

$$Ew_\delta = \frac{\sum_{Mi=M_{inf}}^{M_{up}} N(Mi)Ew_\delta(Mi)Ic_\delta(Mi)}{\sum_{Mi=M_{inf}}^{M_{up}} N(Mi)Ic_\delta(Mi)}$$

where Ew_δ(Mi) is the star equivalent width (paper I) and Ic_δ(Mi), the continuum intensity near Hδ (Kurucz, 1991). Then, the total luminosity at wavelength λ is:

$$L_\lambda = \sum_{Mi=M_{inf}}^{M_{up}} N(Mi)L_\lambda(Mi)$$

where L_λ(Mi) is the stellar luminosity at the given wavelength (L_{Bol}: Schaller et al.,1992;L_V, L_B, L_U:Schmidt–Kaler, 1982). And the total number of Lyman continuum α photons is:

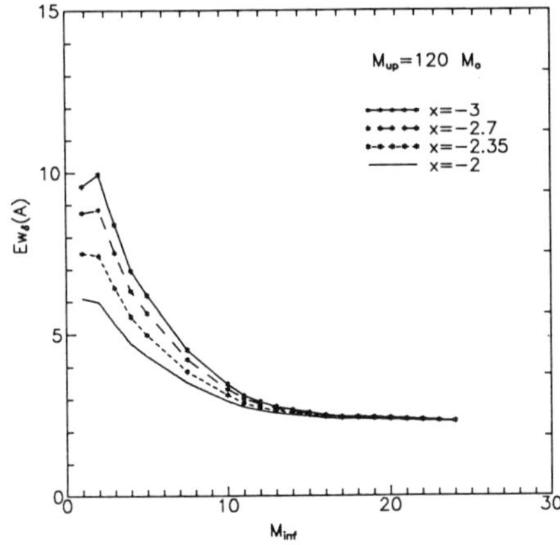

Figure 1a: Variation of the Hδ total equivalent width with the lower mass M_inf and the slope x of the IMF for a zero aged burst

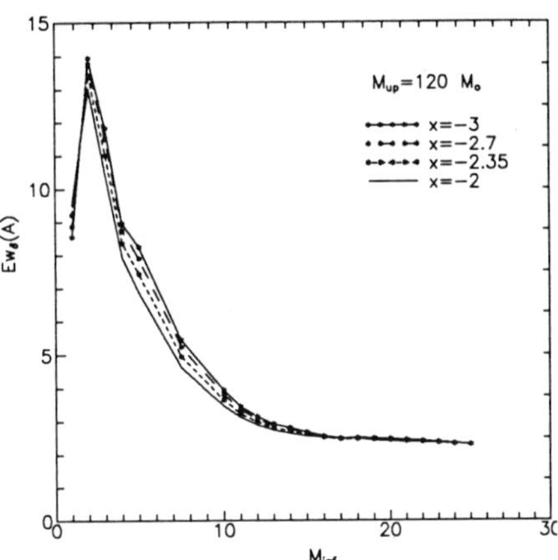

Figure 1b: Variation of the Hδ total equivalent width with the lower mass M_inf and the slope x of the IMF in a 10^{10} years old constant star formation scenari

Fig. 1.   Left: Figure 1a: Variation of the Hδ total equivalent width with the lower mass M_{inf} and the slope x of the IMF for a zero aged burst. Right: Figure 1b: Variation of the Hδ total equivalent width with the lower mass M_{inf} and the slope x of the IMF in a 10^{10} years old constant star formation scenario

$$N_{Lyc} = \sum_{Mi=M_{inf}}^{M_{up}} N(Mi)N_{Lyc}(Mi)$$

where $L_{Lyc}(Mi)$ is the number of the star Lyman α photons (Panagia,1973; Cruz-Gonzales et al.,1974).

3. The hydrogen Balmer lines equivalent widths as a function of the IMF parameters

We want to study the influence of the low mass cutoff M_{inf}, the upper mass M_{up} and the slope x of the IMF on the theoretical equivalent width of the hydrogen Balmer lines. We limit our study to Hδ (hereafter Ew_δ) as Hγ has the same behaviour. We consider here the two extreme cases of a burst at t=0 and a constant star formation lasting up to 10^{10} years. Figure 1a shows Ew_δ as a function of M_{inf} in the burst scenario. According to the IMF, the most numerous stars are those with a mass M close to M_{inf}. Hence for $M_{inf} = 2\ M_\odot$ the values of the equivalent widths derived are maximum because they correspond to stars of spectral type AV,i.e., with the largest equivalent widths. Conversely, for $M_{inf}= 1\ M_\odot$, stars that predominate are of spectral type GV: their values of Ew_δ are much lower. Then,the larger the M_{inf}, the lower the equivalent widths derived until they

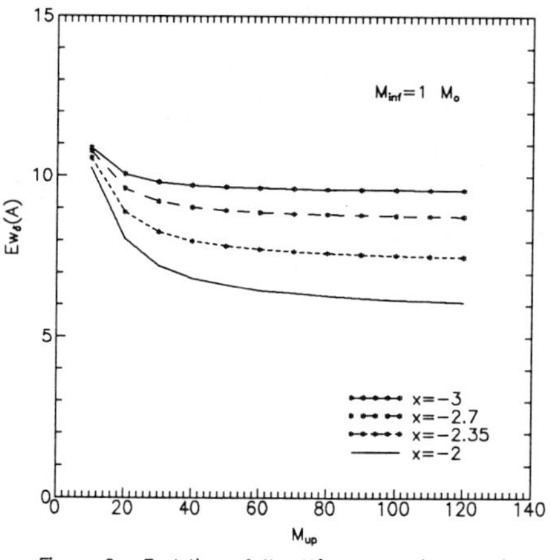

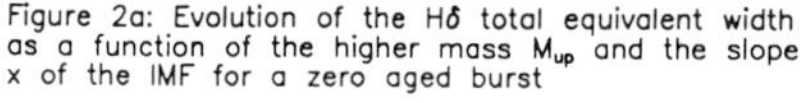

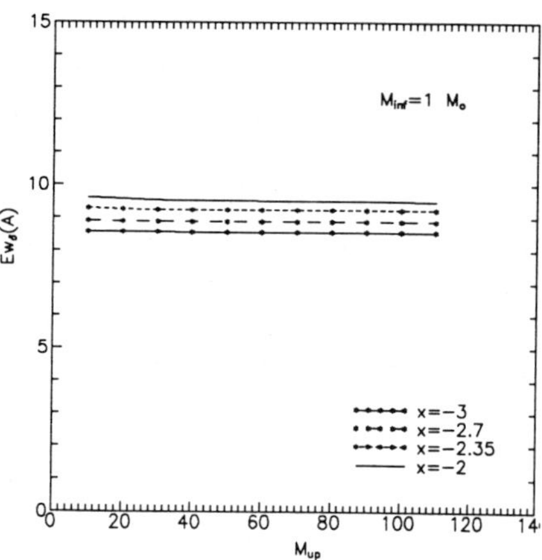

Figure 2a: Evolution of the Hδ total equivalent width as a function of the higher mass M_{up} and the slope x of the IMF for a zero aged burst

Figure 2b: Evolution of the Hδ total equivalent width as a function of the higher mass M_{up} and the slope x of the IMF in a 10^{10} years old constant star formation scenario

Fig. 2.   Left: Figure 2a: Evolution of the Hδ total equivalent width as a function of the higher mass M_{up} and the slope x of the IMF for a zero aged burst. Right: Figure 2b: Variation of the Hδ total equivalent width as a function of the higher mass M_{up} and the slope x of the IMF in a 10^{10} years old constant star formation scenario

reach an asymptotic value for $M_{inf} \geq 15$ M$_\odot$. In that case the equivalent width decrease together with the continuum intensity increase tend to minimize the ratio $\frac{N(Mi)Ew(Mi)Ic(Mi)}{N(Mi)Ic(Mi)}$. The same reasoning applies to figure 1b for the constant star formation case. However we notice that the values of Ew$_\delta$ are generally larger because of the renewal of stars. Figure 2 a & b show the variation of the Balmer equivalent widths with M_{up}. Let's note that massive stars contribute little to the equivalent width derived. As expected in a 10^{10} years constant star formation scenario, massive stars no more influence the value of Ew$_\delta$ because they are drowned in the accumulation of stars of a few solar masses. For a burst, the combination of a smaller equivalent width (for M_{up} = 10 M$_\odot$, Ew$_\delta$ = 7.12 A; for M_{up} = 120 M$_\odot$, Ew$_\delta$ = 2.14 A) and a larger continuum intensity (for M_{up} =10 M$_\odot$, Ic$_\delta$ = 3.9 10^8erg s^{-1} A^{-1};for M_{up} =120 M$_\odot$, Ic$_\delta$ = 14 10^8erg s^{-1} A^{-1}) is responsible for the slight decrease of Ew$_\delta$ when increasing M_{up}. As far as the slope x of the IMF is concerned, we can notice on the previous figures that it plays mainly a role in a burst scenario. In fact, the proportion of stars formed is more important for an IMF with x=-2 than x=-3 and especially massive stars which tend to decrease the equivalent width as we have already seen it. So the values of Ew$_\delta$ will be larger for x=-3. In a constant star formation, this is still true but less obvious as stars of a few solar masses mainly predominate on the main sequence for $\tau = 10^{10}$ years whatever the initial amount of stars of different masses. However, we notice that this trend is reversed for $M_{inf} = 1\ M_\odot$ as there are A5 V stars in the sample, more numerous for x=-2 than for x=-3.

In conclusion, this study stresses out that the equivalent widths are much more sensitive to M_{inf} than to M_{up}, to the slope x of the IMF and even to the stellar formation scenario. In particular, we have derived values of $Ew_\delta = 4.21\text{A}$ and $Ew_\gamma = 4.27\text{A}$ for NGC 604. The application of our model to this brightest HII region of M 33 leads to a lower mass cutoff ranging from 8.25 to 5.10 $M_\odot$ when x varies from -3 to -2, corrobating the idea of a high mass star formation in starburst galaxies (Cananzi-Mathias, 1993)

References

Augarde,R.,Lequeux,J.: 1985, *A&A* **147**, 273

Cananzi,K.,Augarde,R.,Lequeux,J.: 1993, *A&AS* **in press**, paper I

Cananzi- Mathias,K.: 1993, Thèse, Université Paris VII

Cananzi,K.,Augarde,R.,Lequeux,J.: 1993, *A&A* **in preparation**, paper II

Cruz-Gonzales,C., Recillas-Cruz,E., Costero,R., Peimbert,M. and Torres- Peimbert,S.: 1974, In Revista Mexicana de Astronomía y Astrofísica Vol.1, p. 250

Kurucz,R.L. 1991, preprint series No 3348

Leitherer,C: 1990, *ApJS* **73**, 1

Mas-Hess,J.M., Kunth,D.: 1991, *A&A* **88**, 399

Osterbrock,D.E.: 1989, Astrophysics of Gaseous Nebulae and Active Galactic Nuclei; University Science Books, California

Panagia,N.: 1973, *AJ* **78**, 929

Schaller,G., Schaerer,G., Meynet,G., Maeder,A.: 1992, *A&AS* **96**, 269

Schmidt-Kaler,Th.: 1982, Numerical data and Functional Relationships in Science and Technology/ Landolt-Börnstein, Springer Verlag, Berlin- Heidelberg

THE STELLAR CONTENT OF THE ORION OB1 ASSOCIATION

A.G.A. BROWN

Sterrewacht Leiden, P.O. Box 9513, 2300 RA Leiden, The Netherlands

Abstract. We present a photometric investigation, using the VBLUW system, of the stellar content of Orion OB1. Physical parameters ($\log g$, $\log T_{\text{eff}}$) for the stars are derived with the aid of model atmospheres. From these, visual extinctions, absolute magnitudes and distance moduli are derived. The distance moduli are used to determine membership for the stars in each of the subgroups and distances to the subgroups are calculated. The ages of the subgroups are derived through isochrone fitting and the IMF is derived for each subgroup. The energy deposited into the ISM through stellar winds and supernovae is calculated and compared to observed large scale features in the ISM around Orion OB1.

1. Introduction

The study of OB associations is motivated primarily by the fact that they form the fossil record of star formation processes occurring in giant molecular clouds in the galactic plane. The stellar content and the internal kinematics of associations provide valuable information on: the initial mass function, the local star formation rate and efficiency, the velocity distribution of the young stars as a function of mass and position in the system, differential age effects between subgroups in an association, the characteristics of the binary population and the interaction between stars and the surrounding interstellar medium (for a review see eg. Blaauw 1991).

Up to now the study of associations has been hampered by a lack of accurate knowledge of the stellar content as well as the internal kinematics. To remedy this problem the SPECTER consortium was formed at Leiden Observatory. It has been granted observing time on the HIPPARCOS satellite for measuring proper motions of over 10000 candidate members of nearby associations. Candidates were chosen on the basis of spectral type, apparent magnitude and location. A programme to obtain VBLUW photometry for around 5500 candidate stars has been completed. In addition an ESO Key programme aimed at obtaining precise radial and rotational velocities for the stars in Sco–Cen is in progress since 1990 (Hensberge et al. 1990). The results of this programme will provide, in addition to better membership determination, the distribution of rotational velocities ($v \sin i$), information on duplicity and the third component of the space motions. Here we discuss preliminary photometric results for Orion OB1. For results on Scorpio–Centaurus see De Geus et al. (1989) and De Geus (1992).

The Orion OB star complex is an excellent laboratory for studying the questions mentioned above. It contains the four subgroups (1a–1d, Blaauw 1964) of Orion OB1 and the Orion Molecular Cloud complex, the nearest site of active OB star formation. The large scale stellar content was studied extensively by Warren & Hesser (1977, 1978 hereafter WH) using $uvby\beta$ photometry. They derived

Space Science Reviews **66**: 81–84, 1994.
© 1994 *Kluwer Academic Publishers. Printed in Belgium.*

TABLE I

Mean distance modulus and rms spread for the subgroups.

subgroup	WH $\langle dm \rangle$ (mag)	WH rms spread (mag)	VBLUW $\langle dm \rangle$ (mag)	VBLUW rms spread (mag)
a	8.0	0.46	7.9	0.52
b1	8.5	0.53	7.8	0.39
b2	8.2	0.29	7.8	0.45
b3	8.0	0.55	7.8	0.46
c	8.2	0.49	8.0	0.49
d	8.4	0.53	7.9	0.25

distances and ages for the subgroups. During the past fifteen years a wealth of data has been gathered on the interstellar medium surrounding Orion OB1, and the modelling of the effects of early type stars on their surroundings has also improved considerably. So a new investigation into the large scale stellar content of Orion OB1 is justified. Our sample of stars includes all O and B stars in the Orion region and a larger number of late type stars (A, F and G) than were included in WH.

2. Derivation of Physical Parameters

To derive physical parameters for the stars from VBLUW photometry we use the reddening independent colours $[B-U]$ and $[B-L]$, where $[B-U]$ is an indicator of $\log T_{\mathrm{eff}}$ and $[B-L]$ depends mostly on $\log g$. With the aid of Kurucz (1979) atmosphere models one can construct a grid of synthetic colours which is then used to derive $\log T_{\mathrm{eff}}$ and $\log g$ for the stars. These two physical parameters are subsequently transformed into the intrinsic colours of the stars from which visual extinctions follow. Absolute magnitudes are derived from surface gravity and effective temperature with the aid of the Straizys & Kuriliene (1981) calibration tables. For more details see eg. De Geus et al. (1989) and Brown et al. (1993).

3. The Properties of Orion OB1; Distances, Ages and the IMF

Mean distances to the subgroups are calculated using the known member stars (taken from WH). Based on these means and the rms spreads therein the membership criteria for the remainder of the programme stars is established. In Table 1 the derived distance moduli and the rms spreads therein are listed and compared to the values as found by WH.

Subgroup 1b was divided by WH in three parts in order to investigate whether

TABLE II

Ages and total energy output of the subgroups.

subgroup	ages (Myr)			energy output (10^{50} ergs)
	VBLUW	WH	Blaauw	
a	11.4 ± 1.9	7.9	12	26^{+20}_{-13}
b	1.7 ± 1.1	5.1	7	$2^{+8}_{-1}(< 12)$
c	$4.6^{+1.8}_{-2.1}$	3.7	3	13^{+17}_{-10}
d	< 1.0	< 0.5	0	—

there is an increase in distance modulus with right ascension within this subgroup. We find no evidence for such an increase in distance from west to east.

The ages of the subgroups are determined by finding the theoretical isochrone (from stellar evolution models by Schaller et al., 1992) that fits the observations best in the $\log T_{\rm eff}$, $\log g$ plane. The ages are listed in Table 2 together with ages derived by WH and Blaauw (1991). Except for 1b, which we find to be younger than 1c, our results agree with the ages listed in Blaauw (1991). The results are discussed in more detail in Brown et al. 1993.

To determine the stellar IMF for the subgroups of Orion OB1 the present day mass function can be compared to the assumed form of the IMF using the Kolmogorov–Smirnoff test. We find that the IMF for the subgroups in Orion OB1 is best described by a single power law; $\xi(\log M) = AM^{-1.7}$. Where the scaling factor A indicates the richness of the subgroup.

4. Interaction of the Stars and the ISM in Orion

The large scale features in the ISM in the vicinity of Orion OB1 include the Orion molecular cloud, Barnard's loop, and a hole in the HI distribution surrounded by HI shells which extend to 50° below the galactic plane. The whole Orion–Eridanus region shows an enhancement of the background diffuse $H\alpha$ emission. These features can be explained as part of a cavity containing hot ionized gas surrounded by expanding shells of HI (see eg. Reynolds & Ogden, 1979).

To investigate whether the cavity was formed by the combined action of stellar winds and supernovae from the association we calculate the energy output of the subgroups. Three important contributions need to be considered; 1) the winds of main sequence O and B stars, 2) the winds of Wolf–Rayet stars and 3) the supernova explosions which occurred. The total energy output depends on the IMF, the age of the association and the number of high mass stars that were formed.

The energy outputs are computed through Monte Carlo simulations. Synthetic associations are generated with stellar populations drawn from a distribution function corresponding to the IMF. These associations are advanced in time using stellar evolution models. The number of supernovae is monitored and the wind–energy output is calculated. The simulations show that the number of supernova events and the wind–energy output have a large statistical uncertainty. It is assumed that each SN explosion produces 10^{51} ergs of which about 20% will contribute to the kinetic energy of the HI shells (Weaver et al. 1977). Stars more massive than 25 $M_\odot$ are assumed to go through a Wolf–Rayet phase, during which they contribute $3 \pm 2 \times 10^{50}$ ergs. The resulting total energy outputs are listed in Table 2.

The energy output of the subgroups can be compared to the dimensions of the cavity and the expansion velocity of the HI shells. Simple estimates, based on spherical geometry and an assumed constant ambient density of the ISM, show that the energy output of the Orion OB1 stars is sufficient to account for the observed cavity

For more details on the derivation of the ages, the IMF and the energy output we refer to Brown et al. (1993).

References

Blaauw, A.: 1964, *Annual Review of Astronomy and Astrophysics* **2**, 236

Blaauw, A.: 1991, 'OB Associations and the Fossil Record of Star Formation' in C.J. Lada and N.D. Kylafis, ed(s)., *The Physics of Star Formation and Early Stellar Evolution, NATO ASI Series C, Vol. 342*, Kluwer: Dordrecht, 125

Brown, A.G.A., de Geus, E.J., de Zeeuw, P.T.: 1993, submitted to *Astronomy and Astrophysics*

de Geus, E.J., de Zeeuw, P.T., Lub, J.: 1989, *Astronomy and Astrophysics* **216**, 44

de Geus, E.J.: 1992, *Astronomy and Astrophysics* **262**, 258

Hensberge, H., van Dessel, E.L., Burger, M., de Zeeuw, P.T., Lub, J., LePoole, R.S., Verschueren, W., David, M., Theuns, T., de Loore, C., de Geus, E.J., Mathieu, R.D., Blaauw, A., 1990, *The Messenger* **61**, 20

Kurucz, R.L.: 1979, *Astrophysical Journal, Supplement Series* **40**, 1

Reynolds, R.J., Ogden, P.M.: 1979, *Astrophysical Journal* **229**, 942

Schaller, G., Schaerer, D., Meynet, G., Maeder, A.: 1992, *Astronomy and Astrophysics, Supplement Series* **96**, 229

Straizys, V., Kuriliene, G.: 1981, *Astrophysics and Space Science* **80**, 353

Warren, W., Hesser, J.E.: 1977, *Astrophysical Journal, Supplement Series* **34**, 115 (WH)

Warren, W., Hesser, J.E.: 1978, *Astrophysical Journal, Supplement Series* **36**, 497 (WH)

Weaver, R., McCray, R., Castor, J., Shapiro, P., Moore, R.: 1977, *Astrophysical Journal* **218**, 377 (err **220**, 742)

OBSERVATIONS OF THE ATMOSPHERES AND WINDS OF O–STARS, LBVS AND WOLF-RAYET STARS

PAUL A. CROWTHER and ALLAN J. WILLIS
Department of Physics and Astronomy
University College London
Gower Street, London, WC1E 6BT UK

Abstract. This review summarises recent studies of O–stars, Luminous Blue Variables (LBVs) and Wolf-Rayet (WR) stars, emphasising observations and analyses of their atmospheres and stellar winds yielding determinations of their physical and chemical properties. Studies of these stellar groups provide important tests of both stellar wind theory and stellar evolution models incorporating mass-loss effects. Quantitative analyses of O–star spectra reveal enhanced helium abundances in Of and many luminous O–supergiants, together with CNO anomalies in OBN and Ofpe/WN9 stars, indicative of evolved objects. Enhanced helium, and CNO–cycle products are observed in several LBVs, implying a highly evolved status, whilst for the WR stars there is strong evidence for the exposition of CNO–cycle products in WN stars, and helium–burning products in WC and WO stars. The observed wind properties and mass-loss rates derived for O–stars show, in general terms, good agreement with predictions from the latest radiation–driven wind models, although some discrepancies are apparent. Several LBVs show similar mass-loss rates at maximum and minimum states, contrary to previous expectations, with the mechanism responsible for the variability and outbursts remaining unclear. WR stars exhibit the most extreme levels of mass-loss and stellar wind momenta. Whilst alternative mass-loss mechanisms have been proposed, recent calculations indicate that radiation pressure alone may be sufficient, given the strong ionization stratification present in their winds.

Key words: stars:winds – stars:mass-loss – stars:evolution – stars:abundances – stars:Wolf-Rayet – stars:atmospheres

1. O Stars

We review progress in observational and interpretative studies of O–stars, concentrating on determinations of their stellar wind and mass-loss properties and atmospheric chemical composition. More general reviews of this field has been given in Conti & Underhill (1988), whilst Kudritzki & Hummer (1990) have given a review describing progress in quantitative spectroscopy of hot, early–type stars.

1.1. MASS-LOSS RATES AND WIND VELOCITIES

Observational studies over the past decade have used data across the electromagnetic spectrum to characterise the wind properties and mass-loss rates of massive stars, provide tests of radiation-driven wind theory (e.g. Pauldrach *et al*, 1986, 1990), and the mass-loss parameterizations used in stellar evolution models (e.g. Maeder, 1990).

Bieging *et al* (1989) have summarised radio observations at 2, 6 and 20 cm of

Space Science Reviews **66**: 85–103, 1994.
© 1994 *Kluwer Academic Publishers. Printed in Belgium.*

Galactic OB stars. For 7 sources (confined to the most luminous stars, $\geq 10^6$ L$_\odot$) with detected radio emission and confirmed as thermal, free-free emitters, they derive mass-loss rates of $\sim 10^{-5}$ M$_\odot$ yr^{-1}. Leitherer & Robert (1991) report the detection of 4 OB stars at 1.3 mm, finding mm–radio spectral indices of ~ 0.6, consistent with free-free emission expectations, and confirming the radio mass-loss rates. Determinations of $\dot{M}$ for larger stellar samples have generally relied on analyses of UV and/or Hα observations.

Howarth & Prinja (1989) analysed the UV P–Cygni profiles in the IUE spectra of 203 Galactic OB stars, using the radio-derived mass-loss rates to estimate empirical ionization fractions in the stellar winds. They derive a well-defined relation between mass-loss rate and stellar luminosity:

$$\log \left(\dot{M} \ / \ \mathrm{M}_\odot \ \mathrm{yr}^{-1}\right) = 1.69 \log \left(L/L_\odot\right) - 15.4$$

finding good agreement with predictions of finite disk, steady state wind models. In an earlier study of the UV spectra of 16 OB stars, Garmany & Conti (1984) deduced a similar scaling: $\dot{M} \propto L/L_\odot{}^{1.62}$. This relation was used by Leitherer (1988) to calibrate mass-loss rate with observed Hα emission measured in 149 Galactic OB stars, with the wind velocity law derived ($\beta \sim 0.7$) agreeing with the value of $\beta \sim 0.8$ predicted from the wind models of Pauldrach et al (1986). Groenenwegen & Lamers (1989) analysed the UV P–Cygni profiles in 26 O–stars using the SEI method, including allowance for turbulence in the winds, and derived a velocity law with $\beta = 0.68 \pm 0.15$ in good agreement with theory.

Groenewegen & Lamers (1991) find that the empirically–derived ionization ratios in Si IV, C IV and N V, show very large differences (factors of 100-1000) with those predicted in the wind models of Pauldrach et al (1990) and Drew (1989), whilst Groenewegen et al (1989) find that the O–star wind terminal velocities derived from UV profile fitting are systematically lower by about 40% than predicted. They find $v_\infty / v_{\mathrm{esc}} = 2.78 \pm 0.36$ – independent of luminosity class – compared to the value of about 3.9 expected from Pauldrach et al (1990) theory. Similar conclusions are found by Blomme (1990), and also by Prinja et al (1990) who measure v_∞ in a sample of 150 O-stars from the violet limit of saturated troughs in UV P–Cygni absorption profiles, and derive a mean ratio of $v_\infty / v_{\mathrm{esc}} = 2.41$. This discrepancy suggests either some remaining inadequacy in the radiation pressure models or some systematic error in the stellar parameters derived from the evolutionary tracks which are used to determine v_{esc}. To shed light on this question, Lamers & Leitherer (1993) have re–derived mass-loss rates for 28 Galactic OB stars utilising radio and Hα data, finding:

$$\log \left(\dot{M} \ / \ \mathrm{M}_\odot \ \mathrm{yr}^{-1}\right) = 1.738 \log \left(L/L_\odot\right) - 1.352 \log \mathrm{T}_{\mathrm{eff}} - 9.547$$

in good agreement with the scale adopted by Maeder (1990) in computing evolutionary tracks for OB stars. The empirical mass-loss rates are higher by about a factor of two than values predicted in the wind models. Lamers & Leitherer suggest that this discrepancy (particularly evident in the most luminous stars) may be a reflection, at a more modest level, of additional mechanisms (e.g. multiple scattering) required to explain the high mass-loss rates and wind momenta evident in WR stars. Due to the effect of wind structure (e.g. clumping) in O–star winds, uncertainties of ~0.1 dex in derived mass-loss rates need to be admitted.

1.2. WIND STRUCTURE

There is now strong observational (and theoretical) evidence that the winds of O stars are structured and permeated by shocked gas, believed to arise through radiatively–induced instabilities (e.g. Lucy, 1982; Owocki *et al*, 1988).

Non thermal radio emission has been detected in a large fraction ($\geq$ 25%) of O stars (Bieging *et al*, 1989), and proposed by White (1985) to arise in the outer winds through synchrotron emission from relativistic electrons accelerated in these shocks. *Einstein* X–ray observations of O–stars (e.g. Long & White, 1980; Chlebowski *et al*, 1989), with $L_x/L_{bol} \sim 10^7$, show soft spectra (typically $E \leq 0.5$ keV), indicative of emission produced in the outer wind layers, rather than from a base corona, probably due to high temperature shocked gas (Lucy & White, 1980; Lucy, 1982).

The saturated UV stellar wind P–Cygni resonance line profiles in O–stars exhibit extended black absorption troughs, and violet wings, not expected for spherical, monotonic velocity laws. In addition, the unsaturated UV profiles show the (time-dependent) occurrence of Discrete Absorption Components (DACs), which are observed in essentially all O–stars (Prinja & Howarth, 1986; Howarth & Prinja, 1989). Intensive IUE monitoring studies of several individual O–stars (Prinja, 1992) have shown that the DACs are highly variable, initially appear at about 0.3–0.5 v_∞, and progress out to v_∞ whilst becoming narrower. The measured DAC accelerations are slow, crudely corresponding to β = 2–5 type velocity laws, with the reccurrence and development timescales (typically 1— few days) being correlated with $v \sin i$ (Prinja, 1988). Variability in optical O–star wind signatures (e.g. He I λ5876 P–Cygni profiles) has been reviewed by Fullerton (1992), who concludes that the observed changes arise in clumps of material propagating through the winds.

1.3. PHYSICAL AND CHEMICAL PROPERTIES

The optical spectroscopic classification of O–stars, extensively developed by Walborn during the 1970's, has been updated using digital spectrophotometry by Walborn & Fitpatrick (1990), whilst their UV spectral morphology has been described by Walborn *et al* (1985). The development of highly sophisticated, non-LTE model atmosphere codes, including the effects of wind blanketing (Abbott & Hummer, 1985), coupled with the acquisition of very high precision spectral data, has led to tremendous advances in the determination of accurate physical and chemical properties for OB stars. Grigsby *et al* (1992) have presented analyses of 34 OB stars using an independent non-LTE, line-blanketed model atmosphere code. Kudritzki & Hummer (1990) summarise recent determinations of T_{eff}, log g and atmospheric helium abundances for 38 Galactic and 17 Magellanic Cloud OB stars. The results provide important tests and constraints on massive star evolutionary theory, with, *inter alia*, ample evidence for enhanced helium abundances in many stars.

Voels *et al* (1989) carried out detailed non-LTE, wind–blanketed model atmosphere analyses of the optical helium and hydrogen line profiles in four O9.5 stars covering luminosity classes V–Ia. Whilst normal helium abundances were derived for the three lower luminosity stars, an enhanced value of $Y=0.18\pm0.03^*$ was found for α Cam (O9.5 Ia). Voels *et al* concluded that enhanced atmospheric helium and probably also of CNO processed material is likely to be common amongst the Of and O Ia stars. Bohannan *et al* (1990) analysed the optical spectrum of three O4 stars, once again finding enhanced helium ($Y = 0.20\pm0.03$) for the supergiant class (ζ Pup, O4 I(n)f), confirming their earlier results (Bohannan *et al*, 1986). Pauldrach (these proc) confirms the predictions of CNO processed material in Of stars, deriving C/N=0.12 for ζ Puppis, $\sim$25 below solar values! Most recently, Herrero *et al* (1992), in their non-LTE analysis of the optical spectra of 25 Galactic OB stars, derive significant helium enhancements ($Y \geq 0.15$) in many stars, in particular in the rapid rotators and most luminous objects (Of, Ia, Iab stars), whilst the bulk of main sequence and slowly rotating giants and supergiants show solar values. Derived radii, luminosities and masses from spectroscopic analyses, compared with expectations from single star evolutionary models (Maeder, 1990) show significant discrepancies. Evolutionary models do not predict the enhanced He/H and N/C found, whilst the spectroscopically-derived masses are substantially smaller than predicted from the evolutionary tracks.

1.4. OBN STARS

Walborn (1971, 1976) identified the OBN class as a group of O–stars showing anomalously strong nitrogen lines, suggesting evolved objects exhibiting CNO–cycled material. Schonberger *et al* (1988) confirmed this view, from a quantitative non-LTE model atmosphere analysis of the optical and ultraviolet spectra of four OBN stars, compared to two normal stars of similar spectral types. Whilst the two normal stars were confirmed to have solar helium abundances, all four OBN stars were found to exhibit substantial helium enhancements, $Y= 0.15$–0.26. Additionally, the OBN stars showed enhanced nitrogen (typically up by factors of 10–30), with corresponding depletions in carbon, whilst the oxygen abundances were found to be near cosmic, indicating CN equilibrium values. The poor agreement of these results with the predictions of Maeder & Meynet (1987) led Schonberger *et al* to suggest the possibility of homogeneous evolution due to very fast initial rotation, which terminated after about half the hydrogen burning lifetime through angular momentum losses due to stellar wind mass-loss. They also noted the possibily anomalous abundances arising through enhanced mixing and mass exchange in a binary scenario.

---

* $Y= N(He)/[N(H)+N(He)]$

1.5. OFPE/WN9 STARS

Walborn (1977, 1982) first introduced the Ofpe/WN9 subclass, whose members show the presence of both high ionization Of features (N III, He II) plus low ionization WN9 features (N II, He I). Currently ten such stars have been identified in the LMC (Bohannan & Walborn, 1989). Allen *et al* (1990) have detected stars with similar properties at the Galactic Centre and Willis *et al* (1992) have recently discovered the first such star in M33.

Schmutz *et al* (1991) have performed a detailed analysis of the Ofpe/WN9 star R84. They determined $L/L_\odot = 10^{5.7}$, $T_* = 28.5$kK, $\dot{M} = 10^{-4.6}$ $M_\odot$ yr^{-1}, $v_\infty = 400$ km s^{-1} and H/He=2.4 (by number) using the Kiel WR code. From IR spectroscopy, McGregor *et al* (1988) have estimated H/He=2–5 for 5 LMC Ofpe/WN9 stars. Crowther (1993) has recently performed analyses of four such stars using the Hillier WR code and confirms the findings of Schmutz *et al* (1991) that Ofpe/WN9 stars have low temperatures ($\sim$30kK), high luminosities ($L/L_\odot = 10^{5.5-5.9}$), low terminal velocities 300-500 km s^{-1} and atmospheres severely depleted in hydrogen (H/He=1.7–3.5). Nitrogen and carbon abundances were found to be consistent with CN equilibrium values for $Z \sim 0.008$ (N/C=5–35). From the tracks of Maeder (1990) these stars are consistent with stars in the early WNL phase of stellar evolution with initial masses of around $40M_\odot$ and current masses of $\sim 20M_\odot$.

While a link between the Ofpe/WN9 stars and the LBVs has long been suspected, conclusive evidence of a direct connection between the two groups resulted when R127, a prototype Ofpe/WN9 star (Walborn, 1977) quite unexpectedly evolved to a B–type (Stahl *et al*, 1983) and later an A–type supergiant spectrum (Wolf *et al*, 1988).

2. Luminous Blue Variables

The term Luminous Blue Variable (LBV) denotes the most luminous variable blue stars and includes the S Dor, P Cygni type and the Hubble-Sandage variables. Around a dozen LBVs are currently known in the Galaxy and LMC with around 20 more further afield. In the Galaxy there are four well known LBVs, η Car, AG Car, P Cyg and HR Car, with HD 160529 (Sterken *et al*, 1991) and WRA751 (Hu *et al*, 1990) introduced recently. In the LMC, four LBVs are well known (R71, S Dor, R127, HD 269582) with R110 (Stahl *et al*, 1990) and R143 (Parker *et al*, 1993) recent additions.

The LBV definition is very broad and is not based on spectroscopic classification criteria. It is probable that the group contains stars whose evolutionary states are unrelated since it is merely the irregular variability that distinguishes LBVs. These stars lie close to the Humphreys-Davidson instability limit (Humphreys & Davidson, 1979). This limit is an observed luminosity cutoff above which luminous red supergiants are not found. The LBV variability has therefore been linked to the instability which allows the most massive stars to lose sufficient mass to prevent them from becoming red supergiants. This phase is therefore crucial to massive star evolution. Unfortunately the instability mechanism is not well understood.

2.1. VARIABILITY

Three types of photometric variability are observed in LBVs – small scale ($<$ 0.5mag), moderate (0.5–2 mag) and eruptive ($>$2 mag). LBVs typically resemble S Dor type A supergiants at maximum (7–8kK) and P Cyg type B stars at minimum (15–20kK) with some stars proceeding to even earlier spectral type, such as AG Car and R127 (25–30kK). Moderate light changes typically occur on timescales of decades, apparently at constant bolometric luminosity.

The observed lower temperature limit at maximum (7–8kK) is accepted as being due to the decrease in opacity with the onset of hydrogen recombination, causing a drop in radiation pressure (Davidson, 1987). Recently inverse P Cygni profiles have been observed in S Dor and R127 and explained as being due to infalling matter caused by yet lower opacity resulting in less radiation pressure (Wolf, 1992). Schulte-Ladbeck *et al* (1993a) argue that the inverse P Cygni profiles may be attributed to the presence of a circumstellar disk in which redshifted absorption indicates the presence of rotation rather than infalling material. Should this be true for R127 and S Dor, the LBVs may be more closely related to B[e] stars than previously thought (see later). Two exceptions which have crossed the opacity limit are R110, which was a late B supergiant at minimum and had evolved to a G star in 1991 (Wolf, 1992), and Var A in M33 which in 1985/6 showed the spectrum of an M3 supergiant (Humphreys *et al* 1987).

The bolometric magnitude of all known Galactic and LMC LBVs, excluding η Carina, is –9.7±0.9. Recent revisions to M_{bol} for some stars previously considered to have low luminosities such as R71 (M_{bol}=–8.8, Humphreys, 1989) suggest they may not be subluminous (M_{bol}=–9.9, Lennon *et al*, these proc.). Crowther (1993) has determined M_{bol}=–9.3±0.5 for four LMC Ofpe/WN9 stars and Crowther *et al* (these proc) find M_{bol}=–9.0±1.0 for 10 Galactic WN7-9 stars.

While many luminous emission line stars are known, only a few are recognised as LBVs. The major problem relating to the discovery of new LBVs is that the timescale over which variability occurs is generally much longer than that over which spectroscopic and photometric observations have been made. Many stars, such as He 3–519 (Smith *et al*, 1993) and R81 (Wolf *et al*, 1981a) are probably dormant LBVs. The first known LBV, P Cygni itself, would not be classified as an LBV if it were not for the major eruption around 1600.

Evidence for major eruptions in the past are circumstellar shells and nebulae which are associated with most LBVs (Humphreys, 1991) such as the famous Homunculus of η Car (Hillier & Allen, 1992) and the ring nebula around AG Car (Nota *et al*, 1992).

2.2. MASS-LOSS AND WIND VELOCITY

Lamers (1989) gives a detailed discussion of mass-loss in LBVs. In the past it has been widely accepted that the mass-loss rate at maximum is far greater than at minimum. The principal evidence for this resulted from the analysis by Leitherer *et al* (1989) of R71. This star is of early B spectral type at minimum (V=10.9) and early A at maximum (V=9.8) (Wolf *et al*, 1981b). Leitherer *et al*, (1989) determined mass-loss rates a factor of 30 higher at maximum ($10^{-4.7}$ $M_\odot$ yr^{-1}) than at minimum ($10^{-6.2}$ $M_\odot$ yr^{-1}). This situation now appears to be the exception rather than the rule. Major changes in mass-loss are not observed in many other LBVs. For instance, AG Car was observed to be an A–type S Dor star at maximum (V=6.1, Wolf & Stahl, 1982), at which epoch the mass-loss was estimated to be 10^{-4} $M_\odot$ yr^{-1} using the revised distance to AG Car of 6 kpc (Humphreys *et al*, 1989). At intermediate brightness, AG Car is an early B supergiant (V=7.2, Hutsemekers & Kohoutek, 1988). Barlow (1991) estimated $10^{-4.43}$ $M_\odot$ yr^{-1} from IR recombination lines observed in March 1984 (V=7.1) in reasonable agreement with $10^{-4.25}$ $M_\odot$ yr^{-1} found by Smith *et al* (1993) at the extreme minimum state (V=8.1). Stahl (1986) classified AG Car as Ofpe/WN9 although Smith *et al* (1993) suggest WN11 is more appropriate. Leitherer *et al* (1992) found an identical mass-loss rate ($10^{-4.5}$ $M_\odot$ yr^{-1}) to be appropriate for AG Car in Dec 1990 (V=7.9) and Aug 1991 (V=7.4).

In addition to AG Car, the recently discovered LBVs, R110 (Stahl *et al*, 1990) and HD 160529 (Sterken *et al*, 1991) have been found to show similar mass-loss rates at maximum and minimum, with the mass-loss at minimum probably greater than at outburst for R110!

Luminosities and mass-loss rates of LBVs appear to be largely environment independent (Lamers, 1989) although number statistics are of course very low. For instance, Humphreys *et al* (1988) found Var C in M33 to have a bolometric luminosity (M_{bol}=–9.8) and mass-loss rate ($10^{-4.4}$ $M_\odot$ yr^{-1}) which is typical of Galactic and LMC LBVs at maximum.

Terminal wind velocities are found to be dependent on spectral type, with wind velocities typically around 100 km s^{-1} in their A–type S Dor phase, 200 km s^{-1} for the B supergiant phase and >250 km s^{-1} for their hot O–type phase. For example, the terminal velocity of AG Car has been measured to be 166 km s^{-1} in 1981 (V=6.1, Wolf & Stahl, 1982), 205 km s^{-1} in Feb 1977 (V=7.2, Hutsemekers & Kohoutek, 1988) and 250 km s^{-1} in 1990 (V=7.9, Leitherer *et al*, 1992).

Mass-loss rates of LBVs are extremely high during major eruptions, such as the 1840–1860 η Car and 1600–1660 P Cyg events. During such events LBVs typically eject 10^{-2}–10^{0} $M_\odot$ (Lamers, 1989). The mass of the ring nebula associated with AG Car has been determined to be around 4 $M_\odot$ with a dynamical age of 8400 yr (Nota *et al*, 1992). van Genderen & The (1984) showed that the bolometric luminosity of η Car was *not* conserved during the great eruption during which 2 $M_\odot$ was expelled over $\sim$25 years (Davidson, 1989).

Two famous LBVs currently in quiescence are η Car and P Cyg, with mass-loss rates of $\sim 10^{-3}$ $M_\odot$ yr^{-1}(Davidson, 1989) and $10^{-4.67}$ $M_\odot$ yr^{-1} (Barlow, 1991) respectively. The stellar parameters of P Cyg have been determined by Pauldrach & Puls (1990), who found T_*=19.3kK, $L/L_\odot$=$10^{5.9}$, v_∞ =195 km s^{-1} while those of η Car are highly uncertain since the central star is shrouded by dust although stellar temperatures above 30kK are excluded due to the absence of He II emission (Hillier & Allen, 1992). This star is by far the most luminous LBV known (Davidson, 1989), with the revised distance of 3.2kpc to Tr16 (Massey & Johnson, 1993) implying a yet higher luminosity (M_{bol}=$-$12.3).

In summary, mass-loss rates of many LBVs at maximum are no more than those at minimum, within the uncertainties, with only R71 and probably S Dor clearly showing a large variability. The maximum mass-loss rates are also typical of WR stars, with terminal velocities typically a factor of ten lower and a factor of around two larger at minimum than at maximum. Thus, the single scattering limit is not exceeded for LBVs, indicating that radiation driven winds are appropriate (see however Smith *et al*, 1993 for the LBV candidate He 3–519).

There is substantial evidence for non-spherical symmetry in many LBVs, such as R127 (Schulte-Ladbeck *et al*, 1993a), AG Car (Schulte-Ladbeck *et al*, 1993b) and P Cygni (Taylor *et al*, 1991). Thus, mass-loss rates determined under the assumptions of spherical symmetry and homogeneity probably overestimate the true mass-loss rates by typically a factor of two (e.g. Hillier, 1991).

2.3. CHEMISTRY AND EVOLUTIONARY STATUS

It has been only in recent years that estimates of chemical abundances in LBVs have been possible. Helium contents in P Cyg and AG Car have been estimated from recombination lines by Barlow (1991) and found to be severely enhanced relative to solar values (H/He=2 and 4, respectively). An analysis of the optical line spectrum of AG Car by Smith *et al* (1993) using a detailed atmospheric code yielded H/He=2.4 for AG Car. Lennon *et al* (these proc) have recently analysed R71 and determined H/He=2.3, N/He=0.003, N/C=35 demonstrating CN equilibrium values are appropriate for at least one LBV.

Chemical abundances in a subset of LBV nebulae have also been made. Davidson *et al* (1982, 1986) derived abundances in the Homuculus surrounding η Car and found H/He=6, N/H$\sim$0.001, C/N=0.02 and O/N<0.5, demonstrating that η Car is a highly evolved object. Enhanced nitrogen has been found in the P Cyg (N/S=33) and HR Car (N/S=63, N/O$\geq$0.4) nebulae by Johnson *et al* (1992) and Hutsemekers & van Drom (1991) respectively. For the AG Car nebula, Mitra & Dufour (1990) have determined high N/S=63 and N/O=2 ratios, although argued against nitrogen being enhanced (see however de Freitas Pacheco *et al*, 1992).

Hence, abundance estimates imply LBVs are highly evolved objects. It is of use to estimate current stellar masses of LBVs, which can be made using the mass–luminosity relation of Schaerer & Maeder (1992). While this relation is only strictly valid for pure helium WR stars, derived masses are in excellent agreement with those found from independent methods. Stellar masses are found to be 20^{+20}_{-8} $M_\odot$ for the range of bolometric luminosity appropriate for LBVs (~ 200 $M_\odot$ for η Carina!), with initial masses around 40 ± 20 $M_\odot$ from evolutionary tracks. Pauldrach & Puls (1990) derived 23 ± 2 $M_\odot$ for P Cygni using their wind model and Lennon *et al* (these proc) have similarly determined a mass of 20 ± 2 $M_\odot$ for R71. The estimated mass of the P Cygni type eclipsing binary R81 is ~ 25 $M_\odot$ from Stahl *et al* (1987). LBVs are considered to be at an intermediate stage between massive O stars and WR stars in which the outer layers are removed by extreme mass-loss. Severely depleted hydrogen contents and enhanced nitrogen confirm this scenario, although no theory currently explains the mechanism which causes the LBV phenomenon. The long variability timescales inhibit a more complete understanding of LBVs. While reliable estimates of mass-loss rates are now possible using line and continuum transfer codes (e.g. Hamann & Schmutz, 1987; Hillier, 1987a), an explanation for the near constant mass-loss rate between maximum and minimum for some stars (e.g. AG Car) and dramatic differences for others with similar bolometric luminosities (e.g. R71) is not known.

While an Ofpe/WN9 star (R127) has been seen to evolve to an LBV, it is not known whether all Ofpe/WN9 stars evolve through this phase. Also, are all P Cygni type stars dormant LBVs? The slow timescales over which variability occurs does not allow us to currently answer these questions. Finally, the possible presence of disks around LBVs (Schulte-Ladbeck *et al* 1993a) suggests a more intimate connection with the B[e] stars. B[e] stars have a high velocity polar stellar wind, normal for early type supergiants and a low velocity wind in the equatorial region (Zickgraf *et al*, 1986), hot circumstellar dust and occupy a similar location on the H-R diagram to the LBVs (Zickgraf, 1992). Two B[e] stars (Hen S22/LMC, Hen S18/SMC) are spectroscopically variable in the UV (Shore, 1992; Shore *et al*, 1987). Hen S18/SMC is also variable in the optical (He II λ4686, Sanduleak, 1977), shows broad P Cygni profiles of H I, He I, is luminous ($M_{bol} \sim -9.3$), has a mass-loss rate of $>10^{-5}$ $M_\odot$ yr^{-1} (Zickgraf et al, 1989) and an enhanced nitrogen content, although it is correctly classified as a B[e] star due to the presence of narrow [Fe II] and [O I] emission lines. Hillier (1992) remarked that η Car shows many characteristics similar to the B[e] stars – strong broad Balmer emission, narrow Fe II, [Fe II] and a terminal velocity (500 km s^{-1}) similar to Hen S12/LMC (Zickgraf *et al*, 1986), with only the absence of [O I] emission preventing a B[e] classification. In contrast to LBVs, B[e] stars are not known to be photometrically variable, although such a link cannot be excluded due to the slow timescales over which variability occurs.

3. Wolf-Rayet stars

We review recent progress in determinations of the physical, chemical and mass-loss properties of WR stars from observations and analyses of their atmospheres and stellar winds. Advances in overall WR research have been the subject of recent IAU Symposia (IAU 116: de Loore *et al*, 1986; IAU 143: van der Hucht & Hidayat, 1991) and the reviews by Abbott & Conti (1987) and van der Hucht (1992).

3.1. EFFECTIVE TEMPERATURES AND LUMINOSITIES

Considerable advances have been made in recent years in determining temperatures and luminosities for WR stars based on quantitative modelling of observed spectra, using model atmosphere codes suitable for the optically thick, stratified WR atmospheres developed by Hillier (1987a) and Hamann & Schmutz (1987).

Schmutz *et al* (1989) developed a technique whereby large samples of stars can be analysed with minimal computational expense, using measured equivalent widths of selected He I and He II lines and a single continuum band absolute flux. They analysed a subset of the Galactic WR stars, finding typical values for WNL stars of $T_* \sim 35$kK and $L/L_{\odot}=10^{5.5-6.0}$, with the WNE stars spanning a wider range of $T_* \sim 35$–90kK and with $L/L_{\odot}=10^{5.0-5.5}$. This technique has been applied to a further sample of Galactic WN stars by Hamann *et al* (1993a), to Galactic WN and WC stars by Howarth & Schmutz (1992) and to LMC WN stars by Koesterke *et al* (1991).

Tailored analyses of WN stars have been performed for WR6 (WN5; Hillier, 1987b; Hamann *et al*, 1988) and for WR136 (WN6; Hamann *et al* 1993b). Hamann *et al* (1991) and Crowther *et al* (these proc) have, respectively, presented detailed analyses of 4 and 24 Galactic WN stars with surprisingly good agreement between the two groups. Line blanketing by iron group elements remains the principal effect currently neglected in WR models, thus limiting the accuracy of present results (see Schmutz, these proc).

Pure helium analyses have been shown to be inadequate for WC stars (Hillier, 1989; Hamann *et al*, 1992) since the inclusion of metals (carbon and oxygen) is required. Thus far, tailored fitting for only one star has been published (WR111, WC5; Hillier, 1989; Hamann *et al*, 1992) although the Kiel group have recently performed analyses for a sizeable sample of Galactic WCE stars (see Hamann, these proc) which are found to occupy a similar location on the HR diagram to the WNE-s stars.

An important check on WR atmospheric model results can be made using photoionization modelling of their surrounding ring nebulae (Rosa & Mathis, 1990; Esteban *et al*, 1993). Agreement is found to be reasonable for hot WR stars, with lower stellar temperatures derived from nebular analyses for cooler stars, apparently due to the lack of line blanketing in the WR models (see also Smith & Esteban, these proc).

3.2. MASS-LOSS AND WIND VELOCITIES

Observations of the radio emission from WR stars provides, potentially, the most accurate (and least model dependent) determinations of mass-loss rates. In spherical symmetry, assuming thermal free-free emission, the Wright & Barlow (1975) formulation is used to derive $\dot{M}$. As discussed below, new determinations of wind terminal velocities, coupled with re–assessments of the outer wind ionization balance have led to recent revisions in mass-loss rates derived from radio data. Should wind clumping be important, all analyses for single WR stars will overestimate the true mass-loss rate.

3.2.1. Terminal velocities

It is now evident that the true terminal velocities for WR stars are lower than previously inferred from measurements of the maximum violet edge velocity, v_{max}, in UV P–Cygni absorption profiles (Willis, 1982) or from the maximum widths of optical emission lines (Conti *et al*, 1983; Torres *et al*, 1986).

Williams & Eenens (1989) found lower values of ~ 0.7 v_{max} from P–Cygni absorption velocities of the He I 2.058 μm line. From IUE spectra, as for O–stars, Prinja *et al* (1990) measured v_∞ directly from the maximum violet extent of the zero residual intensity in saturated P–Cygni absorption profiles, v_{black} and not v_{max}. For 35 WR stars, they find a mean ratio of $v_\infty / v_{max} = 0.76 \pm 0.12$, in close agreement with the He I 2.058 μm results. Howarth & Schmutz (1992) measured v_∞ from the P–Cygni profiles in He I $\lambda 10830$ in 24 Galactic stars, which are reasonably consistent with the v_{black} measurements.

Hamann *et al* (1993a) estimate v_∞ for 53 Galactic WN stars from optical spectral line profile fits, whilst Koesterke *et al* (1991) using similar techniques derive v_∞ for 19 WN stars in the LMC. They find a good correlation with WN subtype, with v_∞ increasing from about 500 km s^{-1} at WN9 to about 3000 km s^{-1} at WN2 with no significant differences found between Galactic and LMC values.

It is necessary to emphasise that the currently accepted velocity laws (see Hamann, these proc) are not necessarily valid. Constraining the velocity law is difficult since most of the line formation occurs at high velocities, well above the acceleration zone.

3.2.2. Mass-loss rates

Abbott *et al* (1986) presented 4.9 GHz observations of 22 WR stars, showing the majority to be thermal emitters. Barlow *et al* (1981) found IR–radio spectral indices in the range 0.66–0.82 for 12 stars with measured radio fluxes, with a mean value of $\alpha = 0.75 \pm 0.04$, close to that expected for thermal emission. Leitherer & Robert (1991) present measurements at 1.3 mm for three WR stars, yielding a spectral index $\alpha \sim 0.6$ consistent with thermal free–free emission expectations.

For confirmed, or suspected, thermal radio sources, Abbott *et al* (1986) deduced a relatively tight range in WR mass-loss rates of 0.8–8 $\times 10^{-5}$ M$_\odot$ yr^{-1}, with no apparent systematic difference between WN and WC stars, with subtype within each sequence, or between single and binary stars. These results have been revised by Prinja *et al* (1990) and Willis (1991a) who accounted for the appropriate ionization balance and the more recent estimates of v_∞. They found a similar range in mass-loss rate of 1–10 $\times 10^{-5}$ $M_\odot$ yr^{-1}, again confirming the lack of systematic differences with spectral type and binarity.

This scale of WR mass-loss rate has been confirmed by optical spectral line analyses by Schmutz *et al* (1989) and Koesterke *et al* (1991), who also find similar rates for Galactic and LMC WN stars. Howarth & Schmutz (1992), in their spectral analysis of 24 Galactic WR stars derive mean mass-loss rates of: 3.9×10^{-5} (WNE); 2.6×10^{-5} (WNL) and 4.1×10^{-5} (WC), confirming the earlier conclusions of no substantial difference in $\dot{M}$ with spectral type. Additional information on WR mass-loss rates comes from studies of binary systems. For instance, St Louis *et al* (1988) modelled phase–dependent variations observed in optical linear polarization in 10 WR+O binaries, deriving values in the range 1–10 $\times$ 10^{-5} $M_\odot$ yr^{-1}.

Abbott *et al* (1986) suggested a tentative relation based on five WR+O double–lined systems of $\dot{M} \propto M^{2.3}$ which has, generally, not been confirmed by subsequent studies. St Louis *et al* (1988) in their polarization study of WR+O binaries, suggest a rough scaling of $\dot{M} \propto M^{0.8-1.3}$. Hamann *et al* (1993a), in their comprehensive study of Galactic WN stars claim a tentative correlation of mass-loss with stellar luminosity while Howarth & Schmutz (1992) do not find such a relation, although marginal evidence for $\dot{M} \propto M$ for the double–lined binary systems in their sample is found.

Winds of O–stars have mass-loss rates which do not normally exceed the 'single scattering limit' $\dot{M} v_\infty$ /(L/c), while all recent studies of WR stars have shown this to be substantially greater than unity. Hamann *et al* (1993a) derive mean values of $\sim$ 7 for WNL stars, $\sim$ 15 for WNE-w stars, $\sim$ 60 for WNE-s stars. Howarth & Schmutz (1992) determine $\sim$ 4 for WNL stars, $\sim$ 34 for both WNE and WC stars, with similar values found for LMC WN stars by Koesterke *et al* (1991). Rsults from detailed spectral analyses are up to a factor of two greater than those from radio measurements suggesting different degrees of clumping in the optical and radio forming regions. For instance, the mass-loss rate determined for WR136 (WN6) by the detailed line profile analysis of Hamann *et al* (1993b) is 1.4 $\times 10^{-4}$ $M_\odot$ yr^{-1}, while that from radio measurements is 7.6 $\times 10^{-5}$ $M_\odot$ yr^{-1} using observed 4.9 GHz radio fluxes (Abbott *et al*, 1986), appropriate ionization balance and abundances from Hamann *et al*.

These high values of $\dot{M} v_\infty$ /(L/c) have led to considerations that mechanisms other than radiation pressure may be driving the WR mass-loss, such as rotational and magnetic (e.g. Cassinelli, 1991). However, recent calculations by Lucy & Abbott (1993) have indicated that radiatively–driven solutions appropriate for the strong ionization stratification present in WR winds can lead to mass-loss rates with $\dot{M} v_\infty$ /(L/c)$\sim$ 10.

3.3. CLUMPING AND VARIABILITY OF WR WINDS

The formation of shocks and inhomogeneities in winds has been studied by Owocki *et al* (1988) who showed that radiation driven winds are intrinsically unstable. Prinja & Smith (1992) have recently detected DACs in the UV spectrum of WR24 (WN7+abs) with characteristics that suggest some evidence for a common mechanism for intrinsic wind variability in both WR and O–stars. Line and continuum variability is well known for WR stars (Moffat & Robert, 1991, 1992) and substantial evidence points to inhomogeneities from intensive line profile spectroscopy (e.g. Moffat *et al*, 1988). Cherepashchuk *et al* (1984) found that the UV eclipses of WR139 (WN5+O) could be fit with a lower wind density than the IR eclipses, suggesting density enhancements or clumps as the cause. From model line profile fitting, theoretical electron scattering wings appear to be stronger than is observed (e.g. Hillier, 1988) indicating that homogeneity may not be appropriate for some WR stars (Hillier, 1991). In the case of WR binaries, mass-loss rates determined from polarization models (e.g. St-Louis *et al*, 1993) have the advantage over radio and spectral mass-loss estimates that they are not dependent on the assumption of homogeneity.

Schulte-Ladbeck *et al* (1992) have obtained spectropolarimetry for 30 Galactic WR stars of which 4 (all WN subtypes) displayed reduced levels of polarization across emission lines, indicating nonsphericity. Two stars (WR6, WR134) were found to possess globally distorted, axisymmetric winds.

3.4. CHEMICAL COMPOSITION AND EVOLUTIONARY STATUS

As outlined below, most determinations of the composition of WR atmospheres find that they are chemically evolved, with WN stars showing the products of interior H–burning and WC and WO stars He–burning products. However, Bhatia & Underhill (1988) have come to quite different conclusions, suggesting that the atmospheric composition of both WN and WC stars are solar and that WR stars represent pre–Main Sequence stars. Compelling arguments against the latter scenario have been given by Lamers *et al* (1991).

3.4.1. WN stars

Extensive measurements of the H/He ratio in WN stars have been determined by Conti *et al* (1983) for 37 Galactic and 21 LMC stars covering all subtypes. H/He ratios were derived from measurements of the observed He II (n–4) Pickering decrement, where even–n series members coincide with H–Balmer (n–2) lines. Conti *et al* (1983) find most WN stars to be severely H–depleted, with WNL stars exhibiting the largest values (H/H = 0.4 – 8).

Hamann *et al* (1991) and Crowther *et al* (these proc) have confirmed the Pickering decrement results from their non-LTE model atmosphere analyses of Galactic WN stars. A clear temperature effect in the H–abundance is found, with the cooler WN stars ($T_* \sim 35$kK) showing some hydrogen, whilst stars with considerably higher temperatures do not (except WR136, WN6). Zero hydrogen contents are appropriate for most WNE stars, with H/He~1 typical for most WNL stars except for the WN7+abs stars for which Crowther *et al* found H/He=3–4, identical to the value determined for ζ Pup (O4I(n)f) by Bohannan *et al* (1990).

Smith & Willis (1982, 1983) and Nugis (1982) using a crude, Sobolev analysis of spectra of WN stars estimated values of C/N=0.01–0.06 and N/He=0.002–0.02 by number, broadly indicative of CNO–burning products. More recent, non-LTE model atmosphere analyses have largely confirmed these conclusions. For WR6 (WN5), Hillier (1988) determined C/N=0.07 and N/He=0.004 by number, whilst Hamann *et al* (1993b), from a non-LTE analysis of WR136 (WN6) find N/He=0.005 by number. Crowther *et al* (these proc) finds for all WN subtypes studied N/He=0.001–0.005, C/N~ 0.02 and O/N≪0.1, although WR46 (WN3p) is found to have a high oxygen content with O/N~0.2 and C/O<0.2.

An independent method of abundance determination in WR stars is through analyses of their associated nebulae. Rosa & Mathis (1990) and Esteban *et al* (1992) find N and He to be overabundant, with O underabundant, reflecting the products of CNO equilibrium in WR stellar ejecta.

3.4.2. WN/C stars

Conti & Massey (1989) have identified 7 Galactic and 2 LMC stars as having intermediate WN–WC characteristics based on anomalously high C IV λ5801 / He II λ4686 line strengths compared to normal WN stars. For the brightest of these, WR8 (WN6–C4), Willis & Stickland (1990) estimated C/N~ 0.3 and N/He~0.01 by number, from a simple Sobolev analysis, consistent with the star being chemically intermediate between the WN and WC stages. These conclusions have been broadly confirmed by Crowther (1993) who performed a detailed non-LTE analysis of this star deriving N/He=0.005, C/N=2.5 and C/O=3, broadly in line with the theory of Langer (1991).

3.4.3. WC and WO stars

Hydrogen is not detected in the optical, infrared or ultraviolet spectra of WC stars and a zero hydrogen content is normally adopted for this class (Willis, 1991b). Nugis (1982) analysed the C and He optical lines in several WC stars, deriving C/He~0.1–2 (by number), whilst Torres (1988) from a simple recombination analysis of observed optical spectra estimated C/He~0.1–0.8. Smith & Hummer (1988) performed recombination analyses of 17 Galactic WC stars using near–IR spectra deriving C/He ratios in the range 0.04–0.7, with evidence for a clear trend of increasing C/He from WC7 to WC4 subtypes. This was confirmed by Eenens & Williams (1992) who determined C/He=0.007–0.41 for 6 WC5–9 stars although WR146 (WC6) was found to have a very low carbon content (C/He=0.007).

Hillier (1989) and Hamann *et al* (1992) performed a detailed non-LTE model atmosphere analysis of HD 165763 (WC5) adopting C/He=0.4–0.5. Koesterke *et al* (in prep) have carried out analyses for a sample of 30 WC5–8 stars resulting in C/He=0.1–0.5 (by number), although they did not find a trend of C/He with subtype. The C/He ratios derived are clearly indicative of the exposition of He–burning products.

To date the Ne abundance has been derived for only one WC star, γ Velorum (WC8+O9I) for which Barlow *et al* (1988) derived Ne/He$\sim$0.001 by number – close to cosmic. This result conflicts with the expectations from stellar evolution models (e.g. Maeder & Meynet, 1987) which predict a high Ne–abundance in WC stars.

For the WO stars, first identified by Barlow & Hummer (1982), Kingsburgh *et al* (these proc) derive C/He=0.4–0.8 and O/He=0.1–0.4, from recombination analyses of UV and optical spectra of 4 WO stars, yielding C/O ratios between 2.1 and 4.5. Eenens (1993) determined C/He=0.8–2 for 2 WO stars from recombination analyses using IR spectra.

3.4.4. Evolutionary status

Overwhelming evidence now points to WR stars being highly evolved objects. Hydrogen is found to be severely depleted in all WR stars analysed to date. Nitrogen and carbon abundances are consistent with CNO cycle values in WN stars from both atmospheric and nebular analyses. In WC and WO stars extremely high carbon and oxygen abundances are found, clearly reflecting both triple–α and α–capture products.

The standard formation mechanism for single WR stars is that of post-RSG evolution in which mass-loss peels down the hydrogen layers as the O star evolves first to a RSG (or LBV) and then turns blueward. From a comparison of the observed location on the HR diagram of WR stars (Hamann, these proc) with that predicted by single star evolution (e.g. Maeder & Meynet (1987) agreement is found to be poor. A significant population of WR stars is located below the track for an initial mass of $25\,M_\odot$ with no significant difference between results for the Galactic and LMC stars in contrast to that predicted for differing metallicities (Maeder, 1990). The observed low luminosities of many WR stars (Hamann *et al*, 1993a) suggests a formation mechanism other than single star evolution, such as close binary evolution through which the hydrogen atmosphere is lost through Roche lobe overflow (Vanbeveren, 1991, these proc). However, WR stars with possible low mass companions are *not* found to have the lowest stellar luminosities (Hamann, these proc).

For very massive stars, mass-loss can remove the hydrogen envelope before the star can evolve to the red causing evolution to proceed directly from the Of to the WN phase (Conti, 1975). However such luminous WR stars are not observed in our Galaxy. Langer (these proc) finds that pulsational instabilities in massive stars may provide a mechanism for driving extreme mass-loss for lower mass stars. Crowther *et al* (these proc) find the WNL+abs stars have by far the highest hydrogen abundances of the WR stars, with higher hydrogen contents than the LBVs and that extreme Of stars appear to be their immediate precursors.

In summary, while WR stars are clearly evolved objects, their formation channel is currently not well established.

Acknowledgements

We are grateful to Linda Smith and Raman Prinja for providing useful comments on an earlier version of this paper.

References

Abbott, D.C. & Conti, P.S., 1987. *Ann. Rev. Astron. & Astrophys.*, **25**, 113–150.
Abbott, D.C. & Hummer, D.G., 1985. *Astrophys. J.*, **294**, 286.
Abbott, D.C., Bieging, J.H., Churchwell, E. & Torres, A.V, 1986. *Astrophys. J.*, **303**, 239.
Allen, D.A., Hyland, A.R. & Hillier, D.J., 1990. *Mon. Not. R. Astr. Soc.*, **244**, 706.
Barlow, M.J., 1991. In: *Wolf-Rayet Stars and Interrelations with other Massive Stars in Galaxies, IAU Symposium 143*, eds. van der Hucht, K.A. & Hidayat, B., p. 281, Kluwer, Dordrecht.
Barlow, M.J. & Hummer, D.G., 1982. In: *Wolf-Rayet Stars: Observations, Physics, Evolution, IAU Symposium 99*, eds. de Loore, C.W.H. & Willis, A.J., p.387, Reidel, Dordrecht.
Barlow, M.J., Smith, L.J. & Willis, A.J., 1981. *Mon. Not. R. Astr. Soc.*, **196**, 101.
Barlow, M.J., Roche, P.F. & Aitken, D.K., 1988. *Mon. Not. R. Astr. Soc.*, **232**, 821.
Bhatia, A.K. & Underhill, A.B., 1988. *Astrophys. J. Suppl*, **67**, 187.
Bieging, J.H., Abbott, D.C. & Churchwell, E.B., 1989. *Astrophys. J.*, **340**, 518.
Blomme, R., 1990. *Astron. Astrophys.*, **229**, 513.
Bohannan, B. & Walborn, N.R., 1989. *Publ. Astron. Soc. Pac.*, **101**, 639.
Bohannan, B., Abbott, D.C., Voels, S.A. & Hummer, D.G., 1986. *Astrophys. J.*, **308**, 728.
Bohannan, B., Voels, S.A., Hummer, D.G. & Abbott, D.C., 1990. *Astrophys. J.*, **365**, 729.
Cassinelli, J.P., 1991. In: *Wolf-Rayet Stars and Interrelations with other Massive Stars in Galaxies, IAU Symposium 143*, eds. van der Hucht, K.A. & Hidayat, B., p. 289, Kluwer, Dordrecht.
Cherepashchuk, A.M., Eaton, J.A. & Khaliullin, , 1984. *Astrophys. J.*, **281**, 774.
Chlebowski, T., Harnder, F.R. & Sciortino, S., 1989. *Astrophys. J.*, **341**, 427.
Conti. P.S., 1975. *Mem. Soc. Roy. Sci. Liege* , 6eme serie, Tome IX, 193.
Conti, P.S., Leep, E.M. & Perry, D.N., 1983. *Astrophys. J.*, **268**, 228.
Conti. P.S., & Underhill, A.B., 1988. *O stars and Wolf–Rayet Stars*, NASA Sp–497.
Conti, P.S. & Massey, P., 1989. *Astrophys. J.*, **337**, 251.
Crowther, P.A. PhD thesis, University of London, 1993.
Davidson, K., 1987. *Astrophys. J.*, **317**, 760.
Davidson, K., 1989. In: *Physics of Luminous Blue Variables, IAU Colloquium 113*, eds. Davidson, K., Moffat, A.F.J. & Lamers, H.J.G.L.M, p. 101, Kluwer, Dordrecht.
Davidson, K., Walborn, N.R. & Gull, T.R., 1982. *Astrophys. J.*, **254**, L47.
Davidson, K., Dufour, R.J., Walborn, N.R. & Gull, T.R., 1986. *Astrophys. J.*, **305**, 867.
de Freitas Pacheco, J.A., Neto, A.D., Costa, R.D.D. & Viotti, R., 1988. *Astron. Astrophys.*, **266**, 360.
de Loore, C.W.H., Willis, A.J. & Laskarides, P. (eds)., 1986. *Luminous Stars and Associations in Galaxies, IAU Symposium 116*, Reidel, Dordrecht.
Drew, J., 1989. *Astrophys. J. Suppl.*, **71**, 267.
Eenens, P.R.J., 1993. In: *Massive stars: Their Lives in the Interstellar Medium, A.S.P. Conference Series, Vol. 35*, eds. Cassinelli, J.P. & Churchwell, E.B., p. 254.
Eenens, P.R.J. & Williams, P.M., 1992. *Mon. Not. R. Astr. Soc.*, **255**, 227.
Esteban, C., Vlchez, J.M., Smith, L.J. & Clegg, R.E.S., 1992. *Astron. Astrophys.*, **259**, 629.
Esteban, C., Smith, L.J., Vlchez, J.M. & Clegg, R.E.S., 1993. *Astron. Astrophys.*, **272**, 299.
Fullerton, A.W., 1992. In: *Nonisotropic and Variable Outflows from Stars, A.S.P. Conference Series, Vol. 22*, eds. Drissen, L., Leitherer, C. & Nota, A., p. 185.
Garmany, C.D. & Conti, P.S., 1984. *Astrophys. J.*, **284**, 705.

Grigsby, J.A., Morrison, N.D. & Anderson, L.S., 1992. *Astrophys. J. Suppl.*, **78**, 205.

Groenewegen, M.A.T. & Lamers, H.J.G.L.M., 1989. *Astron. Astrophys. Suppl.*, **79**, 359.

Groenewegen, M.A.T. & Lamers, H.J.G.L.M., 1991. *Astron. Astrophys.*, **243**, 429.

Groenewegen, M.A.T., Lamers, H.J.G.L.M. & Pauldrach, A.W.A, 1989. *Astron. Astrophys.*, **221**, 78.

Hamann, W-R. & Schmutz, W., 1987. *Astron. Astrophys.*, **174**, 173.

Hamann, W-R., Schmutz, W. & Wessolowski, U., 1988. *Astron. Astrophys.*, **194**, 190.

Hamann, W-R., Dunnebeil, G., Koesterke, L., Schmutz, W. & Wessolowski, U., 1991. *Astron. Astrophys.*, **249**, 443.

Hamann, W-R., Leuenhagen, U., Koesterke, L. & Wessolowski, U., 1992. *Astron. Astrophys.*, **255**, 200.

Hamann, W-R., Koesterke, L. & Wessolowski, U., 1993a. *Astron. Astrophys.*, **274**, 397.

Hamann, W-R., Wessolowski, U. & Koesterke, L., 1993b. *Astron. Astrophys..* submitted.

Herrero, A., Kudritzki, R.P., Vilchez, J.M., Kunze, D., Butler, K. & Haser, S., 1992. *Astron. Astrophys.*, **261**, 209.

Hillier, D. J., 1992. In: *Atmospheres of Early-Type stars*, eds. Heber, U. & Jeffery, C.S., p. 105, Springer-Verlag.

Hillier, D.J., 1987a. *Astrophys. J. Suppl.*, **63**, 947.

Hillier, D.J., 1987b. *Astrophys. J. Suppl.*, **63**, 965.

Hillier, D.J., 1988. *Astrophys. J.*, **327**, 822.

Hillier, D.J., 1989. *Astrophys. J.*, **347**, 392.

Hillier, D.J., 1991. *Astron. Astrophys.*, **247**, 455.

Hillier, D.J. & Allen, D.A., 1992. *Astron. Astrophys.*, **262**, 153.

Howarth, I.D. & Prinja, R.K., 1989. *Astrophys. J. Suppl.*, **69**, 527.

Howarth, I.D. & Schmutz, W., 1992. *Astron. Astrophys.*, **261**, 503.

Hu, J.Y., de Winter, D., The, P.S. & Perez, M.R., 1990. *Astron. Astrophys.*, **227**, L17.

Humphreys, R.M., 1989. In: *Physics of Luminous Blue Variables, IAU Colloquium 113*, eds. Davidson, K., Moffat, A.F.J. & Lamers, H.J.G.L.M, p. 3, Kluwer, Dordrecht.

Humphreys, R.M., 1991. In: *Wolf-Rayet Stars and Interrelations with other Massive Stars in Galaxies, IAU Symposium 143*, eds. van der Hucht, K.A. & Hidayat, B., p. 485, Kluwer, Dordrecht.

Humphreys, R.M. & Davidson, K., 1979. *Astrophys. J.*, **232**, 409.

Humphreys, R.M., Jones, T.J. & Gerhz, R.D., 1987. *Astron. J.*, **94**, 315.

Humphreys, R.M., Leitherer, C., Stahl, O., Wolf, B. & Zickgraf, F.J., 1988. *Astron. Astrophys.*, **203**, 306.

Humphreys, R.M., Lamers, H.J.G.L.M., Hoekzema, N. & Cassatella, A., 1989. *Astron. Astrophys.*, **218**, L18.

Hutsemekers, D. & Kohoutek, L., 1988. *Astron. Astrophys. Suppl.*, **73**, 217.

Hutsemekers, D. & Van Drom, E., 1991. *Astron. Astrophys.*, **248**, 141.

Johnson, D.R.H., Barlow, M.J., Drew, J.E. & Brinks, E., 1992. *Mon. Not. R. Astr. Soc.*, **255**, 261.

Koesterke, L., Hamann, W-R., Schmutz, W. & Wessolowski, U., 1991. *Astron. Astrophys.*, **248**, 166.

Kudritzki, R.P. & Hummer, D.G., 1990. *Ann. Rev. Astron. & Astrophys.*, **28**, 303–345.

Lamers, H.J.G.L.M., 1989. In: *Physics of Luminous Blue Variables, IAU Colloquium 113*, eds. Davidson, K., Moffat, A.F.J. & Lamers, H.J.G.L.M, p. 135, Kluwer, Dordrecht.

Lamers, H.J.G.L.M. & Leitherer, C., 1993. *Astrophys. J.*, **412**, 771.

Lamers, H.J.G.L.M., Maeder, A., Schmutz, W. & Cassinelli, J.P., 1991. *Astrophys. J.*, **368**, 538.

Langer, N., 1991. *Astron. Astrophys.*, **248**, 531.

Leitherer, C., 1988. *Astrophys. J.*, **326**, 356.

Leitherer, C. & Robert, C., 1991. *Astrophys. J.*, **377**, 629.

Leitherer, C., Schmutz, W., Abbott, D.C., Hamann, W-R. & Wessolowski, U., 1989. *Astrophys. J.*, **346**, 919.

Leitherer, C., Neto, A.D. & Schmutz, W., 1992. In: *Nonisotropic and Variable Outflows from Stars, A.S.P. Conference Series, Vol. 22*, eds. Drissen, C.L. Leitherer & Nota, A., p. 366.

Long, K.S. & White, R.L., 1980. *Astrophys. J.*, **239**, L65.

Lucy, L.B., 1982. *Astrophys. J.*, **255**, 286.

Lucy, L.B. & Abbott, D.C., 1993. *Astrophys. J.*, **405**, 738.

Maeder, A., 1990. *Astron. Astrophys. Suppl.*, **84**, 139.

Maeder, A. & Meynet, G., 1987. *Astron. Astrophys.*, **182**, 243.

Massey, P. & Johnson, J., 1993. *Astron. J.*, **105**, 980.

McGregor, P.J., Hyland, A.R. & Hillier, D.J., 1988. *Astrophys. J.*, **324**, 1071.

Mitra, P.M. & Dufour, R.J., 1990. *Mon. Not. R. Astr. Soc.*, **242**, 98.

Moffat, A.F.J. & Robert, C., 1991. In: *Wolf-Rayet Stars and Interrelations with other Massive Stars in Galaxies, IAU Symposium 143*, eds. van der Hucht, K.A. & Hidayat, B., p. 109, Kluwer, Dordrecht.

Moffat, A.F.J. & Robert, C., 1992. In: *Nonisotropic and Variable Outflows from Stars, A.S.P. Conference Series, Vol. 22*, eds. Drissen, L., Leitherer, C. & Nota, A., p. 203.

Moffat, A.F.J, Drissen, L., Lamontagne, R. & Robert, C., 1988. *Astrophys. J.*, **334**, 1038.

Nota, A., Leitherer, C., Clampin, M., Greenfield, P. & Golimowski, D.A., 1992. *Astrophys. J.*, **398**, 621.

Nugis, T., 1982. In: *Wolf-Rayet Stars: Observations, Physics, Evolution, IAU Symposium 99*, eds. de Loore, C.W.H. & Willis, A.J., p. 131, Reidel, Dordrecht.

Owocki, S.P., Castor, J.I. & Rybicki, G.B., 1988. *Astrophys. J.*, **335**, 914.

Parker, J.W., Clayton, G.C., Winge, C. & Conti, P.S., 1993. *Astrophys. J.*, **409**, 770.

Paudrach, A., Puls, J. & Kudritzki, R.P., 1986. *Astron. Astrophys.*, **164**, 86.

Paudrach, A., Kudritski, R.P., Puls, J. & Butler, K., 1990. *Astron. Astrophys.*, **228**, 125.

Pauldrach, A.W.A. & Puls, J., 1990. *Astron. Astrophys.*, **237**, 409.

Prinja, R.K., 1988. *Mon. Not. R. Astr. Soc.*, **231**, 21P.

Prinja, R.K., 1992. In: *Nonisotropic and Variable Outflows from Stars, A.S.P. Conference Series, Vol. 22*, eds. Drissen, L., Leitherer, C. & Nota, A., p. 167.

Prinja, R.K. & Howarth, I.D., 1986. *Astrophys. J. Suppl.*, **61**, 357.

Prinja, R.K. & Smith, L.J., 1992. *Astron. Astrophys.*, **266**, 377.

Prinja, R.K., Barlow, M.J. & Howarth, I.D., 1990. *Astrophys. J.*, **361**, 607.

Rosa, M.R. & Mathis, J.S., 1990. In: *Properties of Hot Luminous Stars*, ed. Garmany, C.D., p. 135, Astron. Soc. Pac. Conf. Ser 7.

Sanduleak, N., 1977. *Inf. Bull. Var. Stars.*, **1304**.

Schaerer, D. & Maeder, A., 1992. *Astron. Astrophys.*, **263**, 129.

Schmutz, W., Hamann, W-R. & Wessolowski, U., 1989. *Astron. Astrophys.*, **210**, 236.

Schmutz, W., Leitherer, C., Hubeny, I., Vogel, M., Hamann, W-R. & Wessolowski, U., 1991. *Astrophys. J.*, **372**, 664.

Schonberger, D., Herrero, A., Becker, S., Eber, F., Butler, K., Kudritzki, R.P. & Simon, K.P., 1988. *Astron. Astrophys.*, **197**, 209.

Schulte-Ladbeck, R.E., Meade, M.R. & Hillier, D.J., 1992. In: *Nonisotropic and Variable Outflows from Stars, A.S.P. Conference Series, Vol. 22*, eds. Drissen, L., Leitherer, C. & Nota, A., p. 118.

Schulte-Ladbeck, R.E., Leitherer, C., Clayton, G.C., Robert, C., Meade, M.R., Drissen, L., Nota, A. & Schmutz, W., 1993a. *Astrophys. J.*, **407**, 723.

Schulte-Ladbeck, R.E., Clayton, G.C. & Meade, M.R., 1993b. In: *Massive stars: Their Lives in the Interstellar Medium, A.S.P. Conference Series, Vol. 35*, eds. Cassinelli, J.P. & Churchwell, E.B., p. 237.

Shore, S. N., Sanduleak, N. & Allen, D.A., 1987. *Astron. Astrophys.*, **176**, 59.

Shore, S.N., 1992. In: *Nonisotropic and Variable Outflows from Stars, A.S.P. Conference Series, Vol. 22*, eds. Drissen, C.L. Leitherer & Nota, A., p. 342.

Smith, L.F. & Hummer, D.G., 1988. *Mon. Not. R. Astr. Soc.*, **230**, 511.

Smith, L.J. & Willis, A.J., 1982. *Mon. Not. R. Astr. Soc.*, **201**, 451.

Smith, L.J. & Willis, A.J., 1983. *Astron. Astrophys. Suppl.*, **54**, 229.

Smith, L.J., Crowther, P.A. & Prinja, R.K., 1993. *Astron. Astrophys..* in press.

St-Louis, N., Moffat, A.F.J., Drissen, L., Bastien, P. & Robert, C., 1988. *Astrophys. J.*, **330**, 286.

St-Louis, N., Moffat, A.F.J., Lapointe, L., Efimov, Y.S., Shakhovskoy, N.M., Fox, G.K., & Piirola, V. 1993. *Astrophys. J.*, **410**, 342.

Stahl, O., 1986. *Astron. Astrophys.*, **164**, 321.

Stahl, O., Wolf, B., Klare, G., Cassatella, A., Krautter, J., Persi, P. & Ferrari-Toniolo, M., 1983. *Astron. Astrophys.*, **127**, 49.

Stahl, O., Wolf, B. & Zickgraf, F.J., 1987. *Astron. Astrophys.*, **184**, 193.

Stahl, O., Wolf, B., Klare, G., Juttner, A. & Cassatella, A., 1990. *Astron. Astrophys.*, **228**, 379.

Sterken, C., Gosset, E., Juttner, A., Stahl, O., Wolf, B. & Axer, M., 1991. *Astron. Astrophys.*, **247**, 383.

Taylor, M., Nordsieck, K.H., Schulte-Ladbeck, R.E. & Bjorkman, K.S., 1991. *Astron. J.*, **102**, 1197.

Torres, A.V., 1988. *Astrophys. J.*, **325**, 759.

Torres, A.V., Conti, P.S. & Massey, P., 1986. *Astrophys. J.*, **300**, 379.

van der Hucht, K.A. 1992. *Astron. Astrophys. Rev.*, **4**, 123.

van der Hucht, K.A. & Hidayat, B. (eds)., 1991. *Wolf-Rayet Stars and Interrelations with other Massive Stars in Galaxies, IAU Symposium 143*, Kluwer, Dordrecht.

van Genderen, A.M. & The, P.S., 1984. *Space Sci. Rev.*, **39**, 317.

Vanbeveren, D., 1991. *Astron. Astrophys.*, **252**, 159.

Voels, S.A., Bohannan, B., Abbott, D.C. & Hummer, D.G., 1989. *Astrophys. J.*, **340**, 1073.

Walborn, N.R., 1971. *Astrophys. J.*, **164**, 267.

Walborn, N.R., 1976. *Astrophys. J.*, **205**.

Walborn, N.R., 1977. *Astrophys. J.*, **215**, 53.

Walborn, N.R., 1982. *Astrophys. J.*, **256**, 452.

Walborn, N.R. & Fitzpartick, E.L., 1990. *Publ. Astron. Soc. Pac.*, **102**, 379.

Walborn, N.R., Nichols-Bohhi, J. & Panek, R.J., 1985. *NASA Ref. Pub.*, **1115**.

White, R.L., 1985. *Astrophys. J.*, **289**, 698.

Williams, P.E, 1982. In: *Wolf-Rayet Stars: Observations, Physics, Evolution, IAU Symposium 99*, eds. de Loore, C.W.H. & Willis, A.J., p. 73, Reidel, Dordrecht.

Williams, P.M. & Eenens, P.R.J., 1989. *Mon. Not. R. Astr. Soc.*, **240**, 445.

Willis, A.J., 1991a. In: *Wolf-Rayet Stars and Interrelations with other Massive Stars in Galaxies, IAU Symposium 143*, eds. van der Hucht, K.A. & Hidayat, B., p. 265, Kluwer, Dordrecht.

Willis, A.J., 1991b. In: *Evolution of Stars: The Photospheric Abundance Connection IAU Symposium 145*, eds. Michaud, G. & Tutukov, A., p. 195, Kluwer, Dordrecht.

Willis, A.J. & Stickland, D.J., 1990. *Astron. Astrophys.*, **232**, 89.

Willis, A.J., Schild, H. & Smith, L.J., 1992. *Astron. Astrophys.*, **261**, 419.

Wolf, B., 1992. In: *Nonisotropic and Variable Outflows from Stars, A.S.P. Conference Series, Vol. 22*, eds. Drissen, L., Leitherer, C. & Nota, A., p. 321.

Wolf, B. & Stahl, O., 1982. *Astron. Astrophys.*, **112**, 111.

Wolf, B., Stahl, O., de Groot, M.J.H. & Sterken, C., 1981a. *Astron. Astrophys.*, **99**, 351.

Wolf, B., Appenzeller, I. & Stahl, O., 1981b. *Astron. Astrophys.*, **103**, 94.

Wolf, B., Stahl, O., Smolinski, J. & Cassatella, A., 1988. *Astron. Astrophys. Suppl.*, **74**, 239.

Wright, A.E. & Barlow, M.J, 1975. *Mon. Not. R. Astr. Soc.*, **170**, 41.

Zickgraf, F-J., 1992. In: *Nonisotropic and Variable Outflows from Stars, A.S.P. Conference Series, Vol. 22*, eds. Drissen, L., Leitherer, C. & Nota, A., p. 75.

Zickgraf, F.-J., Wolf, B., Stahl, O., Leitherer, C. & Appenzeller, I., 1986. *Astron. Astrophys.*, **163**, 119.

Zickgraf, F-J., Wolf, B., Stahl, O. & Humphreys, R.M., 1989. *Astron. Astrophys.*, **220**, 206.

RADIATION DRIVEN WINDS OF HOT STARS:
THEORY OF O-STAR ATMOSPHERES AS A SPECTROSCOPIC TOOL

A.W.A. PAULDRACH, A. FELDMEIER, J. PULS and R.P. KUDRITZKI*

Universitatssternwarte Munchen, D-81679 Munchen, Scheinerstrae 1, Federal Republik of Germany

Abstract. The status of the continuing effort to construct radiation driven wind models for O-Stars atmospheres is reviewed. Emphasis is given to several problems relating to the fomation of UV line spectra the use of accurate atomic data, the inclusion of EUV radiation by shock heated matter, the simulation of photospheric line blocking.

A new tool for O-star diagnostics is presented. This is based on the use of wind models to calculate synthetic high resolution spectra covering the observable UV region. A comparison with observed spectra then gives physical constraints on the properties of stellar winds and stellar parameters, additionally abundances can be determined.

The astrophysical potential of this method is demonstrated by an application to two Of-stars, the galactic O4f-star ζ-Puppis and the LMC O3f-star Melnick 42. With regard to effective temperatures and gravities, the results from the application of classical methods to the analysis of photospheric lines are only partially verified. Explanations for the shortcomings of classical NLTE methods are discussed.

Key words: stars: atmospheres, early-type, mass-loss, X-rays, fundamental parameters, element abundances

1. Introduction

Although the ultraviolet spectra of O-stars with their hundreds of spectral lines provide a wealth of astrophysically important information about plasma conditions, stellar parameters and abundances, the quantitative analysis of UV O-star spectra concentrates on the few strong stellar wind lines (in particular the resonance lines of CIV, NV, OVI and SiIV). However, numerous other strong wind contaminated lines, especially of the iron group elements (see Fig. 1), are not taken into account. This is dangerous since, due to the usual uncertainty of stellar parameters and abundances (see below) and the complexity of the theory of O-star atmospheres, the uniqueness of an atmospheric model can not be guaranteed by fitting only a few wind lines. This means that an atmospheric model, and the stellar parameters derived from that, can be regarded as correct only if the consistently calculated synthetic high resolution spectrum covering the full observable UV spectral range fits the observed spectrum.

The theoretical tools which are required for this objective are still in an explorative stage. There are two reasons for that:

First, the simplifying assumption of LTE fails completely in O-star atmospheres because of the intense radiation field and, thus, a NLTE treatment of all

* RPK affiliated to Max-Planck-Inst. fur Astrophys., Karl-Schwarzschild-Str. 1, D-85748 Garching bei Munchen, Federal Republik of Germany

Space Science Reviews **66**: 105–125, 1994.

© 1994 *Kluwer Academic Publishers. Printed in Belgium.*

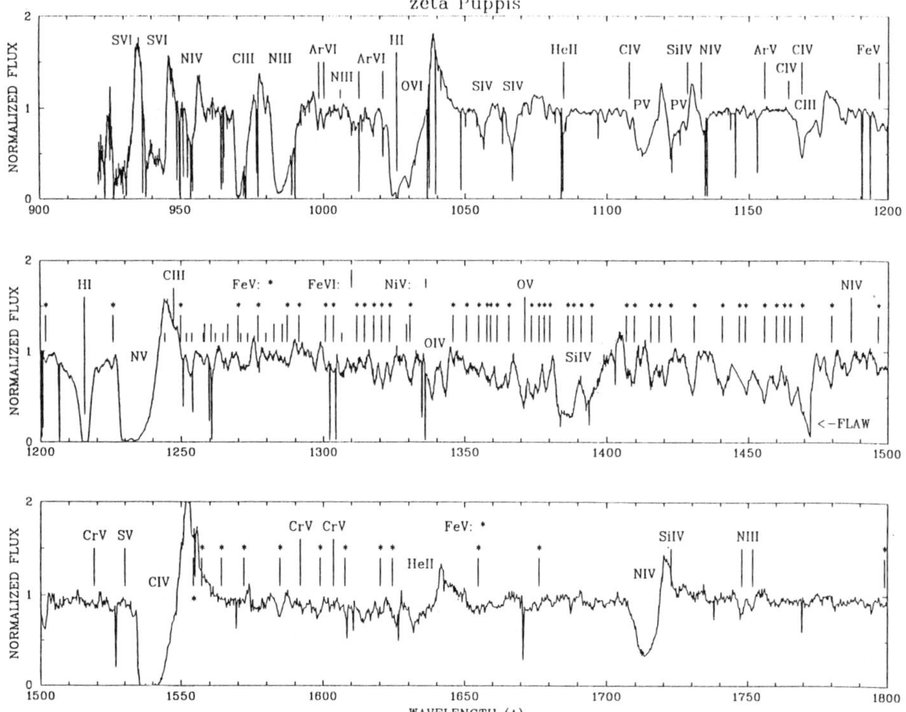

Fig. 1.   Observed Copernicus (900-1500A, Morton and Underhill (1977)) and IUE (1500-1800A, Walborn et al. (1985)) high resolution UV spectrum. A large number of strong and weak winds lines are identified and marked. Note the numerous FeV lines between 1250 and 1500A.

ions including the iron group elements is needed. As has become evident during the past decade, such NLTE calculations, if they are to be used for a quantitative comparison with observations, require very detailed and sophisticated model atoms with a large number of energy levels and transitions and with sufficiently accurate atomic data. Second, the UV line spectrum is strongly affected by the presence of stellar winds. This is true not only for the few strong resonance lines that exhibit P-Cygni profiles but also for the many weak absorption lines observed. In consequence, a hydrodynamic model atmosphere treatment that includes the effects of stellar winds is needed for theoretical spectrum synthesis calculations in the ultraviolet.

Usually the basic parameters of O-stars, consisting of effective temperature and gravity, are determined by an analysis of photospherical H and He lines (cf. Kudritzki and Hummer, 1990). These parameters enter in the hydrodynamical calculations as a starting point. Hence, it is important to note that these lines are affected by stellar winds as well (cf. Gabler et al. 1989). This effect can be clearly seen in Fig. 2 where the observed H_γ line profile of the two Of-stars (ζ-Puppis

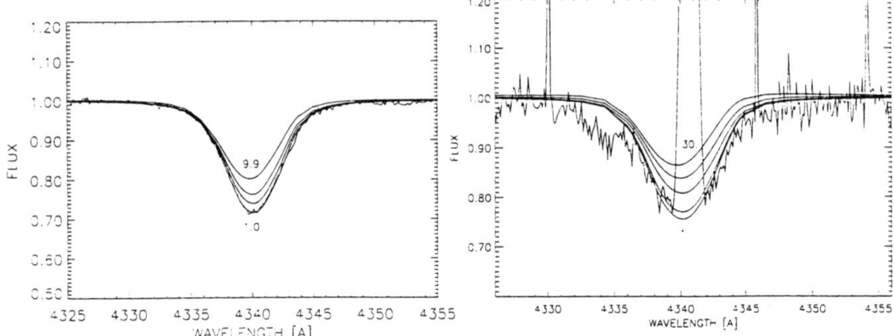

Fig. 2.   Observed and calculated Hγ profiles of ζ Puppis (Left - a sequence of $\dot{M} = 1.0, 3.2, 5.5, 9.9 \cdot 10^{-6} M_\odot/yr$ was adopted for the various models) and Melnick 42 (right - $\dot{M} = 1.0, 3.3, 10., 20., 30. \cdot 10^{-6} M_\odot/yr$ was adopted). From Sellmaier et al. (1993).

and Melnick 42) are compared to a sequence of "unified models" (photospherical NLTE models which include spherical extension and stellar winds) with different mass loss rates. The effect of wind contamination fills up the absorption profile with increasing mass loss rate. To fit the observed profile a higher value for the surface gravity is required (cf. Sellmaier et al. 1993). However, as is shown in Fig. 2, the influence of wind contamination can just be quantified if the hydro-dynamical structure and hence the mass loss rate is known; but this requires a knowledge of the stellar parameters in advance. Although the procedure is straightforward in the case the value for the mass loss rate is not too high (for ζ-Puppis with $\dot{M} = 3 - 5 \ 10^{-6} M_\odot/yr$ an enhancement of 0.1-0.15 dex is required for the surface gravity, Sellmaier et al. , 1993), almost nothing can be predicted in cases where the value for the mass loss rate is highly uncertain. In sect. 3 it will be shown that Melnick 42 belongs to this class of objects where only a lower limit for the basic parameters can be obtained from a photospheric analysis. We also review in that section a recently developed step towards fully quantitative UV-spectroscopy of O-stars based on NLTE radiation driven wind models, and we point out that this kind of spectroscopy can fill the diagnostic gap shown to exist for O-stars like Melnick 42, i.e. objects with high and uncertain mass loss rates. For this purpose the concept and the status of the computational method is summarized in sect. 2.

2. The Theory of Radiation Driven Winds

The basis of our model calculations is the concept of homogeneous, stationary and spherically symmetric radiation driven winds (where the driving mechanism is the line scattering of the photospheric UV radiation field by metal ions in the expanding and hence Doppler-shifted atmosphere) that describes correctly the

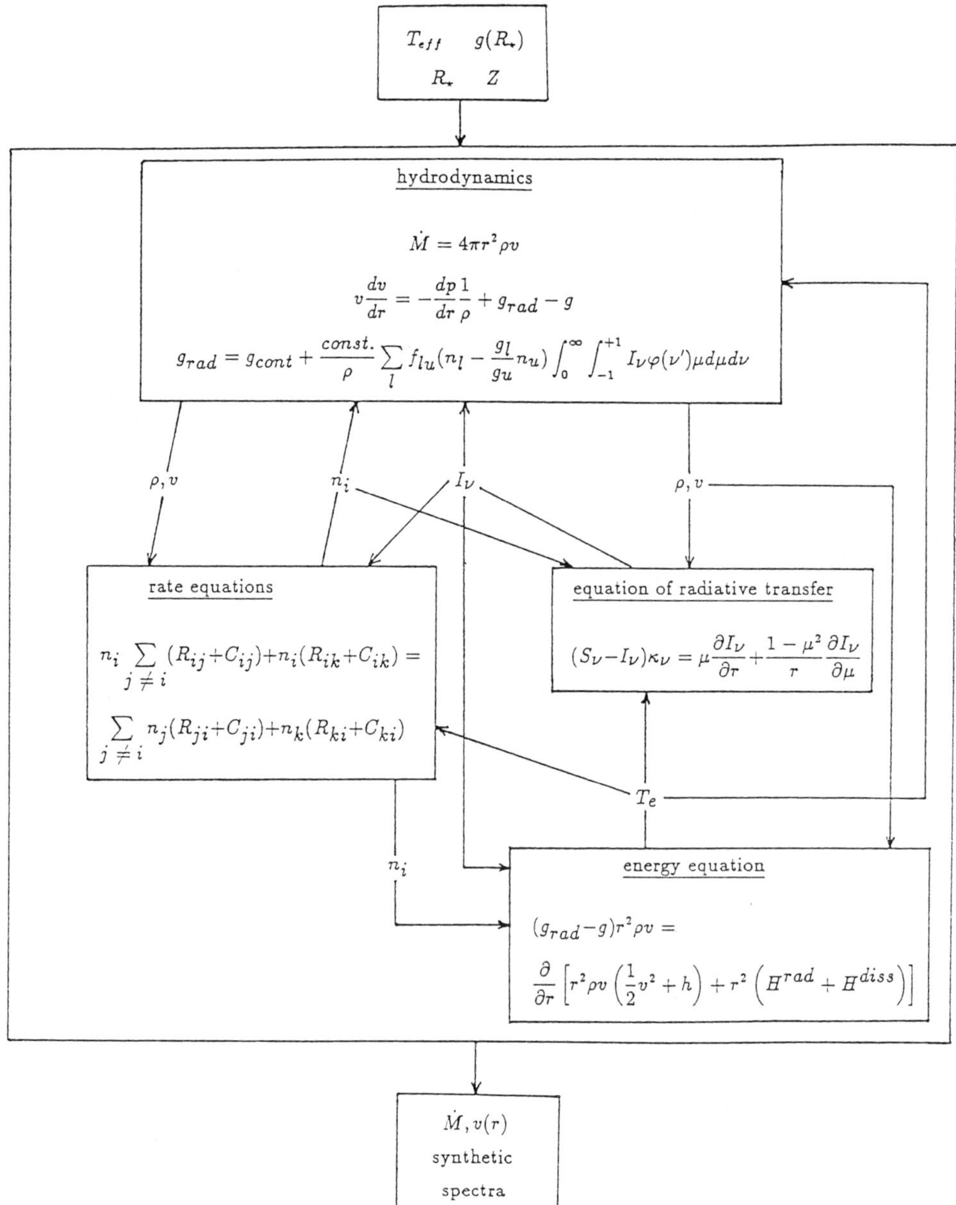

Fig. 3.   Schematic sketch of the non-linear system of integro-differential equations that form the basis of stationary radiation driven wind theory (see text).

time average mean of most spectral features in the UV (see below). This paper is intended to summarize the status of radiation driven wind models and not to give a comprehensive review of the wind theory of hot stars. Before we start to describe the theory in its present form it should not be forgotten that the basic ideas of the theory and the first attempt at its solution was the fundamental work of Lucy and Solomon (1970) and that the pioneering step in the formulation of the theory in a self-consistent manner was performed by Castor, Abbott and Klein (1975) and Abbott (1982). Although their approaches where only qualitative due to many simplifications, the theory was only further developed owing to their promising results.

The concept and the status of this development are sketched in Fig. 3. (essential steps of this building have been described and performed by Pauldrach, Puls and Kudritzki (1986, Paper I), Pauldrach (1987, Paper III), Puls (1987, Paper IV), Pauldrach and Herrero (1988), Pauldrach et al. (1990, Paper VII), Pauldrach et al. (1990, Paper IX) and Pauldrach et al. (1993, Paper XII).

To calculate a wind model, the stellar parameters T_{eff} (effective temperature), log g (logarithm of photospheric gravitational acceleration), R_* (photospheric radius defined at a pre-specified Thomson optical depth) and Z (abundances) have to be specified. Then the stationary hydrodynamic equations are solved in spherical symmetry (r is the radial coordinate, ϱ the mass density, v the velocity, p the gas pressure and $\dot{M}$ the rate of mass-loss). The crucial term is the radiative acceleration g_{rad} that has contributions from continuous absorption and scattering (g_{cont} - Thomson scattering, bound-free and free-free absorption of all elements considered (see below) are taken into account) and the line absorption. The calculation of the line acceleration is performed by summing the contributions of more than 200000 lines (see Paper XII). For each line the oscillator strengths f_{lu}, the statistical weights g_l, g_u and the occupation numbers n_l, n_u of the lower and upper level enter together with the frequency and angle integral over the specific intensity I_ν and the line broadening function φ_ν accounting for the Doppler effect.

The occupation numbers are determined by the *rate equations* containing collisional (C_{ij}) and radiative (R_{ij}) transition rates. It is important to note that the hydrodynamical equations are coupled directly with the rate equations. The velocity field enters into the radiative rates while the density is important for the collisional rates and the equation of particle conservation. On the other hand, the occupation numbers are crucial for the hydrodynamics since the radiative line acceleration dominates the equation of motion.

In addition, the radiation field determined by the *equation of transfer* is coupled with the hydrodynamics (radiative line acceleration) and the rate equations (radiative rates).

The temperature is, in principle, determined by the energy equation, which depends on h, the free enthalpy, H^{rad} the radiative flux and H^{diss} the energy flux generated by dissipative processes. Since we showed in Paper XII that the emergent spectrum is insensitive to the wind temperature structure used in our code we assumed a temperature structure on the basis of radiative equilibrium where bound-bound and bound-free opacities have been treated in NLTE and the effects of adiabatic cooling have been included (cf. Gabler, 1992).

The iterative solution of the total system of equations then yields the hydro-dynamic structure of the wind – including *mass-loss rate* and *terminal velocity* (v_∞) together with synthetic spectra (see Puls and Pauldrach (1990) and Paper XII) that can be compared with observations.

Since our treatment of O-star atmospheric models was recently described comprehensively in Paper XII, we mention only the crucial points which either will have important consequences or still imply some uncertainties for our model calculations.

2.1. ATOMIC MODELS

For detailed NLTE spectrum snthesis calculations, accurate atomic data are required. We are presently replacing the simplified atomic models (cf. Paper III, Paper IX) of the 149 ionization stages of 26 elements considered. This has already been done for the most important ionization stages, where more energy levels (comprising a total of 5000) and transitions (comprising 25000 bound-bound transitions and 20000 individual transition probabilities of low-temperature dielectronic recombination) were included for the following ions: H I; He I, II; C III, IV, V; N III, IV, V, VI; O IV, V, VI; Si IV; P V, VI; S VII; Ar VI, VII, VIII; Fe IV, V, VI ,VII, VIII; Ni V, VI, VII, VIII. To implement these improvements we utilised and modified the program SUPERSTRUCTURE (Eissner et al. 1974, Nussbaumer and Storey 1978), which uses the configuration interaction approximation to determine wavefunctions and radiative data.

2.2. RADIATIVE RATES AND RADIATION TRANSFER

i)   For the calculation of the radiative bound-bound transition probabilities R_{ij} the Sobolev-approximation is used in the entire atmosphere. As this might be a poor approximation in the subsonic region of the atmospheric layers where the continuum is formed, we presently implement improvements such as the Sobolev plus continuum method (Hummer and Rybicki, 1985; Puls and Hummer, 1988) or the comoving frame method (for applications in stellar wind dynamics, see Puls, 1987).

ii)   Low-temperature dielectronic recombination is included in the approximation described by Mihalas and Hummer (1973).

iii) The spherical transfer equation which yields the continous radiation field at up to 900 frequency points at every depth point including the deepest layers where the radiation is thermalized and the diffusion approximation is applicable is correctly solved, but without line opacities. Hence, the effects of photospheric EUV line blocking - due to metal ions in the spectral region between 228A and 911A - on the ionization and excitation of levels are treated separately in a realistic but still approximate way (see section 3.).

iv)   The emission from shocks arising from the non-stationary, unstable behaviour of radiation driven winds (cf. Lamers et al., 1982; Prinja and Howarth, 1986; Henrichs, 1986; Ebbets, 1982; Bieging et al., 1989) is additionally taken into account in the rate equations and the radiative transfer. This source was incorporated in a preliminary way on the basis of an approximate calculation of the shock emission coefficient (see section 3.).

v)   Using the cross-sections from Deltabuit and Cox (1972) the K-shell absorption was included for C, N, O, Ne, Mg, Si and S in the radiative transfer and Auger- ionization was taken into account in the rate equations (see Hunsinger and Pauldrach, 1994).

This present approach to the theory of O-star atmospheres is obviously not free from approximations. However, a detailed comparison with the observations can demonstrate its reliability.

3. Wind models: towards detailed UV line diagnostics of O-stars

The objectives of a detailed comparison are twofold. The primary aim in a first step concerns the investigation of wind physics in order to find physical constraints on the properties of stellar winds and in order to prove our method. If this step has been successfully completed we will try to determine stellar parameters and abundances by means of UV spectral synthesis. Concerning the first point the galactic O4f-star ζ-Puppis has been chosen, since excellent high resolution UV spectra including the important EUV spectral range have been obtained with the Copernicus satellite (see Fig. 1), and accurate flux measurements at radio-, IR- and X-ray wavelengths are available. The method is then applied to the O3 If*/WN6 − A (Walborn et al., 1993) star Melnick 42-observed with HST by Heap et al. (1991)-in the 30 Dorados complex of the LMC, which is a suitable candidate due to the uncertainty of its stellar parameters (see section 1- for a more detailed discussion see Paper XII) and its extreme nature.

It is a very interesting result that hydrostatic NLTE analyses yielded almost identical basic parameters for ζ-Puppis (T_{eff} = 42000K, log g = 3.5 - Kudritzki et al. (1983), Bohannan et al. (1986), Voels et al. (1989) - $R/R_\odot$ = 19) and MK 42 (T_{eff} = 42500K, logg = 3.5 - Heap et al. (1991) - $R/R_\odot$ = 28), whereas the UV spectra of these objects look rather different. This points to the necessity of a quantitative treatment of the full UV-spectrum. *The strategy of our procedure is as follows:*

i)   Starting from an adopted value of T_{eff} and an estimate of R_* and abundances (Z) the observed value of the terminal velocity is fitted by a sequence of models with varying log g. This gives $\dot{M}$ and the *surface gravity* which can be compared to the value obtained from the "unified models".

ii) Then, a model grid is calculated and the corresponding synthetic UV high resolution spectra are compared to the observed spectrum in order to get *constraints on the wind physics* or *constraints on the basic parameters* T_{eff}, R_*.

iii) In the test phase it is also worthwhile to check the dynamics by a comparison of the predicted- and the observed mass loss rate; the latter should be

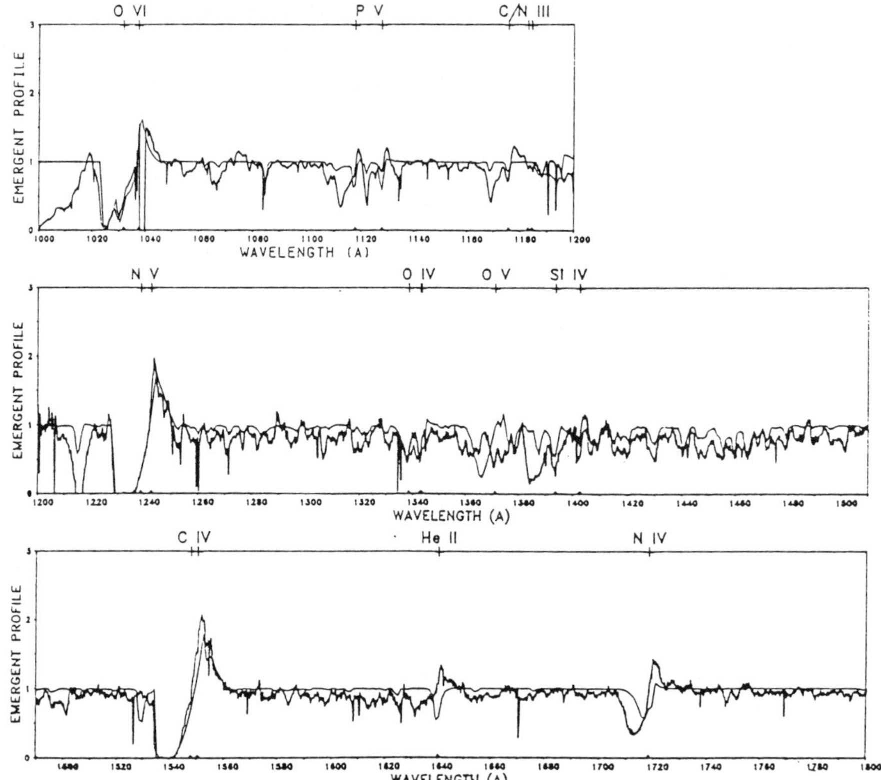

Fig. 4. Calculated and observed UV spectrum for ζ-Pup. The calculated spectrum belongs to model 1.

determined from radio data or H_α.

iv)　In the next step the observed- and systhetic spectrum of the final model is compared again for a precise determination of *abundances*.

This procedure provides the diagnostic tool for the determination of complete sets of stellar parameters consisting of *R, L, M, Z* and hence *distance*.

3.1. ANALYSIS OF THE UV SPECTRUM OF ζ-PUPPIS

In order to verify our method and to show the influence of the crucial points of our treatment metioned in section 2, a series of models will be investigated in the following. For all of these models we adopt a value of $T_{\text{eff}} = 42000K, R_*/R_\odot = 19$ and $v_\infty = 2260km/s$ (Groenewegen et al. 1989; Kudritzki et al. 1992 - Paper X). The different assumptions are summarized in Table 1.

3.1.1. Model 1

We start with a spectrum synthesis calculation where the assumptions of paper IX have been adopted - *simplified atomic models,*Kurucz (1979) *LTE-fluxes* are used to simulate the *EUV line blocking* shortword of the He II groundstate edge, constant temperature in the wind part ($T(r) \approx T_{eff}$) *EUV and X-ray radiation* by shock heated matter *is neglected* and solar abundances are assumed.

TABLE I

Assumptions characterising five different models for ζ-Pup.

model	atomic models	blocking	T_{eff}	shocks	Z	log g	$\dot{M}$ $(10^{-6}\,M_{\odot}/\mathrm{yr})$
1	simplified	LTE	$T_e = T_{eff}$	no	solar	3.5	3.6
2	*improved*	LTE	$T_e = T_{eff}$	no	solar	3.63	5.1
3		*(N)LTE*	*NLTE*	no	solar	3.63	5.1
4				*yes*	solar	3.63	5.1
5					*spectr.*	3.63	5.1

With these approximations a reasonable value for the mass loss rate is obtained (the actual value obtained from the observed radio flux is in between $3 - 5\ 10^{-6} M_{\odot}/\mathrm{yr}$ - see discussion in Paper XII) and a surface gravity of log g = 3.5. Although this value coincides with the result from the hydrostatic analysis, it is at least 0.1 dex smaller than the value obtained from the "unified model" (see sect. 1) and thus incorrect. The comparison between the observed and the synthetic spectrum (Fig. 4) shows good agreement for the strong resonance lines of CIV, NV and OVI-the latter indicates that the problem of "superionization" (cf. Lamers and Morton (1976), Cassinelli and Olson (1979), Hamann(1980), Olson and Castor (1981)) can be solved without any other source of ionization (cf. Paper III, Paper IX), but the rest of the spectrum is riddled by striking discrepancies. Among others the FeV lines disagree completely. Hence we have to note that the wind physics is obviously not yet correctly described.

Following the strategy outlined above we will now investigate how far the improved atomic models can influence these mainly negative results.

3.1.2. Model 2

As an example of the new atomic data sets the Grotrian diagram of the most crucial ionization stage in stellar wind calculations-FeV-is displayed in Fig 5. The reasons for the importance of FeV are threefold: 1. FeV shows numerous spectral features in the UV, 2. most of the almost 4000 spectral lines which are available from this atomic model contribute to the photospheric EUV blocking, and 3. up to 50% of the line force in the lower part of the wind-where $\dot{M}$ is fixed-is due to FeV. A change in the opacity-especially the iron opacity-owing to the more accurate and complete atomic data will, therefore, influence not only the spectrum calculation but also the dynamics. As is verified in Fig. 6 far more spectral lines contribute to the line force for model 2. This leads not only to a

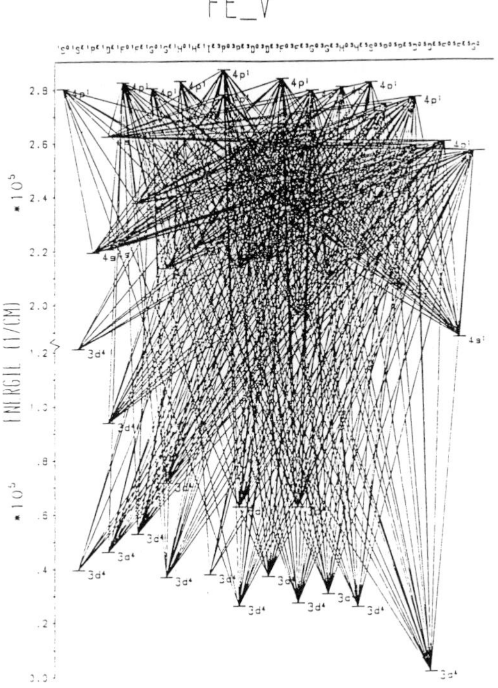

Fig. 5. Grotrian diagram of the atomic model of Fe V.

significantly larger value for the mass-loss rate ($\dot{M} = 5.1\ 10^{-6}M_{\odot}$/yr),but also to an increased value for the surface gravity (log g = 3.63), which is now exactlythe value obtained by the "unified model".

Concerning the synthetic spectrum of this model the comparison shows some improvements (Fig. 7), e.g. the FeV lines between 1420 to 1480 A are reasonable now, but there are still striking discrepancies. In particular the low ionization stages CIII, NIII, SiIV, and HeII (λ 1640A) disagree completely-instead of characteristic P-Cygni profiles only photospheric components are predicted.

3.1.3. Model 3

Since the ground state edges of these stages (SiIV - 274.4A, NIII - 261.4A, CIII - 258.9A) are located just longward of the HeII edge (227.8A), where photospheric line opacities are significant, the reason for this failure was attributed to insufficient line blocking in this frequency range (Paper VII, Paper IX). However, in Paper XII it turned out that the radiative ionization rates of these species are also strongly influenced by the continuum opacity shortward of the HeII - edge. Too strong an ionization is prevented only if the opacity shortward of 227A is consid-

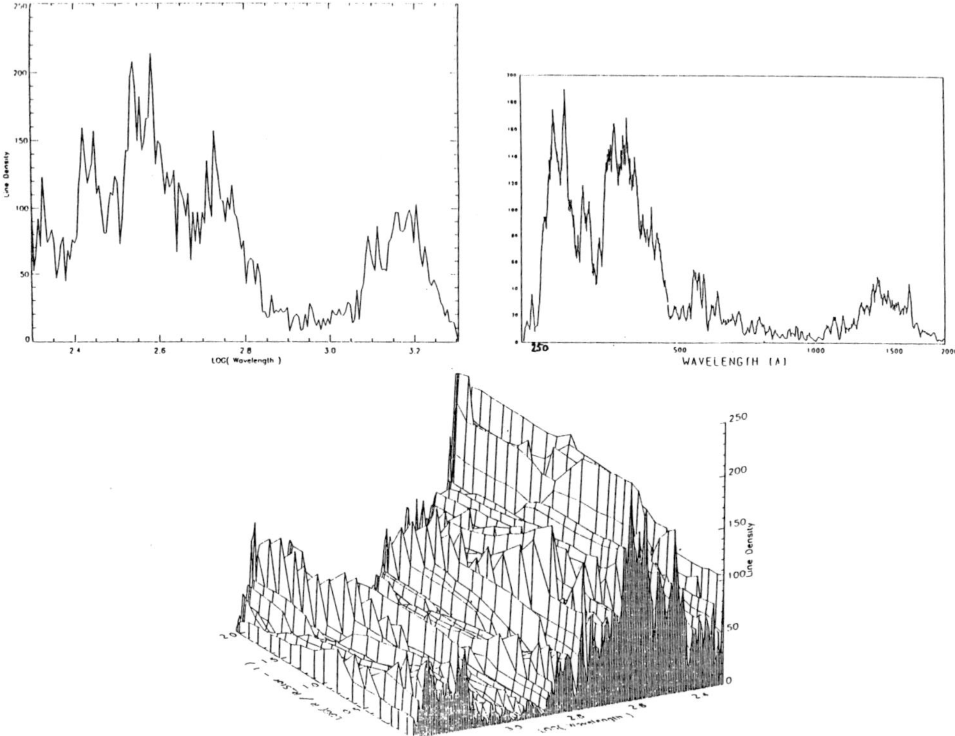

Fig. 6. Projection of line density (number of lines with opacity larger than Thomson-opacity in the maximum Doppler-interval) - upper left, model 2; upper right, model 1 - and line density versus radius and wavelength - below, model 2 - of ζ-Pup.

erably large, so that the radiation field is reduced due to the optical thickness. In Paper XII it was shown that this situation can only be obtained by recombination of HeIII to HeII in the outer wind layer; and this requires not only a *more realistic treatment of photospheric EUV line blocking*,but also a *wind temperature structure* which drops significantly in the outer wind layers. This is the case for *radiative equilibrium* calculations by Gabler (1992).

As the correct treatment of the blocking influence of all metal lines in the entire sub- and supersonically *expanding atmosphere* has not yet been realized (cf. Puls and Pauldrach, 1990), up to now one has had to rely on approximations. So far, Monte Carlo simulations have been performed (Abbott and Lucy (1985), Lucy and Abbott (1992), Schmutz and Schaerer (1992)). However, these calculations suffer from the treatment of line opacities-instead of solving the correct rate equations in NLTE, inadequate gaseous nebula like formula have been used. On the other hand emergent fluxes obtained from conventional hydrostatic NLTE model atmospheres for which the metal line opacity has been included in LTE (Fig. 8) turned out to be not too bad *an approximation*, since our wind models

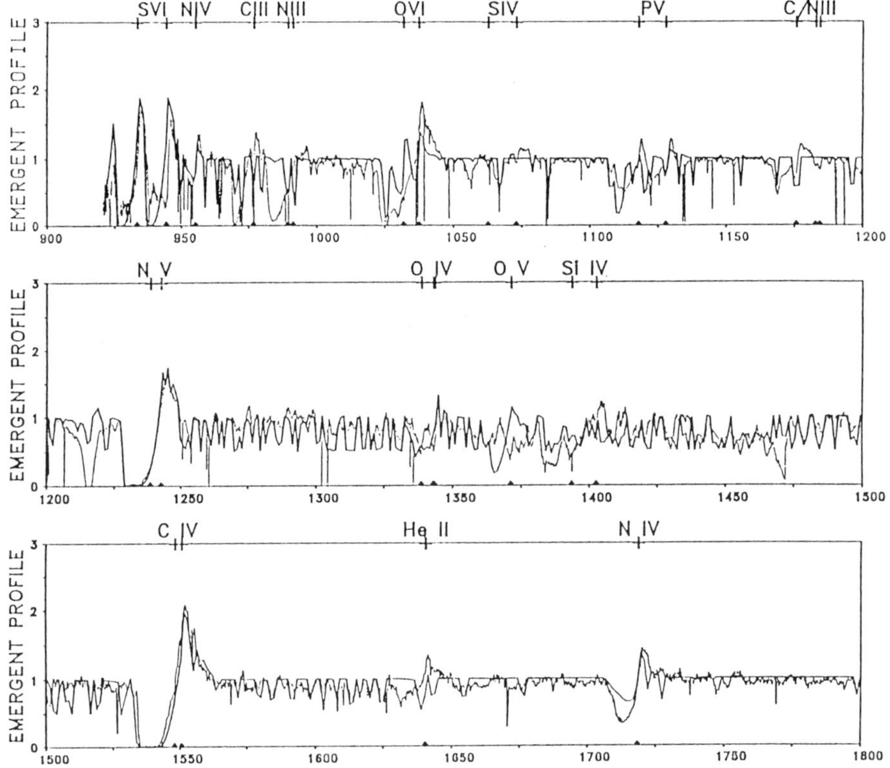

Fig. 7. Calculated and observed UV spectrum for ζ-Pup. The calculated spectrum belongs to model 2. From Paper XII.

showed that the iron like ions dominating the blocking are indeed almost in LTE in the region where the blocking takes place (cf. Paper XII).

Fig. 9 shows the synthetic spectrum of a model calculated with this kind of blocking, reducing the flux significantly shortward of 300A, and with T(r) according to Gabler (1992). This model proved indeed that helium recombined in the outer wind layers. As a consequence of both effects the situation has clearly improved regarding CIII, NIII, SiIV, NIV, OV and HeII 1640, and almost all FeV lines are now represented well. However, these improvements are accompanied by increased deficiencies for high ionization stages like NV and OVI, so that we are again faced with the problem of "superionization". But, as was shown in Paper XII, this problem is solved by accounting for the EUV radiation field of shocks.

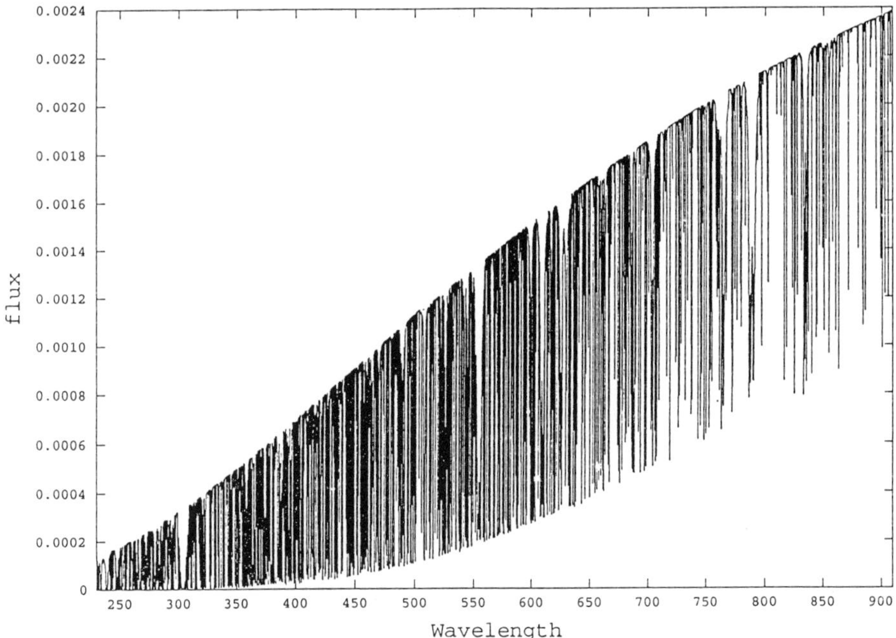

Fig. 8. The emergent flux $(erg/(cm^2 sHz))$ of a photospheric model of ζ-Pup. in the range 227A- 911A. The influence of blocking due to line opacities if shown. This reduces the radiation field drasically in this frequency regime. From Paper XII.

3.1.4. Model 4

The present scenario for producing the soft X-ray emission of O stars (detected by Seward et al. (1979) and Harnden et al. (1979) where L_x/L_{Bol} is typically 10^{-7} - Chlebowski et al. (1989)) is to assume that they arise from shock instabilities in the stellar wind. This picture was guided by the finding of Lucy and Soloman (1970) that radiation driven winds are inherently unstable. Hence, non-stationary features must show up leading to shocks and the X-rays are explained by radiative losses of the post shock regions (Lucy and White (1980), Lucy (1982)). Although Krolik and Raymond (1985) had already pointed out that shock-heated matter radiates also in the ultraviolet, only the effects of Auger-ionization caused by X-rays were studied to investigate the influence on ionization (Cassinelli and Olson (1979), Olson and Castor 1981), Cassinelli and Swank 1983, Waldron (1980)). This mechanism is, however, probably of secondary importance (cf. Paper IX), whereas enhanced *direct ionization by EUV shock radiation* has important effects (cf. Paper XII).

A consideration of this process requires a theoretical investigation of time dependent radiation-hydrodynamics which describes the creation and developement of shocks (cf. Owocki, Castor and Rybicki (1988), Feldmeier (1993)). From this kind of calculations is turned out that the wind is not inherently unstable as

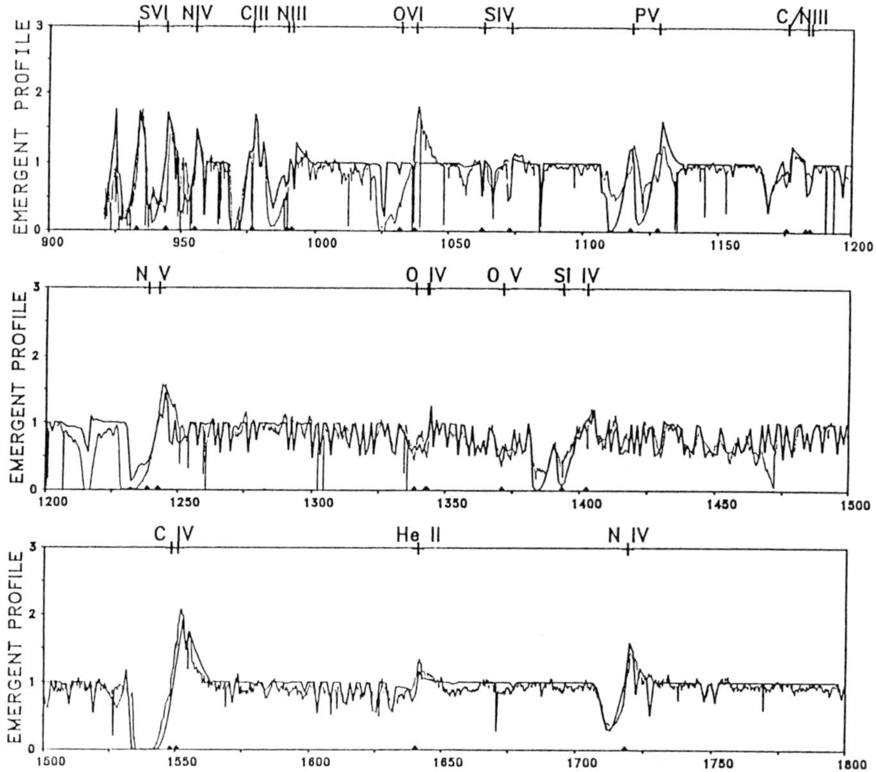

Fig. 9. Calculated and observed UV spectrum for ζ-Pup. The calculated spectrum belongs to model 3. From Paper XII.

was suggested earlier, but is unstable against small perturbations. This is shown in Fig. 10a where the temporal development of $\dot{M}$ is plotted versus radius. After about 10 hours - from model start - the wind settles down to a pronounced periodic response to the perturbations which are described by sound waves with an amplitude of 1% in density and a period of 5000s. Although the spatial variation of the velocity seems to be in contrast to the stationary picture (Fig. 10b), the gross wind properties like the time averaged velocity structure (over 1 hour) and the mass distribution (Fig. 10c) turn out to be in quite good agreement with those of the stationary models. Moreover, since the shock distance is much larger than the shock cooling length in the accelerating port of the wind, the picture of a stationary "cool wind" with embedded randomly distributed shocks is apporporiate. Fig. 10c also shows that there is only a small amount of high velocity material (the spikes) which gives a filling factor not much larger than $f^{sh} \sim 10^{-2}$, and from the height of the spikes jump velocities $u^{sh} = 300 - 500 km/s$ and, hence, immediate post shock temperatures of $T^{sh} = 1 - 3 \cdot 10^{6} K$ are deduced. The reliability of

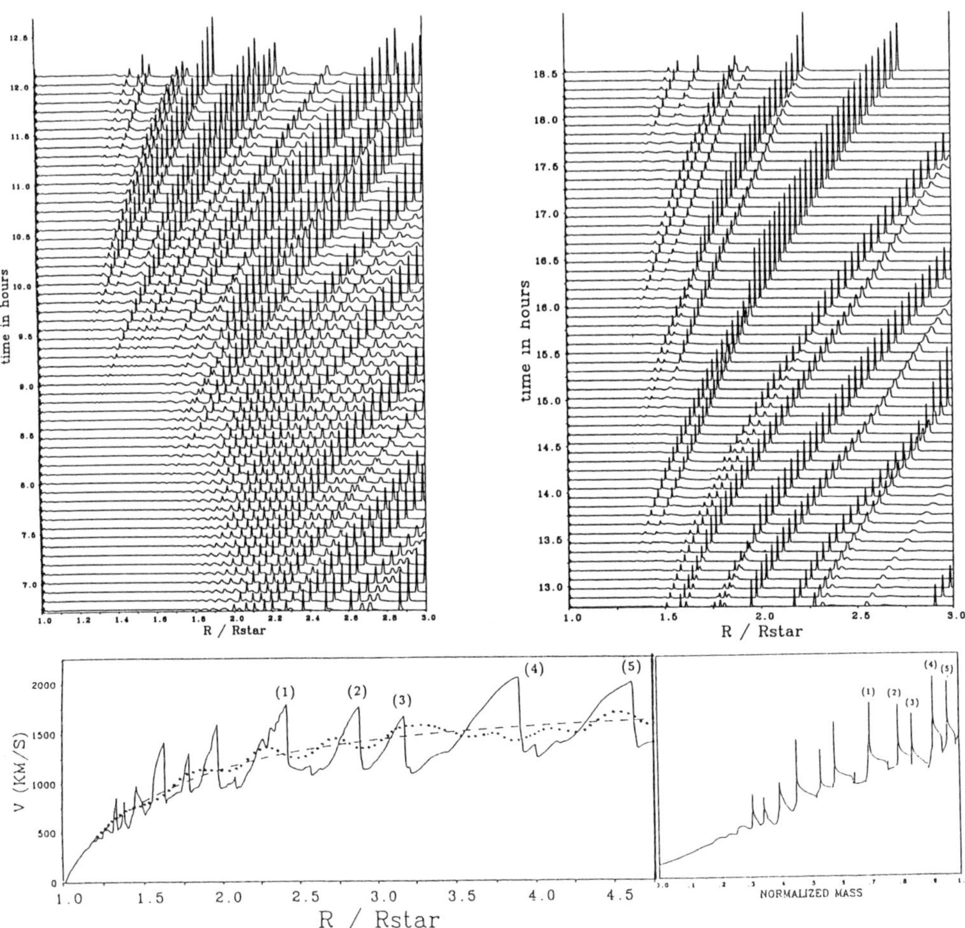

Fig. 10.   Temporal development of $\dot{M}$ versus radius (upper part - Fig. 10a). Note the merging shells on the right panel. Velocity at a later time step plotted versus radius (lower panel left - Fig. 10b; stationary model - dashed line; time average over 1 hour - dotted line) and normalized mass (lower panel right - Fig. 10c). Pronounced features are indicated by numbers.

these calculations can be demonstrated by a comparison to ROSAT-observations.

On the basis of this picture the radiative transfer has been solved including shock emission - where the volume emission coefficient (Λ_ν) of an X-ray plasma is calculated using the Raymond and Smith (1977) code and T^{sh} and f^{sh} enter as fit parameters -, K shell absorption and absorption by the "cool wind" (cf. Hillier et al., 1993). A detailed modelling of the observed X-ray spectrum of ζ-Puppis revealed that the fitted jump velocities ($n^{sh} = 290 - 520 km/s$) and filling factors ($f^{sh} = 1.8 - 2.1 \cdot 10^{-2}$) agree quite well with the results of the time

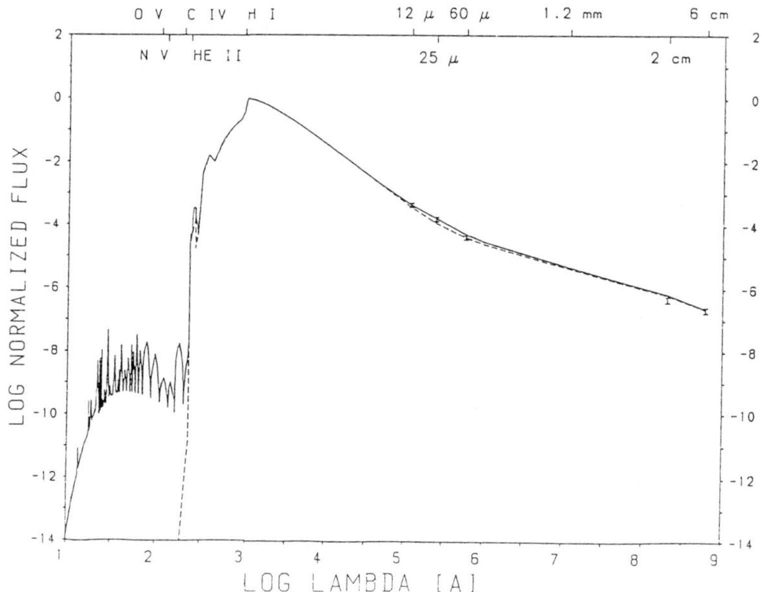

Fig. 11. Energy distribution of two atmospheric models - including shock emission (fully drawn), whithout (dashed) - of ζ-Pup.

dependent calculation, and that the best fit occurs when the cool wind opacity of model 3 is used, i.e. the model where He-recombination takes place. It is further important to note that the jump velocity which fits the energy port below the K-shell edges (u^{sh} = 290km/s) is surprisingly equal to the "turbulence velocity" (v_{turb} = 290±70km/s, Groenewegen et al., 1989). Hence, the primary contribution of shock radiation to the ionization equilibrium by direct ionization is obviously characterized by the shock temperature resulting from u^{sh}_{max} = v_{turb} (cf. Paper XII).

Fig. 11 shows the emergent energy distribution calculated for model 4, where an ad hoc approach for the spatial behaviour of the jump velocities has been adopted and K-shell absorption, Auger-ionization and Λ_ν calculated using the Raymond and Smith code have been included (cf. Paper XII, Hunsinger and Pauldrach, 1994). The influence of shock emission shortward of the HeII edge is significant and reveals the importance of direct ionization due to EUV and soft X-ray photons. This not only affects the high ionization stages - NV, OVI -, but also leads to synthetic lines which reproduce the observations almost perfectly (cf. Fig. 12).

3.1.5. Model 5

At this step we regard the wind physics as correctly described, and the stellar parameters as determined and verified. Hence, the stellar mass $M/M_\odot$ = 55.6

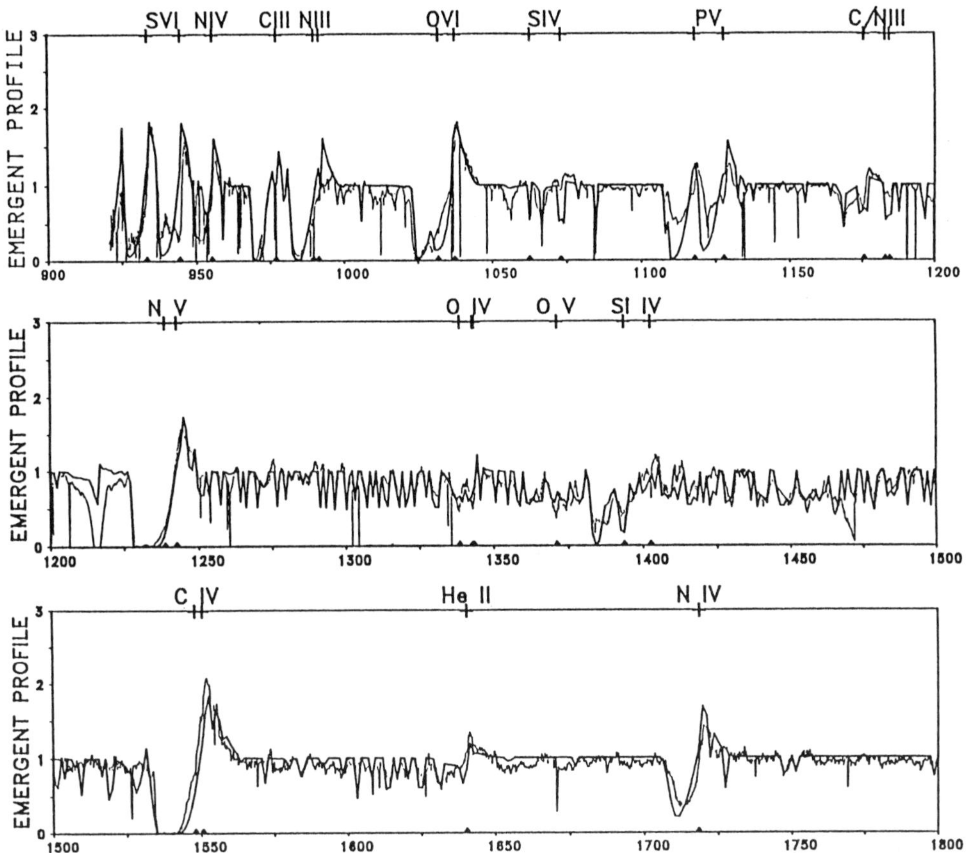

Fig. 12.   Calculated and observed UV spectrum for ζ-Pup. The calculated spectrum belongs to model 5.From Paper XII.

obtained from the derived value of the surface gravity can be compared to the mass obtained from stellar evolution theory-$M/M_\odot$ = 68.5 (Maeder, 1990). A difference of only 20% in this case does not support the "mass discrepancy problem" (cf. Groenewegen et al. (1989), Herrero et al. (1992)). The remaining item is the *determination of abundances*. Here we have to consider that ζ-Pup. must be regarded as an evolved object-photospheric analyses showed an enhanced He-abundance (cf. Kudritzki et al. (1983), Voels et al. (1989)). This is supported by the fact that the CIII resonance line of model 3 is too strong and the NIII resonance line is too weak (cf. Fig. 9 - note that both lines are physically affected in the same way). We have, therefore, calculated a model 5 where the CIII resonance line was fitted-yielding $\epsilon_C = 0.35\epsilon_{C,\odot}$ ($\epsilon_{C,\odot}$ is the solar number fraction)-and consistent predictions from evolutionary tracks have been taken for the other elements $-\epsilon_N = 8.0\epsilon_{N,\odot}$, $\epsilon_O = 0.75\epsilon_{O,\odot}$, $n_{He}/n_H = 0.12$ (Maeder, 1990). As is shown in Fig. 12, the observed spectrum is fitted quite well apart from minor differences (a completely independent fit resulted in $\epsilon_N = 4.0\epsilon_{N,\odot}$, $\epsilon_O = 1.0\epsilon_{O,\odot}$, $\epsilon_P = 0.6\epsilon_{P,\odot}$, $n_{He}/n_H = 0.1 - 0.2$). Therefore, we conclude that

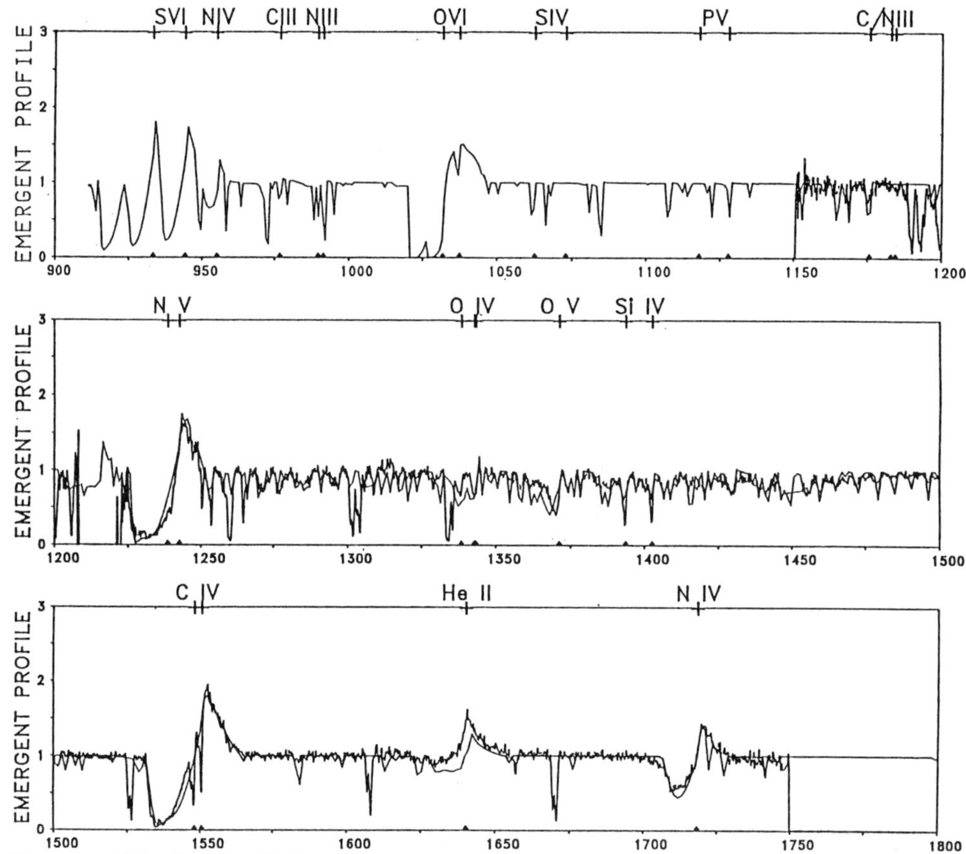

Fig. 13. Calculated and observed (HST high resolution) spectra for Melnick 42. Note that the stellar parameters have been determined using only the tools of UV diagnostics.

our spectrum synthesis technique does already allow the determination of abundances, especially the Fe-abundance can be derived with high precision due to its strong influence on the dynamics which in turn changes the synthetic spectrum considerably.

3.2. ANALYSIS OF THE UV SPECTRUM OF MELNICK 42

The UV analysis carried out in Paper XII by incorporating all the improvements discussed above led to the conclusion that the effective temperature ($T_{eff} = 50500K$) and the surface gravity ($\log g = 3.8$) deduced from the observed terminal velocity ($v_\infty = 3000km/s$) are much higher than those values derived by Heap et al. (1991) who used hydrostatic atmosphere models ($T_{eff}^{hyd.} = 42500K, \log g^{hyd.} = 3.5$). This conclusion relies on two aspects. 1. The synthetic UV spectrum calculated on the basis of the parameters from Heap et al. showed serious descrepancies compared to the HST observations. 2. A quite reasonable fit of the whole UV spectrum was achieved by a calculation based on a grid of self-consistent mod-

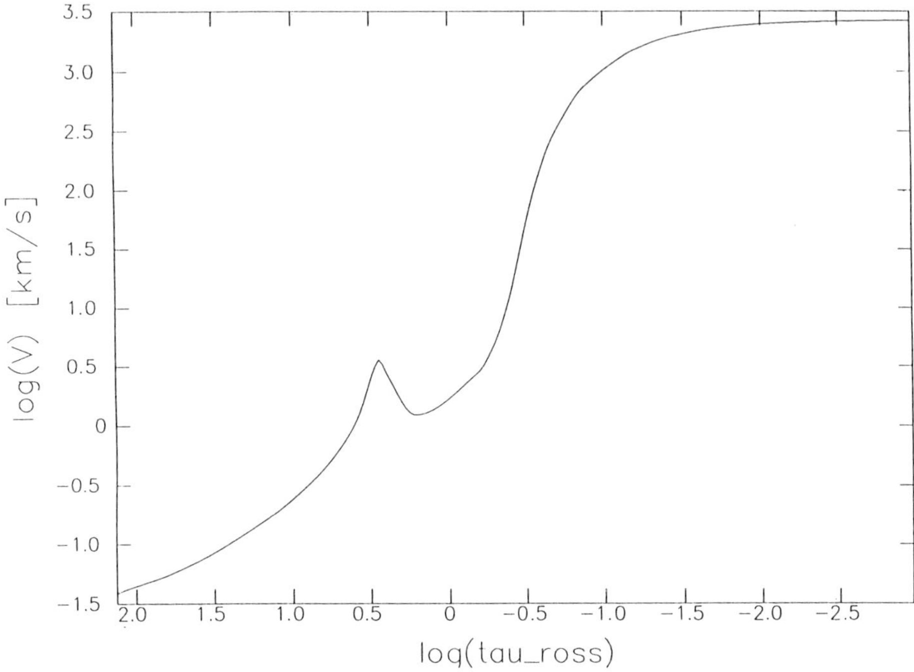

Fig. 14.   Velocity versus the Rosseland-optical depth. The influence of the continuous acceleration in the photospherical region is clearly displayed.

els for Melnick 42 (cf. Fig. 13), where T_{eff} was obtained from a comparison of the FeV/IV lines, and the SiIV and CIII line. Furthermore, a much higher mass (150.4$M_\odot$), luminosity (log $L/L_\odot$ = 6.6) and mass-loss rate ($\dot M$ = 27. $\cdot\,10^{-6}M_\odot$/yr) were derived consistently, and abundances a factor of four smaller than solar were obtained. Moreover, an extrapolation of Maeder's (1990) evolutionary tracks showed that the corresponding mass of 155$M_\odot$ agrees well with our value. Hence, MK42 does not show indications of the "mass discrepancy" either.

However, one problem remains: The failure of the photospheric analysis. Although it is well known that photospheric lines might be wind contaminated (cf. sect. 1), the effective temperatures derived from hydrostatic atmosphere models are commonly regarded as correct, since they are mainly deduced from the purely hydrostatic HeI λ 4471A line. The presence of this line, with roughly 250mA equivalent width, constrained T_{eff} to a value not higher than 42500K. However, this argument was already weakened in Paper XII, where it was shown that due to the different density structure-compared to hydrostatic calculations-caused by the extremely high $\dot M$ of MK42, the EUV blocking influences even deeper regions where the HeI line is formed. It was also shown that the line has a reasonable strength (125mA) when blocking is included in the calculations, but the line still turned out to be too weak. Hence, we started to investigate a second aspect. This concerns the influence of the continuous acceleration (g_{cont} - cf. sect. 2) due to Thomsom scattering and bound-free and free-free absorption on the photospheric density structure. It turned out from these calculations that the influence of g_{cont} on the velocity and, hence, density structure is considerable at an optical depth of τ_{ROS} 1 − 5 for extremely high values of the mass-loss rate ($\dot{M} > 10 \cdot 10^{-6} M_{\odot}/yr$). In the case of Melnick 42 the velocity increases again to a value of 20% of the sound velocity at τ_{ROS} 3 (cf. Fig. 14). Therefore, an optical depth larger than 10 is required to match the approximation of hydrostatic equilibrium. As the He I line is formed at much smaller depths its spectral shape and, hence, its equivalent width is certainly influenced by this effect; calculations allowing a quantitative investigation of this finding will be forthcoming soon (we note here that the dynamics were only slightly affected by g_{cont} - less than 10% - in the considered range of $\dot{M}$; we note further that the effect decreased considerably if the contribution of the metals was neglected for g_{cont}).

4. Acknowledgements

We wish to thank our colleagues S.M. Haser, Dr. K. Butler, Dr. M. Lennon, C. Kronberger, Dr. R. Gabler for helpful discussions and Dr. D. Lennon for carefully reading the manuscript. This research was supported by the Deutsche Forschungsgemeinschaft in the "Gerhard Hess Programm" under grant PA 477/1-1.

References

Abbott, D.C.: 1982, *Astrophys. J.* **259**, 282
Abbott, D.C. and Lucy, L.B.: 1985, *Astrophys. J.* **288**, 679
Bieging, J.H., Abbott, D.C., Churchwell, E.B.: 1989, *Astrophys. J.* **340**, 518
Bohannan, B., Abbott, D.C., Voels, S.A., Hummer, D.G.: 1986, *Astrophys. J.* **308**, 728
Cassinelli, J.P., Olson, G.L.: 1979, *Astrophys. J.* **229**, 304
Cassinelli, J.P., Swank, J.H.: 1983, *Astrophys. J.* **271**, 681
Castor, J., Abbott, D.C., Klein, R.: 1975, *Astrophys. J.* **195**, 157
Chlebowski, T., Harnden, F.R.,Jr., Sciortione, S.: 1989, *ApJ.* **341**, 427
Ebbets, D.C.: 1982, *Astrophys. J.* suppl. **48**, 399
Eisner, W., Jones. M., Nussbaumer, H.: 1974, *Comput. Phys. Commun.* **8**, 270
Feldmeier, A.: 1993, 'Thesis', *Ludwig-Maximilians-Univ.,Muenchen* ,
Gabler, R., Gabler, A., Kudritzki, R.P., Puls, J., Pauldrach, A.W.A.: 1989, *Astron.Astrophys.* **226**, 162
Gabler, R.: 1992, 'Thesis', *Ludwig-Maximilians-Universitaet, Muenchen* ,
Groenewegen, M. A. T., Lamers, H. J. G. L. M., Pauldrach, A. W. A.: 1989, *Astron. Astrophys.* **221**, 78
Hamann, W.R.: 1980, *Astron. Astrophys.* **84**, 342
Harnden, F.R., Jr., Branduardi, G., Elvis, M., Gorenstein, P., Grindlay, J., Pye, J.P., Rosner, R., Topka, K., and Vaiana, G.S.: 1979, *Astrophys. J. (Letters)* **234**, L51
Heap, S. R., Altner, B., Ebbets, D., Hubeny, I., Hutching, J. S., Kudritzki, R. P., Voels, S. A., Haser, S., Pauldrach, A. W. A., Puls, J., Butler, K.: 1991, *Astrophys. J. Letters* **377**, L29
Henrichs, H.F.: 1986, *Pub. Astron. Soc. Pacific* **98**, 48

Herrero, A., Kudritzki,R.P., Vilches,J.M., Kunze,D., Butler,K., Haser,S.: 1992, *Astron.Astrophys.* **261**, 209

Hillier, D.J., Kudritzki,R.P., Pauldrach,A.W.A., Puls,J., Baade,D., Schmitt,J.M.: 1993, *Astron. Astrophys.* **in press,**

Hummer, D.G., Rybicki, G.B.: 1985, *Astrophys. J.* **293**, 258

Hunsinger, J., Pauldrach, A.W.A.: 1994, *Astron.Astrophys.* **in prep.,**

Krolik, J.,H. Raymond, J.C.: 1985, *Astrophys. J.* **298**, 660

Kudritzki, R.P., Simon, K.P., Hamann, W.R.: 1983, *Astron. Astrophys.* **118**, 245

Kudritzki, R.P., Hummer, D.G.: 1990, *Annual Rev. Astron. Astrophys.* **28**, 303

Kudritzki, R. P., Hummer, D. G., Pauldrach, A. W. A., Puls, J., Najarro , F., Imhoff, J.: 1992, *Astron., Astrophys.* **257**, 655(Paper X)

Kurucz, R.L.: 1979, *Astrophys. J. Suppl.* **40**, 1

Lamers, H.J.G.L.M., Morton, D.C.: 1976, *Astrophys. J. Suppl.* **32**, 715

Lamers, H.J.G.L.M., Gathier, R., and Snow, T.P.: 1982, *Astrophys. J.* **258**, 186

Lucy, L.B., Solomon, P.: 1970, *Astrophys. J.* **159**, 879

Lucy. L.B., White, R.: 1980, *Astrophys. J.* **241**, 300

Lucy, L.B.: 1982, *Astrophys. J.* **255**, 286

Lucy, L.B., Abbott, D.C.: 1992, *Ap. J.* **405**, 738

Maeder, A.: 1990, *Astron. Astrophys. Suppl. Ser.* **84**, 139

Mihalas, D., Hummer, D.G.: 1973, *Astrophys. J.* **179**, 872

Morton, D.C., Underhill, A.B.: 1977, *Ap. J. Suppl.* **33**, 83

Nussbaumer, H., Storey, P.J.: 1978, *Astron.Astrophys.* **64**, 139

Olson, G.L., Castor, J.I.: 1981, *Astrophys. J.* **244**, 179

Owocki, S.P., Castor, J.I., Rybicki, G.B.: 1988, *Astrophys. J.* **335**, 914

Pauldrach, A.W.A., Puls, J., Kudritzki, R.P.: 1986, *Astron. Astrophys.* **164**, 86(Paper I)

Pauldrach, A.W.A.: 1987, *Astron. Astrophys.* **183**, 295 (Paper III)

Pauldrach, A.W.A., Herrero, A.: 1988, *Astron. Astrophys.* **199**, 262

Pauldrach, A. W. A., Kudritzki, R. P., Puls, J., Butler, K.: 1990, *Astron. Astrophys.* **228**, 125-154 (Paper VII)

Pauldrach, A. W. A., Puls, J., Gabler, R., Gabler, A.: 1990, *Boulder-Munich workshop, ed. C.D. Garmany, San Francisco : Astronomical Society of the Pacific Conference Series* **7**, 171 (Paper IX)

Pauldrach, A. W. A., Kudritzki, R.P., Puls, J., Butler, K., Hunsinger,J.: 1993, *Astron. Astrophys.* **in press,** Paper XII

Prinja, R.K., Howarth, I.D.: 1986, *Astrophys. J. Suppl.* **61**, 867

Puls, J.: 1987, *Astron. Astrophys.* **184**, 227 (Paper IV)

Puls, J., Hummer, D.G.: 1988, *Astron. Astrophys.* **191**, 87

Puls, J., Pauldrach, A.W.A.: 1990, *in Astronomical Society of the Pacific Confe rence Series 7, "Boulder Munich Workshop "*. ed. C.Garmany **203**,

Raymond, J.C., Smith, B.W.: 1977, *Ap.J. Suppl.* **35**, 419

Raymond, J.C.: 1988, *in 'Hot Thin Plasmas in Astrophysics', ed. R. Pallavicini, Kluwer Academic Publishers* **3**,

Schmutz,W., Schaerer,D.: 1992, *Lecture Notes in Physics* **401**, 409

Sellmaier,F., Puls,J., Kudritzki;R.P., Gabler,R., Gabler,A., Voels,S.: 1993, *Astron. Astrophys.* **273**, 533

Seward, F.D., Forman, W.R., Giacconi, R., Griffiths, R.E., Harnden, F.R., Jones, C., Pye, J.P.: 1979, *Astrophys. J. (Letters)* **234**, L55

Voels, S.A., Bohannan, B., Abbott, D.C., Hummer, D.G.: 1989, *Astrophys. J.* **340**, 1073

Walborn, N.R., Nichols-Bohlin, J., Panek, R.J.: 1985, *"IUE Atlas of O-Type Spectra from 1200 to 1900 Angst. ", NASA Reference Publication* **1155**,

Walborn, N.R., Ebbets,D.C., Parker,J.W., Nichols-Bohlin, J., White,R.l.: 1993, *Ap.J. Letters, in press* ,

Waldron,W.L.: 1984, *Ap.J.* **282**, 256

THE CHEMICAL COMPOSITION OF B-SUPERGIANT ATMOSPHERES

D.J. LENNON

Universtatssternwarte Munchen, D-81679 Munchen, Scheinerstrasse 1, Federal Republic of Germany

Abstract. We present new estimates of He/H and CNO abundance ratios in the atmospheres of a selection of B2 supergiants which imply that the C/N ratio in the most luminous Ia stars is close to its equilibrium value. The is also some evidence for more moderate CN abundance anomalies in the B2Ib and B2II supergiants. These results, together with other recent work, imply that the effects of the CNO bi-cycle on the composition of B-supergiant atmospheres are most severe for the more luminous and massive stars. Furthermore, studies of LMC B-supergiants indicate that a small fraction of these very luminous stars are nitrogen weak. This picture is qualitatively consistent with theoretical predictions whenever massive stars perform blue loops in the HR diagram, returning from a red supergiant phase to become core helium burning blue supergiants with atmospheres contaminated by nuclear processed material.

Key words: B-supergiants – abundances – evolution

1. Introduction

Supergiants have always been considered as key tests of stellar evolution, for example the ratio of blue to red supergiants has for a long time been regarded as an important diagnostic of massive star evolution (see Stothers (1991) for a discussion of this and other points). In fact the very existence of some B-supergiants poses a problem for all evolutionary calculations since they all predict a post-main-sequence gap which is not evident from the observations (Tuchman & Wheeler 1990). Of course the presence of this gap may be masked by observational uncertainties related to the use of photometry or spectral classification in the assignment of effective temperatures in combination with binarity. Thus the existence of this feature is still something of an open question. Nevertheless, abundances remain a very important diagnostic and can provide critical constraints for the evolutionary calculations. For example, the extent of helium enrichment can discriminate between mixing processes on the main-sequence (through rotation) and during the RSG phase (dredge up) while the C/N/O ratios can give information on the extent to which the material has been processed by the CN and ON cycles of the CNO bi-cycle and an indication of dilution effects. Some of these points have been discussed recently by Maeder (1987) and Langer (1991,1992) where it is found that enrichment by CNO processed matter should be more severe for the more massive and luminous stars.

Carbon and nitrogen anomalies in B-supergiants are well known and have led to the introduction of BN and BC classifications (Walborn 1976). Note that in this paper we are concerned only with supergiant BN/BC stars; we assume that main-sequence BN/BC stars arise through some other mechanism, perhaps through

Space Science Reviews **66**: 127–135, 1994.

© 1994 *Kluwer Academic Publishers. Printed in Belgium.*

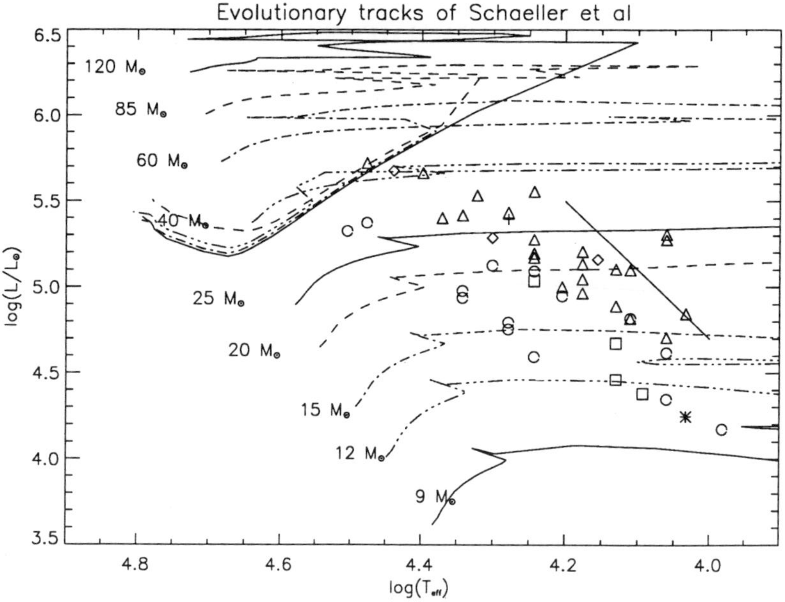

Fig. 1. HR diagram showing positions of B-supergiants observed by Lennon et al (1992,1993) and for comparison, the evolutionary tracks of Schaller et al (1992) as well as the position of the Fitzpatrick & Garmany ledge. Symbols represent the different luminosity classes as follows; pluses - Ia$^+$, triangles - Ia, diamonds - Iab, circles - Ib, squares - II.

binary interaction. In his paper, Walborn suggested that nitrogen weak or BC supergiants have in fact normal composition while the morphologically normal B-supergiants are nitrogen rich. Thus by implication the nitrogen strong BN supergiants should have even greater nitrogen enhancements. In fact the number of B-supergiants for which detailed abundance analyses have been performed is rather small, most of our information being derived from differential analyses, qualitative comparisons of their spectra, or indirectly through analyses of SN ejecta (such as SN1987A). This information is well summarised in the review by Walborn (1988) and for the remainder of this paper we review results from more recent programs concerning both galactic and LMC B-type supergiants. In addition we present some new results for CNO abundances in B-type supergiants.

2. B-supergiants in the Milky Way

In Fig. 1 we show the positions in the HR diagram of the 46 galactic B-supergiants observed in the recent survey of Lennon et al (1992,1993). Also shown for illustrative purposes are the evolutionary tracks of Schaller et al (1992) as well as the approximate position of the Fitzpatrick & Garmany (1990) ledge. As shown by Lennon et al (1993), the strengths of the C II and N II appear to be anti-correlated implying that some CN processing (at least) has taken place. Further, this effect

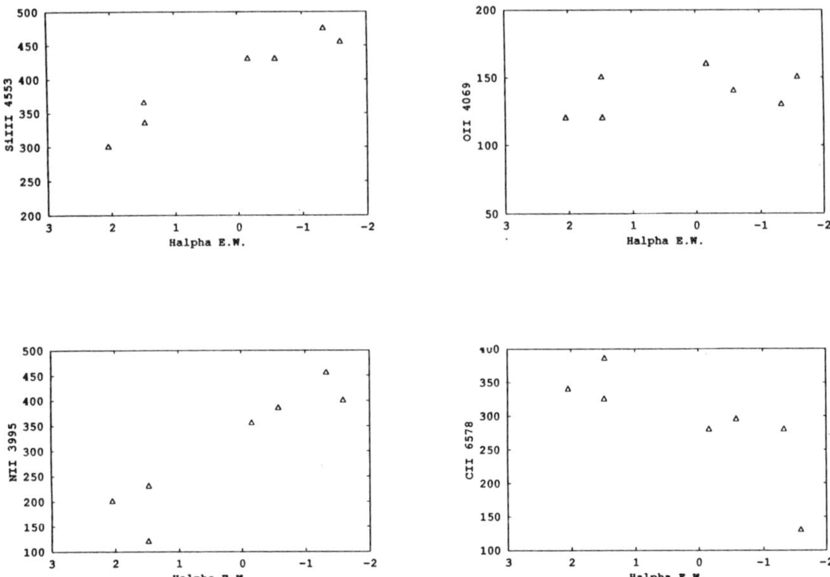

Fig. 2. Line strength variations as a function of Hα equivalent width for B2 supergiants for Si (top left), O (top right), N (bottom left), and C (bottom right).

appears to be more pronounced for the most luminous stars. We illustrate this effect here at spectral type B2, where in Fig. 2 we show the luminosity dependence (as indicated by the strength of Hα) of a number of metal lines. We can see that the N II lines show a strong positive luminosity dependence but note that other species, such as Si III, Mg II also show such a dependence. Thus this alone is not sufficient evidence from which to deduce CN processing. However C II shows the opposite behaviour which is to be expected if carbon has been converted into nitrogen. (The alternative is to invoke NLTE effects such that they effect carbon in a sense different from all other elements.) These variations are dramatically illustrated in Fig. 3 where we show the spectral region centered on Hα which includes both carbon and nitrogen lines. The almost sudden appearance of the N II lines in the Ia supergiants and the weakening of the C II doublet are clearly shown. Below we attempt to quantify these results, concentrating on spectral type B2 and looking at this luminosity dependence in more detail.

2.1. HELIUM

Consider first the helium abundance, line strengths for the helium lines have been computed using NLTE models as described by Lennon et al (1991) for helium fractional abundances by number of 0.1, 0.2 and 0.3. In Fig. 4 we have compared the computed helium line strengths with the measurements of Lennon et al (1993). We note here that the effective temperatures of the B2 supergiants can be derived from the silicon lines rather conveniently by realising that for these stars the Si II

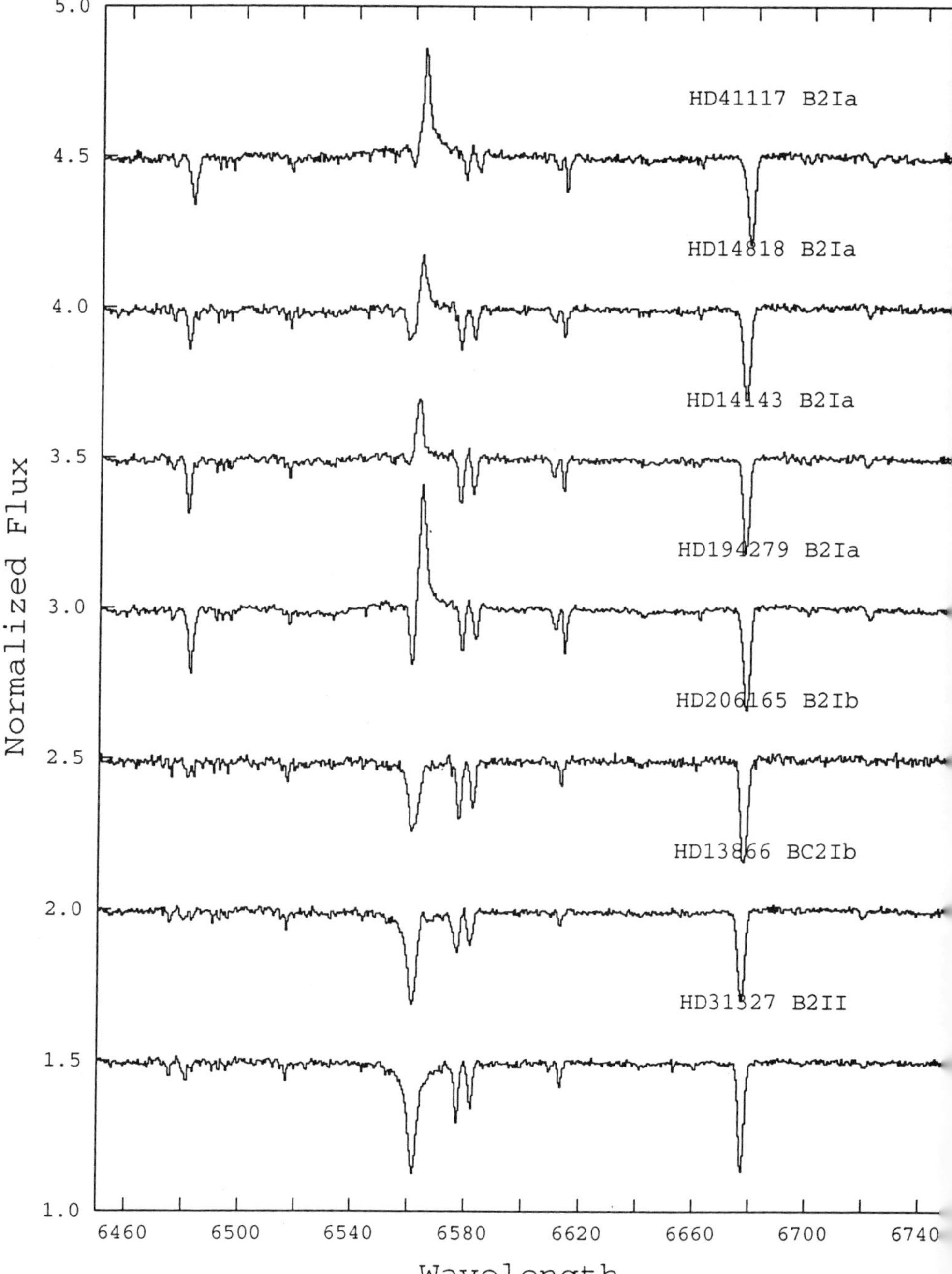

Fig. 3. Variation of N II and C II line strengths in B2 supergiants in the vicinity of Hα. The main features are, from left to right, N II λ6482, Hα, C II λλ6578 and 6582, N II λ6610, a DIB at λ6613 and He I λ6678. Note the sudden strengthening of the N II features at luminosity class Ia.

TABLE I

CNO abundance ratios, by number, for B-supergiants.

Object	Sp. type	C/N	O/N	Source
HD 41117	B2 Ia	0.02	1.51	Present work
HD 14818	B2 Ia	0.10	2.00	Present work
HD 14143	B2 Ia	0.08	1.17	Present work
HD 194279	B2 Ia	0.05	2.04	Present work
HD 206165	B2 Ib	1.38	4.90	Present work
HD 13866	BC2 Ib	3.39	47.9	Present work
HD 31327	B2 II	1.07	5.01	Present work
B-stars		2.45	7.40	Gies & Lambert (1992)
HD 51309	B3 II	1.20	2.75	Gies & Lambert (1992)
HD 52089	B2 II	1.95	3.39	Gies & Lambert (1992)
HD 91316	B1 Ib	0.32	1.48	Gies & Lambert (1992)
HD 198478	B3 Ia	0.52	1.23	Gies & Lambert (1992)
HD 206165	B2 Ib	1.17	4.27	Gies & Lambert (1992)
ζ Per	B1 Ib	1.25	–	Massa et al (1991)
HD 93840	BN1 Ib	0.03	--	Massa et al (1991)
HD 96248	BC1 Iab	2.5	15.8	Dufton (1972)

and Si IV line strengths are falling off sharply. From consideration of NLTE line formation calculations this happens at an effective temperature of approximately 20 000 K and does not depend strongly on silicon abundance, microturbulence or He/H ratios. We therefore adopt this effective temperature, recognizing that there is an inherent uncertainty of about $\pm 1\,000$ K, and derive surface gravities from fitting the Balmer Hγ and Hδ lines. This results in values of log g ranging from 2.15 to 2.60 for the luminosity classes Ia to II. Accordingly, we adopt values of 2.25 and 2.50 for the comparison of helium line strengths shown in Fig. 4. From this we can see, in general terms, that while He I λ4471 implies a helium abundance of 0.3-0.4, λ4026 implies a value around 0.1-0.2. Thus while an enhanced helium abundance may be deduced for normal B-supergiants, one cannot arrive at a precise estimate of this quantity using these models (see also Lennon et al 1991).

2.2. CNO ELEMENTS

Turning now to the CNO elements, in order to derive abundances, NLTE curves of growth were computed for all the CNO lines tabulated by Lennon et al (1993).

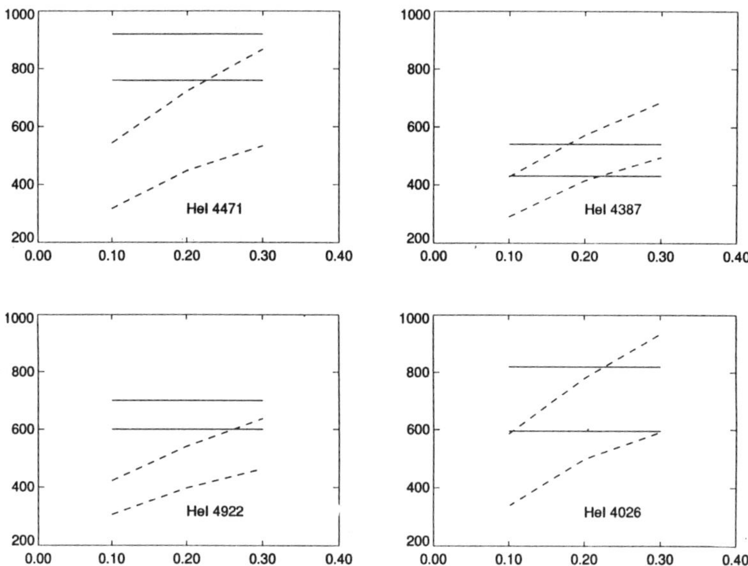

Fig. 4.   He I line strengths (mA) compared to NLTE calculations for various helium fractional abundances by number. The dashed curves represent NLTE calculations for a log g of 2.25 (lower) and 2.5 (upper) while solid lines brackett the observed values for Ia (lower) and II (upper) B2 supergiants.

The calculations were carried out for a range of microturbulent velocities and helium abundances at an effective temperature of 20 000 K and log g values of 2.25 and 2.50. In practice the uncertainties in the derived abundances for a given star are dominated by the microturbulent velocity. Here we adopt values of 10 and 15 km/s for the luminosity class Ib-II and Ia supergiants respectively, although for some Ia stars there are indications that an even higher value would be more appropriate. With this in mind, it is clear that a comparison of absolute abundances between Ia and Ib-II supergiants probably contains systematic errors. We therefore expect the abundance ratios to be more robust towards uncertainties in the microturbulence and the results of this analysis are shown in Table I, where we also show the implied abundance ratios derived for main-sequence B stars (Gies & Lambert 1992). In the derivation of these abundance ratios we have assumed a helium fractional abundance of 0.3, however changing this to 0.1 leads to only small changes in the results. It is clear from this table that, even assuming large uncertainties due to the assumptions mentioned above, there is evidence for very significant departures from normal CNO abundance ratios in the Ia supergiants. In fact the C/N ratio is close to its equilibrium value for these objects while the O/N ratio implies that some processing due to the ON cycle may have occurred. The Ib-II supergiants on the other hand exhibit evidence for only moderate CN processing, while the BC supergiant HD13866 seems to have

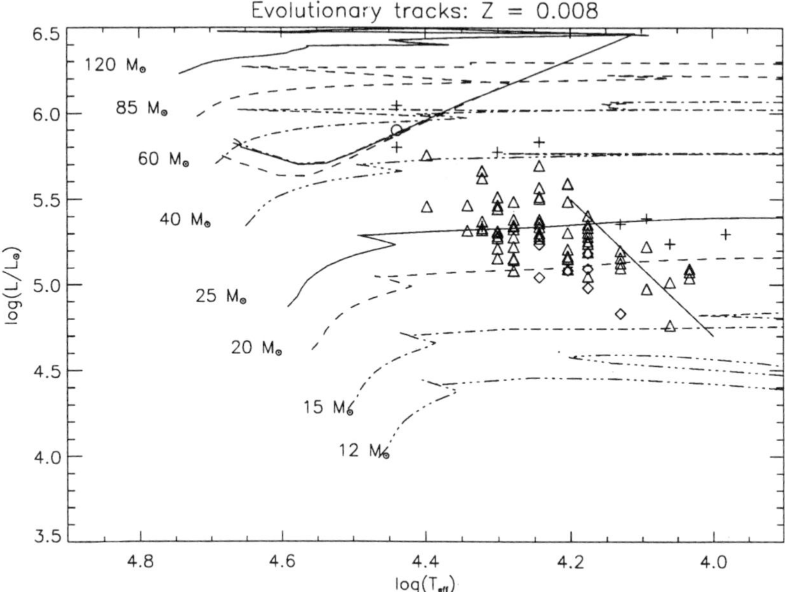

Fig. 5.   HR diagram of Fitzpatrick's (1991) sample of LMC supergiants, for comparison with Fig. 1.

a peculiar oxygen abundance.

Also in Table I, we summarize some results from the literature, including for comparison LTE results for another BC supergiant (HD96248) also found to have a rather high O/N ratio (Dufton 1972). We note that these results reinforce the general trend found here, in that the most luminous supergiants at a given spectral type appear to exhibit the greatest degree of contamination by CNO processed material (see also Kudritzki et al (1987) and Lennon et al (1991)). This result, if confirmed by detailed modelling for individual stars, confirms one of the general predictions of stellar evolution calculations discussed in section 1, namely that the most massive and luminous supergiants in the post-RSG scenario should have the strongest CNO processing signature. There are some remaining problems however, in particular the helium fractional abundance adopted here (0.3 by number) is much higher than is generally predicted by theory. It is therefore of great importance that this parameter be derived more precisely, especially in view of the fact that one can have N enrichment without significant helium enrichment under certain conditions (see Langer 1991 for example).

3. B-supergiants in the LMC

The Large Magellanic Cloud, being the nearest external galaxy and containing a rich population of young stars, is the ideal laboratory in which to study these objects, since uncertainties in distance are significantly reduced. In an important

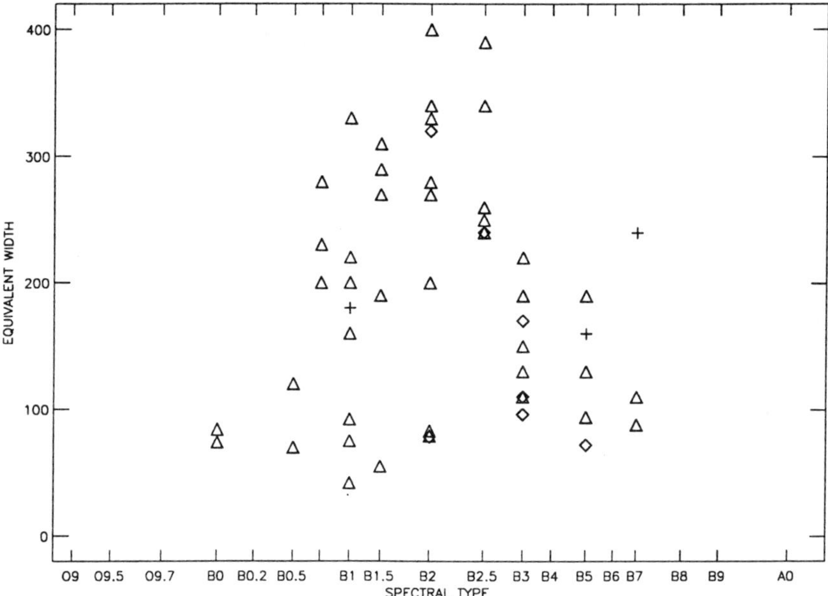

Fig. 6. Fitzpatrick's (1991) derived nitrogen line strengths as a function of spectral type. Compare this with Fig. 14 of Lennon et al (1993).

paper, Fitzpatrick (1991) surveyed approximately 100 B-type supergiants in the LMC at intermediate resolution and measured line strengths for a number of strong absorption lines. It was found that the spread in nitrogen line strengths at a given spectral type was unusually large. We illustrate both this sample of stars and the nitrogen results in Fig. 5 and Fig. 6 for comparison with the Milky Way data. Notice an important selection effect, the LMC stars are much more luminous on average than the galactic sample. Nevertheless, the spread in the nitrogen line strengths is enormous, this is particularly evident at spectral types B1-B2. Subsequent high resolution observations of a subset of these stars by Fitzpatrick & Bohannan (1993) has confirmed that, in a morphological sense, nitrogen weak stars in this sample are also carbon strong. Furthermore, they found some evidence for a correlation between the carbon and oxygen features. The conclusion drawn by the authors is that the nitrogen weak stars, constituting about 10-20% of their sample, are the pre-RSG objects with normal nitrogen abundances while the majority of morphologically normal supergiants are nitrogen rich. Thus we have evidence to support another prediction of the 'blue loop' scenario, namely that most B-supergiants should then be post-RSG helium burning stars.

4. Conclusions

In a qualitative sense, the studies of Milky Way and LMC B-supergiants are in agreement with stellar evolution calculations in which massive stars are allowed to perform blue loops in the HR diagram. Both the preponderance of nitrogen rich stars and the luminosity dependence of this contamination predicted by such calculations are observed. Some questions remain however. The nitrogen enrichment of the B2 Ib-II supergiants is less secure than that of the Ia stars, while the large helium abundance implied by some He I lines in the latter is difficult to reconcile with theoretical predictions which tend to be somewhat lower. Clearly more precise estimates are needed and, given the importance of the microturbulent velocity noted above, a high priority is the inclusion of velocity fields for the NLTE calculations. Concerning the position of these stars in the HR diagram, if they really are post-RSG stars then blue loops must extend further bluewards than is commonly predicted. On the other hand recent analyses of A-supergiants by Venn (these proceedings), imply normal nitrogen abundances for these objects. So the interpretation of the still patchy pattern of abundances in luminous blue supergiants is not yet clear. Finally, extending this type of study to supergiants in the SMC and to less luminous supergiants in the LMC will give important constraints on stellar evolution over a wide range of metallicities.

Acknowledgements

I wish to thank the BMFT for financial support under grant number 010R90080 and Rolf Kudritzki, Ed Fitzpatrick, Philip Dufton and Alan Fitzsimmons for many useful discussions.

References

Dufton, P.L.: 1972, *A&A* **16**, 301
Fitzpatrick, E.L.: 1991, *PASP* **103**, 1123
Fitzpatrick, E.L. and Bohannan, B.: 1993, *ApJ* **404**, 734
Fitzpatrick, E.L. and Garmany, C.D.: 1990, *ApJ* **363**, 119
Gies, D.R. and Lambert, D.L.: 1992, *ApJ* **387**, 673
Kudritzki, R.P. et al: 1987, 'ESO Conference and Workshop Proceedings No. 26' in Danziger, I.J., ed(s)., *SN1987A*, ESO:Garching, 39
Langer, N.: 1991, *A&A* **243**, 155
Langer, N.: 1991, *A&A* **252**, 669
Langer, N.: 1992, *A&A* **265**, L17
Lennon, D.J., Kudritzki, R.-P., et al: 1991, *A&A* **252**, 498
Lennon, D.J., Dufton, P.L. and Fitzsimmons: 1992, *A&AS* **94**, 569
Lennon, D.J., Dufton, P.L. and Fitzsimmons: 1993, *A&AS* **97**, 559
Massa, D. : 1991, *A&A* **242**, 188
Maeder, A.: 1987, *A&A* **173**, 247
Schaerer, D., Meynet, G., Maeder, A. and Schaller, G.: 1993, *A&AS* **98**, 523
Schaller, G., Schaerer, D., Meynet, D. and Maeder, A.: 1992, *A&AS* **96**, 269
Tuchman, Y. and Wheeler, J.C.: 1990, *ApJ* **363**, 255
Walborn, N.R.: 1976, *ApJ* **205**, 419
Walborn, N.R.: 1988, '108th IAU Coll.' in Nomoto, K., ed(s)., *Atmospheric diagnostics of stellar evolution*, Springer-Verlag:Berlin, 70

INTRINSIC PARAMETERS OF MASSIVE OB STARS*

A. HERRERO

Instituto de Astrofsica de Canarias, E-38200 La Laguna, Tenerife,Spain

Abstract. We present the results of our observations of stars of type O5 and earlier and show that inclusion of the line blocking between 228 and 912 A solves the problem found by Herrero et al. (1992) in the determination of their stellar parameters. We study the influence of the line blocking and other effects on the mass and helium discrepancies and show that the first one is reduced by the use of spherical, non hydrostatic model atmospheres and that the second one is probably due to exposure of CNO material.

Key words: spectroscopy – stars:early types – mass – helium – line blocking

1. Introduction

Accurate parameters of massive hot stars are needed today to extend our present understanding of these and related objects in many astrophysical fields: in the theory of radiatively driven winds, (see Pauldrach et al., 1993), where they provide boundary values; in the study of the interstellar medium, where hot stars determine the energy and momentum inputs and the chemical yields (Leitherer et al., 1992; Maeder, 1993); and in extragalactic studies, where due to their luminosities they play a key role in the comparison of galaxies and where they could be used in the future as standard candles. However, presently is in the theory of stellar structure and evolution where accurate stellar parameters could have a major influence in the future research as a consequence of the so called *mass* and *helium discrepancies*. Herrero et al. (1992, hereafter Paper I) have shown that the masses derived from the spectroscopic analyses and the wind theory are systematically lower than those derived from the evolutionary theory (see Paper I for details on how these masses are derived and for a discussion of the associated errors). These differences, that sometimes amount to more than a factor of two, constitute the so called *mass discrepancy*. In Paper I it has also been shown that the helium abundances derived from the spectroscopic analyses do not agree with the predictions of the evolutionary theory. Nearly all O supergiants show helium enrichment, which cannot be explained by present evolutionary models.

The unavoidable conclusion is that one or both theories have to be corrected and improved, because both of them try to represent the same object. In the present paper we present results of our analysis to determine which improvements in the model atmospheres could account for the described discrepancies, or could at least reduce them, and attempt to find independent additional evidence that could indicate the direction of future efforts.

---

* The INT is operated on the island of La Palma by the RGO in the Spanish Observatorio de El Roque de los Muchachos of the Instituto de Astrofsica de Canarias

Space Science Reviews **66**: 137–145, 1994.
© 1994 *Kluwer Academic Publishers. Printed in Belgium.*

There is however another effect found in Paper I that could be relevant for the mentioned discrepancies, and that in any case is of great importance for the theory of stellar atmospheres and for the accuracy of the derived parameters, the central point of our work. For the two hottest stars in Paper I, two O5 stars, we found that the stellar parameters derived using the He I singlet lines at 4388 and 4922 A were different from those derived using the triplet line at 4471 A. The main effect was in the temperature, that could change about 5 000 K (temperatures derived with the triplet line being higher), i.e., much more than the error bars would allow. The change in gravity was simply that demanded by the change in temperature. No change was found in the derived helium abundance.

The described effect is important for two reasons: first, it makes it impossible to speak of accurate parameters for the early types (O5 and earlier), as we do not know the reason for the difference nor do we know any reason to trust one line more than the others, as they all are in these types weak lines that should form in photospheric layers well described by present plane parallel, hydrostatic (PPH) models; second, the effect could be present in a smaller grade in later types, where usually only the singlet lines are used to determine the stellar parameters, that could then need a correction if these lines are the ones giving the wrong parameters.

To investigate the effect and its cause we have observed a number of stars of type O5 and earlier and have analyzed them using the models described in Paper I with the improvements described in the next section. These analyses are also used to determine if the mass and helium discrepancies extend above spectral type O6.

Section 2 describes the observations. In Sect. 3 we describe the improvements made to resolve the temperature problem in O5 types and in Sect. 4 we look for the effects of mass loss and sphericity. The next section is dedicated to a brief discussion of the role of distance and rotation. Section 6 is dedicated to the possibility that the stars are contaminated with CNO burning products. Conclusions are presented in Sect. 7.

2. Observations

The observations, carried out with the Isaac Newton 2.5 m. telescope at the Observatory of El Roque de los Muchachos in La Palma during two different observational runs, were similar to those described in Paper I, except that during the second run (August 1992) we used the large EEV 5 CCD chip, resulting in a larger spectral range (about 400 A). The first run (September 1991) was as in Paper I. The Intermediate Dispersion Spectrograph was used with the H 2400 B grid and the 235 mm camera, which resulted in a spectral resolution of 0.6 A/pixel, measured on the Cu–Ar calibration lamp. The observed objects are listed in Tab. 1.

The reduction of the data was made using own software, based on programs developed at the Munich Observatory. We follow the standard procedure: bias subtraction, flat field division, spectrum extraction and continuum rectification. The S/N ratio is about 200.

3. Spectral Analysis and Line Blocking Effect

The model grid from Paper I was used to analyse the objects. All objects in which it was possible to identify the singlet lines showed the effect described in the introduction.

Recently Pauldrach et al. (1993) have shown the importance of the line blocking in the wind for the ionization structure and the emergent flux. To look for the effect of this line blocking on the photospheric lines we have included in our calculations the line list between 912 and 228 A used by Pauldrach et al. (1993), and the corresponding model atoms. We have included all the transitions in the H and He line formation problem.

We first investigate the effect of the line blocking on a model appropriate for O5 stars. We have chosen the model with $42\,500$ K, $\log g = 3.70$ and $\epsilon = 0.09$. Figure 1 shows the profiles of the H_γ line with and without inclusion of the line blocking. We can see that the effects are very small and in comparison with the observations, where the broadening due to rotation has to be included, only the small difference in the wings will remain significant. It is interesting to note that the new wings are deeper, which means that the $\log g$ derived with the inclusion of the line blocking will be lower, thus increasing slightly the mass discrepancy.

The same changes are seen on the He I 4471 and the He II 4541 lines, indicating that the ionization equilibrium of He is not altered by the inclusion of the line blocking. However, the singlet lines He I 4922 and 4388 (the last one shown in Fig. 2) are strongly affected by the additional opacities. The new lines are much deeper, thus fitting the observed profiles at higher temperatures and removing the discrepancy found in Paper I. All He I lines indicate now the same temperature.

Why does the line blocking only affect the singlet lines significantly? The reason lies in the atomic structure of He I . The lower level of the singlet lines, the $2p^1P^0$ level, is connected to the ground level of He I through a radiative transition at 584 A. In absence of line blocking this line lies in a region where the only important opacity source is electron scattering, far from the continuum edges of H I and He I . When the new opacities are added the global opacity at this wavelength is strongly increased, and the line formation region is changed to upper layers. The triplets are not altered because they are not connected to the ground level by radiative transitions, and the ionization equilibrium is not changed because the integrals of the continuum rates are dominated by values close to the edges, where the new opacities are added to the strong opacities of H I and He I, and thus have only a small effect on the global opacity.

We have calculated for different models the changes produced in the stellar

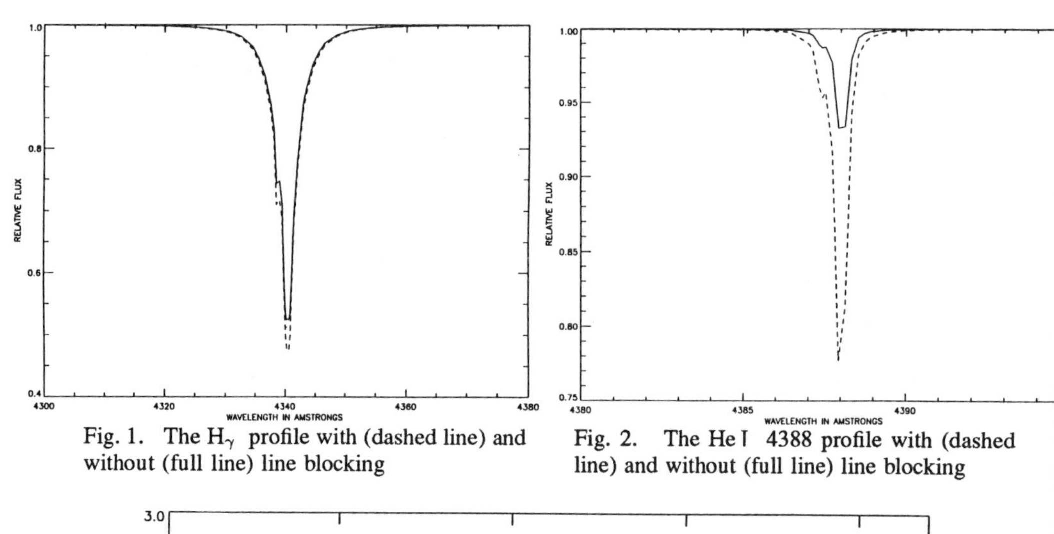

Fig. 1. The H$_\gamma$ profile with (dashed line) and without (full line) line blocking

Fig. 2. The He I 4388 profile with (dashed line) and without (full line) line blocking

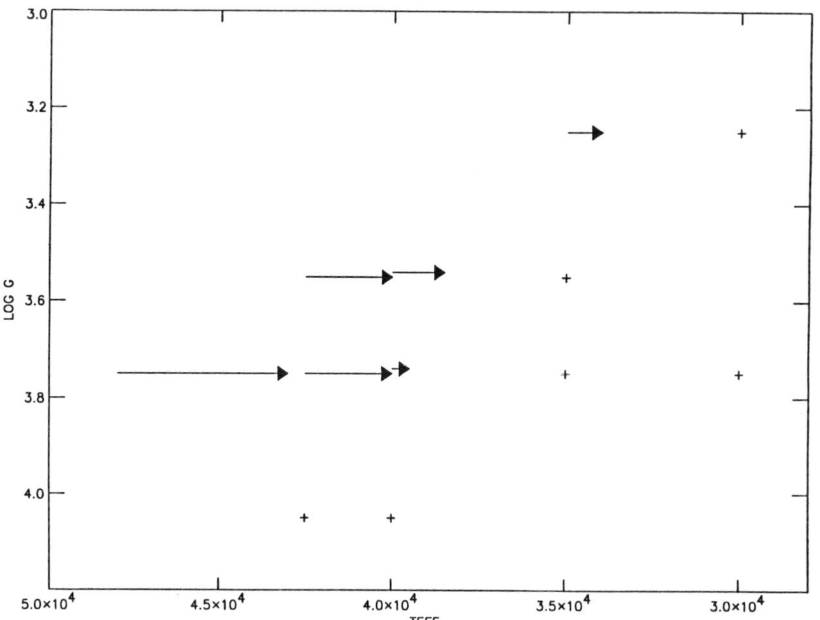

Fig. 3. Effect of the line blocking on the stellar parameters. The arrows begin at the correct parameters and end at the parameters determined using the HeI singlet lines in absence of line blocking. Models starting at 40 000, 3.75 and 40 000, 3.55 have slightly displaced for clarity. Crosses represent models for which no changes were found

parameters due to this effect. In Fig. 3 we see that the effect is on the temperatures, which are now higher. The effect depends on both T_{eff} and $\log g$, and is larger at higher temperatures and lower gravities. Below 40 000 K the effect is important only for very low gravities.

Using our former model atmospheres and line formation codes (see Paper I) we have used the displacements vectors from Fig. 3 to stimate the stellar param-

TABLE I

Observed stars and their estimated stellar parameters. Temperatures are given in thousands of Kelvin, velocities in km/s, masses and radii in solar units.

STAR	Sp.Type	$T_{\rm eff}$	$\log g$	ϵ	Vsin i	$M_{\rm s}$	R
Cyg OB2#7	O3 If	51.5	3.70	0.15	105	50.4	16.5
HD 15570	O4 If$^+$	49.0	3.50	0.25	105	69.1	24.3
HD 15558	O5 III f	48.0	3.75	0.07	120	97.7	21.8
HD 15629	O5 V((f))	47.0	3.75	0.07	90	41.7	14.2
Cyg OB2#8A	O5.5 If	45.0	3.55	0.15	95	68.5	22.9
HD 14947	O5 If	43.5	3.45	0.15	140	28.1	16.1

eters of the observed O5 stars, which are given in Tab. 1. These are provisional parameters, as we still do not have a grid of models including the line blocking. But again, the mass and helium discrepancies are found for stars with low gravities, in full agreement with the results found for types later than O5.

Placing the stars from Paper I on the $\log g$- $T_{\rm eff}$ diagram we see that many of them lie in the region where a correction to the derived stellar parameters is expected. We have estimated this correction and have calculated new spectroscopic masses to see if the mass discrepancy is reduced. The result is that only a small effect is present, and that this effect goes in the direction of increasing the mass discrepancy. No effect on the helium discrepancy was found, because the derived helium abundance is not affected.

4. Unified Models

We have used the models of Gabler et al. (1989) to determine displacement vectors due to the effects of sphericity and mass loss by comparing with PPH models. The main effect is on the derived gravity, which is now larger by 0.1–0.15 dex. The changes in temperature are smaller than the uncertainties quoted in Paper I. These corrections, however, are dependent on the stellar parameters themselves, and for very large mass loss rates ($> 10^{-5}$ M_0/yr) even larger changes may be expected.

The correction to the gravity changes the derived spectroscopic mass, so that the mass discrepancy is reduced by about a factor of two. This means that the spectroscopic and evolutionary masses now agree within the errors, but still the evolutionary masses are systematically larger than the spectroscopic ones by about 30 per cent typically.

Unlike the mass discrepancy, the helium discrepancy is not affected by the introduction of sphericity and mass loss, because the derived helium abundance does not change.

5. On the Effect of Distance and Rotation

In the derivation of the spectroscopic radii and masses the distance plays an important role (see Paper I) through the absolute magnitude. Accurate distances are one of the most difficult astrophysical problems and recently Garmany and Stencel (1992) have determined new distances to open clusters which differ from those of Humphreys (1978) in 0.5 mag typically. As we used these last distances in Paper I we have recalculated the spectroscopic masses with both distance scales.

The results is that the mass discrepancy is only weakly dependent on the distance. For example, for HD 190 864, one of the stars in Paper I, a change of 0.6 mag. in the distance, as taken from Garmany and Stencel, changes the mass discrepancy from 20 to 22 solar masses (and increasing it!). The reason is that a change in the distance produces a change in the absolute magnitude and thus a change in the derived spectroscopic mass (see Paper I), but it also produces a change of the same sign in the luminosity and thus in the mass we read out of the evolutionary tracks.

The effect seen for HD 190 864 is representative of all other stars (the sign depending on the sign of the distance change). We thus conclude that changes in the distance will not solve the mass discrepancy.

If we represent the mass discrepancy against the helium discrepancy for all stars in Paper I no correlation is seen. However, when we do not plot the stars with large projected rotational velocities we see that all stars with large helium discrepancies show a large mass discrepancy, whereas the rapid rotators showed large helium discrepancies without large mass discrepancies. Whereas this result may have different interpretations, we will only mention here that it is compatible with the picture presented by Schonberner et al. (1988), who proposed that the fast rotators could evolve first to the left of the ZAMS and then after homogeneous evolution and strong mixing, due to a braking of its rotation, to the right, reaching thus a position close to the ZAMS, with high gravities and abundances corresponding to advanced evolutionary stages.

6. The CNO Content

If the helium discrepancy is due to exposure at the surface of processed material we expect that it is accompanied by CNO burning products, i.e., we expect that it correlates with N enrichment and C depletion, the effect on O being smaller. Thus we have tried to determine the CNO abundances for stars in which we could measure the relevant lines. Unfortunately we found a strong dependence of the derived abundances on the stellar T_{eff}, indicating that the calculations should be improved (for example, test calculations showed that the line blocking described in Sect. 3 also has an influence on the line formation of metals, and thus in the derived abundances. Calculations are currently being performed to clarify this

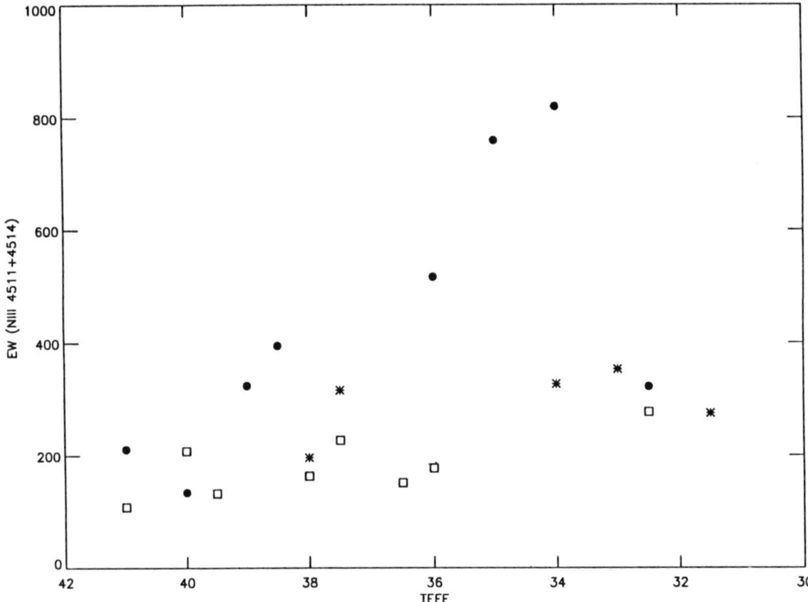

Fig. 4.   Equivalent width of N III 4511+4514 against temperature. Open squares are stars with normal He abundance, asteriscs are stars with intermediate He abundance and filled circles are stars with high He abundance

effect).

. However we still may study the correlation of the helium discrepancy with the equivalent widths of the C and N lines. Figure 4 is a plot of the equivalent width of the N III 4511+4514 lines against the temperature, for stars with different helium abundances. We inmediately see that stars with higher derived helium abundances also have larger equivalent widths, supporting the suggestion that we are actually seeing CNO processed material at the stellar surface. This is again supported by Fig. 5 where we have plotted the equivalent widths of the mentioned N III lines versus that of the C III 4069+4187 lines. We see that the star with enhanced helium abundance occupy the region of low C III and large N III equivalent widths.

7. Conclusions

We have observed and analyzed some stars of spectral type O5 and earlier, a region where only a few analyses are available. We find that to analyze them the line blocking between 912 and 228 A has to be included in the calculations, because if it is neglected we could make important errors in the derived temperatures (and smaller errors in the gravity). The effect is larger for higher temperatures and lower gravities, and is important also for stars cooler than O5 with low gravities. The effect also solves the difficulty found in Paper I for the

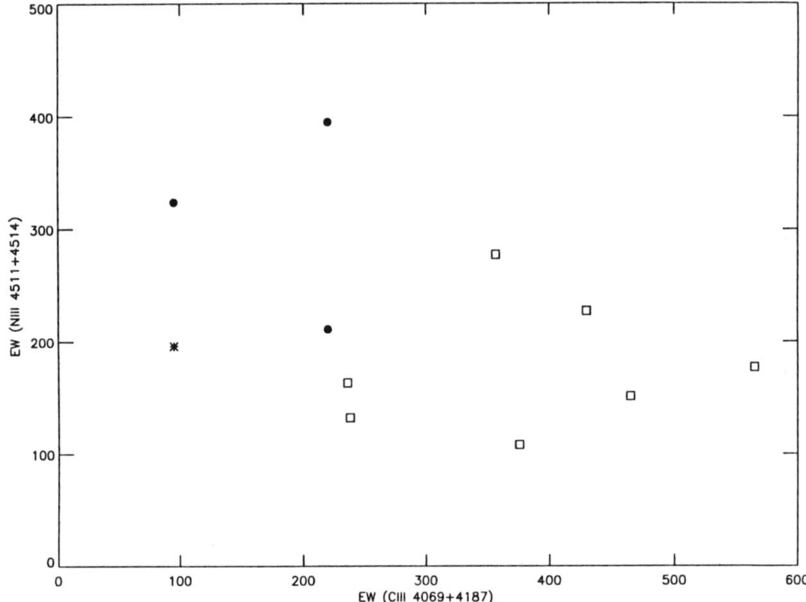

Fig. 5. Equivalent width of N III 4511+4514 versus C III 4069+4187. Symbols are the same as in Fig. 4

determination of the stellar parameters of O5 stars.

We have studied the influence of this effect on the mass discrepancy, and have found that it cannot explain it. The mass discrepancy cannot be explained either by uncertainties in the distance. However, we have found that it is reduced if we use models with sphericity and mass loss, being the errors in the gravity of the PPH models larger for stars with larger mass loss rates.

None of the effects mentioned above changes the derived abundance, and thus they cannot explain the helium discrepancy. We have found that the fast rotators appear to follow a different behaviour with respect to the mass and helium discrepancy than the rest of the stars, perhaps due to a different evolutionary history.

We have shown that there is evidence favouring the interpretation of the helium discrepancy as due to the exposure of CNO processed material at the stellar surface. However, a confirmation by direct determination of metal abundances requires improvements in the current calculations.

Acknowledgements

I would like to thank S. Becker for the data needed for the calculations of the line blocking, and R. Gabler for unified models to compare with PPH ones. I would like to thank A. Pauldrach for encouraging me to consider the line blocking effect on photospheric lines. This work has been partly supported by the Alexander von Humboldt Stiftung and the Ministerio de Educacion y Ciencia through the Acciones Integradas Hispano-alemanas

References

Gabler, R., Gabler, A., Kudritzki, R.P., Puls, J., Pauldrach, A., 1989, A&A, 226, 162

Garmany, C.D., Stencel, R.E., 1992, A&AS, 92, 211

Herrero, A., Kudritzki, R.P., Vilchez, J.M., Kunze, D., Butler, K., Haser, S., 1992, A&A, 261, 209 (Paper I)

Humphreys, R.M., 1978, ApJS, 38, 309

Leitherer, C., Robert, C., Drissen, L., 1992, ApJ, 401, 596

Maeder, A., 1993, A&A 264, 105

Pauldrach, A., Kudritzki, R.P., Puls, J., Butler, K., Hunsinger, H., 1993, A&A, in press

Schonberner, D., Herrero, A., Butler, K., Becker, S., Eber, F., Kudritzki, R.P., Simon, K.P., 1988, A&A, 197, 209

LINE BLANKETING BY IRON GROUP ELEMENTS IN NON-LTE MODEL ATMOSPHERES OF HOT STARS

STEFAN DREIZLER

Dr.-Remeis-Sternwarte, Astronomisches Institut der Universitat Erlangen-Nurnberg, Sternwartstr. 7, D-96049 Bamberg, Germany

and

KLAUS WERNER

Institut fur Theoretische Physik und Sternwarte der Universitat Kiel, D-24098 Kiel, Germany

Abstract. We report on our recent progress in modeling non-LTE atmospheres of O-stars including blanketing by lines from the iron group elements. The numerical method to account for the huge number of atomic levels and line transitions is presented. Results of exploratory model calculations examining the effects on the temperature structure, the hydrogen and helium line profiles and UV/EUV fluxes are discussed.

Key words: stars: early-type – stars: atmospheres – Radiative Transfer – Line: formation – Methods: numerical – Ultraviolet: stars

1. Introduction

In the recent years model atmospheres for O stars were improved substantially. The Kiel/Bamberg group and now also a group at Goddard works to improve the "classical" NLTE models (plane-parallel, static) to a degree that is comparable to the state-of-the-art in LTE modeling, i.e., to include as much metal opacities as possible. On the other hand the Munich group works successfully accounting for the effects of stellar winds by the construction of so-called unified model atmospheres (Pauldrach et al. 1986, Gabler et al. 1989).

The realistic treatment of iron group elements with millions of line transitions in NLTE model atmospheres of hot stars was a long standing problem. So far, line blanketing has been treated realistically only under the assumption of *Local Thermodynamic Equilibrium* (LTE) using Opacity Distribution Functions or Opacity Sampling techniques (Kurucz 1979, 1991, Gustafsson et al. 1975, Sneden et al. 1976). These models have proven to satisfactorily represent stars of spectral type B or later. For O stars this approach is not sufficient due to the failure of the assumption of LTE at high temperatures and due to the intense radiation fields.

In the next section we describe the numerical method which allows the inclusion of a huge number of line transitions in NLTE model atmosphere calculations. Some applications and results are given in the third section.

Space Science Reviews **66**: 147–152, 1994.
© 1994 *Kluwer Academic Publishers. Printed in Belgium.*

2. Computational method

The pioneering work of Auer & Mihalas (1969) set the stage for the first successful computation of NLTE model atmospheres and many spectroscopic analyses of hot stars were performed successfully since then. Their *Complete Linearization* (CL) solution technique became a robust and widely-used standard tool. However, inherent capacity limits of this method in its original formulation only allowed for the calculation of models with relatively simple compositions. By the development of new numerical techniques in the last decade NLTE model atmospheres including other elements than only hydrogen and helium became available. With the *Accelerated Lambda Iteration* (ALI) technique (Werner & Husfeld 1985) and with several refinements during the last few years accounting for more and more elaborate atomic model atoms (H, He, C, N, O, Ne, Mg, Si, see e.g. Dreizler 1993) was possible. The next step, inclusion of opacities from the iron group elements, was begun last year and first results for hot stars are published (Dreizler & Werner 1993).

An alternative numerical approach (*multi-frequency/multi-gray*, MF/MG) was developed by Anderson (1985) who for the first time computed fully metal line blanketed NLTE model atmospheres for cool stars (1989). We adapted his ideas to construct generic model atoms for the iron group elements and implemented the statistical treatment of the opacities into our ALI code. Finally, a third numerical method, regarded as a hybrid ALI/CL approach, has been introduced by Hubeny & Lanz (1993). This method is also capable to account for metal line blanketing in NLTE model atmospheres.

NLTE model atmospheres require a simultaneous solution of the radiative transfer and statistical equilibrium equations in order to achieve a self-consistent solution for the radiation field and the atomic level populations. For numerical reasons the total number of equations can not exceed a few hundred. This is by far too less if detailed atomic models with some hundred levels and thousands of line transitions are included in the calculations. The methods differ in how this severe limitation can be circumvented or at least be relaxed. The ALI method avoids the explicit solution of the radiative transfer equations by solving them implicitly with the statistical equilibrium by iteration. The MF/MG method reduces the number of radiative transfer equations by a pre-integration in frequency in a combination with variable Eddington factors.

Nevertheless, these methods are without further modifications still not powerful enough when iron group elements are to be treated, because of the huge number of atomic levels and line transitions involved. Approximations have to be used to reduce the explicit number of atomic levels and the frequency points to represent the line transitions. This requires the introduction of mean levels ("superlevels", or "bands") that reduces the $\approx 10^5$ atomic levels to a manageable amount, typically in the order of hundred. The basic approximation here is that the population among all levels combined to such a superlevel obeys Boltzmann statistics. Mean excitation energies and statistical weights for these super-levels are calculated with this presumption. However, departures from LTE for these mean levels and for the ionisation equilibria are accounted for. The second approximation concerns the treatment of line opacities. A detailed description of opacities during model atmosphere calculations would require a number of frequency points that is simply impossible to deal with, even with the fastest computers. This difficulty is not specifically inherent to the NLTE problem but it is also encountered in LTE. In principle two successfully used LTE techniques are known to reduce the number of frequencies and both can be generalized to the NLTE case as was shown by Anderson (1989, 1991). One technique is the well known Opacity Distribution Function (ODF) approach, the other one is called Opacity Sampling (OS) technique. The very reason why neither ODF nor OS techniques can be applied in the NLTE case without modification is the following. In LTE atomic populations n_l and opacities κ_ν are simply depending on two local state variables only (e.g. temperature and pressure) and can be computed from the Saha-Boltzmann equations: $\kappa_\nu = \kappa_\nu(\mathrm{T,p})$. The time-consuming construction of ODF tables for example can therefore be performed once and for all in advance of any model atmosphere calculation.

On the other hand, the populations in NLTE are strongly dependent on the (non-local) radiation field J_ν which enters through the statistical equilibrium equations. The radiation field, which itself is determined by level populations of all atmospheric layers, is not known in advance, prohibiting the construction of opacity tables at the outset. One has to go back one step. Instead of computing the opacity arising from some particular levels i and j, $\kappa_\nu = n_l \sigma_{ij}$ (where σ_{ij} is the photon cross-section), one computes only the σ_{ij} in advance. After having constructed super-levels one can generate suitable profiles of "super-lines" between those levels. These complex line profiles are weighted sums of individual line profiles originating from all individual levels that are comprised by the super-levels. Instead of opacities, it are these profiles that are computed in advance of model calculations. They are essentially dependent on atomic quantities only. Those complex line profiles may span a wide range of frequency and have a complicated shape. A detailed representation requires, occasionally, several 10^5 frequency points. In analogy to the LTE case they are reduced either by redistribution into small frequency intervals in order to give a smooth profile (corresponding to the ODF approach) or they are sampled on a coarse frequency grid (OS approach). A direct comparison between both approaches in NLTE calculations is desirable but still lacking. The results presented here were obtained with models computed by the OS approach. Further details on the model construction can be found in Dreizler & Werner (1993).

3. Applications and results

A series of NLTE models for various types of hot evolved stars was calculated to investigate the effects on the atmospheric stuctures, line profiles, UV and soft X-ray fluxes. Representative element abundances for each spectral type in question were assumed for H, He, C, N, O, and Fe group elements. Abundance ratios within the iron group were always kept at the solar values (Holweger 1979). Detailed standard model atoms for H and He as well as simplified versions of our C,N,O models (Werner & Heber 1991) are used for the calculations. Atomic data for the iron group elements are taken from Kurucz (1991). Up to now about 130 000 line transitions are included where the corresponding levels were identified in laboratory experiments. However, the complete line list of Kurucz (1991) contains also millions of predicted transitions, which will be included in future calculations.

A few calculations were performed to study the influence of the metal line blanketing on the atmospheric structures as well as the effect on hydrogen and helium line profiles commonly used to determine gravities and effective temperatures. Comparing blanketed with unblanketed temperature stratifications we found a cooling of the outer regions and a backwarming of the continuum forming layers. However, if C, N, and O are already included in the calculations, additional line blanketing by iron group elements has only a small effect. The outer layers are dominated by the strong C, N, and O resonance lines; additional line blanketing therefore leads to no significant change. The continuum forming layers have, however, a higher temperature due to the backwarming. The effect on hydrogen and helium line profiles is in general not very drastic. Balmer lines become slightly broader in the wings, the effect is decreasing at higher series members. He II lines become deeper and broader, the strongest effect can be seen in the He II 4686 A line. These effects are, however, too small to explain difficulties in the analysis of hydrogen rich CSPN encounterd by Napiwotzki (1992). He I lines become slightly weaker, we found that the singlet lines are more sensitive than the triplet lines. These results might change if the complete line list of Kurucz (1991) is included in the model atom or if non-solar compositions for the iron group elements are considered.

Since the self-consistent treatment of the metal line blanketing of iron group elements is very time consuming it might be tempting or necessary to employ inconsistent models using LTE population numbers for the iron group elements and ignoring the back-reaction of the opacities on the atmospheric structure. Since the ionisation equilibrium of iron is far from LTE, the ionisation equilibrium is shifted to higher stages, changing the blanketing effect in certain frequency ranges quite drastically. Sensitive ionisation equilibria of other elements might therefore be affected by unrealistic blocking as can be seen in Dreizler & Werner (1993).

After the discussion of the effects of metal line blanketing on the atmospheric structure and line profiles of other elements we will now outline the applications

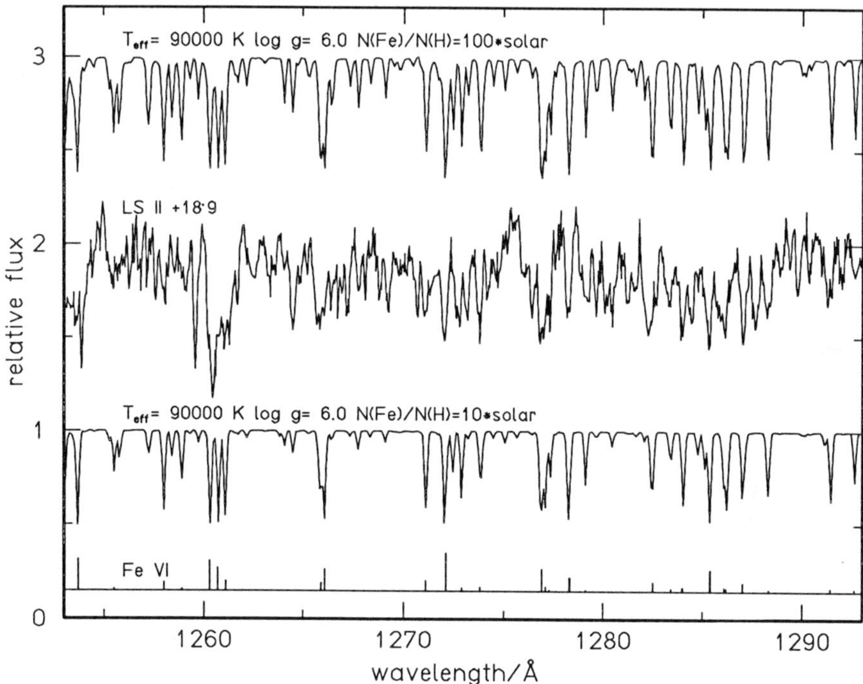

Fig. 1.   Synthetic spectra from line blanketed NLTE models with T_{eff}=90 000 K, $\log g$ =6, and solar abundances except for iron group elements which are set to 100 resp. 10 times the solar value. **Middle:** IUE high resolution spectrum of the sdO star LS II 18°9.

of the interpretation of the iron lines themselves. UV and X-ray spectra of many hot stars are dominated by iron lines. Line blanketing by iron group elements have therefore to be used if *IUE, HST, EUVE,* and *ROSAT* data of hot stars shall be interpreted. Iron lines are most useful temperature indicators, in particular in very hot stars for which the He I/II ionisation balance cannot be used since the He I lines are absent. Instead we can use the Fe V/VI/VII balances for the determination of the effective temperature. Lines from iron group elements can also be used to determine the abundances of these elements in order to decide if abundance pattern are influenced by nuclear burning, radiative levitation, or dust fractionation. Example interpretations of UV and EUV spectra of the DA white dwarfs G191-B2B, RE 2214-492, and RE 0632-377 as well as the sdO star BD+28°4211 analysed with our models are given in Dreizler & Werner (1993) and Holberg et al. (1993). As an application we present here the UV spectrum of a hot helium-poor sdO star, LS II 18°9. Its optical spectrum is very similar to BD+28°4211 so that a similar effective temperature and gravity can be presumed, the iron lines however, are much stronger in LS II 18°9 as already pointed out by Schonberner & Drilling (1985). In Fig. 1 we compare a part of the IUE spectrum with the synthetic spectrum of a line blanketed model atmosphere with T_{eff}= 90 000 K and $\log g = 6.0$ and solar abundances, except for iron group elements which are set to ten and hundred times the solar abundances, respectively. The agreement is quite promising, it has to be remarked that the effective temperature and gravity are not yet fine tuned. The iron abundance is definitely larger than solar, a value between ten and hundred times the solar abundance seems to be realistic. A detailed analysis of LS II 18°9 will be presented in the near future.

Acknowledgements

S. Dreizler and K. Werner are supported by the Deutsche Forschungsgemeinschaft under grants He 1365/16-1 and We 1312/6-1, respectively. The authors thank Thomas Meier (Bamberg) for reducing the IUE spectrum of LS II 18°9.

References

Anderson L.S. 1985, ApJ 298, 848

Anderson L.S. 1989, ApJ 339, 558

Anderson L.S. 1991, in Stellar Atmospheres: Beyond Classical Models, NATO ASI Series C, Vol. 341, eds. L. Crivellari, I. Hubeny and D.G. Hummer, Kluwer, Dordrecht, p. 29

Auer L.H., Mihalas D. 1969, ApJ 158, 641

Dreizler S. 1993, A&A 273, 212

Dreizler S., Werner K. 1993, A&A in press

Gabler R., Gabler A., Kudritzki R.P., Puls J., Pauldrach A. 1989, A&A 226, 162

Gustafsson B., Bell R.A., Eriksson K., Nordlund A. 1975, A&A 42, 407

Holberg J.B., Barstow M.A., Chen A., Dreizler S., Marsh M.C., O'Donoghue D., Sion E.M., Tweedy R.W., Vauclair G., Werner K. 1993, ApJ in press

Holweger H. 1979, in Les Elements et leurs Isotopes dans l'Universe, Universite de Liege, Inst. d'Astrophysique, p. 117

Hubeny I., Lanz T. 1993, IAU Coll. 138, ASP Conf. Series 44, 98

Kurucz R.L. 1979, ApJS 40, 1

Kurucz R.L. 1991, see reference Anderson 1991, p. 441

Mendez R.H., Kudritzki R.P., Herrero A., Husfeld D. 1988, A&A 190, 113

Napiwotzki R., 1992, in The Atmospheres of Early Type Stars, Lecture Notes in Physics, Vol. 401, eds. U. Heber and C.S. Jeffery, Springer, Berlin, p. 310

Pauldrach A., Puls J., Kudritzki R.P. 1986, A&A 164, 86

Schonberner S., Drilling J.S. 1985, ApJ 290, L49

Sneden C., Johnson H.R., Krupp B.M. 1976, ApJ 204, 281

Werner K. Husfeld D. 1985, A&A 148, 417

Werner K., Heber U. 1991, see reference Anderson 1991, p. 341

A COMPARISON BETWEEN OBSERVED AND PREDICTED MASS-LOSS RATES AND WIND MOMENTUM OF O STARS

CLAUS LEITHERER
Space Telescope Science Institute
Baltimore, Maryland

and

HENNY J. G. L. M. LAMERS
SRON Laboratory for Space Research and Astronomical Institute
Utrecht, The Netherlands

Abstract. Empirical mass-loss rates were derived for 28 luminous O stars from radio fluxes and Hα equivalent widths. Comparison with theoretical values predicted by the theory of radiatively driven winds reveals a discrepancy of 0.30 ± 0.05 dex, with the theoretical values being too low. We show that there is not only a mass-loss discrepancy but also a momentum flux discrepancy. The theoretically predicted momentum fluxes are too low by 0.17 ± 0.04 dex. This discrepancy is independent of the adopted stellar mass. We demonstrate that the momentum discrepancy in the most luminous O stars is comparable to the one found in the least extreme Wolf-Rayet stars. We suggest that the physical reason for the break-down of the theory in Wolf-Rayet stars and the most luminous O stars may be related.

Key words: O stars – mass loss – radiatively driven winds

1. Two Decades of Mass-Loss Studies of ζ Puppis

The O4f star ζ Pup arguably deserves the credit of being the star with the largest number of theoretical and empirical mass-loss determinations. ζ Pup was the benchmark test for the original theory of radiatively driven winds by Castor et al. (1975). It was the first O star for which a detailed study of the stellar wind was made on the basis of UV spectra (Lamers & Morton 1976). Subsequently it was the target of ultraviolet (e.g. Garmany et al. 1981; Howarth & Prinja 1989), Hα (e.g. Conti & Frost 1977; Leitherer 1988a; Lamers & Leitherer 1993), infrared (Barlow & Cohen 1977; Lamers & Waters 1984; Lamers et al. 1984), and radio (Abbott et al. 1980; Bieging et al. 1989; Leitherer & Robert 1991) observations to derive empirical mass-loss rates. Surprisingly, theoretical and empirical mass-loss determinations agree rather well. The average from 26 studies found in the literature over the past 20 years is $\dot{M} \approx 5 \times 10^{-6} M_\odot yr^{-1}$, with a spread of about a factor of 2. The latest theoretical studies by Kudritzki et al. (1992) also favor this value.

Does this indicate perfect agreement between theoretical and empirical O-star mass-loss rates in general? Not necessarily, because in many studies the mass-loss rate derived from the radio flux of this star was used as a scaling factor to adjust for uncertainties in the analysis of the observations. We note that the error bars of individual mass-loss determinations are large — a factor of two is common, and

Space Science Reviews **66**: 153–161, 1994.

© 1994 *Kluwer Academic Publishers. Printed in Belgium.*

the stellar parameters can always be adjusted to simulate agreement, in particular when dealing with one object only. The latter problem may be overcome by increasing the sample size, but even then inconclusive results have been found (cf. Leitherer 1988b).

The terminal velocity of the stellar wind v_∞ is the second fundamental wind parameter which may flag discrepancies between theory and observations. Groenewegen et al. (1989) and Blomme (1990) found that the observed values are $\sim$40% *lower* than theoretical ones. This result is difficult to dismiss as the observational errors are negligibly small. One explanation could be that the adopted stellar masses are in error because v_∞ is expected to scale with v_{esc}. Alternatively, the theory may make wrong predictions for v_∞. If this is the case, one might expect the theoretical prediction for $\dot{M}$ to be wrong as well. The wind theory describes the conversion of radiative into kinetic *momentum*, and if a discrepancy exists for v_∞, the same may hold for $\dot{M}$.

2. A Homogenous Sample of O Stars with $\dot{M}$ Determinations

Recently, Howarth & Brown (1991) obtained precise radio fluxes of O stars for which high-resolution ultraviolet line profiles were available as well. This provides the unique opportunity to derive mass-loss rates *and* velocity laws (including v_∞) in the same stars. In order to increase the sample and to check consistency between radio and Hα mass-loss rates, we also added O stars with high-quality Hα observations. A total of 28 stars ranging from types O3 to O9.5 could be included in our sample. Table 1 gives a summary. The program stars occupy the uppermost part of the HRD. They are representative of the most massive, luminous population of O stars. In contrast, less luminous O stars close to the main-sequence which comprise the bulk of the stars in a Salpeter-type mass spectrum are hardly represented.

Mass-loss rates were derived taking into account the velocity information from the ultraviolet. We largely followed the procedures described by Bieging et al. (1989) for the radio and by Leitherer (1988a) for Hα. Typical errors are $\pm$0.20 dex. In most cases the limiting quantities for the accuracy of the mass-loss rates are the distances and the errors of the radio measurements themselves. Our sample is the largest homogenous set of radio mass-loss rates ever published. In Table 1 we indicate if $\dot{M}$ was derived from the radio ('R') or Hα ('H'). In those cases where both methods were available we list the radio value, which is generally more reliable. Comparison of radio and Hα rates in those objects suggests no significant difference between the two methods.

3. A Comparison: Observations versus Models

We calculated mass-loss rates and wind velocities predicted by the theory of radiatively driven winds via the scaling relations of Kudritzki et al. (1989). In

TABLE I

Observed and predicted $\dot{M}$ and v_∞

HD	Type	$v_{\infty\ obs}$ [$km\ s^{-1}$]	$\dfrac{v_{\infty\ pred}}{v_{\infty\ obs}}$	$\log \dot{M}_{obs}$ [$M_\odot yr^{-1}$]	Method	$\log \dfrac{\dot{M}_{pred}}{\dot{M}_{obs}}$
			Class I and f			
93129A	O3 I	3050 ± 60	1.452	$-4.88^{+0.18}_{-0.18}$	H	-0.42
15570	O4 I	2600	1.335	$-5.33^{+0.18}_{-0.18}$	R	-0.19
66811	O4 I	2200 ± 60	1.700	$-5.62^{+0.15}_{-0.15}$	R	-0.24
190429A	O4 I	2300 ± 70	1.556	$-5.16^{+0.18}_{-0.18}$	H	-0.47
14947	O5 f	2300 ± 70	1.465	$-5.32^{+0.18}_{-0.18}$	H	-0.55
210839	O6 I	2100 ± 60	1.509	$-5.68^{+0.15}_{-0.15}$	R	-0.18
151804	O8 I	1600 ± 70	1.456	$-5.00^{+0.17}_{-0.17}$	R	-0.43
152408	O8 I	960 :	–	$-4.87^{+0.15}_{-0.15}$	R	-0.74
188001	O8 I	1800 ± 70	1.478	$-5.38^{+0.18}_{-0.18}$	H	-0.46
149404	O9 I	2450	0.914	$-4.91^{+0.15}_{-0.15}$	R	-0.64
30614	O9.5 I	1550 ± 60	1.516	$-5.41^{+0.18}_{-0.18}$	R	-0.51
37742	O9.5 I	2100 ± 150	1.052	$-5.60^{+0.15}_{-0.15}$	R	-0.14
149038	O9.5 I	1750 ± 100	1.343	$-5.67^{+0.18}_{-0.19}$	H	-0.25
152424	O9.5 I	1760	1.250	$-5.26^{+0.17}_{-0.17}$	R	-0.43
37128	B0 Ia	1500 ± 150	1.407	$-5.39^{+0.16}_{-0.16}$	R	-0.52
mean			1.39 ± 0.05			-0.41 ± 0.05
std. dev.			0.20			0.18
			Class II and III			
15558	O5 III	3350 ± 200	1.072	$-5.61^{+0.19}_{-0.23}$	H	-0.08
190864	O6.5 III	2450 ± 150	1.465	$-5.88^{+0.18}_{-0.20}$	H	-0.33
24912	O7.5 III	2400 ± 100	1.467	$-5.89^{+0.18}_{-0.19}$	H	-0.63
36861	O8 III	2400 ± 150	1.420	$-6.20^{+0.21}_{-0.41}$	H	-0.31
37043	O9 III	2450 ± 150	1.208	$-6.50^{+0.18}_{-0.18}$	R	$+0.27$
57061	O9 II	1960	1.306	$-5.20^{+0.15}_{-0.15}$	R	-0.42
36486	O9.5 II	2000	1.245	$-5.97^{+0.15}_{-0.15}$	R	$+0.10$
mean			1.31 ± 0.06			-0.20 ± 0.13
std. dev.			0.15			0.31
			Class V			
46223	O4 V	2800 ± 60	1.625	$-5.85^{+0.18}_{-0.20}$	H	-0.30
164794	O4 V	2950 ± 150	1.376	$-5.62^{+0.19}_{-0.20}$	H	-0.08
15629	O5 V	2900 ± 70	1.421	$-5.77^{+0.18}_{-0.20}$	H	-0.30
46150	O5 V	2900 ± 200	1.417	< -5.88	H	–
47839	O7 V	2300 ± 200	1.757	$-6.30^{+0.21}_{-0.41}$	H	-0.28
149757	O9 V	1500 :		$-7.41^{+0.16}_{-0.16}$	R	$+0.36$
mean			1.52 ± 0.08			-0.12 ± 0.14
std. dev.			0.16			0.28

Table 1 we show how the theoretical and observed terminal velocities and mass-loss rates compare. The predicted velocities $v_{\infty\,pred}$ are on average about 40% larger than observed, a result known before. As for the mass-loss rates, we find $\log(\dot{M}_{pred}/\dot{M}_{obs}) = -0.30 \pm 0.05$ for the whole sample. Supergiants have the largest discrepancy with $\log(\dot{M}_{pred}/\dot{M}_{obs}) = -0.41 \pm 0.05$. No significant difference between theory and observations is detected in main-sequence stars.

What is the reason for the discrepancy? One might suspect that systematic errors in the adopted stellar parameters could introduce a bias. We investigated the effect of varying the fundamental stellar parameters and found no significant changes in $\dot{M}_{pred}/\dot{M}_{obs}$. The stellar luminosity is the most crucial parameter for the determination of $\dot{M}_{obs}$ and $\dot{M}_{pred}$ but it affects both the observational *and* the theoretical values in a rather similar way. Therefore $\dot{M}_{pred}/\dot{M}_{obs}$ is hardly affected by possible uncertainties in the stellar luminosity. Sellmaier et al. (1993) and Schaerer & Schmutz (1993) found that the surface gravities of the most luminous O stars have been underestimated in the past. This immediately translates into an upward revision of the *spectroscopic* masses. We note that the masses adopted here are the *evolutionary* masses, which are not affected. Moreover, we will show below that the $\dot{M}$ and v_{∞} discrepancy cannot be solved simultaneously by varying the stellar mass.

We checked the validity of Kudritzki et al.'s scaling relations with respect to fully self-consistent wind models. Kudritzki et al.'s relations are only first-order approximations to a complete model using average line-force multipliers, and one might expect that systematic errors are possibly introduced. Pauldrach et al. (1990) solved the full radiative transfer and hydrodynamics of the wind for a grid of O stars. We calculated wind parameters for Pauldrach et al.'s grid in exactly the same way we did for our present sample. No significant difference was found between the results of the scaling relations and the full theory.

What are the observational uncertainties? We note that many program stars have been detected at multiple wavelengths. The slope of their radio spectrum is in excellent agreement with the prediction for a homogeneous, stationary outflow at constant velocity. Inhomogeneities could in principle be a reason for concern, as free-free radiation is sensitive to spatial density variations with a resulting overestimate of $\dot{M}$. It is very unlikely that such conditions exist in the winds of O stars. Local density enhancements in the wind are observed as 'discrete absorption components' in ultraviolet absorption lines (Prinja 1992). These absorptions are *not* observed at velocities $\lesssim 700\ km\ s^{-1}$, or (with a standard velocity law) at distances $\lesssim 1.5\ R$. This is the region where Hα emission typically originates. Since recombination radiation is rather sensitive to inhomogeneities, the presence or absence of density variations should be immediately noticeable. $\dot{M}$ derived from Hα (where clumps are not observed in the ultraviolet) and from mm- to cm-radiation (at distances larger than several stellar radii, where clumps are detected in the ultraviolet) is not significantly different. This makes it unlikely that clumping introduces a significant bias in the observed mass-loss rates. We recall that there is also a v_∞ discrepancy. It is difficult to imagine how clumping would affect a velocity measurement.

4. The *Momentum* Discrepancy

$\dot{M}_{pred}$ is lower and $v_{\infty\ pred}$ is higher than observed, and the $\dot{M}$ discrepancy is highest in evolved stars. Herrero et al. (1992) derived stellar masses via a non-LTE analysis which were systematically lower than expected from evolutionary models. The difference was most pronounced for evolved O supergiants and vanished for main-sequence stars. If this result suggests errors in the evolutionary masses, an explanation for the $\dot{M}$ and v_∞ discrepancy would be found. We adopted evolutionary masses when calculating $\dot{M}_{pred}$ and $v_{\infty\ pred}$. If the actual masses were lower, higher theoretical mass-loss rates and lower terminal velocities would follow.

Although $\dot{M}_{pred}$ and $v_{\infty\ pred}$ increase and decrease, respectively, with lower mass, the momentum flux $(\dot{M}v_\infty)_{pred}$ is virtually independent of M. As an example we list the corresponding values for ζ Pup. Keeping all stellar parameters fixed (except for the mass), we find the results listed in Table 2.

TABLE II

Theoretical momentum flux vs. mass for ζ Pup

M	$\log \dot{M}_{pred}$	$v_{\infty\ pred}$	$\log(\dot{M}v_\infty)_{pred}$
$[M_\odot]$	$[M_\odot yr^{-1}]$	$[km\ s^{-1}]$	$[g\ cm\ s^{-2}]$
120	−5.89	5448	28.65
100	−5.83	4833	28.65
80	−5.76	4128	28.66
60	−5.64	3161	28.66
40	−5.36	1663	28.66

Despite the factor-of-3 variation of M, $(\dot{M}v_\infty)_{pred}$ remains constant. The increase of $\dot{M}_{pred}$ is exactly compensated by a decrease of v_∞. In Figure 1 we show a comparison between $(\dot{M}v_\infty)_{pred}$ and $(\dot{M}v_\infty)_{obs}$ for all program stars. Interestingly, the theoretical momentum fluxes are tightly correlated with the stellar luminosity: $(\dot{M}v_\infty)_{pred} \approx 0.3L/c$. The theory predicts that about 30% of the available radiative momentum flux is converted into kinetic momentum flux, with little dependence on other stellar parameters. The observed momentum fluxes are higher. Some stars are close to — or even above — the single-scattering limit for the momentum transfer. We find that $\log((\dot{M}v_\infty)_{obs}/(\dot{M}v_\infty)_{pred}) = 0.17 \pm 0.04$, suggesting there exists a *momentum discrepancy* between theory and observa-

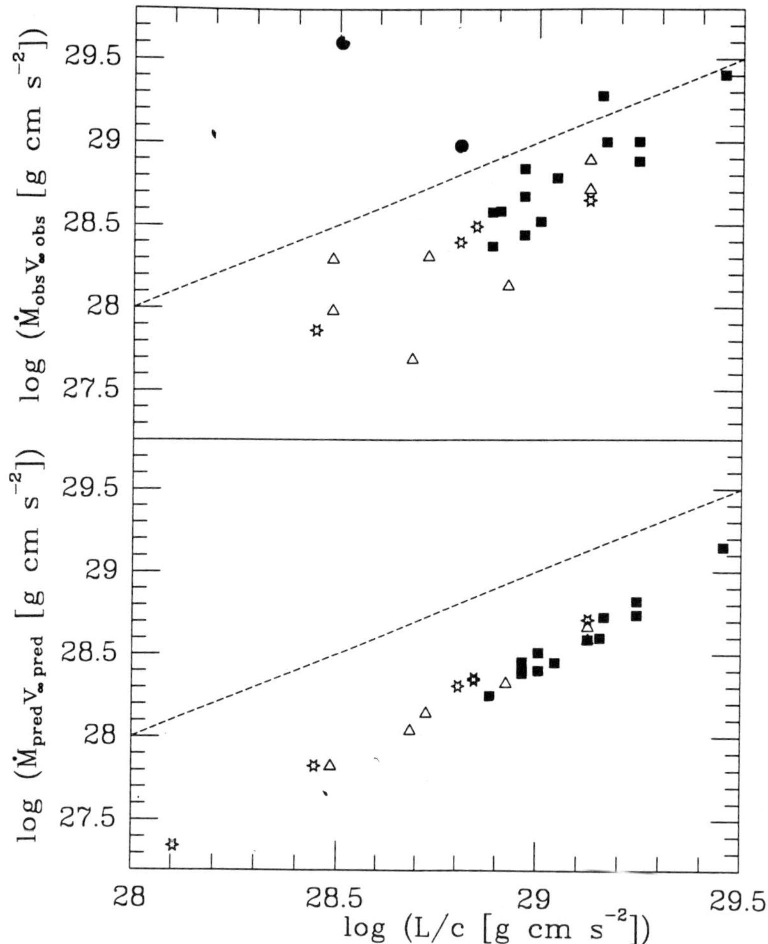

Fig. 1.   Observed (top) and theoretical (bottom) momentum fluxes. Squares: luminosity class I; stars: II, III; triangles: V. Dashed line indicates the single-scattering limit. The theoretical values reveal a momentum deficit of the wind theory.

tions for the most luminous stars. Since the momentum flux is independent of the adopted stellar mass, this discrepancy cannot be caused be errors in the adopted stellar masses.

5. The O-Wolf-Rayet Connection

We searched for correlations between stellar parameters and $\dot{M}_{pred}/\dot{M}_{obs}$ and $v_{\infty\,pred}/v_{\infty\,obs}$. No significant correlation with simple stellar parameters, such as mass, luminosity, or T_{eff} was found. There is, however, a clear tendency for $\dot{M}_{pred}/\dot{M}_{obs}$ to become larger with higher wind densities. (Wind density is defined here generically as $\dot{M}_{obs}/(4\pi R^2 v_{\infty\,obs})$. This is the density at a distance

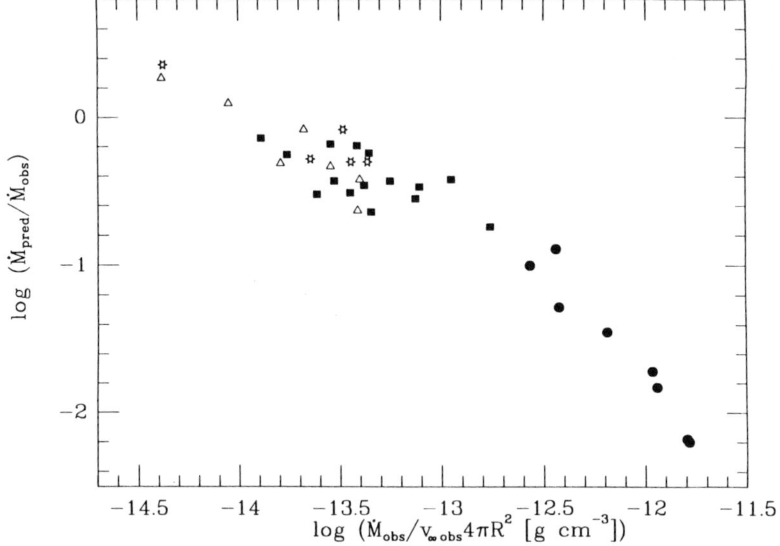

Fig. 2.   Mass-loss discrepancy versus wind density for O stars and 8 WNL stars (WNL: filled circles; otherwise same symbols as in Figure 1). The most extreme O stars form an extension of the least extreme WNL stars.

of 1.5 R if the wind has a β-type velocity law with $\beta = 0.70$).

This trend becomes clearer after inclusion of WNL stars. We selected all WN7, WN8, WN9 stars with reliable parameters from Schmutz et al. (1989). It is well-known (e.g. Cassinelli 1991) that the wind theory in its present form is not capable of describing typical Wolf-Rayet winds. However, the most extreme O stars (such as HD152408) have wind conditions not too different from the least extreme WNL stars. We calculated theoretical wind parameters for the WNL sample in the same way we did for the O sample. We do not expect the relations of Kudritzki et al. (1989) to hold for Wolf-Rayet stars but comparison of the results for the most extreme O stars and the least extreme WNL stars may bear the clue for the resolution of the momentum discrepancy.

$\dot{M}_{pred}/\dot{M}_{obs}$ versus wind density for the O- and Wolf-Rayet-stars is plotted in Figure 2. A clear correlation is discernible. The discrepancy increases with increasing wind density. $\dot{M}_{pred}/\dot{M}_{obs}$ becomes smaller for more luminous O stars and reaches up to a factor of 10^{-2} for some of the WNL stars. Note the smooth transition from O stars with the densest winds to the WNL stars with the least dense winds.

The results of Figure 2 are strongly suggestive of a common physical origin of the mass-loss (and momentum) discrepancy in O- and Wolf-Rayet-stars. This is not unexpected in view of the rather similar wind properties prevailing in the O/WR transition domain. Lucy & Abbott (1993) found that the observed ionization stratification in Wolf-Rayet winds can significantly increase the theoretical efficiency of the momentum transfer by multiple scattering as compared to the constant ionization case. The smooth transition from WNL to O stars found here may hint at a similar mechanism responsible in O-star winds.

Kudritzki (private communication and this meeting) found that the wind theory in its current version lacks a significant number of important driving lines. In particular, highly ionized Fe lines are incomplete, leading to a deficit in the line force. This deficit had previously been discovered by Schmutz & Schaerer (1992). Inclusion of the missing lines may at least partially resolve the momentum discrepancy in O stars. In fact, initial results (Pauldrach, this meeting) indicate an upward revision of the theoretical mass-loss rates by up to a factor of 2.

6. Conclusions and Further Directions

The study of a homogeneous sample of mass loss rates from O-stars shows that the observed momentum of stellar winds is larger than predicted. The discrepancy shows a trend with wind-density. The late-WN stars follow this same trend. This suggests that the mechanism that is responsible for the momentum problem in WR stars is already operating in O-stars. Evolutionary calculations based on the *predicted* mass-loss rates will be significantly in error.

The latest improvements of the wind theory reported at this meeting by R. Kudritzki and A. Pauldrach indicate a partial (if not complete) solution of the momentum deficit. If additional driving lines due to iron are included in the models, the theoretical momentum flux increases. A detailed comparison of the next generation of models with the observations will be required to test if the theory of radiatively driven winds will eventually be able to make truly *quantitative* predictions.

7. Acknowledgements

Support for this work was provided by NASA through grant number GO-3663.03-91A from the Space Telescope Science Institute, which is operated by the Association of Universities for Research in Astronomy, Inc., under NASA contract NAS5-26555. H. J. G. L. M. L. gratefully acknowledges support by the STScI Collaborative Visitor Fund.

References

Abbott, D. C., Bieging, J. H., Churchwell, E., & Cassinelli, J. P.: 1980, *ApJ* **238**, 196
Barlow, M. J., & Cohen, M.: 1977, *MNRAS* **213**, 737
Bieging, J. H., Abbott, D. C., & Churchwell, E.: 1989, *ApJ* **340**, 518
Blomme, R.: 1990, *A&A* **229**, 513
Cassinelli, J. P.: 1991, in IAU Symp. 143, Wolf-Rayet Stars and Interrelations with Other Massive Stars in Galaxies, K. A. van der Hucht & B. Hidayat, eds. (Kluwer: Dordrecht), 289
Castor, J. I., Abbott, D. C., & Klein, R. I.: 1975, *ApJ* **195**, 157
Conti, P. S., & Frost, S. A.: 1977, *ApJ* **212**, 728

Garmany, C. D., Olson, G. L., Conti, P. S., & Van Steenberg, M. E.: 1981, *ApJ* **250**, 660

Groenewegen, M. A. T., Lamers, H. J. G. L. M., & Pauldrach, A. W. A.: 1989, *A&A* **221**, 78

Herrero, A., Kudritzki, R. P., Vlchez, J. M., Kunze, D., Butler, K, & Haser, S.: 1992, *A&A* **261**, 209

Howarth, I. D., & Brown, A. B.: 1991, in IAU Symp. 143, Wolf-Rayet Stars and Interrelations with Other Massive Stars in Galaxies, K. A. van der Hucht & B. Hidayat, eds. (Kluwer: Dordrecht), 315

Howarth, I. D., & Prinja, R. K.: 1989, *ApJS* **69**, 527

Kudritzki, R. P., Hummer, D. G., Pauldrach, A. W. A., Puls, J., Najarro, F., & Imhoff, J.: 1992, *A&A* **257**, 655

Kudritzki, R. P., Pauldrach, A., Puls, J, & Abbott, D. C.: 1989, *A&A* **219**, 205

Lamers, H. J. G. L. M., & Leitherer, C.: 1993, *ApJ* **412**, 771

Lamers, H. J. G. L. M., & Morton, D. C.: 1976, *ApJS* **32**, 715

Lamers, H. J. G. L. M., & Waters, L. B. F. M.: 1984, *A&A* **136**, 37

Lamers, H. J. G. L. M., Waters, L. B. F. M., & Wesselius, P. R.: 1984, *A&A* **134**, L17

Leitherer, C.: 1988a, *ApJ* **326**, 356

Leitherer, C.: 1988b, *ApJ* **334**, 626

Leitherer, C., & Robert, C.: 1991, *ApJ* **377**, 629

Lucy, L. B., & Abbott, D. C.: 1993, *ApJ* **405**, 738

Pauldrach, A. W. A., Kudritzki, R. P.,Puls, J., & Butler, K.: 1990, *A&A* **228**, 125

Prinja, R. K.: 1992, in Nonisotropic and Variable Outflows from Stars, L. Drissen, C. Leitherer, & A. Nota, eds. (Brigham Young University: Provo), 167

Schaerer, D., & Schmutz, W.: 1993, *A&A* , in press

Schmutz, W., & Schaerer, D.: 1992, in The Atmospheres of Early-Type Stars, U. Heber & C. S. Jeffery, eds. (Springer: Berlin), 409

Schmutz, W., Hamann, W.-R., & Wessolowski, U.: 1989, *A&A* **210**, 236

Sellmaier, F., Puls, J., Kudritzki, R. P., Gabler, A., Gabler, R., & Voels, S. A.: 1993, *A&A* **273**, 533

LTE AND NLTE ABUNDANCES IN A-SUPERGIANTS
A TEST OF THEIR EVOLUTIONARY STATUS

KIM A. VENN

Univ. of Texas at Austin, Dept. of Astronomy, RLM 15.308, Austin, TX, 78712, USA

1. Motivation

The A-type supergiants occupy an interesting region on the HR-diagram, where theories of the evolution of 10–20 solar mass stars differ. The differences between the scenarios depend on the input assumptions for several physical parameters, *e.g.*, treatments for convection parameters and mass loss rates. A review of the predictions of evolution calculations for 10–30 solar mass stars is given by Fitzpatrick & Garmany (1991).

If an A-supergiant has evolved through the red supergiant phase, then deep surface convection is predicted to mix gas from the hydrogen-burning layers to the observable photosphere (the first dredge-up). Since hydrogen-burning is dominated by the CNO-cycle in these stars, then the surface CNO abundances may be altered in a discernible way.

As a dissertation project, I have calculated the CNO and metal abundances in a sample of bright A-supergiants (A0-F0). Preliminary NLTE results for 22 stars are presented here. In an attempt to avoid severe problems with departures from LTE, I have chosen low luminosity A-supergiants, types Ib-II, although four Iab and Ia stars are included. Many of these stars are MK standards, some are members of open clusters, and a few are known binaries.

2. LTE Analysis

Abundance results are presented for 22 massive ($\sim$ 10–15 $M_\odot$), Population I, Galactic, A-supergiants. Equivalent widths of weak carbon, nitrogen, oxygen, and metal lines have been collected from high resolution (about 0.1 Å/pix), high signal-to-noise (about 100) CCD spectra taken at the Mc-Donald Observatory. Atmospheric parameters (T_{eff} and gravity) have been determined from spectroscopic indicators (Hγ line profiles, ionization equilibrium of elements such as Mg I/Mg II, and in some cases $\log\epsilon$(Fe) versus χ).

An initial analysis of the data set has been made assuming LTE and adopting ATLAS9 line-blanketed model atmospheres (includes the new Kurucz ODFs). From this analysis, the A-supergiants have roughly solar metal abundances (*i.e.*, [M/H]= 0.0 $\pm$ 0.3).

Space Science Reviews **66**: 163–168, 1994.

© 1994 *Kluwer Academic Publishers. Printed in Belgium.*

Abundance results for the light elements, C, N, and O, suggest extensive mixing of CNO-processed gas. However, the N abundances are often much larger than is typically predicted for the first dredge-up ([N]~0.5). Also, the N abundances appear correlated with temperature as shown in Figure 1, whereas this is not clearly seen for any other element. This would imply a systematic error in the abundance of this element. NLTE corrections for N I have been estimated as large in A-F supergiants (~ -0.5 to -1.0) by Sadakane, Takeda & Okyudo (1993) and Luck & Lambert (1985).

3. NLTE Nitrogen and Carbon Analyses

NLTE corrections for nitrogen have been carried out using the detailed statistical equilibrium code developed at Kiel University (Steenbock & Holweger 1984). This code has been updated by Dr. M. Lemke to include the new opacity distribution functions from Kurucz (1991).

We have compiled atomic data for neutral and singly ionized nitrogen to develop a model atom to be used with the Kiel code (Lemke & Venn 1994). The model comprises 88 levels of N I, which includes all known terms of N I to n=8 at 0.11 eV below the ionization limit (14.53 eV). An additional 5 lower levels of N II are included up to excitation energy 25.97 eV. A total of 111 line transitions are treated explicitly. Transition probabilities and photoionization cross-sections are from the Opacity Project (OP) calculations. Multiplet structure has been neglected. Collisional excitation cross-sections have been taken from the formulae of Auer & Mihalas (1973). Collisional ionization cross-sections from the formula of Mihalas (1978). This input data has been examined to determine the sensitivity of the departure coefficients and subsequent NLTE corrections (see Table 1). *We consider this model atom basically complete, although there may be small adjustments after we complete the calibration of Vega's N I spectrum.*

The NLTE corrections for N I lines range from -0.2 to -1.4 depending on equivalent width and atmospheric parameters, *e.g.*, even weak lines have large corrections in early A-supergiants. The corrected N abundances are independent of temperature as seen in Figure 1, and appear roughly solar.

NLTE corrections for C have been calculated using the Stürenburg & Holweger (1990) neutral carbon model atom with the Kiel code. This atom comprises 83 levels of C I and the 5 lowest terms of C II. A total of 63 line transitions are treated explicitly; transition probabilities were compiled from the literature. Photoionization cross-sections were taken from Hofsaess (1979), with the exception of the three lowest levels (from Hofmann *et al.* 1983, and Mathisen 1984). Electron collision cross-sections were calculated from the Drawin (1967) formula. The sensitivities of the input data for this model have not yet been tested for A-supergiants, and therefore *all NLTE corrections for the carbon abundances should be regarded as preliminary.*

The NLTE C corrections range from −0.2 to −2.0 depending mostly on the equivalent width (weak lines are hardly effected). However, the new C abundances appear to be related to temperature as seen in Figure 1. This may indicate a sensitivity in the atomic data of the carbon model in these extreme stellar environments. Also, the corrected C abundances are quite low ($<[C]>\sim -0.6$), much more depleted than first dredge-up predictions ($[C]\sim -0.15$).

4. Evolutionary Status

The corrected nitrogen abundances appear to be roughly solar, which would imply that these stars have not undergone the first dredge-up mixing event, and may have evolved directly from the main-sequence. There is also a hint of *partial* mixing of CN-processed gas, as found by Gies & Lambert (1992) and Lennon (this conference) for the B-supergiants; this is because $[N]<0.5$ (the first dredge-up prediction) but often slightly above solar.

The N/C ratios are more similar to the first dredge-up predictions, however, due to the large (uncertain) C depletions; see Figure 1. The solar-like N abundances could be in agreement if the metallicities are less than solar by ~ 0.2 to 0.5 dex since [N/Fe] would then be larger than solar. The lower limit correction in metallicity (0.2 dex) is within the uncertainties of the LTE analysis for most stars, but the upper limit (0.5 dex) is too large to be due to uncertainties, and therefore this possibility is unlikely.

In Figure 2, the N/C ratios range from solar to beyond first dredge-up abundances when plotted against N/Fe. Even though the uncertainty for any one star on this plot is large (note the estimated error bar), the majority of the NLTE abundances are near the first dredge-up values. However, if the C NLTE corrections are systematically too large, then the stars range only from solar to about the first dredge-up values. Therefore, mixing of CN-processed gas is indicated, but not necessarily first dredge-up abundances.

5. Conclusions

Departures from LTE have an important effect on the level populations of the nitrogen atom in A-supergiant atmospheres, and must be considered when determining the photospheric N abundance.

The carbon atom may be significantly effected by departures from LTE in the atmospheres of early A-supergiants, but this result is preliminary and our NLTE C corrections may be too large.

The evolutionary status of the A-supergiants continues to be difficult to determine due to uncertainties in the NLTE calculations for C and N. Also, the uncertainties for any individual stellar abundance are quite large, such that discussions of the amount of CN-mixing are difficult. At present however, the N NLTE results of near solar abundances suggests that these stars may have *not* undergone the first dredge-up.

Acknowledgements

I would like to thank Michael Lemke for excellent instructions and advice on working in NLTE, and Drs. David Lambert and Chris Sneden for many helpful discussions and support. I also wish to acknowledge an International Travel Grant from the American Astronomical Society to attend this meeting, and additional travel funds from the McDonald Observatory Stars Group.

TABLE I

NLTE and LTE Uncertainties for N I

		HD 87737	HD 34578	HD 36673
T_{eff} (K) :		9800	8300	7500
Log g :		2.0	1.8	1.15
number of N I lines included :	NLTE	8	9	8
	LTE	3	5	1
σ (line-to-line scatter) :	NLTE	± 0.09	± 0.16	± 0.14
	LTE	± 0.12	± 0.27	± 0.0
$\Delta\, T_{eff} = +200K$:	NLTE	+0.02	−0.03	−0.07
	LTE	+0.07	+0.05	−0.08
$\Delta\, logg = -0.2$:	NLTE	0.00	−0.02	−0.04
	LTE	+0.04	+0.05	−0.04
$\Delta\, \xi = -1\ kms^{-1}$:	NLTE	+0.06	+0.07	+0.15
	LTE	+0.09	+0.11	+0.11
Photo. and Coll. Ion. X-sections :				
OP vs Hydrogenic		+0.09	+0.03	+0.01
Coll. Exc. X-sections :				
allowed trans., $f_{ij}*3$		+0.17	+0.07	+0.04
allowed trans., Drawin form.		−0.06	−0.04	+0.02
forb. trans., $\Omega*10$		+0.03	−0.03	−0.01
forb. trans., Drawin form.		+0.08	+0.04	−0.01
input $log\epsilon(C) = 8.0$:		−0.15	−0.06	+0.08
ATLAS6 ODF :		+0.07	+0.01	0.00

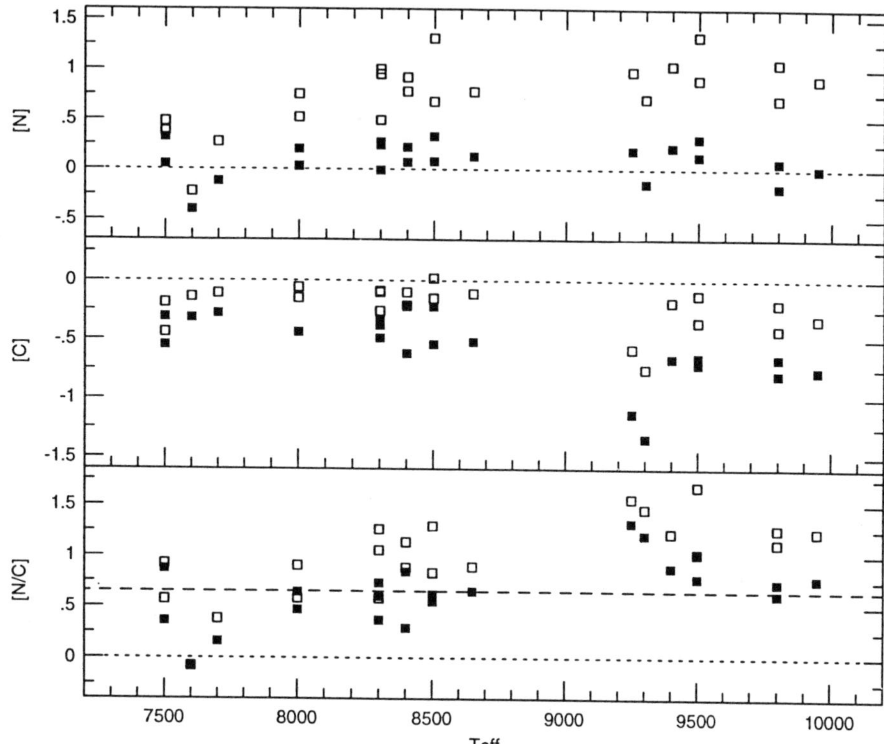

Fig. 1.   LTE (*hollow squares*) and NLTE (*filled squares*) calculations of the C and N abundances versus temperature. Dotted lines represent solar abundances. The dashed line represents the approx. first dredge-up abundances

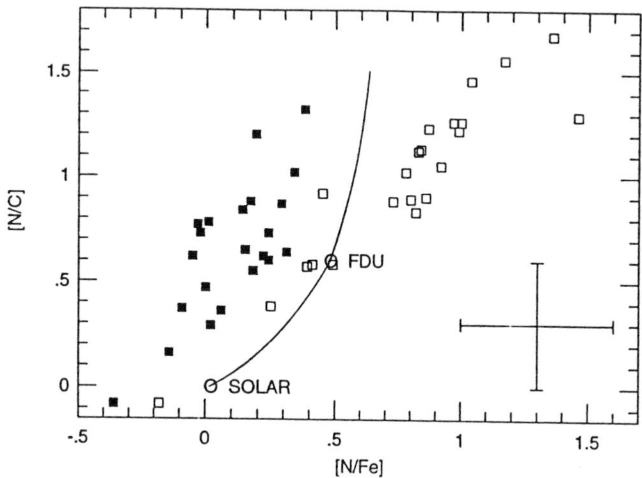

Fig. 2. N/C vs N/Fe for A-supergiants in LTE (*hollow squares*) and NLTE (*filled squares*). The solar and first dredge-up abundances are indicated along the loci of CNO-processed gas abundances. The error bar indicates the typical random uncertainties per star.

 K. A. VENN

References

Auer L.H., Mihalas D.: 1973, *ApJ* **184**, 151
Drawin H.W.: 1967, 'Collision and Transport Cross-Sections, Association Euratom',
    C.E.A., EUR-CEA-FC-383.
Fitzpatrick E.L., Garmany C.D.: 1991, *ApJ* **363**, 119
Gies D.R., Lambert D.L.: 1992, *ApJ* **387**, 673
Hofmann H., Saha H.P., Trefftz E.: 1983, *A&A* **126**, 415
Hofsaess D.: 1979, *Atomic Data and Nuclear Data Tables* **24**, 285
Kurucz R.: 1991, private communication.
Lemke M., Venn K.A.: 1994, *ApJ*, in prep.
Luck R.E., Lambert D.L.: 1985, *ApJ* **298**, 782
Mathisen R.: 1984, 'Photo Cross-Sections for Stellar Atmosphere Calculations' Inst. of
    Theoretical Astrophysics, Univ. of Oslo, Publ. Series No. 1.
Mihalas D.: 1978, *Stellar Atmospheres, Second Edition*, W.H. Freeman and Company: San
    Francisco
Sadakane K., Takeda Y., Okyudo M.: 1993, *PASJ*, in press.
Steenbock W., Holweger H.: 1984, *A&A* **130**, 319
Stürenberg S., Holweger H.: 1990, *A&A* **237**, 125

A SPECTROSCOPIC ANALYSIS OF B STARS IN THE SMC CLUSTER
NGC 330

D.J. LENNON
Universtatssternwarte Munchen, D-81679 Munchen, Scheinerstrasse 1, Federal Republic of Germany

PAOLO A. MAZZALI and F. PASIAN
Osservatorio Astronomico di Trieste, Via G.B.Tiepolo, 11, I-34131 Trieste, Italy

P. BONIFACIO
Scuola Internazionale di Studi Avanzati (SISSA), Strada Costiera 11, I-34131 Trieste, Italy

and

V. CASTELLANI
Istituto di Astronomia, Universita di Pisa, Piazza Torricelli, I-56100 Pisa, Italy

Abstract. Medium resolution (2A/px) but high s/n spectra of approximately twenty of the brightest blue stars in the young open cluster NGC 330 in the SMC have been analyzed in order to determine their atmospheric parameters and the evolutionary status. Stellar parameters are determined by comparison with LTE and NLTE model atmosphere calculations and an HR diagram constructed. Luminosities of the sample stars lie in the range $4.0 < \log(L_*/L_\odot) < 5.0$ and spectral types between O9 and late-B. The stars in our sample appear to define 4 groups: main-sequence B-stars (B2-B4), B-supergiants (B4) in a blue-loop phase of evolution, a small number of blue stragglers (O9-B0 near main-sequence stars) and a group of luminous giants (B1-B2) which reside in the so-called 'post main-sequence gap' of the HR diagram. Furthermore, we have confirmed spectroscopically the very high incidence of Be stars in this cluster. Finally the almost complete absence of metal lines (at this resolution) is in keeping with the expected very low metallicity of the SMC.

Key words: NGC330 − B-stars − abundances − evolution

1. Introduction.

Star clusters in the Magellanic Clouds have come under intense scrutiny in recent years offering as they do a number of excellent advantages compared to galactic clusters; well determined distances, very populous, low reddening, small angular size facilitating CCD photometry and a range of metallicities. NGC330 is one of the better studied young clusters in the Small Magellanic Cloud (SMC) with photometric studies by Arp (1959), Robertson (1974), Carney et al (1985), Balona (1992) and Grebel & Richtler (1992). Further studies of Hα emission in the cluster have revealed the presence of a large number of main-sequence B-stars with Balmer emission (Feast 1972, Grebel et al 1992). There have also been recent estimates of the metallicity based upon spectroscopy of the brightest K and F supergiants (Spite et al 1991) and one B-giant (Reitermann et al 1990). Despite this, precise estimates of the stellar parameters for the B-stars in this cluster, which are necessary for a detailed comparison with evolutionary calculations, are unavailable. Note that the estimates based upon UV continuum fitting (Caloi et al 1993) make an assumption concerning the surface gravity or reddening.

2. Observations and analysis

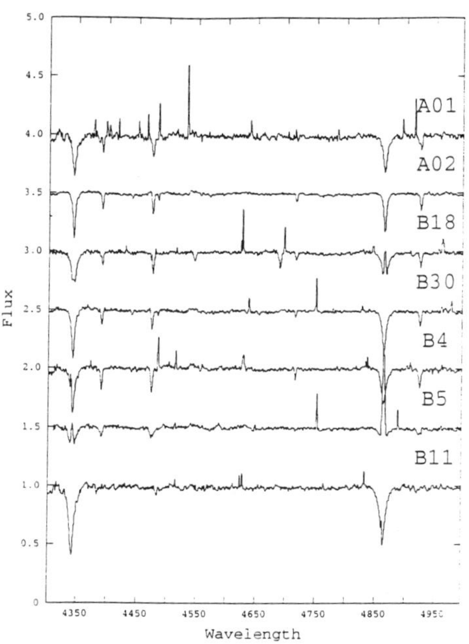

Fig. 1.   Sample EFOSC spectra of B-stars in NGC330, for spectral types and stellar parameters see Table I.

In October 1992 approximately 20 B-stars in NGC330 were observed by us using the 3.6m telescope and EFOSC at a resolution of approximately 2A the data covering almost the complete visible wavelength range. The s/n obtained was typically 100, even for the fainter V=15.5 B-stars and these data were reduced in the Midas environment. In our preliminary analysis we have used Kurucz LTE model atmospheres deriving atmospheric parameters from fits to the hydrogen Balmer line profiles, to the He I lines (for a few hotter stars we also used the He II lines) and in addition using the UV energy distribution to constrain the model fit. We emphasize here that the surface gravities are constrained primarily by fitting the wings of the Balmer lines. In most of our targets, Hα emission is present and we estimated values of the rotational value from this feature. Our results are summarized in Table I.

3. Discussion

In Fig. 1 we have constucted a physical HR diagram and, for comparison, we have also plotted the positions of the evolutionary tracks of Schaller et al (1992)

TABLE I

B-stars observed in NGC330 together with derived parameters and estimated spectral types. Values of v_{rot} are derived from Hα emission.

STAR	T_{eff} (K)	$\log g$	M_{bol}	$\log(L/L_\odot)$	Type	v_{rot}(kms^{-1})
A01	29000± 1000	4.25± 0.20	-7.10	4.74	B0.5 Ve	125
A02	16000± 1000	2.50± 0.20	-7.29	4.86	B4 Iab/b	–
B04	25000± 1000	3.90± 0.20	-5.82	4.23	B1.5 IVe	180
B05	22000± 1000	3.50± 0.25	-5.63	4.16	B2 III/IVe	300
B06	22000± 1000	3.50± 0.30	-5.78	4.21	B2 IIIe	300
B07	22000± 3000	3.70± 0.30	-5.46	4.09	B2 III/IVe	160
B11	12000± 2000	3.25± 0.50	-4.15	3.56	B7 II/III	–
B12	22000± 2000	3.50± 0.30	-5.80	4.22	B2 IIIe	300
B13	22000± 2000	3.60± 0.40	-5.43	4.07	B2 III/IVe	160
B16	10000± 1500	2.60± 0.50	-5.29	4.02	B9 Ib/II	–
B18	32000± 1000	4.50± 0.20	-6.48	4.49	B0 Ve	180
B21	22000± 1000	3.00± 0.25	-6.71	4.59	B1.5 II/IIIe	320
B22	20000± 1000	3.20± 0.20	-6.56	4.53	B2 IIe	150
B24	25000± 3000	3.90± 0.50	-6.28	4.42	B1 IVe	150
B28	30000± 3000	4.40± 0.40	-6.10	4.34	B0 Ve	125
B30	20500± 1000	3.25± 0.25	-6.67	4.57	B2 II	–
B35	25000± 5000	3.50± 0.50	-6.19	4.38	B1.5 III/IVe	400
B37	18000± 1000	2.60± 0.20	-7.27	4.81	B3 Ib	–

for a metallicity of z=0.001. We can see immediately that there are a number of problems in interpretation of this figure. If we assume that B37, A02 and the B21, B30 and B22 group are evolved post red supergiant stars (the blue loops perhaps being bluer than is indicated by the tracks), then we have a number of blue straggler stars (A01 and B18). In fact a preliminary analysis of A01 indicates that this star is helium rich, which would be consistent with a prolonged main-sequence lifetime due to some mixing process (rotationally induced perhaps). The group of stars around B05 would appear from their stellar parameters to be main-sequence, although a high proportion have Hα emission, yet appear to be too red for the tracks considered here. Stars B11 and B16 are considered cluster members in previous work however their positions are difficult to reconcile with the apparent youth of NGC330, these may be field stars in the SMC. Finally, we note that almost all the stars considered here (see Table I) have Balmer emission, even some stars designated as being without Hα emission from photometric studies (Grebel et al 1992) exhibit weak, but definite, emission in our spectrograms. Further high resolution observations are highly desirable, in particular for the determination of metal abundances in these stars.

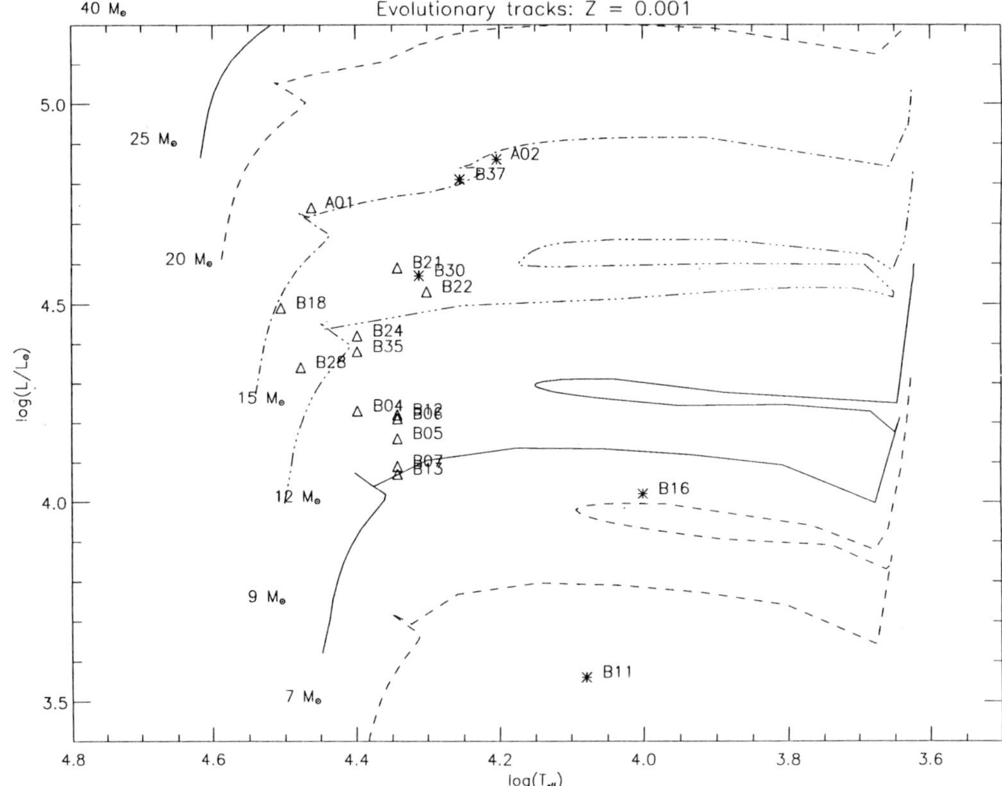

Fig. 2. HR diagram for NGC330, evolutionary tracks from Schaller et al (1992) for z=0.001. Asterisks denote stars with no clear Balmer emission.

Acknowledgements

DJL wishes to thank the BMFT for financial support under grant number 010R90080.

References

Arp, H.: 1959, *AJ* **64**, 254
Balona, L.A.: 1992, *MNRAS* **256**, 425
Caloi V., Cassatella, A., Castellani, V., Walker, A.: 1993, *A&A* **271**, 109
Carney, B.W., Janes, K.A., Flower, P.J.: 1985, *AJ* **90**, 1196
Feast, M.W.: 1972, *MNRAS* **159**, 113
Grebel E.K., Richtler, T.: 1992, *A&A* **253**, 359
Grebel E.K., Richtler, T., de Boer, K.S.: 1992, *A&A* **254**, L5
Reitermann, A., Baschek, B., Stahl, O., Wolf, B.: 1990, *A&A* **234**, 109
Robertson, J.W.: 1974, *A&AS* **15**, 261
Schaller, G., Schaerer, D., Meynet, D. and Maeder, A.: 1992, *A&AS* **96**, 269
Spite, F., Richtler, T., Spite, M.: 1991, *A&A* **252**, 557

HYDRODYNAMIC ATMOSPHERE MODELS FOR HOT LUMINOUS STARS II. METHOD AND IMPROVEMENTS OVER UNIFIED MODELS

D. SCHAERER

Geneva Observatory , CH-1290 Sauverny, Switzerland

and

W. SCHMUTZ

Institut fur Astronomie, ETH Zentrum, CH-8092 Zurich, Switzerland

Abstract. In a paper submitted to A&A we present the first line blanketed hydrodynamic models of spherically expanding atmospheres of hot stars. This paper is complementary to the submitted paper. Here, we emphasize the advantages and the weak points of our approach and we present additional technical aspects.

The models are characterised by a simultaneous solution of the equation of motion, the non–LTE populations of H and He, and radiation transfer in a line blanketed atmosphere. The entire domain from the optically thick photosphere out to the terminal velocity of the wind is treated. The radiative forces are evaluated consistently with the depth-dependent radiation field, taking into account multiple scattering by metal lines and line overlap.

Key words: Stars: atmospheres – Stars: mass loss – Stars: early-type

1. Method

Four stellar parameters describe our models: T_{eff}, R, the mass M, and the chemical composition. A model atmosphere is computed in three steps, which are iterated for convergence.

1) Simultaneous solution of line blanketed radiative transfer and statistical equilibrium of non–LTE populations in the comoving frame.

2) Multi-line transfer to determine line blanketing factors averaged over a given wavelength and compute the radiative forces (Monte Carlo simulation).

3) Simultaneous solution of the equation of motion and temperature structure to yield the density structure and velocity field (Hydrodynamic code).

The results shown in this paper are from a model with stellar parameters similar to that of ζ Puppis (Paper I, model A): T_{eff}= 42000 K, $R/R_{\odot} = 17$, $\log L/L_{\odot} = 5.91$, $[Y] = 0.2$. The resulting wind has v_{∞}= 2200 km s^{-1} and $\dot{M}$= 3. 10^{-6} M$_{\odot}$yr^{-1} for $M = 35\ M_{\odot}$.

1.1. STRENGTHS OF THE MODEL

The difference to conventional atmosphere models for expanding atmospheres (e.g. Hamann & Schmutz 1987) is the inclusion of line blanketing. The line blanketing is incorporated by intensity–weighted effective mean scattering and absorptive opacity coefficients, which are evaluated by a Monte–Carlo (MC) technique for a wavelength band of 50 A (see Sect. 1.2). These then account for the effects of hundreds or thousands of spectral lines within this band.

Space Science Reviews **66**: 173–177, 1994.

© 1994 *Kluwer Academic Publishers. Printed in Belgium.*

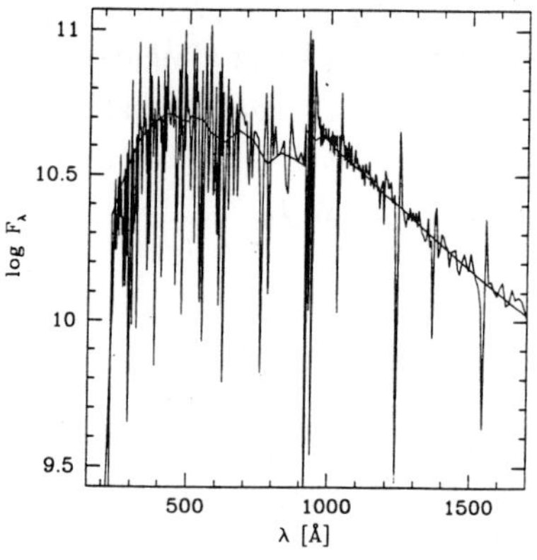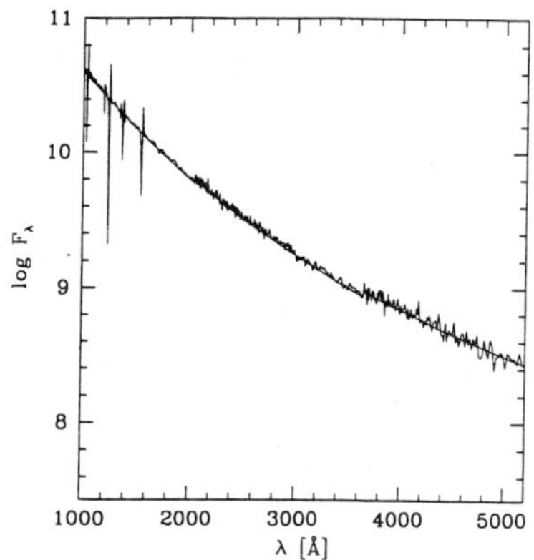

Fig. 1. Logarithm of the astrophysical flux from the line blanketed non–LTE model (continuum flux) compared to the MC simulation including $\approx$ 40000 metal lines.

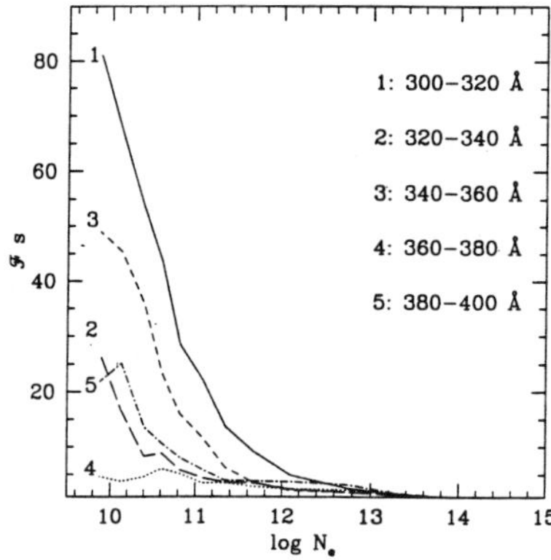

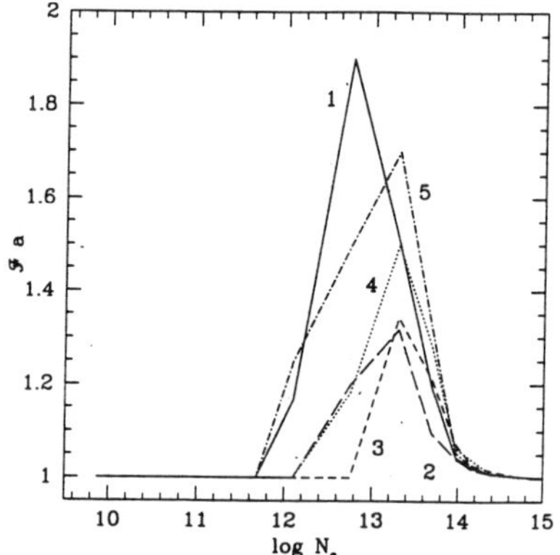

Fig. 2. Scattering (left panel) and absorptive opacity factors (right) as a function of depth (electron density) and wavelength band in the interval 300-400 A (see text).

The MC calculation performs a formal solution of the equation of transfer in the differentially expanding flow, allowing for scattering by electrons and lines, line absorption, and continuum emission/absorption processes. The MC emergent UV flux is shown in Fig. 1 together with the continuum flux obtained from the non–LTE calculation. This shows the perfect agreement between the two completely independent methods for the radiation transfer.

1.2. BLANKETING FACTORS

An essential result of the MC calculation are the evaluation of mean opacity coefficients, which allow us to treat line blanketing. For the strongly blanketed interval 300-400 A, Fig. 3 shows the factors $\mathcal{F}_s(N_e, \lambda) = \frac{\mathcal{N}_s(N_e,\lambda)+\mathcal{N}_{ls}(N_e,\lambda)}{\mathcal{N}_s(N_e,\lambda)}$, and $\mathcal{F}_a(N_e, \lambda) = \frac{\mathcal{N}_c(N_e,\lambda)+\mathcal{N}_{la}(N_e,\lambda)}{\mathcal{N}_c(N_e,\lambda)}$, where $\mathcal{N}_s$ and $\mathcal{N}_{ls}$ are the numbers of electron and line scatterings, respectively, and $\mathcal{N}_c$ and $\mathcal{N}_{la}$ are the numbers of continuum and line absorption processes (the latter treated according to Puls & Hummer

1988). These ratios are counted as a function of wavelength and electron density.

Besides the continuum radiation field also the line transfer is affected by line blanketing. Since the lines sample a much smaller interval than the band width used to evaluate the mean blanketing factors, our treatment is correct only to first order. Test calculations with smaller band width (20 A instead of 50 A) showed no significant change in the shape and strength of the line profiles. However, we note that by assuming either that there is no blanketing for an individual spectral line ($\mathcal{F}_{s,a}(N_e, \lambda) = 1$) or, on the other extreme, assuming a much larger blanketing factor ($\times 10$) for say, the HeII Ly_α transition $\lambda 304A$, the results would be changed non-negligibly. But even with a much stronger blanketing for the HeII Ly_α we did not find changes large enough to affect the ionisation condition of our model. The ionisation condition at large distances from the star is determined by ionisation from the ground state by the far UV radiation. Therefore, we question the recent suggestion by Hillier et al. (1993) that recombination in the wind of ζ Puppis takes place because of the blanketing effect on the continuum at 304 A.

1.3. WEAKNESS OF THE MODEL

In the MC calculation we use the non–LTE opacities and emissivities. For the metals however, assumptions about ionisation and excitation (see Paper I) have to be made. For the ionisation, which is more important than excitation, we adopt a modified nebular approximation, which takes non–LTE conditions in the stellar wind into account. The formula used to calculate the ionisation equilibrium is (cf. Schmutz 1991):

$$\frac{N_{j+1} N_e}{N_j} = ((1 - \zeta)W + \zeta)W \left(\frac{T_e}{T_R}\right)^{1/2} \left(\frac{N_{j+1} N_e}{N_j}\right)^{Saha-Boltz.}_{T_R(r,IP)} \tag{1}$$

The radiation temperature $T_R(r, IP)$ (at location r and for the ionisation potential IP of ion j) is obtained by inversion of Eq. 1 using the ionisation ratios of the elements computed in the non–LTE atmosphere (i.e. H and He). ζ is the fraction of recombinations going directly to the ground state (cf. Abbott & Lucy 1985, hereafter AL85), and $W = 1 - \frac{1}{4}F/J$ the dilution factor defined by Schmutz et al. (1990).

The advantage of this method is that it overcomes the severe underestimate of the ionisation by other nebular approximations (Abbott 1982, AL85, Schmutz et al. 1990). This can be seen from Fig. 3, which shows the deviations of the ionisation fractions of Ne, Ar and Fe from Eq. 11 of AL85. A comparison with detailed statistical equilibrium calculations of Pauldrach (1987) shows good agreement. Similarly the ionisation fractions of CIII, CIV, and NV computed with Eq. 1 (see Paper I, Table 4), agree well with a model where we have perfomed statistical equilibrium calculations for H, He, C, N, and O.

An empirical evaluation of the the ionisation and excitation conditions by comparing the predicted UV flux from the MC calculation with observations (Paper

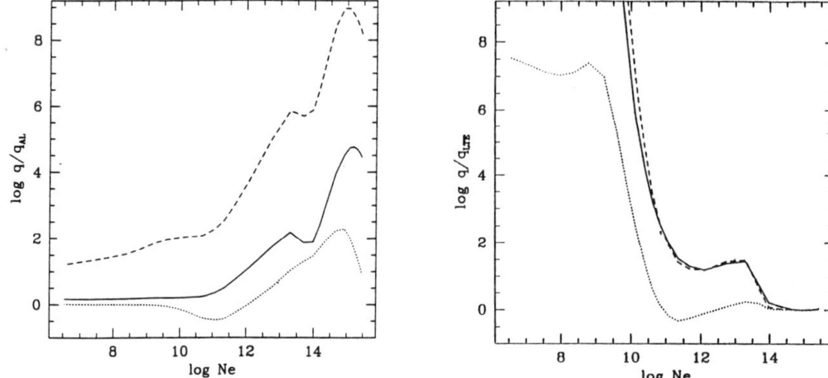

Fig. 3.   Left: Logarithm of the ionisation fraction q_i computed with Eq. 1 using the depth dependent T_R (see text) divided by the value obtained with the nebular approximation from AL85 (Eq. 11) using $T_R = \text{const} = 36000$ K as in their standard model. Right: Same as a but divided by the LTE value. The electron density is used as a depth parameter, where $\tau_R = 1$ is at $\log N_e \approx 14.5$. The curves are given for FeVI (solid), ArVI (dashed), and NeIV (dotted). The small dip at $\log N_e \approx 14$ is due to a decrease of the ionisation temperature $T_R(\text{He}^+)$ (cf. Paper I, Fig. 14).

I) shows basic agreement of the main features. However, since our calculations agree with other theoretical predictions (Pauldrach 1987; Drew 1989), our results are subject to the same disagreements pointed out by Groenewegen & Lamers (1991) between observed and theoretical ionisation fractions.

2. Differences to the unified models

The two major differences to non–LTE model atmospheres from the Munich group are the following:

1. The effects of line blanketing on the radiation transfer and statistical equilibrium equations are included in the atmosphere calculations — this is a significant improvement.

2. The ionisation and excitation equilibrium of the metals are specified by a modified nebular approximation, which takes non–LTE conditions in the stellar wind into account — this is a disadvantage of our approach but in Sect. 2.2 we have shown that this simplification is roughly correct.

   In contrast to the so called "unified models" by Gabler et al. (1989) we do not take fixed values for the line force parameters k, α, δ. Instead, in our models the radiative force is evaluated using the depth–dependent radiation field of the model atmosphere. We also account for multiple scattering and line overlap and solve for the dependence of the radiative forces on the density structure, ionisation rate and the velocity field (see Paper I).

   Since the inclusion of line blanketing has a significant effect on the statistical equilibrium equations and consequently on the predicted line profiles, we conclude that our new approach results in an overall improvement of the atmosphere models for hot stars.

References

Abbott D.C., 1982, ApJ 259, 282
Abbott D.C., Lucy L., 1985, ApJ 288, 679
Drew J.E., 1989, ApJS 71, 267
Gabler R., Gabler A., Kudritzki R.P., Puls J., Pauldrach A., 1989, A&A 226, 162
Groenewegen M.A.T., Lamers H.J.G.L.M., 1991, A&A 243, 429
Hamann W.-R., Schmutz W., 1987, A&A 174, 173
Hillier D.J., Kudritzki R.P., Pauldrach A., Baade D., Cassinelli J., Puls J., Schmitt J.H.M.M., 1993, A&A 276, 117
Pauldrach A., 1987, A&A 183, 295 125
Puls J., Hummer D.G., 1988, A&A 191, 87
Schaerer D., Schmutz W., 1993, A&A submitted (Paper I)
Schmutz W., 1991, in "Stellar Atmospheres: Beyond Classical Model s", Eds. Crivellari, L., Hubeny, I., Hummer, D.G., NATO ASI Series C, Vol. 341 , p. 191
Schmutz W., Abbott D.C., Russell R.S., Hamann W.-R., Wessolowski U ., 1990, ApJ 355, 255

HUBBLE SPACE TELESCOPE SPECTROSCOPY OF MASSIVE HOT STARS IN THE MAGELLANIC CLOUDS AND M31

S.M. HASER, D.J. LENNON, R.-P. KUDRITZKI* and J. PULS
Universitatssternwarte Munchen, D-81679 Munchen, Scheinerstrasse 1, Federal Republic of Germany

N.R. WALBORN and L. BIANCHI**
Space Telescope Science Institute, 3700 San Martin Drive, Baltimore, MD 21218, USA

and

J.B. HUTCHINGS
Dominion Astrophysical Observatory, NRC of Canada, 5071 West Saanich Road, Victoria, B.C., Canada, V8X4M6

Abstract. Using the Hubble Space Telescope (HST) and the Faint Object Spectrograph (FOS) high signal to noise spectrograms were obtained for 15 OB stars in the Magellanic Clouds *** , three of which are of spectral type O3. The data cover the spectral region from 1150 A – 2300 A with a resolution of $\Delta\lambda/\lambda \approx 1$ A. One O8.5 supergiant, OB78#231, in M31 [‡]is also included in this work. These data are a substantial improvement on previous high resolution IUE observations in the Magellanic Clouds (Walborn et al. 1985 and references therein) because of the smaller aperture and the much better signal to noise ratio, while no high resolution UV spectra of O stars in M31 have been obtained before. In this paper we discuss various morphological aspects of the spectra, concerning metallicity and the stellar winds, compared to galactic analogues.

Key words: Hot stars – extragalactic UV spectroscopy – metallicity

1. Introduction and procedure

The presented dataset of OB-stars, although still incomplete (three of the hottest objects in the SMC have not yet been observed successfully), will provide a means of determining the dependence of mass loss rate upon metallicity for these most luminous stars using the theory of radiation driven winds (see Pauldrach et al. 1993 and references therein, Kudritzki et al. 1992), the metallicity being a most influential parameter affecting the evolution of such massive stars. In this paper we will discuss the spectral morphology via comparison with previous IUE observations of galactic O-stars (data from Howarth & Prinja 1989). For this purpose the following steps of data processing must be performed beforehand:

* SMH & RPK affiliated to Max-Planck-Inst. fur Astrophys., Karl-Schwarzschild-Str. 1, D-85748 Garching bei Munchen, Federal Republic of Germany
** affiliated to Osservatorio Astronomico di Torino, 10025 Pino Torinese (TO), Italy
*** The data were obtained for the HST Cycle 1 program "The Physics of Massive O-Stars in different Parent Galaxies". PI.: R.-P. Kudritzki, co-I's: H.G. Groth, D. Husfeld, K. Butler, S. Voels, J. Puls, A.W.A. Pauldrach, N.R. Walborn, S.R. Heap, B. Bohannan, P.S. Conti, C.D. Garmany, D.G. Hummer, and D. Baade
‡ This spectrum is from the HST GO program P2581. PI.: L. Bianchi, co-I's: J.B. Hutchings, P. Massey, H.J.G.L.M. Lamers, R.-P. Kudritzki

Space Science Reviews **66**: 179–182, 1994.
© 1994 *Kluwer Academic Publishers. Printed in Belgium.*

TABLE I

The sample of SMC/LMC OB stars. Melnick 42 has been observed with the Goddard High Resolution Spectrograph (Heap et al. 1991). Given are: spectral classification (Walborn 1977, 1982, 1983, except for Sk -66° 172, which is reclassified as O3 due to its O v P-Cygni Profile), the terminal velocities v_∞, heliocentric radial velocities v_{rad}, and interstellar H I column densities.

		sp. classif.	v_∞ (km s^{-1})	v_{rad}(km s^{-1})	N_{HI} (cm^{-2})
LMC	Sanduleak -65° 21	B0 Ia	1600	281	$3.5 \cdot 10^{20}$
	Sanduleak -66° 100	O6 III	2150	309	$6.0 \cdot 10^{20}$
	Sanduleak -66° 172	O3 V	3250	303	$2.0 \cdot 10^{21}$
	Sanduleak -67° 166	O4 If	1900	259	$5.0 \cdot 10^{20}$
	Sanduleak -67° 167	O4 Inf$^+$	2150	290	$5.2 \cdot 10^{20}$
	Sanduleak -67° 211	O3 V	3750	264	$1.1 \cdot 10^{21}$
	Sanduleak -68° 41	B0.5 Ia	1050	239	$4.5 \cdot 10^{20}$
	Sanduleak -68° 137	O3 III	3400	273	$3.5 \cdot 10^{21}$
	Sanduleak -70° 69	O4 V	2600	283	$6.5 \cdot 10^{20}$
	Melnick 42	O3 If/WN	3000	261	$6.0 \cdot 10^{21}$
SMC	AV 232	O7 Iaf$^+$	1400	164	$1.3 \cdot 10^{20}$
	AV 238	O9 III	1200	240	$3.0 \cdot 10^{21}$
	AV 243	O6 III	2050	210	$3.4 \cdot 10^{21}$
	AV 388	O4 V	2100	187	$1.9 \cdot 10^{21}$
	AV 488	B0.5 Ia	1300	193	$1.8 \cdot 10^{21}$
	NGC 346#1	O4 III	2650	155	poor signal

The FOS spectra must be corrected for the **radial velocity** Doppler shift. The heliocentric radial velocities, obtained from optical spectra, are given in the table. They are more accurate than those obtained from the C III λ2297 and He II λ2306 lines in the FOS data due to the comparably large instrumental profile of the FOS (see below).

A **degradation** to $\lambda/\Delta\lambda \approx 1400$ (the resolving power of the FOS for the 1"x0.25" aperture) **of the IUE data** is necessary to compare them with the FOS spectra. First, the slope of the blue edges of the P-Cygni profiles becomes shallower and the black troughs of the very strong lines can become desaturated slightly. Second, the shape of the pseudo continuum depends on the resolution when blends of many weak photospheric lines are present, because flux maxima present in stars with small $v \sin i$ are smeared out. However, less crowded regions can be identified to serve as reference points for all stars (for example, the "islands" at 1680 A and 1500A). "Internal consistence" within the sample is achieved by using the same reference points where possible, and the cubic spline representing the continuum should be as smooth as possible. The interstellar Lyman-α damping profile was fitted in the same process, thus also determining the interstellar H I coulumn density and reconstructing the N v P-Cygni profile.

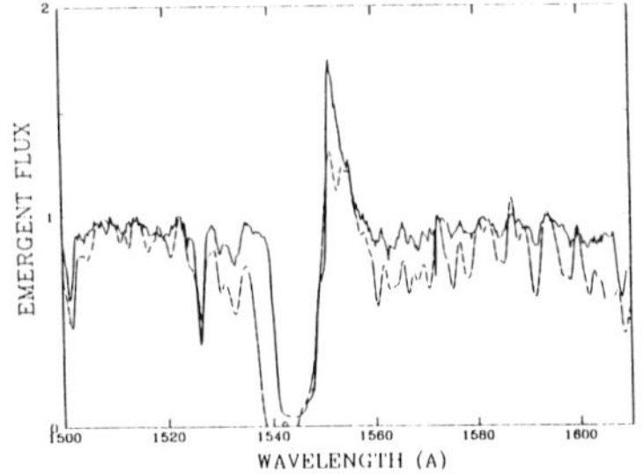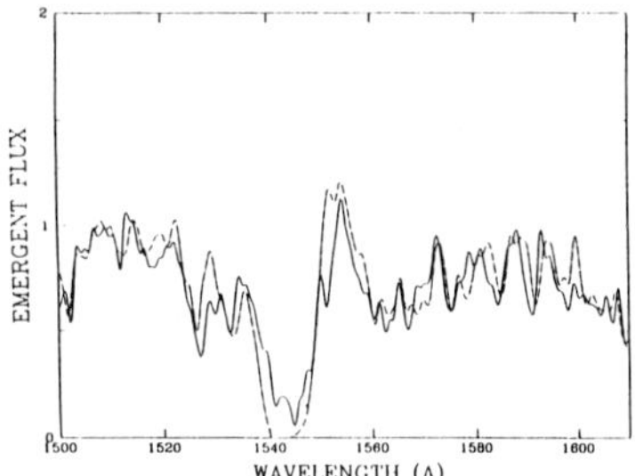

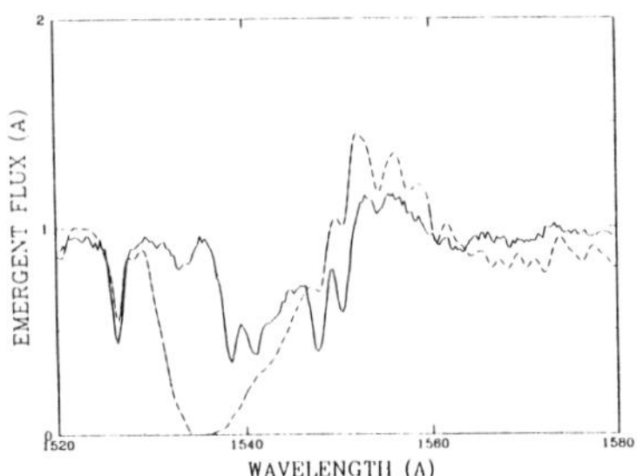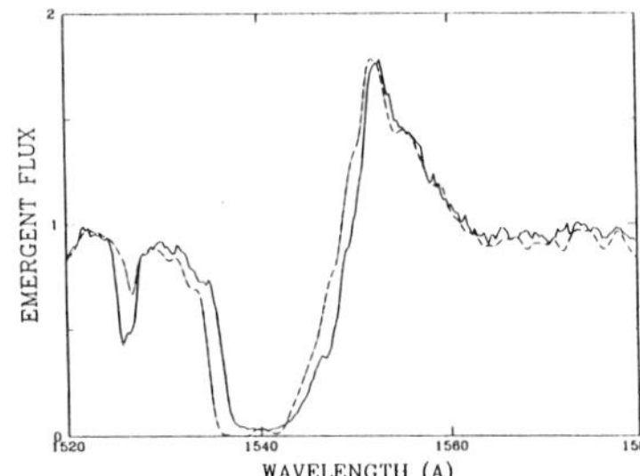

Fig. 1.   Left: full: AV 232 (SMC), dash-dotted: HD 188001 (Galaxy). The desaturation of the FOS data is due to an incomplete background subtraction. Right: full:OB78#231 (M31), dash-dotted: HD 112244 (Galaxy). See text for further explanation.

Fig. 2.   Left: C IV P-Cygni profiles of AV 388 (SMC, full) and 9 Sgr (Galaxy, dash-dotted). Note the extremely weak line in AV 388. Right: the same for Sk -67° 167 (LMC, full) and ζPup (Galaxy, dash-dotted). See text for further explanation.

2. Spectral morphology and conclusions

The strength of the **"pseudo photospheric"** metal line spectrum (mainly Fe/Ni IV and Fe/Ni V, so called because also wind affected, see Kudritzki 1992) is used as a direct, qualitative indicator of metallicity. Fig. 1 shows the comparison of AV 232 (O7 Iaf⁺) in the SMC with HD 188001 (O7.5 Iaf) in the Galaxy. The continuum depression at $\lambda > 1510$ A, typical for late O stars, pronounced in the latter, nearly absent in the former, mirrors the metal poorness of the SMC in the UV by the weakness of the Fe IV lines.

Fig. 1 compares OB78#231 (O8.5 I(f), see Hutchings et al. 1992) in M31 with HD 112244 (O8.5 Iab(f)). No significant difference in the metal line spectrum is found, implying that the metallicity of OB78#231 is not much different from galactic. This observation allows, for the first time, a direct comparison of a high resolution UV spectrum of a – hopefully – single O-star in a different *spiral* galaxy.

Comparisons of the LMC stars with their galactic analogues show, that the photospheric spectra are similar, with only slightly stronger lines found in the galactic stars. Thus in the LMC the metals are less depleted.

In the SMC spectra *all* but one (C IV of AV 232) of the **P-Cygni profiles** of N V, Si IV, and C IV are *un*saturated (see Fig. 2). The lower absorber density can be due to both the lower element abundance and/or the smaller mass loss rate (caused by the smaller metallicity). Also the terminal velocities, given by the blue edges of the P-Cygni profiles (see Haser et al., this proceeding), are lower on avarage (Table I). In the LMC objects these two effects are less evident, especially in the supergiant spectra, where many of the resonance lines are saturated. However, excepting the O3 stars, the terminal velocities are never larger than in the galactic comparison stars, with a slight tendency towards smaller values (see Fig. 2).

Acknowledgements

We thank Ian D. Howarth and Raman K. Prinja for kindly providing us with their sample of 203 IUE spectra of galactic O-stars. SMH acknowledges a grant of the Max-Planck-Institut fur Astrophysik, DJL is grateful for support from the Bundesminister fur Forschung und Technologie under grant 010 R 90080. This work was also supported by the Deutsche Forschungsgesellschaft in the "Gerhard Hess Programm" under grant PA 477/1-1.

References

Heap, S.R., Altner, B., Ebbets, D., Hubeny, I., Hutchings, J.B., Kudritzki, R.-P., Voels, S.A., Haser, S.M., Pauldrach, A.W.A., Puls, J., Butler, K.: 1991, *ApJ* **377**, L29

Howarth, I.D., Prinja, R.K.: 1989, *ApJS* **69**, 527

Hutchings, J.B, Bianchi, L., Lamers, H.J.G.L.M., Massey, P., Morris, S.C.: 1992, *ApJ* **400**, L35

Kudritzki, R.-P.: 1992, *A&A* **266**, 395

Kudritzki, R.-P., Lennon, D.J., Becker, S.R. et al. in: 'Science with the Hubble Space Telescope': ESO Conference and Workshop Proceedings No. 44, eds. P. Benvenuti and E. Schreier, *1992*, p. 279

Science with the Hubble Space Telescope Garching279

Pauldrach, A.W.A., Kudritzki, R.-P., Puls, J., Butler, K., Hunsinger, J.: 1993, *A&A* , in press

Prinja, R.K., Barlow, M.J., Howarth, I.D.: 1990, *ApJ* **361**, 607

Walborn, N.R.: 1977, *ApJ* **215**, 53

Walborn, N.R.: 1982, *ApJ* **254**, L15

Walborn, N.R.: 1983, *ApJ* **265**, 716

Walborn, N.R., Nichols-Bohlin, J., Panek, R.J.: 1985, 'IUE Atlas of O-type spectra from 1200 to 1900 A', *NASA reference publication* , 1155

STELLAR WINDS OF MASSIVE STARS IN M31

LUCIANA BIANCHI *
Space Telescope Science Institute, 3700 San Martin drive, Baltimore, MD21218, USA

and

JOHN HUTCHINGS
DAO 5071 West Saanich road., Victoria,B.C. V8X 4M6, Canada

Abstract. We have obtained the first UV high resolution spectra of hot luminous stars in M31 with the FOS on *Hubble Space Telescope*. The spectra, combined with optical spectroscopic and photometric observations, enable us to study their stellar winds and photospheric parameters. We derive mass-loss rates and velocity laws from the wind line profiles, with the SEI method, as well as information on abundances. The wind lines and photospheric spectra are compared with galactic stars of the same spectral type.

The spectra analyzed so far indicate that the stars have mass–loss rates comparable or slightly lower than galactic stars of the same spectral type, but possibly different velocity laws in their winds. The spectra of two stars are discussed here.

Key words: O stars – mass loss – radiatively driven winds – external galaxies

1. Introduction

Only with the advent of the *Hubble Space Telescope* it became possible to obtain high resolution UV spectra of massive stars in M31, a spiral galaxy similar to our own Milky Way. In the past, many galactic hot stars, and the brightest MC stars were observed at high resolution in the UV with IUE, to investigate properties of the winds.

We obtained the first high resolution UV spectra of hot supergiants in M31 with the Faint Object Spectrograph on the *Hubble Space Telescope* (five stars to date, from one GO and one GTO program of L.Bianchi and J.Hutchings, respectively). They are the first detailed observations of luminous hot stars in M31, that allow quantitative analysis of the lines and comparison with similar objects in our galaxy and the MC's. The purpose was to further investigate the dependence of winds and photospheres of hot stars on the global properties of their host galaxies. Hot star are important as they reflect current star formation activity, and because of the amount of kinetic and thermal energy injected by their winds into the interstellar medium (ISM). Moreover, the stars are probes of the ISM in the M31 halo.

In a previous paper (Hutchings et al 1992) we presented the two first stars observed. Here the spectrum of a B1.5 supergiant in M31 is presented, which is yet unpublished. We also briefly re-discuss one of the previously published spectra.

---

* On leave from Osservatorio Astronomico di Torino, Italy

Space Science Reviews **66**: 183–186, 1994.

© 1994 *Kluwer Academic Publishers. Printed in Belgium.*

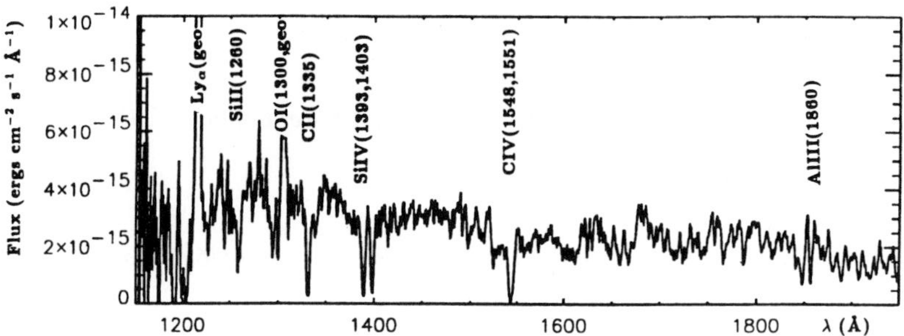

Fig. 1. The FOS spectrum of star 277 in the range shortwards of 1950 A. Note that the emissions of Ly$_\alpha$ and OI are of geocoronal origin.

2. Star OB78-277

Star 277 in the OB78 region of M31 is a bright isolated object, classified as B1I by Humphreys, Massey and Freedman (1990). Hutchings et al (1987) and Bianchi et al (1990) discussed its optical and IUE low resolution spectra. Our optical data taken at KPNO, and UV data presented here suggest a spectral type of B1.5Ia.

In Figure 1 we show a portion of the FOS spectra of star 277, in the range of the main wind lines. The main features are identified. The FOS spectra are affected by non-negligible diffuse background, that is not removed by the standard pipeline reduction and had to be carefully evaluated and subtracted case by case by us before analyzing the spectra. The FOS spectra cover the range 1100-3300 A, allowing also the analysis of the continuum flux in the UV. This analysis, that will be published in detail elsewhere (Bianchi et al, in preparation), includes also optical CCD photometry, and yields a temperature of T_{eff}=20000 K, M_V=-7.26 and log $L/L_\odot$=5.6, in good agreement with the spectral classification.

The SiIV and CIV resonance lines show P Cygni profiles with asymmetric absorption but no emission. In addition, CII 1335 and AlIII 1860 show P Cygni profiles, confirming the spectral type between B1 and B2. Fitting the CIV and SiIV lines with the SEI method, we find a terminal velocity of 700 km s^{-1}, comparable or somewhat lower than similar galactic stars. From the SEI analysis we also derive the optical depth of these lines, that can be used to derive the mass loss rate when the abundances and the ionization structure of the wind are known. As for the abundances, we can reasonably assume that they are close to solar within the uncertainties of the todate existing information. Empirical determinations of ionization fractions for stars later than B0 however almost do not exist, even for galactic stars, nor reliable models that can be used(see Bianchi et al1993 for details). Fortunately, we also have the H$_\alpha$ profile of star 277, obtained by Artemio Herrero at the WHT, in collaboration with the Munich group. The analysis of this line with the SEI method, assuming a photospheric profile from a galactic star of the same type but lower luminosity, gives a very

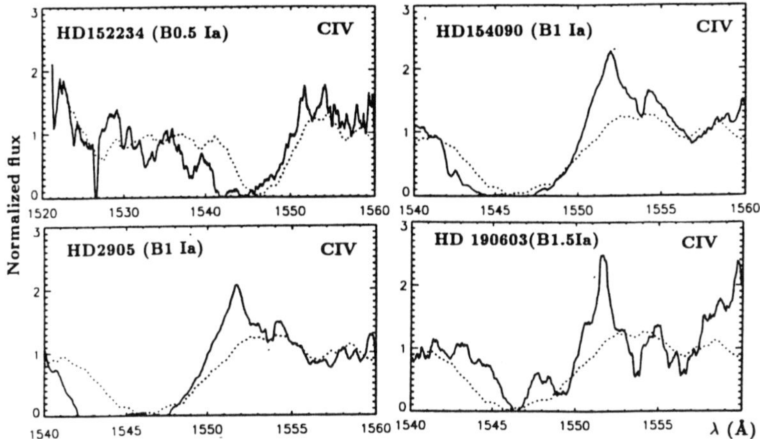

Fig. 2.   The CIV profile of star 277 (dotted line) is compared to CIV profiles of galactic stars observed with IUE. Star 277 has weaker emission in all cases, and smaller velocity in all cases except for HD190603

good fit to the observed profiles with the same velocity derived from the UV lines, and yields $\dot{M}$=1. 10^{-6} $M_\odot yr^{-1}$. We can compare this value with predictions and empirical relations derived for galactic stars. Extrapolating the relation of Lamers and Leitherer (1993) for O and B0 supergiants we would expect $\dot{M}$=2.2 10^{-6} $M_\odot yr^{-1}$ for our object. From the relation of Vyverman et al. (1994, this volume) valid for O and B stars, we predict $\dot{M}$=3.4 10^{-6} $M_\odot yr^{-1}$. This indicates that the mass loss rate of the first B supergiant observed in M31 is rather similar to its galactic counterparts, or slightly lower.

It is also interesting to directly compare of the line profiles of the M31 star with galactic stars of similar type. This is shown in Figure 2. The galactic stars were observed with IUE, at higher resolution but worse S/N. The IUE spectra have been smoothed to obtain better signal, and resolution comparable to the FOS. However the actual FOS resolution is compromised by the HST Point Spread Function, that makes the comparison difficult. The emission component seems generally weaker in star 277 of M31 than in galactic stars, possibly indicating difference in the velocity law.

3. Star 231 in OB78

We published the FOS spectra of this O8.5I(f) star in two previous papers (Hutchings et al 1992, Bianchi et al 1992). The wind lines were analysed with the SEI method, and the mass loss rate derived with the formula of Groenewegen and Lamers (1989), adopting empirical ionization fractions adequate for a galactic star of the same spectral type. Unfortunately wrong numbers were printed in those papers for the final results. The correct number, for the same SEI analysis and ionization fractions is instead log $\dot{M}$ ≈ -5.9 $M_\odot yr^{-1}$, close to what one would expect for the spectral type.

4. Future work

More observing time have been assigned to us with HST and with optical telescopes to significantly extend this study to more hot objects in M31, and also to
massive stars in the other nearby galaxy M33, that we have extensively studied in
the optical and with the *Ultraviolet Imaging Telescope (UIT)* , and to the farthest
possible extent with IUE (e.g. Bianchi et al 1991).

In our analysis so far we adopted solar abundances and galactic ionization
fractions for differential comparison of the M31 stars with galactic objects. In fact,
the existing literature indicates that the abundances are solar at the positions of our
stars in M31 (these results are mainly based on HII regions) and visual comparison
of the FOS spectra with spectra of galactic stars indicate some similarities of
the photospheric lines (see Haser *et al.,* this book). Our final goal is also the
determination of the abundances for the individual stars, by analysing the FOS
spectra with the code elaborated by the group of Prof. Kudritzki. The FOS data
in fact allow for the first time analysis of the abundances in single objects of
M31, and of the ISM in different lines of sight in this external galaxy. For this
reason, our targets were selected at different galactocentric distances in M31.

5. Acknowledgements

We would like to thank Dr. Claus Leitherer for very valuable comments, and Dr.
Artemio Herrero and Dr.Rolf Kudritzki for giving us the H_α spectrum.

References

Bianchi,L., Hutchings,J., Massey,P.: 1991, *A&A* **249**, 14
Bianchi,L., Hutchings,J., Lamers,H., Massey,P., Morris,S., De Francesco,G. 1992, in "Science with
    the Hubble Space Telescope", Benvenuti and Scrherier eds.,p.361
Groenewegen,M., and Lamers, H.: 1989, *A&ASuppl* **79**, 359
Humphreys,R., Massey,P., Freedman,W.: 1990, *AJ* **99**, 84
Haser,S., Lennon,D., Kudritzki,R., Walborn,N., Bianchi,L., Hutchings,J.: 1993, *this book*
Hutchings, J., Massey,P., Bianchi,L.: 1987, *ApJ* **322**, L79
Hutchings,J., Bianchi,L., Lamers,H., Massey,P., Morris S.: 1992, *ApJ* **400**, L35
Lamers,H., Leitherer,C.: 1993, *Ap.J* **412**, 771

O-STAR WINDS IN THE MAGELLANIC CLOUDS AND THE MILKY WAY

A Modified Empirical Analysis Of UV P-Cygni Profiles

S.M. HASER, J. PULS and R.-P. KUDRITZKI*

Universitatssternwarte Munchen, D-81679 Munchen, Scheinerstrasse 1, Federal Republik of Germany

Abstract. UV P-Cygni profiles of OB-stars in the Magellanic Clouds (observed with HST), and the galaxy (observed with IUE) are analyzed empirically using a line formation procedure similar to the one described by Lamers et al. (1987). The assumption of a constant microturbulence v_{turb} throughout the wind is dropped and replaced by a radially increasing turbulence parameter $v_{\text{turb}}(v)$, thus improving the fit for the emission peaks substantially, and at the same time avoiding the need for a justification of extremely supersonic turbulence in the vicinity of the wind's sonic point. The Sobolev optical depth is determined iteratively at fixed velocities in the wind, which removes the bias introduced by the choice of a specific parameterization function. Where it was possible and necessary a full photospheric spectrum was used to illuminate the wind line. The terminal velocities v_∞ are are found to be largest in the Galaxy, smallest in the Small Magellanic Cloud, and intermediate or similar to galactic in the Large Cloud.

Key words: Hot star winds – UV P-Cygni profiles

1. Introduction

For a faster computation of P-Cygni profiles originating in spherically symmetric and strict monotonic expanding stellar winds, Hamann (1981) suggested how to overcome one of the shortcomings of a pure Sobolev method by computing the intensity integral exactly, involving a widening of the resonance zone ("turbulence", see below), while evaluating the source function by Sobolev theory. Lamers et al. (1987, LCSP) have developed this method as an efficient tool ("SEI") for computing P-Cygni profiles.

The method uses several input parameters to specify the properties of the stellar wind. These are: the velocity field, here: $v(r) = (1 - b/r)^\beta$, $b \approx 0.99$, $\beta \approx 0.5 \ldots 2$, the "turbulence velocity" $v_{\text{turb}}(r)$, which mimics a stochastic motion of the absorbers, and the Sobolev optical depth $\tau(v) = \frac{\pi e^2}{m_e c} f_{\text{lu}} \lambda_{\text{lu}} n_{\text{l}}(v) \frac{R_*}{v_\infty} \frac{1}{dv/dr}$, which can be parameterized in terms of a line strength $k(v)$: $\tau(v) = k(v) \frac{1}{r^2 v} \frac{1}{dv/dr}$. (For $k(v) = const$ this implies a constant ionization fraction. All quantities are given in units of either R_* or v_∞). The quantities resulting from a line fit are then: the wind's terminal velocity v_∞, the absorber column density $N_{\text{col}}(v_1, v_2) = v_\infty \frac{m_e c}{\pi e^2} \int_{v_1}^{v_2} \tau(v)\, dv$, and, in the case of

---

* SMH & RPK affiliated to Max-Planck-Inst. fur Astrophys., Karl-Schwarzschild-Str. 1, D-85748 Garching bei Munchen, Federal Republik of Germany

Space Science Reviews **66**: 187–190, 1994.

© 1994 *Kluwer Academic Publishers. Printed in Belgium.*

a resonance line, the product of mass loss rate $\dot{M}$ and ionization fraction $q(v)$:

$$\dot{M} q(v) = 8.74 \cdot 10^{-19} \frac{1+4Y_{\text{He}}}{f_{\text{lu}} \lambda_{\text{lu}} A_{\text{Elem}}} k(v) R_* v_\infty^2.$$

In this paper we present an independent code, which uses modifications of the above method concerning the "turbulence"-parameter and the parameterization of the optical depth. For the application to the observations an incident photospheric profile, which covers the *complete* frequency range of the P-Cygni profile, is used in most cases. The terminal velocities of 16 OB stars in the Magellanic Clouds are given, and the results are compared to those of a large sample of galactic stars.

2. Parameterizations

Parameterization of "turbulence": In their analysis of galactic OB-stars Groenewegen & Lamers (1989, GL) used the method described by LCSP with a *constant* $v_{\text{turb}} \approx 0.05 \ldots 0.15 v_\infty$ *throughout the wind.* In a NLTE analysis of photospheres of hot stars stars Becker & Butler (1992) showed, that photospheric turbulence is smaller than the sound speed (see also Kudritzki 1992). Thus a jump from subsonic to extremely supersonic turbulence near the wind's sonic point would be implied by the use of a large v_{turb}, which would be difficult to justify. We therefore use a radially dependent turbulence parameterization of the form: $v_{\text{turb}} \propto v(r)$ (first suggested by Olson 1981), with $v_{\text{turb}}(r = 1) \approx 2v_{\text{sound}}$ and $v_{\text{turb}}(v_\infty) \approx 0.05 \ldots 0.15 v_\infty$. The application to some 130 galactic O stars showed that a large turbulence velocity is *not needed* at low outflow velocities, but still is required at large $v(r)$. This parameterization results in better fits for the position of the emission peaks in almost all cases.

Parameterization of the Sobolev optical depth: The Sobolev optical depth $\tau(v)$ (see Sec. 1) is usually parameterized as some power of $v(r)$ (Castor & Lamers 1979), Hamann 1980, GL, Howarth & Prinja 1989 (HP)). But regardless of the chosen parameterization function a bias is always introduced by *any* choice for it. Thus the best solution is a *parameter free representation* of $\tau(v)$. This can be achieved by the following iteration scheme for $\tau(v)$, when v_∞, β, and $v_{\text{turb}(v)}$ are known from a previous fit "by hand": First, one selects n observed profile points P_k^{obs}, $k = 1 \ldots n$, $n \approx 4 \ldots 8$. The goal is a best fit to these n points. Starting at an optically very thin value for $\tau(v)$ an automatic variation $\Delta\tau(v_j)$, $j = 1 \ldots n$ at n equidistant velocity points (spanning the whole v-range) is performed. This results in a Jakobi-Matrix

$$\mathcal{A} = \left(\frac{\Delta P_k^{\text{obs}}}{\Delta \tau_j} \right)_{j,k=1\ldots n}$$

The correction $\delta\tau$ for τ_{old} is given by the iteration condition that $\mathcal{A} \cdot \delta\tau - \mathbf{P}^{\text{obs}} = 0$. After $3 \ldots 5$ iterations convergence is usually achieved. Such few observed points are sufficient because with the previous fit by hand much information has already been extracted. Since the matrix equation is ill conditioned in most cases, the technique of the "Singular Value Decomposition"

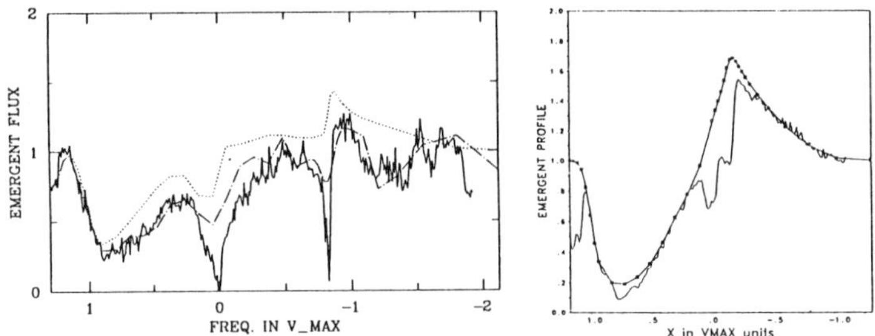

Fig. 1.   Left: The influence of the photospheric profile on the fit for the Si IV line of HD 190429A (O4 If$^+$). Dash-dotted: with the photospheric profile of 9 Sgr (O4 V); dotted: $I_{Core} = 1$, other parameters identical. Right: the C IV profile of Sk -67° 221 (O3 V) in the LMC. The computed profile has been convolved with a gaussian of FWHM = 1 A to approximately account for the resolution of the FOS.

(see Press et al. 1988, and references therein) is applied here.

3. Photospheric profiles

The incident photospheric radiation is distorted by absorption lines, not only of the parent ion, but also of blends of many weak additional metal lines like Fe/Ni IV, Fe/Ni V and others. This has strong effects on the shape of the resulting P-Cygni profiles and therefore changes the extracted result, sometimes severely. For a weak line like Si IV in an O4 I star, the derived ion column densities are much larger for $I_{Core} = 1$, since then the photospheric absorption is attributed to the wind line only (see Fig. 1). In the case of strongly saturated lines sometimes only the input of a photospheric absorption spectrum allows one to fit the shortward wing of the emission peak *without* a large v_{turb} at low velocities (see Sec. 2).

If available, a wind free photospheric spectrum of a star with similar spectral type should be used. This is easy to do for Si IV due to the luminosity effect of this line (Walborn et al. 1985). For N V and C IV the difficulty arises that most stars show these as wind lines. One then has to decide whether the photospheric spectrum of a hot subdwarf (early stars), a late main sequence star, or a constant continuum has to be used.

4. Results

From application to a sample of some 120 galactic O stars and 16 objects in the Magellanic Clouds we found that v_∞ is the quantity best determined with the above method. (For reasons of saving space, the values are listed in Haser et al., this proceeding, the other parameters will be discussed in a future paper.) The numbers agree in general with those obtained by other authors (GL, Prinja, Barlow & Howarth 1990 (PBH)), but individual differences can be large. For example: HD 14947 (v_∞ = 2350 km/s (this work), 2300 km/s (GL), 1885 km/s

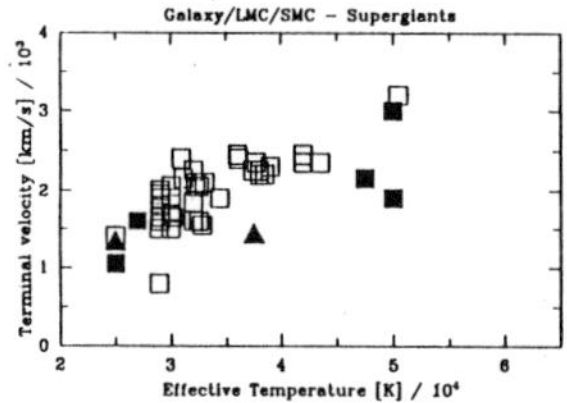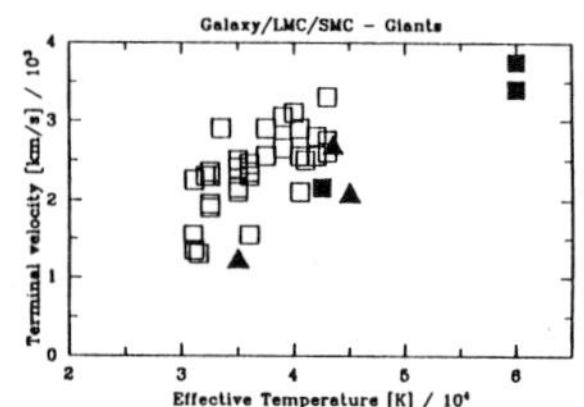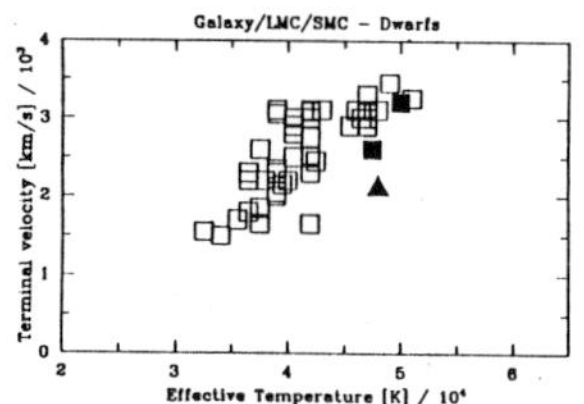

Fig. 2. v_∞ as function of effective temperature for luminosity classes I, III, and V (left to right). light squares: Galaxy, filled squares: LMC, filled triangles: SMC

(PBH, they used a different criterion). We find, that for comparable stars in the three galaxies, the overall tendency exists for v_∞ to be smallest in the SMC, while in the LMC the values are similar to or slightly less than those in the Milky Way, where they are largest. For the O3 stars the situation is different. The LMC objects show larger v_∞ than galactic ones, but the sample is very small (only 4 objects in our galaxy, 4 in the LMC, none yet in the SMC).

For future work, we intend to combine the results with optical analyses, which give the stellar parameters, the mass loss rates from H_α and v_∞, and the helium abundance Y_{He}, to derive ionization fractions $q(v)$, and to compare these with the predictions of the theory of radiation driven winds as described by Pauldrach et al. (1993).

Acknowledgements: This work was supported by the Max-Planck-Institut fur Astrophysik and the Deutsche Forschungsgesellschaft in the "Gerhard Hess Programm" under grant PA 477/1-1.

References

Becker, S.R., Butler, K.: 1992, *A&A* **265**, 647

Castor, J.I., Lamers, H.J.G.L.M.: 1979, *ApJS* **39**, 481

Groenewegen, M.A.T., Lamers, H.J.G.L.M.: 1989, *A&AS* **79**, 359

Hamann, W.-R.: 1981, *A&A* **93**, 353

Howarth, I.D., Prinja, R.K.: 1989, *ApJS* **69**, 527

Kudritzki, R.-P.: 1992, *A&A* **266**, 395

Lamers, H.J.G.L.M., Cerruti-Sola, M., Perinotto, M.: 1987, *ApJ* **314**, 726

Olson, G.L.: 1982, *ApJ* **255**, 267

Pauldrach, A.W.A., Kudritzki, R.-P., Puls, J., Butler, K., Hunsinger, J.: 1993, *A&A* , in press

Press, W.H., Flannery, B.P., Teukolsky, S.A., Vetterling, W.T.: 1988, *in 'Numerical Recipes', Cambridge University Press* **p. 52**, ff

Prinja, R.K., Barlow, M.J., Howarth, I.D: 1990, *ApJ* **361**, 607

ON THE RELATION BETWEEN THE MASS LOSS RATE AND THE STELLAR PARAMETERS OF OB-TYPE STARS

K. VYVERMAN, W. VAN RENSBERGEN

Astrofysisch Instituut, Vrije Universiteit Brussel, Pleinlaan 2, B-1050 Brussel, Belgium

and

D. VANBEVEREN

Dept. of Physics, Vrije Universiteit Brussel, Pleinlaan 2, B-1050 Brussel, Belgium

Abstract. We discuss mass loss relations for OB-type stars as a function of luminosity, effective temperature, and mass. We conclude that a simple first order linear regression relation is as good as any other more sophisticated relation, with the advantage that the simple form consumes much less computer time when used in evolutionary codes.

1. Introduction

We want to have at our disposal a reasonably good mass loss relation for OB-type stars, for use in evolutionary computations. For these types of stars, it is assumed that the unique mechanism for mass loss is described by the radiatively driven wind theory. There exists a theoretical relation (Kudritzki et al. 1989) which expresses the $\log \dot{M}$ as a linear function of $\log L/L_\odot$, $\log M/M_\odot$, and $\log T_{\text{eff}}$.

2. The model

Using observed mass loss rates of 271 stars from all over the Hertzsprung-Russell diagram, De Jager et al. (1988) computed a twenty-terms Chebyshev polynomial expansion describing the dependence of $\dot{M}$ on L and T_{eff}.

There are some remarks to be made about this expression:

— There is no physical reason for using Chebyshev-polynomials.

— Performing multiple regressional analysis upon the sample of stars listed in De Jager et al. (1988), it appeared that a number of these twenty coefficients are not significant.

Using only the O- and B-type stars from the tables in De Jager et al. (1988), we computed the relation

$$-\log\left(-\dot{M}\right) = 8.332 - 1.699 \log \frac{L}{L_\odot} + 1.547 \log T_{\text{eff}},$$

with a standard error of estimate of 0.32 dex. This relation is much less time-consuming for use in an evolutionary code. $\dot{M}$ is expressed in $M_\odot yr^{-1}$. All three terms are now significant, and the coefficient of correlation between the observed and the calculated mass losses of the stars in the sample is about 0.884, which is

Space Science Reviews **66**: 191–192, 1994.

© 1994 *Kluwer Academic Publishers. Printed in Belgium.*

similar to the performance of the formula derived by De Jager et al. (1988) –in the case of OB-type stars–.

3. Adding mass...

In consideration of the physical background, we added a term in $\log M/M_\odot$ to our model, using interpolated masses from recent evolutionary tracks published by the Geneva group (Schaller et al. 1992).

This however, did not significantly improve the mass loss regression, since we found a strong correlation ($r = 0.963$) between $\log M/M_\odot$ and $\log L/L_\odot$ for the stars in our sample.

4. Conclusions

The above three-terms relation can safely be used to calculate mass loss from O- and B-type stars in a time-efficient manner. As can be seen in Figure 1 however, the stars in our sample are not homogeneously distributed over the hydrogen-burning region in the Hertzsprung-Russell diagram. One should therefore be careful when using our relation outside the indicated region. A warning that applies to any $\dot{M}$ relation published so far...

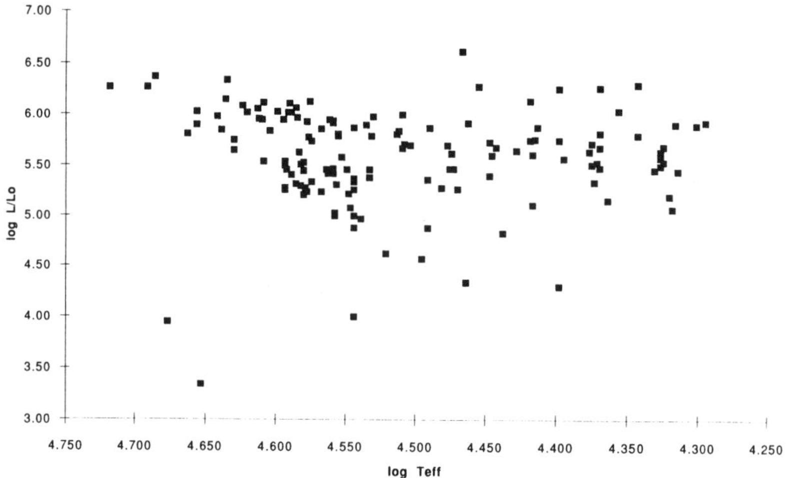

Fig. 1.   The sample of OB-type stars in the Hertzsprung-Russell diagram.

References

De Jager, C., Nieuwenhuijzen, H., Van der Hucht, K. A. 1988, A&AS, 72, 259
Kudritzki, R. P., Pauldrach, A., Puls, J., Abbott, D. C. 1989, A&A, 219, 205
Schaller, G., Schaerer, D., Meynet, G., Maeder, A. 1992, A&AS, 96, 269

WIND ASYMMETRIES IN MASSIVE STARS

R.E. SCHULTE-LADBECK

Department of Physcis and Astronomy, University of Pittsburgh, U.S.A.

G.C. CLAYTON

Center for Astrophysics and Space Astronomy, University of Colorado, U.S.A.

C. LEITHERER, L. DRISSEN, C. ROBERT and A. NOTA

Space Telescope Science Institute, Baltimore, U.S.A.

and

J. WM. PARKER

Goddard Space Flight Center, Greenbelt, U.S.A.

Abstract. We are in the process of surveying the linear polarization in luminous, early-type stars. We here report on new observations of the B [e] stars S 18 and R 50, and of the Luminous Blue Variables HR Car, R 143, and HD 160529. Together with previously published data, these observations provide clear evidence for the presence of intrinsic polarization in 1 B[e] star (HD 34664) and in 5 LBVs (η Car, P Cyg, R 127, AG Car, and HR Car). The data indicate that anisotropic stellar winds are a common occurrence among massive stars in these particular evolutionary stages. For such stars, mass-loss rates estimated using the assumption of a spherical, homogeneous and stationary outflow may be in error.

Key words: Luminous Blue Variables – B [e] stars – polarization

1. Introduction

Luminous Blue Variables (LBVs) are supergiants which are located in the HR diagram very near the empirical upper luminosity boundary of stars recognized as the Humphreys-Davidson limit (Humphreys & Davidson 1979). A famous example is η Carinae, the most luminous star known in the Galaxy. In LBVs, the mass loss can occur in short, eruptive episodes, leading to the formation of shells and ring nebulae (Humphreys 1989). Although radiation pressure is suspected to play at least some role, the principal driving mechanism for these outflows has not yet been identified.

The LBVs are considered to be a very important phase in the evolution of all massive stars. Only about 31 stars in the local group of galaxies are known to be LBVs, which implies that this is a very short-lived phase. Maeder (1989) distinguishes two evolutionary scenarios involving LBVs:

For 50 $M_\odot$ < M < 120 $M_\odot$ O - Of - (BSG) - LBV - WR - SN

For 40 $M_\odot$ < M < 50 $M_\odot$ O - Of - (BSG) - LBV - OH/IR - WR - SN

The upper part of the HR diagram is also populated by the "slash" stars, stars of mixed Of/WN classification (Bohannan & Walborn 1989), and the B[e] supergiants (Zickgraf 1990). Notice that these stars are not considered in the theoretical evolutionary chain by Maeder, although relationships have now been

Space Science Reviews **66**: 193–198, 1994.

© 1994 *Kluwer Academic Publishers. Printed in Belgium.*

established observationally. The minimum-brightness spectral types of R 127 and AG Car are those of slash stars (Stahl et al. 1983, Stahl 1986). It is thus possible that Of/WN stars are quiescent LBVs. Schulte-Ladbeck & Clayton (1993) recently found that the B[e] star HD 34664 displays signatures of an LBV star, hence B[e] stars and LBVs might also be more closely related evolutionary phases than previously thought.

2. Polarimetric observations

Polarimetry of LBVs is a powerful tool for probing the nature of mass loss on very small spatial scales which cannot be studied through normal imaging techniques or coronography (Schulte-Ladbeck et al. 1993, Schulte-Ladbeck, Clayton & Meade 1993). η Car has long been known to be polarized very near the star with a PA that is parallel to the minor axis of the homunculus nebula (Visvanathan 1967). Due to the large size of the homunculus nebula, it is even possible to map the polarized, scattered light of the resolved nebula as a function of wavelength (Warren-Smith et al. 1979, Briggs & Aitken 1985, Meaburn, Wolstencroft & Walsh 1987). Polarization observations of P Cyg were reported by Hayes (1985), Lupie & Nordsieck (1987), and Taylor et al. (1991) and can be interpreted with a spherical or only slighly flattened wind which is permeated with time variable, electron opacity enhancements. The resolved P Cyg nebula appears spherical and clumpy when observed with coronography (Clampin et al. 1993a). Observations of two LBVs, R 127 (Schulte-Ladbeck et al. 1993) and AG Car (Schulte-Ladbeck, Clayton & Meade 1993), and of the B[e] star possibly turning LBV HD 34664 (Schulte-Ladbeck & Clayton 1993) have shown that their stellar winds are axisymmetric within a few R_*. The position angles of the flattened stellar winds of AG Car and R 127 indicate a relationship with their resolved nebulosities recorded with coronography (Nota et al. 1992, Clampin et al. 1993b).

We here report new polarimetric observations of the B[e] stars S 18 and R 50 and of the LBVs HR Car, R 143, and HD 160529. The data were obtained with either the CCD spectropolarimeter at the 3.9-m Anglo-Australian Telescope (AAT) or with the PISCO filter polarimeter attached to the 2.2-m telescope of the European Southern Observatory. We also obtained PISCO observations of a large sample of slash stars. The results of that work will be published in a separate paper (Drissen et al. 1994).

S 18. The polarization data of S 18 are displayed in Figure 1. In general, the observed polarization of a star is the vector sum of interstellar polarization (ISP) and intrinsic polarization in the star. We have overplotted the data in Fig. 1 both wth a best fit using the Serkowski law of ISP in the formulation of Wilking, Lebowsky & Rieke (WLR, 1982), and with an eyeball-fit constant. The data are equally well described by either, and it is difficult to decide whether the polarization is completely interstellar, or partially intrinsic in nature. A previous observation of S 18 was reported by Magalhaes (1982). He finds a V-band polar-

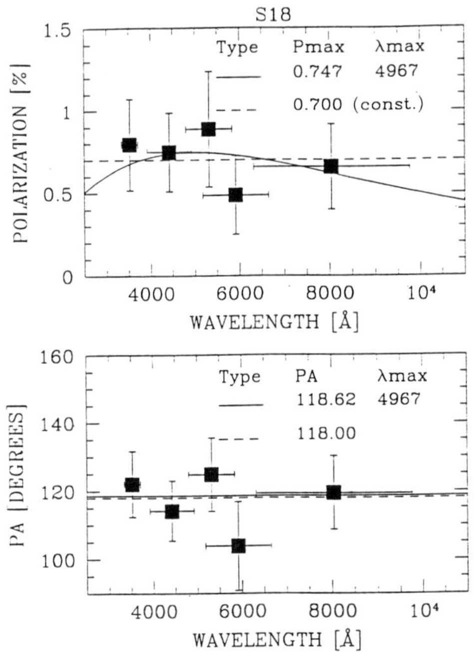

Fig. 1.   UBVRI polarimetry of S 18. The solid line represents a fit with a Serkowski-type (WLR) law for the ISP; the parameters of that fit, Pmax, λmax, and PA are indicated. The dashed line shows P, PA for the hypothesis of a constant polarization.

ization of 0.18% $\pm$ 0.08% at a position angle (PA) of 58°. While our V-filter observation by itself is not significantly different from Magalhaes', we can establish from the fit with a constant that the overall level of observed polarization is higher and the PA is different in our data. Thus, S 18 is a new candiate for variable intrinsic polarization. If a variable polarization in S 18 could be firmly established, this would further strengthen the link between the B[e] and LBV classes.

R 50. Our observations of R 50 are displayed in Figure 2, together with three about equally good fits for the ISP. R 50 was also observed by Magalhaes, and his and our V-filter datum agree within the errors. The observed polarization of R 50 is inistinguishable from ISP.

HD 160529. Filter polarimetry of HD 160529 is shown in Figure 3, together with the best-fit ISP. This star has been observed repeatedly in surveys of the ISP and has been used as a high-polarization standard (Serkowski 1974, Dolan & Tapia 1986 and ef.). It shows a variation of PA with wavelength from the optical into the IR. This could be due to the interstellar medium, or it could indicate an intrinsic component (Dolan & Tapia 1986). Even if HD 160529 has a component of intrinsic polarization, this will be extremely difficult to separate from the substantial ISP on this sightline. However, we caution against the use of this star as a polarimetric standard until further measurements of its polarization

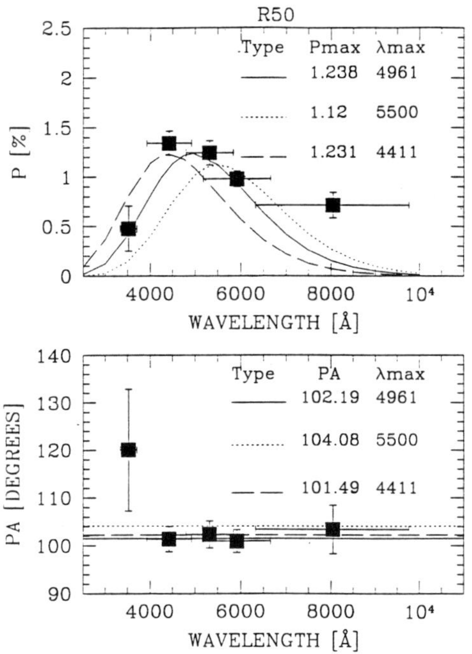

Fig. 2. UBVRI polarimetry of R 50. Three about equally good solutions for fits with the WLR law are overplotted.

have been carried out.

R 143. We do not display our spectropolarimetric observation of R 143 here. Although the data show a significant change of the polarization across the Hα line with respect to the continuum, there was an unfortunate cosmic-ray hit on our spectra. After cosmic-ray removal, a significant change of polarization at the line remained. However, we do not feel confident in the line data until we can obtain a repeat observation. We compared our data with the filter observations of Clayton, Martin & Thompson (1983). In our data, simulted box-shaped filters centered at the central wavelengths of their filters yield P = 3.30% ± 0.03%, PA = 69°.6 at 0.44 μm, P = 3.24% ± 0.01%, PA = 71°.6 at 0.59 μm. There is a 4.5σ difference in one Stokes parameter comparing our to one of their data sets, which probably does not indicate a real variation. More data are needed to establish an intrinsic polarization component in R 143.

HR Car. The change of polarization at Hα is significant in our data (not shown here). This clearly indicates that HR Car possesses an anisotropic wind. In order to establish the wind geometry, spherical or axisymmetric, we need to obtain more data. We note that Bandiera et al. (1989) published high-resolution profiles of Hα and He I which evidence a complex wind structure. On one occasion, the Hα profile was double peaked and closely resembled that of the B[e] star HD 34664. New unpublished coronographic data of Clampin (private communication) show

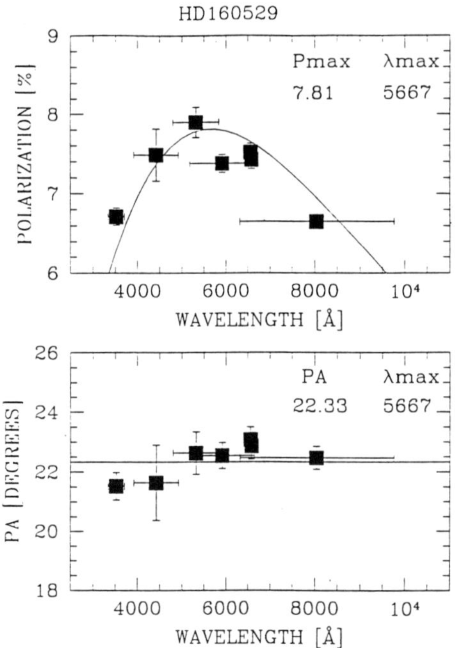

Fig. 3.   Polarimetry in UBVRI, Hα, [NII] filters of HD 160529 (see Schulte-Ladbeck et al. 1993 for filter definitions). Overplotted is the best-fit solution with the WLR law.

3. Discussion

The LBVs show a high incidence of being intrinsically polarized. With the exception of HD 160529, which has a high ISP, and WRA 751 and He 3-519, two recently identified LBVs which we have not observed with polarimetry, all galactic LBVs display intrinsic linear polarization. In the well-studied cases, this polarization is found to be variable as a function of time. The outflows are inhomogeneous. The time-averaged geometry of the wind of P Cyg is spherical, but that of η Car, R 127 and AG Car is axisymmetric. The resolved nebulae around these 4 stars have geometries that reflect those of their winds, i.e., symmetric wind - symmetric nebula, asymmetric wind - asymmetric nebula. We have not enough data to establish the wind geometry of HR Car, but the recent detection of an axisymmetric nebula makes this an interesting object for future polarimetric studies. Multi-wavelength polarimetric observations of the B[e] stars indicate that their winds may be time-dependent. While B[e] stars are considered to have axisymmetric stellar winds as a class, they were not thought to be variable. The possible polarimetric variability of S 18 suggests that the B[e] stars should be monitored more closely in the future, especially with long-term photometry. Such observtions might reveal that B[e] stars are just another phase in the life of a luminous, blue variable.

4. Acknowledgements

We thank M.R. Meade for her continuing assistance with the data analysis. A. Vaidya helped with the preparation of the figures. This work was supported by NAGW-2977.

References

Bandiera, R., Focardi, P., Altamore, A., Rossi, C., Stahl, O.: 1989, *Ap&SS* **157**, 279
Bohannan, B., Walborn, N.R.: 1989, *PASP* **101**, 520
Briggs, G.P., Aitken, D.K.: 1985, *Proc. ASA* **6**, 145
Clampin, M., Robberto, M., Nota, A., Paresce, F., Staude, J.: 1993, *in "Circumstellar Media in the Late Stages of Stellar Evolution"*, conference guide, abstract D8
Clampin, M., Nota, A., Golimowski, D.A., Leitherer, C., Durrance, S.T.: 1993b, *ApJL* **410**, 35
Clayton, G.C., Martin, P.G., Thompson, I.: 1983, *ApJ* **265**, 194
Drissen, L., et al.: 1994, *in preparation* **n.a.**, n.a.
Dolan, J.F., Tapia, S.: 1986, *PASP* **98**, 792
Hayes, D.P.: 1985, *ApJ* **289**, 726
Humphreys, R.M., Davidson, K.: 1979, *ApJ* **232**, 409
Humphreys, R.M.: 1989, *Ap&SS* **157**, 3
Lupie, O.L., Nordsieck, K.H.: 1987, *AJ* **92**, 214
Maeder, A.: 1989, *Ap&SS* **157**, 15
Magalhaes, A.M.: 1992, *ApJ* **398**, 286
Meaburn, J., Wolstencroft, R.D., Walsh, J.R.: 1987, *A&A* **181**, 333
Nota, A., Leitherer, C., Clampin, M., Greenfield, P., Golimowski, D.A.: 1992, *ApJ* **398**, 621
Schulte-Ladbeck, R.E., Clayton, G.C.: 1993, *AJ* **106**, 790
Schulte-Ladbeck, R.E., Clayton, G.C., Meade, M.R.: 1993, *ASP Conf. Ser.* **35**, 237
Schulte-Ladbeck, R.E., Leitherer, C., Clayton, G.C., Robert, C., Meade, M.R., Drissen, L., Nota, A., Schmutz, W.: 1993, *ApJ* **407**, 723
Serkowski, K.: 1974, *in "Planets, Stars and Nebulae studied with Photopolarimetry"*, edt. T. Gehrels, U. Arizona Press, 135
Stahl, O., Wolf, B., Klare, G., Cassatella, A., Krautter, J., Persi, P., Ferrari-Toniolo, M.: 1983, *A&A* **127**, 49
Stahl, O.: 1986, *A&A* **164**, 321
Taylor, M., Nordsieck, K.H., Schulte-Ladbeck, R.E., Bjorkman, K.S.: 1991, *AJ* **102**, 1197
Visvanathan, N.: 1967, *MNRAS* **135**, 275
Wilking, B.A., Lebowsky, M.J., Rieke, G.H.: 1982, *AJ* **87**, 695 (WLR)
Warren-Smith, R.F., Scarrott, S.M., Murdin, P., Bingham, R.G.: 1979, *MNRAS* **187**, 761
Zickgraf, F. J.: 1990, *NATO ASI Ser.* **316**, 245

NLTE ANALYSIS OF HOT BINARIES

K.P. SIMON and E. STURM

Institut fur Astronomie und Astrophysik
der Universitat Munchen, Scheinerstr. 1., 81679 Munchen, Germany

Abstract. We present quantitative spectroscopic NLTE analyses of the components of well detached early type binaries (DH Cep, Y Cyg, V453 Cyg, and CW Cep). The position of the stars in the $\log L - \log T_{\text{eff}}$ diagram is discussed. We find significantly higher temperatures for the components of Y Cygni from spectral analysis by means of unblanketed NLTE model photospheres than those given by the orbit analysis. Therefore the comparison with evolutionary tracks yields larger masses. The spectroscopic temperatures of V453 Cygni and CW Cephei agree with the orbit data, but the evolutionary tracks point to larger masses also. However, if we account for some 2000K lower effective temperatures due to line blanketing, the luminosities, temperatures and masses of all stellar components are in good agreement, except for the case of DH Cep.

1. Introduction

The determination of absolute dimensions, luminosities and temperatures is one of the major problems in the study of early–type stars. These data provide the basis of any consistency check of the theory of stellar structure and evolution. While absolute dimensions are of low accuracy for single stars with poorly known distances, photospheric parameters like effective temperature and surface gravity can now be determined with unprecedented accuracy of a few percent. In order to determine photospheric parameters of massive stars with known absolute dimensions, we performed spectral analyses of several well–detached, hot binaries.

2. Disentangling of composite spectra and NLTE-analysis

The component spectra were reconstructed from a set of composite spectra at different phases by means of the disentangling method, a novel method introduced by Simon and Sturm (1992,1994). The algorithm also yields precise radial velocities, even at phases, where the lines are not properly resolved (e.g. at conjunction). This is a great advantage for the construction of radial velocity curves, the determination of γ values and orbit eccentricities. The reconstructed spectra of the components do not suffer from the disadvantages of composites (e.g. line overlap) and can be analysed like single star spectra. This technique makes it possible to determine the photospheric parameters gravity, helium abundance and effective temperature of both components with the same accuracy as for single stars and independent of the analysis of the photometric orbit. In the case of eclipsing binaries the results of the spectral analysis are an important possibility to check the accuracy of the orbit analysis. Another important application of the spectroscopic analysis of disentangled spectra is the determination of absolute

Space Science Reviews **66**: 199–202, 1994.

© 1994 *Kluwer Academic Publishers. Printed in Belgium.*

dimensions of variables which do not exhibit eclipses. Radii and masses can be determined by means of an analysis of the ellipsoidal light curve if light ratio and temperatures can be fixed. Sturm and Simon (1994) applied this method to the O-type binary DH Cephei. The spectroscopic analyses presented here are performed in the same manner. Results of the spectral analyses are presented in Tab. I (small corrections of the fit parameters are possible due to a planned revision of the reduction of the spectra). Typical errors of the spectral analyses are

TABLE I

Photospheric parameters T_{eff} [K], g [cm s^{-2}] and the light ratio in the visual l as determined from spectroscopy. The orbit parameters are taken from: Y Cyg: Giuricin et al. (1980), V453 Cyg: Popper and Hill (1991), CW Cep: Clausen and Giminez (1991). The orbit analysis of the ellipsoidal variable DH Cep was performed by Sturm and Simon (1994), who used temperatures determined by spectral analysis. The dimensions of the system components are given in solar units. L^* denotes the corrected luminosity and M_{evol} the mass from comparison to evolutionary tracks.

star	orbit					spectroscopy			
(l)	M	R_p	$\log L$	T_{eff}	$\log g$	T_{eff}	$\log g$	M_{evol}	$\log L^*$
DH Cep	29.4	8.3	5.39	44000	3.95	44000	3.95	42.1	5.39
(0.84)	25.0	7.8	5.30	43000	4.03	43000	4.03	36.7	5.30
Y Cyg	17.0	6.6	4.31	27000	4.03	35000	4.26	20.4	4.76
(0.92)	16.3	5.2	4.27	29600	4.22	35200	4.24	20.0	4.57
V453 Cyg	13.9	8.5	4.69	29500	3.72	29900	3.80	18.8	4.71
(0.39)	10.7	5.3	4.18	28200	4.02	30600	4.06	14.8	4.32
CW Cep	11.8	5.5	4.24	28300	4.03	30000	4.05	14.5	4.34
(0.77)	11.1	5.0	4.12	27700	4.09	28000	4.10	12.9	4.14

$\Delta T_{\mathrm{eff}} = \pm 1500$K and $\Delta \log g = \pm 0.15$. The spectroscopic temperatures of both components of Y Cygni were found to be much higher than the temperatures determined by Giuricin et al. (1980). The effective temperature must be significantly higher than 30000K, because the HeII lines at 4686A, 4542A and 4200A are quite strong. Our analysis yields a gravity higher than 4.2 for Y Cygni, which is slightly larger than that of the ZAMS at that temperature. However, the ZAMS is well within the error of $\log g$. The normal helium abundance of 0.09 is within the error ($\Delta \epsilon_{\mathrm{He}} = \pm 0.02$) for all components except the primary of V453 Cyg. For this star our spectra imply slightly enhanced helium.

3. Discussion

Vrancken et al. (1991) discussed the dynamic masses of the primaries of early–type binaries in comparison to masses from evolutionary tracks using data given by Popper (1980). They concluded that the masses are in good agreement. This statement still holds, though in the meantime the orbits and dynamical masses

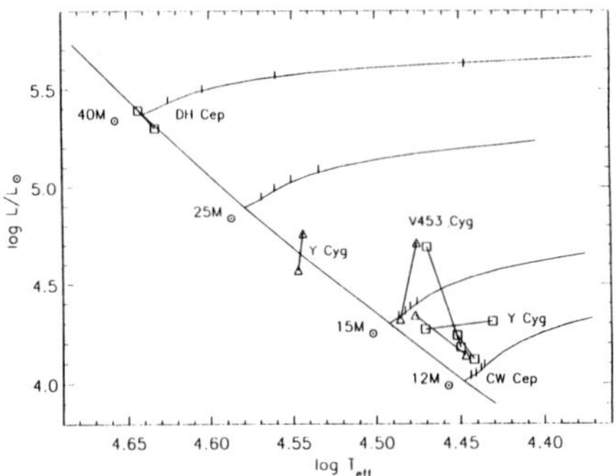

Fig. 1.   Position of the stellar components in the $\log L$ - $\log T_{eff}$ diagram. The squares denote the orbital data and the triangles the position according to the spectral analysis. The ZAMS and the tracks are computed by Schaller et al.(1992) for a metallicity of $Z = 0.02$. Note the large temperature difference of the spectroscopic analysis and the orbit for Y Cygni. For DH Cephei the determination of orbit parameters is not independent from the spectral analysis.

of V453 Cygni and CW Cephei have been improved by Popper and Hill (1991) and Clausen and Gimenez (1991), respectively. On the other hand, significant discrepancies have been encountered in the course of an analysis of 25 galactic OB-stars by Herrero et al.(1992), when masses determined from spectroscopy in conjunction with known distances have been compared with those from evolutionary tracks.

In Table I the absolute dimensions (M, polar Radius R_p, and L) of the orbit analyses are given. In Figure 1 the positions of all eight components are shown. The comparison of the positions of the stellar components in the $\log L - \log T_{eff}$ diagram with evolutionary tracks shows differences in masses, which are decreasing with temperature. However, if we rely on the spectral analysis, we shall derive the luminosity from the spectroscopically determined temperature. Therefore we corrected the luminosity from the orbit by $4\log(T_{spectrum}/T_{orbit})$. This value ($L^*$), as well as the mass ($M_{evol}$) from comparison with the tracks of Schaller et al. (1992) is given in Table I also. Because the spectroscopic temperatures are larger than the orbital temperatures we obtain larger luminosities L^* and in turn the mass differences increase.   Beside the mass discrepancy Herrero et al.(1992) detected several cases of increased helium abundance in the photospheres under study. However, both problems are significant only far from the ZAMS. This is a strong hint to some mixing process neglected in the evolutionary theory. The stars' positions on and close to the ZAMS should not be altered by mixing, unless the underlying mechanism changes the stellar structure significantly, for instance by changing the size of the convective core. If the evolutionary theory should predict wrong ZAMS positions, another explanation is needed.

On the other hand, the mass discrepancy may be caused by systematically too high spectroscopic temperatures. Using line blanketed NLTE atmospheres, the temperatures are expected to be some 2000K lower. This will remove the encountered mass discrepancies of Y Cygni, V 453 Cygni and CW Cephei perfectly. The components of Y Cygni will be close enough to the ZAMS after this correction so that the system is still young enough for not having undergone orbit circularization. However, after this correction the spectroscopic temperatures of the components of Y Cygni are still much higher than the temperatures from orbit analysis of Giuricin et al. (1980). The components of V453 Cygni do not fit to a common isochrone. Possibly this system is the result of a mass exchange event. The hottest star under study, DH Cephei, would need a lower temperature of some 37000K in order to remove the mass discrepancy. Both components would then fit perfectly to the evolutionary tracks corresponding to the masses from the orbit. The orbital solution has not been found independently from the spectral analysis. However, this mass discrepancy cannot be removed, even if we assume much lower temperatures than those determined by spectral analysis.

We conclude, that the determination of masses from analysis of the orbit in conjunction with spectral analysis is most reliable for stars close to the ZAMS and for temperatures below $\approx$ 40000K, if the effects of line blanketing are taken into account.

References

Clausen, J.V., Gimenez, A.: 1991, *A&A* **241**, 98

Giuricin, G., Mardirossian, F., Mezzetti, M.: 1980, *A&AS* **39**, 255

Herrero, A., Kudritzki, R.P., Vilchez, M., Kunze, D., Butler, K.: 1992, *A&A* **261**, 209

Popper, D.M.: 1980, *ARA&A* **18**, 115

Popper, D.M. and Hill, G.: 1991, *AJ* **101**, 600

Schaller, G., Schaerer, D., Meynet, G., Maeder, A: 1992, *A&AS* **96**, 269

Simon, K.P., Sturm, E.: 1992, 'Disentangling of composite spectra' in P.J. Grosbl and R.C.E. de Ruijsscher, ed(s)., *Proceedings of the 4th ESO/ST-ECF Data Analysis Workshop*, ESO:Garching, 91

Simon, K.P., Sturm, E.: 1994, *A&A* **280**, in press

Sturm, E., Simon, K.P.: 1994, *A&A* **281**, in press

Vrancken, M., van Rensbergen, W., Vanbeveren, D.: 1992, 'A comparison of orbital masses of early-type binary components and masses predicted by stellar evolution' in U. Heber and C.S. Jeffery, ed(s)., *The Atmospheres of Early-Type Stars*, Springer:Berlin, 24

EFFECTIVE TEMPERATURES AND SURFACE GRAVITIES OF EARLY TYPE STARS

C. JORDI, E. MASANA, F. FIGUERAS, J. TORRA and R. ASIAIN
Departament d'Astronomia i Meteorologia, Universitat de Barcelona
Avda. Diagonal, 647. E-08028 Barcelona, Spain

1. Introduction

Intermediate band photometry is a powerful tool to determine physical parameters of early type stars. Among others, Moon & Dworetsky (1985) (hereinafter MD) built a grid that relates effective temperatures and surface gravities to Stromgren photometric indices (c_o, β) up to 20,000K on the basis of line-blanketed LTE models of stellar atmospheres from Kurucz (1979). Recently, Castelli (1991) (hereinafter Cs) and Napiwotzki *et al* (1993) (hereinafter NSW) extended this grid to higher temperatures and proposed new interpolations.

We present a comparison of these new algorithms based on a large sample of stars. These stars are the ones with complete Stromgren photometry in the compilation by Hauck & Mermilliod (1990), classified as "early group" (as defined by Stromgren, 1966) using the algorithm by Figueras *et al* (1991), but without peculiarities according to the criteria by Philip *et al* (1976). There are about 3700 early type stars with $11,000K \leq T_{eff} \leq 30,000K$ and $2.5 \leq \log g \leq 4.5$.

2. Comparison between algorithms of Moon (1985), Castelli (1991) and Napiwotzki *et al* (1993)

Moon (1985) fitted two polynomial expressions to the MD grid for early type stars: one in the range of effective temperatures 11,000-15,000K, and the other in the range 15,000-20,000K. In addition, he used the expressions of Balona (1984) to compute effective temperatures and surface gravities for the stars hotter than 20,000K.

Cs and NSW noticed discrepancies between the published values and those computed using Moon's code, especially in the boundary regions of the polynomial fits. Moreover to present new fits that better reproduce the published grid, these authors extended it up to 25,000K (Cs) and 30,000K (NSW) avoiding the use of Balona's formulae, which produces an important discontinuity with the grid.

Figure 1 shows, as expected, non-significant ΔT_{eff} differences between NSW and Moon procedures in the range $11,000K \leq T_{eff} \leq 15,000K$, maximum differences of about 400K in the range $15,000K \leq T_{eff} \leq 20,000K$, and a systematic difference of about 1200K between the extended values and those given by

Space Science Reviews 66: 203–206, 1994.

© 1994 *Kluwer Academic Publishers. Printed in Belgium.*

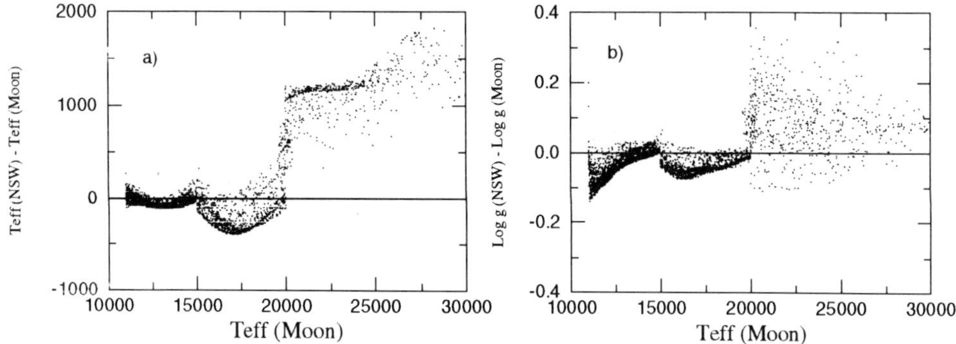

Fig. 1. New algorithm of Napiwotzki *et al* (1993) compared with the Moon (1985) code: a) T_{eff}, b) log g (the correction proposed by NSW was not applied).

Balona's formulae. Marginal differences that appeared in the boundary regions of T_{eff}=11,000K and T_{eff}=15,000K were removed by iterating the computation of the parameters in the codes.

Cs and NSW procedures show the same general behaviour. The largest differences appear between the two extended grids for stars with $T_{eff} \geq 20,000$K. As a whole, the differences are small and well in the range of observational and fitting errors.

NSW derived a correction that depends on temperature (Δlog g=-2.9406+0.7224 log T_{eff}) to match the gravities computed from the grid with those deduced from spectrograms of Balmer lines. This correction, not included in figure 1, will be discussed later.

North & Kroll (1989) (hereinafter NK) compared the physical parameters obtained from Geneva photometry with those computed using Moon's code. The systematic differences in effective temperatures between NSW (or Cs) and Moon procedures found here are very similar to those pointed out by NK, thus suggesting that they are a consequence of the polynomial fits of Moon and the inaccuracy of the Balona (1984) expressions. Thus, there is a good agreement between the effective temperatures obtained from Stromgren and Geneva photometries provided that the new algorithms are used.

When comparing log g determinations, NK found a linear relation with decreasing slopes as the temperature increases (see Table II in NK). This tendency is reversed in the range 18,000-25,000K probably due to the use of Balona's formulae from 20,000K on. When comparing the algorithms of NSW (or Cs) and Moon there is a clear trend for the stars hotter than 20,000K, giving

log g(NSW) = 1.246 log g(Moon) - 0.849

log g(Geneva) = 0.683 log g(NSW) + 1.336.

The correction proposed by NSW to the computed log g increases the scatter but it does not change the slope of the regression lines. Thus, systematic differences in surface gravities obtained from Stromgren and Geneva photometries still

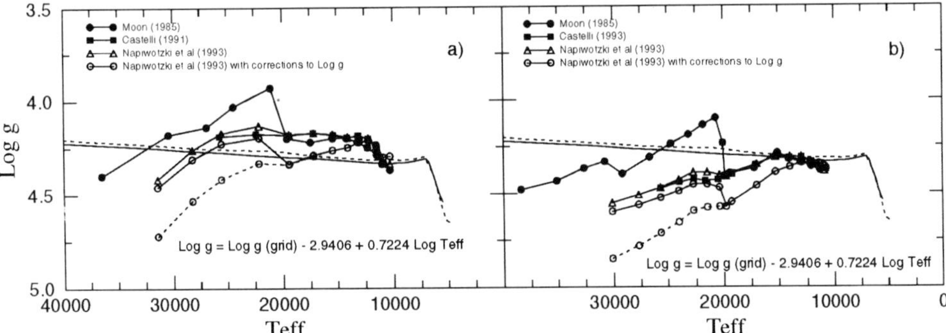

Fig. 2.   Comparison between observational and theoretical ZAMS on the plane (T_{eff}, log g): a) Crawford (1978), b) Balona & Shobbrook (1984). Theoretical ZAMS by Schaller *et al* (1992) and Claret & Giménez (1992) are shown as dashed and solid lines, respectively

remain.

3. Application to observational ZAMS

Observational ZAMS on the plane (c_o, β) of Crawford (1978) and Balona & Shobbrook (1984) have been translated onto the plane (T_{eff}, log g) through the codes and plotted in figure 2, together with theoretical ZAMS of recent evolutionary models for pop I stars (Schaller *et al*, 1992; Claret & Gimenez, 1992). The differences between the observational ZAMS are mainly in the range 0.010-0.015 mag in the β index, which yields differences in log g in the range 0.10-0.30 dex.

The extensions of the grid (NSW and Cs) agree and prevent the discontinuity at 20,000K between the grid of MD and Balona's expressions.

As mentioned above, NSW deduced a correction to the gravities computed from the grid. It can be seen that this correction substantially changes the slope of the observational ZAMS in respect to the theoretical. The sample used to derive this correction basically contained stars with T_{eff} lower than 20,000K (only one star had T_{eff}=21,000K). More recently, Napiwotzki (1993) compared the log g computed from the grid with those obtained by Grigsby *et al* (1992) from high-quality spectroscopic data and NLTE fully line-blanketed model atmospheres. The sample used contained stars in the range of temperatures from 20,000 to 30,000K. A new correction Δlog g=+0.5962-0.1227 log T_{eff} was derived for this range. In figure 2, the observational ZAMS are also plotted after applying different Δlog g corrections for temperatures lower or higher than 20,000K. This produces a discontinuity of 0.10 dex at 20,000K and, as pointed out by the author, a smooth transition should be adopted. As a whole, these corrections increase the surface gravity between 0.05 and 0.15 dex, significantly decreasing the computed ages of the stars. On the other hand, Castelli (1993) did not find such differences in the computed gravities. For all these reasons, further work is needed in order to confirm the application of these corrections.

The ZAMS defined by Crawford (1978) has a slope similar to the theoretical ZAMS in the range of 11,000-25,000K the differences being about 0.10-0.12 dex, which means differences of 0.006-0.009 mag in the β index. For higher temperatures it seems to give slightly larger gravities, which means the β index is also slightly larger. Taking into account the uncertainty of the calibration for the hottest stars already mentioned by the author, there is good agreement between this ZAMS and the theoretical.

On the other hand, the ZAMS defined by Balona & Shobbrook (1984) resembles the theoretical in the range 11,000-15,000K (c_o in the range 0.85-0.40 mag) and for higher temperatures the difference in gravity increases with increasing temperature, reaching about 0.3 dex at 30,000K (0.016 mag in the β index). This seems to indicate that the lower envelope defined by the authors is slightly low at these temperatures.

Acknowledgements

This work has been supported by the DGICYT and the CICYT under contracts PB91-0857 and ESP93-1020-E respectively.

References

Balona L.A. 1984, MNRAS 211, 973
Balona L.A. and Shobbrook R.R. 1984, MNRAS 211, 375
Castelli F. 1991, A&A 251, 106
Castelli F. 1993, private communication
Claret A. and Gimenez A. 1992, A&AS 96, 255
Crawford D.L. 1978, AJ 83, 48
Figueras F., Torra J. and Jordi C. 1991, A&AS 87, 319
Grigsby J.A., Morrison N.D. and Anderson L.S. 1992, ApJS 78, 205
Hauck B. and Mermilliod M. 1990, A&AS 86, 107
Kurucz R.L. 1979, ApJS 40, 1
Moon T.T. 1985, Comm. Univ. London Obs. No. 78
Moon T.T. and Dworetsky M.M. 1985, MNRAS 217, 305
Napiwotzki R. 1993, private communication
Napiwotzki R., Schonberner D. and Wenske V. 1993, A&A 268, 653
North P. and Kroll R. 1989, A&AS, 78, 325
Philip A.G.D., Miller T.M. and Relyea L.J. 1976, Dudley Obs. Report No. 12
Schaller G., Schaerer D., Meynet G. and Maeder A. 1992, A&AS 96, 269
Stromgren B. 1966, ARA&A 4, 433

THE ATMOSPHERIC COMPOSITION, EXTINCTION AND
LUMINOSITY OF THE LBV R71

D.J. LENNON, D. WOBIG and R.-P. KUDRITZKI*

Universtatssternwarte Munchen, D-81679 Munchen, Scheinerstrasse 1, Federal Republic of Germany

and

O. STAHL

Landessternwarte Konigstuhl, D-6900 Heidelberg, Federal Republic of Germany

Abstract. ESO 3.6m Caspec spectra of the LMC luminous blue variable (LBV) taken at minimum have been analysed using NLTE model atmospheres and line formation calculations to derive atmospheric parameters and chemical composition. Using the silicon ionization balance and the hydrogen Balmer lines we derive T_{eff}= 17250, log g = 1.80 and a microturbulent velocity of 15-20 km/s. The analysis yields abundance ratios by number of approximately 0.43 for He/H, 0.03 for C/N and 0.14 for O/N, implying that enrichment of the atmosphere by processed material has taken place. We have re-evaluated the reddening of R71 using IUE low resolution data and published UBVRIJHKL photometry and derive a value for A_V of 0.63. We also construct an extinction curve using archive IUE data for mid-B LMC supergiants and show that the extinction is anomalous; the 2175A bump being almost absent and the far UV rise very pronounced. A comparison of our model flux in the V-band with the observed (dereddened) V magnitude and the D.M. of the LMC (18.45), implies that the bolometric magnitude or R71 is -9.9. This is significantly higher than the value of -9.0 usually adopted for R71 and suggests that this object may not in fact be a 'subluminous' LBV.

Key words: stars:R71 – stars:LBV – stars:abundances – stars:extinction

1. Observations and analysis.

Information on the composition of the outer layers of Luminous Blue Variables (LBVs) has been derived mainly through inference via analyses of their ejecta (Crowther & Willis, these proceedings). In this paper we present photospheric abundances for the LBV R71 obtained directly from an analysis of its minimum phase optical spectrum. R71 lies in the Large Magellanic Cloud (LMC) and is one of the best studied S Doradus variables since Thackery (1974) showed the star to be of this type, oscillating between spectral types B2.5Iaep and A1eq. Estimates of its luminosity have tended to assert a rather low value for this object (Wolf et al 1981) in comparison with other LBVs, a peculiarity we also investigate.

The observational data consisted of ESO 3.6m Caspec spectrograms (blue and red) and IUE low resolution data, all taken at mimimum phase. We classified the optical (absorption) spectrum as B2.5Ia using Fitzpatrick's (1991) atlas of LMC B-supergiants. NLTE plane parallel model atmospheres and line formation calculations for H, He, C, N, O, Mg and Si were performed and T_{eff} and log g were

* Max-Planck-Inst. fur Astrophys., Karl-Schwarzschild-Str. 1, D-85784 Garching bei Munchen, Federal Republic of Germany

Space Science Reviews **66**: 207–210, 1994.

© 1994 *Kluwer Academic Publishers. Printed in Belgium.*

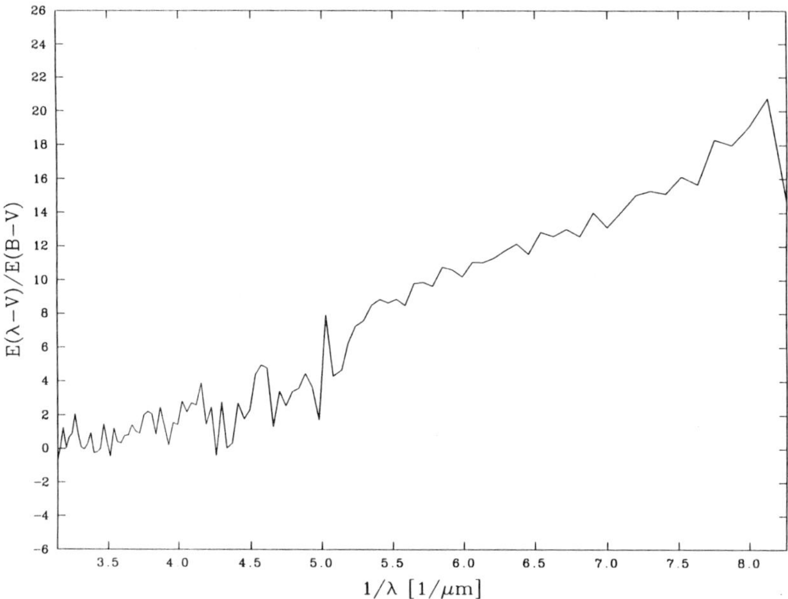

Fig. 1.   Sample extinction curve for R71 obtained using the LMC B2.5Ia supergiant Sk-68 40.

derived from the Hydrogen Balmer line fits and the silicon ionization balance. Microturbulence was used as a fitting parameter here and also for abundances of the other metals, the results being summarized in Table I.

Attempts to derive T_{eff} from UBVRIJHKL photometry and IUE fluxes by fitting the dereddened (using the LMC standard curve) flux distribution were unsuccessful. To investigate this we used low resolution IUE data for unreddened LMC B2.5 Ia supergiants to derive extinction curves for R71. This resulted in curves which show almost no evidence for the 2175A bump and a very steep far-UV rise, atypical for the LMC (Fig.1). In particular, depression of the UV continuum shortward of about 2000A is due to this enhanced extinction, it does not appear be caused by additional line blocking peculiar to this LBV since the (unbinned) extinction curves show no evidence of structure which one would expect to see if such were the case. Thus previous work standard extinction laws will have resulted in too low a value for the effective temperature. Normalizing our model flux distribution to the near-IR observed values implies a total extinction in the V-band (A_V) of approximately 0.63 magnitudes.

Using our estimate of the value of A_V and comparing our model fluxes with the observed V magnitude we derive the ratio of the distance to the radius. Our assumed distance modulus to the LMC of 18.45 then yields the radius which together with our values of T_{eff} and $\log g$ give us the mass and luminosity. We thus arrive at a value for the bolometric magnitude of -9.9, more in line with the values derived for other LBVs. (The bolometric correction then follows from the

TABLE I

Parameters of R71 as derived from NLTE model atmosphere analysis. Abundances in square brackets are given on the scale

$$\varepsilon_X = \frac{N_X}{N_H + 4N_{He}}$$

effective Temperature T_{eff}	17000 - 17500 K
Gravity $log\,g$	1.75 - 1.80
Helium abundance $y_{He} = \frac{N_{He}}{N_{He}+N_H}$	0.3
rotational Velocity $v \cdot sin\,i$	$51 \pm 6\,km/s$
Absorption in the V Magnitude A_V	$0.63 \pm 0.04\,mag$
absolute Magnitude M_V	$-8.2 \pm 0.1\,mag$
Bolometric Correction $B.C.$	$-1.69 \pm 0.1\,mag$
Bolometric Magnitude M_{bol}	$-9.89 \pm 0.1\,mag$
Radius R	$95 \pm 4\,R_\odot$
Mass M	$20 \pm 2\,M_\odot$
initial Mass M_{ZAMS}	$40 - 45\,M_\odot$
Luminosity $log\frac{L}{L_\odot}$	5.85 ± 0.04
Microturbulence v_T	$18 \pm 3\,km/s$

ABUNDANCES:	
Silicon $log\,\varepsilon_{Si}$	$-4.68 \pm 0.18\,[-4.96]$
Nitrogen $log\,\varepsilon_N$	$-3.28 \pm 0.29\,[-3.56]$
Oxygen $log\,\varepsilon_O$	$-4.12 \pm 0.14\,[-4.40]$
Carbon $log\,\varepsilon_C$	$-4.82 \pm 0.15\,[-5.10]$
Magnesium $log\,\varepsilon_{Mg}$	$-4.23 \pm 0.15\,[-4.51]$

absolute visual magnitude, see Table I.)

2. Discussion

The results of our analysis are summarized in Table I. The abundance ratios for R71 need to be compared with those of normal OB-stars in the LMC. Thus, from Rolleston (this meeting) we have for main-sequence B-stars approximately C/N= 0.5 and O/N= 7.0 compared with values of 0.03 and 0.14 derived here for R71. We also obtain He/H= 0.43. It thus appears that the surface material in R71 has undergone significant CNO processing, the small O/N ratio implying that the ON cycle has had an important influence on the current composition. These abundance ratios also indicate a greater degree of contamination in comparison to SN1987A, C/N= 0.13 and O/N= 0.63, which may be expected due to R71's greater mass. We have shown that the extinction of R71 is peculiar for the LMC. The absence of a

2175A bump and the very steep far-UV rise may be explained by a lack of carbon in the constituent material, implying that the extinction may be due primarily to silicates, and by a predominantly small-sized grain distribution. Given that R71 is an IRAS point source, evidence that it is surrounded by a gas and dust envelope (Wolf & Zickgraf 1986), and that more recently this circumstellar dust emission has been found to display prominent silicate emission bands (Roche et al 1993), an obvious explanation is that this extinction is caused locally by dust formed from the carbon deficient ejecta of R71. Finally, our re-evaluation of the luminosity of R71 removes one obstacle to the interpretation of LBVs in general. It has been widely argued that the cause of the instability of these objects is connected with their proximity to the Humphreys-Davidson (HD) luminosity limit in the HR diagram. R71 was conspicuous in that its previous assumed luminosity was well below the horizontal part of the HD limit. However these estimates were based on the assumptions of either a lower effective temperature than is derived here or of little internal LMC extinction and thus the the bolometric magnitude has hovered around -9.0 making R71 rather sub-luminous for an LBV. Our revised bolometric magnitude now puts this star slightly above this limit, in fact it now lies more in line with other well known LBVs such as S Doradus and P Cygni.

Acknowledgements

I wish to thank the BMFT for financial support under grant number 010R90080 and Ed Fitzpatrick and John Hillier for useful discussions.

References

Fitzpatrick, E.L.: 1991, *PASP* **103**, 1123
Roche, P.F., Aitken, D.K. and Smith, C.H.: 1993, *MNRAS* **262**, 301
Thackery, A.D.: 1974, *MNRAS* **168**, 221
Wolf, B.,Appenzeller, I. and Stahl, O.: 1981, *A&A* **103**, 94
Wolf and Zickgraf: 1986, *A&A* **164**, 435

THE VARIABILITY OF ETA CAR: A TOOL FOR THE LBV PHENOMENON

A. DAMINELI NETO
Instituto Astronomico e Geofsico da USP, Av. Miguel Stefano 4200, Sao Paulo, Brazil

R. VIOTTI and A. CASSATELLA
Istituto di Astrofisica Spaziale, CNR, Via Enrico Fermi 21, 00044 Frascati, Italy

and

G.B. BARATTA
Osservatorio Astronomico, Via del Parco Mellini 84, 00136 Roma

Abstract. We discuss the historical light curve of the most peculiar superluminous star η Car, and the spectroscopic variations during the last 100 y. After the nova–like spectral evolution following the 1889 light maximum, the star underwent many shell episodes which were characterized by a large fading of the higher ionization emission lines. We describe the most recent 1992 event when the He I and [N II] emission lines nearly faded out, and a broad P Cygni absorption appeared in the H I and He I lines. A recurrence time of about 5 years is suggested from the times of the spectroscopic episodes and the IR light curve. The results are discussed in the light of possible models.

Key words: η Car – Luminous blue variables – Shell phases – Stellar wind

1. The light history of η Car

The historical light curve of η Car can be divided into three main phases: the *bright phase* of the first half of the past century, the *deep fading* of 1856-1870, and the currently ongoing *minimum phase*. The light variation of η Car during the 19th century was described in detail by Innes (1905), and discussed among others by Gratton (1963), Andriesse et al. (1978), and van Genderen and Thé (1984). Andriesse et al. (1978) in particular found that the deep fading was associated with an efficient process of dust formation in the stellar envelope which started in the middle of the 19th century, as a result of an increase of the mass loss rate. Recently Polcaro and Viotti (1993), from the analysis of old, so far neglected, observations of the last century, discovered that this fading phase must have been far from regular, with possible bright luminosity peaks around 1860-62, followed by a sharp fading in 1865. According to Polcaro and Viotti, this behaviour is consistent with the fact that during this period of time the star was subject to huge mass ejection followed by the formation of circumstellar dust clouds which irregularly masked stellar light. After the large fading, the star remained at minimum until the present time, with a 1^{m} luminosity maximum in 1889, and a very slow brightening expecially during 1940-50 (O' Connell 1956).

Since 1893 the star underwent a nova-like evolution (Hoffleit 1933, Whitney 1952), which should be associated with the formation of a dense expanding

Space Science Reviews **66**: 211–214, 1994.

© 1994 *Kluwer Academic Publishers. Printed in Belgium.*

envelope, followed by its gradual dissipation. Since then, the η Car spectrum has remained up to the present mainly constant with strong H and Fe^+ emission lines.

2. The shell episodes

In more recent times a number of transient events, or *shell episodes*, have been observed which were characterized by marked spectral variations. The first episode was recorded in 1948–49 by Gaviola (1953) and described in detail by Viotti (1968), who found a general decrease of the emission line equivalent widths, from a factor two for the Fe ɪɪ lines up to a factor $\sim$7–10 for the highest ionization lines of He ɪ, [Fe ɪɪɪ] and [N ɪɪ], while [Ne ɪɪɪ] completely disappeared. Similar episodes were discovered by Rodgers and Searle (1967) in 1964–1965, and by Zanella et al. (1984) in December 1981. This latter episode was also observed in the UV with the IUE satellite. In the June 1981 spectrum there was a large decrease of the N ɪɪɪ] λ1750 intercombination multiplet, and of the Fe ɪɪ fluorescence lines at $\lambda\lambda$1785–8 and $\lambda\lambda$2507–8. The December 1982 IUE spectrum indicates that this shell episode has ended.

η Car underwent a new shell episode in 1992 which was recorded with the 1.6 m telescope of the Brazilian National Astronomical Laboratory (LNA). From a continuous spectroscopic monitoring in the optical and near–IR we found that in June 1992 the metastable lower level He ɪ λ1082.9 nm line, which normally displays a very strong emission peak and high velocity P Cygni absorptions (Damineli et al. 1993), had largely decreased down to near disappearance (Fig.1). A similar diminuition was found in the other He ɪ emission lines, and in the [N ɪɪ] doublet near Hα. In the meantime the hydrogen lines had changed by a lesser amount, indicating that the episode was mostly focused on the higher energy lines. A careful analysis of all the previous observations has shown that the episode started fairly long ago, probably around 1989, with a gradual fading of the He ɪ λ1082.9 nm intensity and a more rapid decrease during March–May 1992. At present the star is in a phase of gradual recovery. We have also noted a significant decrease of He ɪ λ1082.9 nm in the January 1987 near–IR spectrum of η Car discussed by Maillard et al. (1992). It should be noted that the IR light curve has shown two maxima at the time of the 1981 and 1987 episodes and a slight increase during 1990–1991 (Whitelock 1992).

The above discussed shell episodes should be important probes of the structure of the η Car wind, but in the absence of a consistent model of the η Car atmosphere, we could only make a qualitative interpretation of the phenomenon. We know that both low and high energy lines with the same velocity range have been identified in the optical and UV spectrum of η Car (e.g. Viotti et al. 1989), revealing a wide ionization range in all parts of the expanding envelope. Damineli et al. (1993) have shown that the He ɪ λ1082.9 nm is also formed in the envelope's outskirts. Probably, an increase of the mass loss from the central star could have caused an increase of the wind opacity to the higher energy photons, which

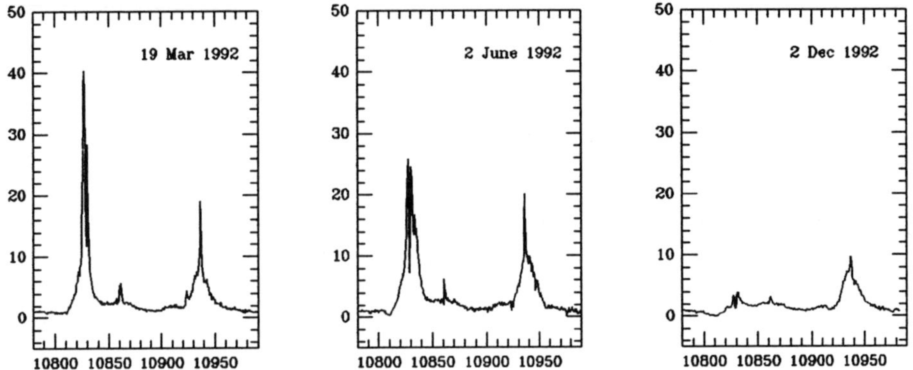

Fig. 1.   The 1992 shell episode of η Car in the near-IR with the large decrease of the He I λ1082.9 nm emission line. Ordinates are fluxes normalized to the continuum; LNA 1.6 m telescope.

resulted in a decrease of the mean wind ionization.

A crucial point would be the knowledge of the duration of the shell episodes, which appears to be of the order of several months to 1–2 y and their repetition time if any. Many other shell episodes similar to those here described may probably have occurred during the last years, but they have not been recorded because of the lack of a continuous monitoring of the spectrum of η Car, or are still hidden in some observatory archives. The IR light curve of η Car combined with the times of the (few) observed shell episodes are consistent with a recurrence time of about 5 y.

3.   The nature of η Car and its fate

According to Andriesse et al. (1978), η Car is a very luminous star (5×10^6 $L_\odot$) which is suffering a huge mass loss (0.075 $M_\odot$ y^{-1}). In the wind matter is accelerated up to velocities $\geq$1000 km s^{-1} (Damineli et al. 1993). At some distance from the star dust condenses from the wind material. The process of dust formation should have started after the bright light maxima of the past century, probably as the result of an increase of the star's mass loss. The circumstellar dust envelope at present absorbes most of the stellar radiation and reemits it in the IR. Andriesse et al. (1978) have shown that the present total (radiative plus mechanical) power of the star is the same as during the bright phase (outside the three main light maxima), so that *the big luminosity fading of η Car occurred at constant bolometric luminosity*. Therefore the deep luminosity decrease has to be ascribed to an increase of the mass loss rate, followed by a kind of *dust* *catastrophe*, probably similar to that occurring in other astrophysical objects, such as novae and R CrB variables.

At present the activity of η Car is enlivened by many (possibly recurrent) shell episodes, which indicate that wind structure is strongly variable, probably as a consequence of a transient increase of the mass outflow, as also suggested by the near-IR maxima at the time of the 1981, 1987 and 1992 episodes (Whitelock 1992). However, the process of dust condensation in the envelope of η Car will probably sometime end and consequently, the stellar radiation, which is at present emitted in the IR, should start to emerge again in the visual. The process could be quite fast, since a decrease of the mass loss rate could result in a lowering of the wind opacity. This should increase the rate of dust grain destruction by the stellar radiation, which produces a further decrease of the wind opacity. This chain reaction might develop fastly and end with a kind of an *inverse dust catastrophe*, following which η Car will trace back the light curve of the past century, and become again a very bright star.

We are grateful to M. Friedjung for comments. A. Damineli Neto acknowledges a contribution from FAPESP.

References

Andriesse C.D., Donn B.D., Viotti R.: 1978, *MNRAS* **185**, 771.
Damineli Neto A., Viotti R., Baratta G.B., de Araujo F.X.: 1993, *A&A* **268**, 283.
Gaviola E.: 1953, *ApJ* **118**, 233.
Gratton L.: 1963, in *Star Evolution*, L. Gratton ed., Academic Press, New York, 297.
Hoffleit D.: 1933, *Harvard Bulletin* **893**, 111.
Innes R.T.A.: 1903, *Cp. A.* **9**, 75B.
Kulczycky A.: 1865, *Connaissance des temps*, Bureau des Longitudes, Paris, August, 42.
Maillard J.P., Viotti R., Altamore A.: 1992, in *High Resolution Spectroscopy with the VLT*, M.-H. Ulrich ed., ESO Conf. Workshop Proc. **40**, 179.
O' Connell D.J.K.: 1956, *Vistas in Astronomy* **2**, 1165.
Polcaro V.F., Viotti R.: 1993, *A&A* **274**, 807.
Rodgers A.W., Searle L.: 1967, *MNRAS* **135**, 99.
van Genderen A.M., The P.S.: 1984, *Space Science Reviews* **39**, 317.
Viotti R.: 1968, *Mem. Soc. Astr. It.* **39**, 105.
Viotti R., Rossi L., Cassatella A., Altamore A., Baratta G.B.: 1989, *ApJS* **71**, 983.
Whitelock P.A.: 1992, *SAAO Report* **1991**.
Whitney C.A.: 1952, *Harvard Bull.***921**, 8.
Zanella R., Wolf B., Stahl O.: 1984, *A&A* **137**, 79.

THE NATURE OF THE LUMINOUS BLUE VARIABLE AG CARINAE

R. VIOTTI, A. CASSATELLA and V.F. POLCARO
Istituto di Astrofisica Spaziale, CNR, Via Enrico Fermi 21, 00044 Frascati, Italy

G.B. BARATTA
Osservatorio Astronomico, Via del Parco Mellini 84, 00136 Roma

A. DAMINELI NETO
Instituto Astronomico e Geofsico da USP, Av. Miguel Stefano 4200, Sao Paulo, Brazil

C. ROSSI
Istituto Astronomico, Universita La Sapienza, Via Lancisi 29, 00161 Roma

and

M. BARYLAK
IUE Observatory, Villafranca Satellite Tracking Station, 28080 Madrid, Spain

Abstract. After many years of permanence at minimum, the luminous blue variable AG Car started in mid 1990 a new brightening phase. We review the spectroscopic variations of the star since 1949, and discuss the nature of the circumstellar nebula. We give evidence that, like in η Car, also in AG Car dust is continuously condensing from the stellar wind. We suggest that the star could be partially reddened by the circumstellar dust, which could affect the estimates of the stellar distance. An extended H II *halo* is present outside the ring nebula, which should be associated with the wind of a previous cooler evolutionary stage of AG Car.

Key words: AG Car – Luminous blue variables – Dust envelope – Stellar wind

1. The spectral variations of AG Carinae

AG Car, together with η Car and P Cyg, is one of the few well established galactic Luminous Blue Variables (LBVs), which are believed to be a short-living (10^4-10^5 y) phase of the high mass star evolution, immediately preceding the WR phase. AG Car is of particular interest for its large photometric and spectroscopic changes during the last decade which were followed by many investigators, and might represent a unique chance to study the nature of LBVs. During 1981-1985 the star underwent a large luminosity fading from V=6^m to 8^m, which was accompanied in the visual and UV by a deep temperature change from $\sim 10000°$K to $\sim 30000°$K (Wolf & Stahl 1982; Viotti et al. 1984; Stahl 1986, Viotti et al. 1993). Viotti et al. (1984) showed that the visual fading after 1981 was accompanied by a corresponding increase in the UV flux, so that the total bolometric luminosity remained constant. This is an important result which was later found in other LBVs. Fig.1 illustrates the variation of the UV energy distribution of AG Car between August 1981 and June 1983, showing the large decrease of the long wavelength flux, and the simultaneous increase of the far-UV flux. The more recent may 1993 IUE observations disclosed a far–UV spectrum similar to that of June 1983 suggesting that the star had not yet reached its lowest temperature

Space Science Reviews **66**: 215–218, 1994.

© 1994 *Kluwer Academic Publishers. Printed in Belgium.*

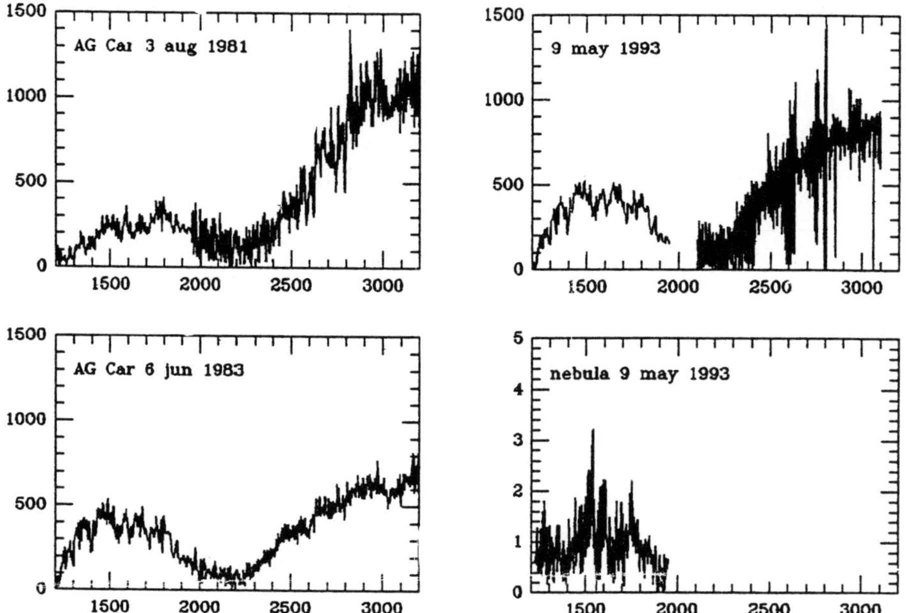

Fig. 1.   The ultraviolet spectrum of AG Car in August 1981, June 1983, and May 1993 (low and high resolution), and of the nebula in May 1993.

phase.

2. The dusty circumstellar nebula and halo

A key characteristics of AG Car is its expanding ring nebula which should have been ejected by AG Car itself about 10^4 y ago. Viotti et al. (1993) found that the optical spectrum of some regions near the star is very similar to that of AG Car itself, which suggests that it partially is dust scattered spectrum of the central star. The presence of dust grains near AG Car was first suggested by Viotti et al. (1988) from the study of the UV spectrum of the nebula, and confirmed by the far-IR KAO observations of McGregor et al. (1988). Paresce and Nota (1989) discovered an inner dusty jet-like feature with a possible helical structure. Smith (1991) and Nota et al. (1992) studied the nebula dynamics and found that the radial velocities are consistent with a hollow expanding asymmetrical shell. It should be noted that a marked deceleration of the stellar wind in the immediate star's surroundings is suggested by the much lower motions (~100 km s^{-1}, Smith 1991) of the nebula a few arcsec from AG Car.

A main problem of AG Car is the chemical composition of the star and of its nebula, since the current evolutionary models indicate that LBVs should be overabundant in N, and underabundant in CO. In fact, a N/C overabundance in the star spectrum could be suggested by the strength of the N II lines during the hotter phase (Caputo and Viotti 1970), but Viotti et al. (1993) have remarked the presence of weak O II and C II lines. Actually, no composition estimate can be done from the stellar spectrum until a consistent atmospheric model of AG Car would be developed. A better estimate can be done from the ring nebula spectrum, which however refers to matter ejected by the star a few 10^3 y ago. Mitra and Dufour (1990) derived a high N/CO ratio in the nebula, whereas de Freitas Pacheco et al. (1992) found that nitrogen is overabundant by one order of magnitude, and oxygen deficient by at least a factor six.

Viotti et al. (1993) have recently discovered outside the ring nebula the presence of an extended H II, which should be associated with the residual of the stellar wind of a previous, probably cooler, evolutionary phase of AG Car.

In the outer halo the nitrogen abundance is probably normal, suggesting that it is unprocessed matter ejected in a previous phase of AG Car. It should noted that the intensity ratio of the Hα, [N II] and [S II] lines in the ring and halo are those typical of PN and H II regions, respectively.

Dust signatures have been found in different parts of the circumstellar environment of AG Car, even in regions very close to the star. This led us to conclude that, *even in present times, dust is continuously condensing from AG Car's wind.* Since carbon and oxygen are among the main components of different types of condensate, their possible underabundance in the stellar wind is difficult to reconcile with the presence of dust in the nebula, unless different types of condensate are considered. In AG Car the dust condensation is probably favoured by the breaking of the outflowing wind into denser cloudlets, as proposed for the similar object η Car (Damineli et al. 1993). This process could be associated with the deceleration of the expansion velocity discussed above. Actually, the "jet-like" structure discovered by Paresce and Nota (1989) and the evidence of a high dust density in the "anti-jet" position point to the possibility of an axisymmetric enhancement of the dust concentration. The jet itself could be composed by many badly resolved cloudlets. Whatever the mechanism of dust condensation can be, it would reduce the CO abundance of the ejected gas, and be at least partially responsible of the anomalous abundance of the ring nebula.

3. A possible model of AG Carinae

Our data show that the ring nebula is surrounded by a much less dense, apparently spherically symmetric H II region which can be considered a residual of the stellar wind of a previous evolutionary phase of the star. Indeed, the nitrogen abundance in this halo might be normal. The measure of the expansion velocity of the halo is crucial to understand whether it was ejected by a red supergiant. The outer boundary of the ring appears very sharp, from both the morphological and spectroscopic points of view, which might suggest the presence of a physical discontinuity, probably a *shock front.* In any case the transition between the phases when the outer halo and the inner ring were formed represents a discontinuity in the star evolution. All these observational facts cannot be explained by any simple model of the AG Car nebula, nor to the presence of a close companion. On the contrary, they point to a complex interaction of different velocities in the wind, associated with different evolutionary phases of the central object. These interactions can also explain the velocity discontinuity between the stellar wind and the inner nebula.

A last evolutionary consideration follows from the determination of the star's distance. The AG Car's reddening of $E_{B-V}\sim0.7$ seems to favour a value of ~6 kpc (Humphreys et al. 1989, Hoekzema et al. 1992). If this is true, the corresponding large absolute luminosity ($M_{bol}\sim-10.8$) would imply, following the current models, an evolutionary track not reaching the RSG phase. However, the large reddening could be at least partially attributed to a local reddening, due to absorption from circumstellar dust in the nebula as well as in the extended halo. An even larger extinction could arise from the innermost dusty regions of the nebula discussed above. Viotti (1971) suggested that AG Car is at the same distance of the Carina complex. Thus, its absolute luminosity ($M_{bol}=-8.3$, Viotti et al. 1984) and mass should be definitely lower than that of the other LBVs. In this case the star can have experienced an RSG phase and would be now in an evolutionary phase that has not been previously recognized.

References

Caputo F., Viotti R. 1970, A&A 7, 266
de Freitas Pacheco J.A., Damineli Neto A., Costa R.D.D., Viotti R. 1992, A&A 266, 360
Damineli Neto A., Viotti R., Baratta G.B., de Araujo F.X.: 1993, A&A 268, 283
Hoekzema N.M., Lamers H.J.G.L.M., Genderer A.M. 1992, A&A 257, 118
Humphreys R.M., Lamers H.J.G.L.M., Hoekzema N., Cassatella A. 1989, A&A 218, L17
McGregor P.J. et al. 1988, ApJ 329, 874
Mitra P.M., Dufour R.J. 1990, MNRAS 242, 98
Nota A. et al. 1992, ApJ 398, 621
Paresce F., Nota A. 1989, ApJ 341, L83
Smith L.J. 1991, *Wolf Rayet stars and Interrelation with Other Massive Stars*, in IAU Symp. 143, van der Hucht K.A. & Hidayat B. (eds.), Kluwer, Dordrecht, p.385
Stahl O. 1986, A&A 164, 321
Viotti R. 1971, PASP 83, 170
Viotti R. et al. 1984, in: *Future of Ultraviolet Astronomy based on Six Years of IUE Research*, NASA CP-2349, Washington, p.231
Viotti R., Cassatella A., Ponz D., The P.S. 1988, A&A 190, 333
Viotti R., Polcaro V.F., Rossi C. 1993, A&A 276, 432
Wolf B., Stahl O. 1982, A&A 112, 111

THE EVOLUTIONARY STATUS OF ETA CAR BASED ON HISTORICAL AND MODERN OPTICAL AND INFRARED PHOTOMETRY

A.M. VAN GENDEREN
Leiden Observatory

M. DE GROOT
Armagh Observatory

and

P.S. THE
Astronomical Institute, Amsterdam

Abstract. We present a photometric study of the luminous early-type star η Car. The star's secular brightness increase is due largely to the expansion of the Homunculus. η Car is a LBV with a possible B0.5V companion. The combined effect of Balmer-line emission, and of warm and cold dust radiation explains the star's other light variations.

1. Introduction

η Carinae is a strong emission-line source surrounded by a pure reflection nebula, the Homunculus. It is one of the brightest IR sources in the sky. Its vital statistics are as follows: distance = 2.8 kpc; $L = 6 \ 10^6 \ L_\odot$; $T_{eff} = 30000$ K; $M \simeq 100 - 150 \ M_\odot$.

2. Observations

We used photometric data from different observers covering the period 1935 - 1992 to reconstruct η Car's long-term light curve (de Vaucouleurs and Eggen, 1952; Feinstein, 1967; Feinstein and Marraco, 1974; van Genderen and The, 1984; Manfroid *et al.*, 1991, 1993). As an example, Walraven VBLUW photometry covering the period 1974 - 1991 is shown in Figure 1.

Quite apart from several short-term variations in brightness, the light curve shows a secular rise which is shown in Fig. 2. Observations were averaged over each observing season, corrected for aperture effects, and transformed to Johnson V (V_J). This gives a homogeneous light curve for the period 1952 - 1992. The 1935 - 1952 period is covered by photographic observations transformed to V_J taken from O'Connell (1956).

The secular rise in brightness since 1952 cannot be explained by evolutionary effects because these predict $\Delta V = 0.^m025$/century (Maeder and Meynet, 1987), whereas the observed rise is $2.^m46$/century.

Space Science Reviews **66**: 219–223, 1994.
© 1994 *Kluwer Academic Publishers. Printed in Belgium.*

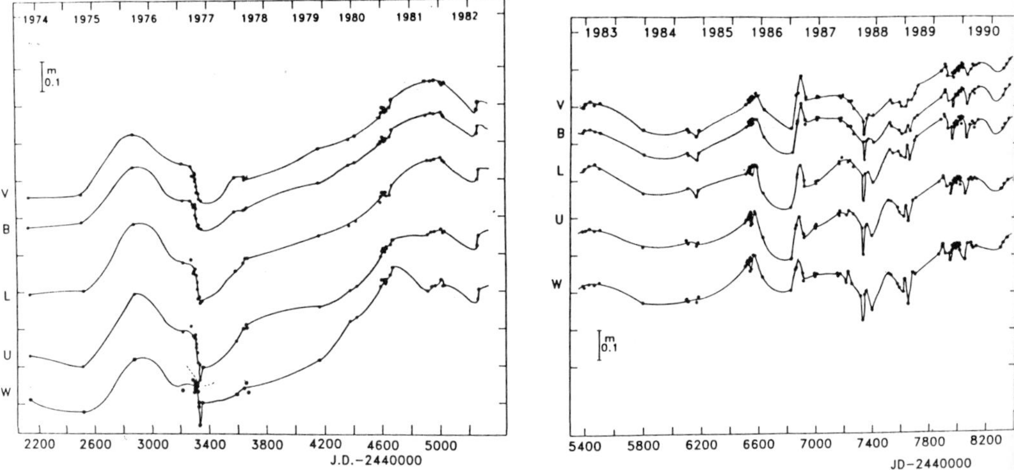

Fig. 1.   Walraven VBLUW light curves, 1974 - 1991. Note (i) the different kinds of long and short time scale variations; (ii) the gradually rising mean brightness (secular light variation).

The expansion of the Homunculus at $\sim$ 750 km s^{-1} causes a decrease of the circumstellar extinction (CSE) which can be calculated assuming that each layer expands homologously. The resulting light curve is also shown in Fig. 2. The CSE curve fits the observations very well. However, considering the colour variations of η Car, while these go in the right direction (i.e. η Car is getting bluer) they do show some abnormalities which are probably due to the varying influence of emission lines (see below).

η Car also shows S Dor-type light variations with some differences as follows: (i) η Car has outbursts on a shorter time scale of 1 to 3 years only; (ii) η Car's outbursts have smaller amplitude of $\sim 0.^m2$ only; (iii) η Car's outbursts show flux excess in the L and U bands. We conclude that η Car is either an S Dor variable or an LBV, with some additional source of variation in the ultraviolet.

3. Tentative Explanation

Due to continuous mass loss of the LBV, a nearby emission region is replenished continuously but at a variable rate. This emission region surrounds either the LBV or a companion. In the latter case a hot spot may be part of the system. An optically thin disk will radiate mostly in spectral lines with a $T_{eff} < 10000$ K. An optically thick disk will radiate mostly in the continuum with a $T_{eff} > 10000$ K. Thus, it is possible that the physical state of the emission region is changing as a function of the amount of infalling matter.

Bath's (1979) model of η Car has an accreting massive main-sequence star in a binary with $L_{disk} \simeq 10^{29} - 10^{32}$ if $\dot{M} = 10^{-6} - 10^{-2}$ M$_\odot$ yr^{-1}. This is similar

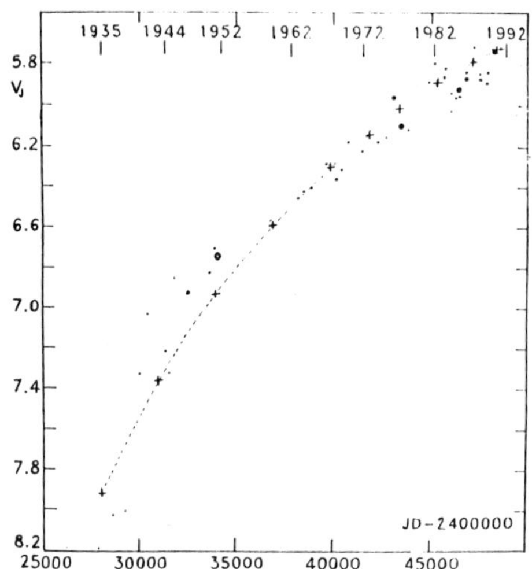

Fig. 2. η Car's secular light variation; 1962 - 1992: seasonally averaged photoelectric photometry; $\diamond$: m_{vis} from de Vaucouleurs and Eggen (1952); pre-1952 dots: adapted photographic observations of O'Connell (1956); +: light curve due to the decrease in circumstellar extinction (CSE) according to the text.

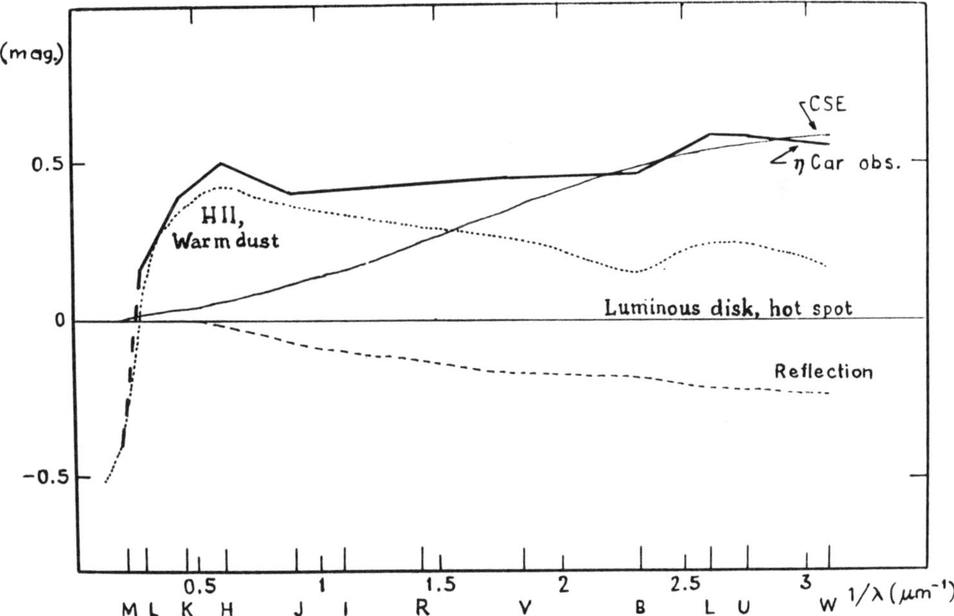

Fig. 3.   Amplitude (A) verus $1/\lambda$ for η Car during the period 1971-1992 (fat line, broken at $\lambda > 3\mu$m). The dotted curve represents the increase in amplitude caused by the central emission regions.

to our above figures for η Car with the observed variations in the L, U and W bands of $\sim 3\ 10^{29}$ W fallin within the above predicted range.

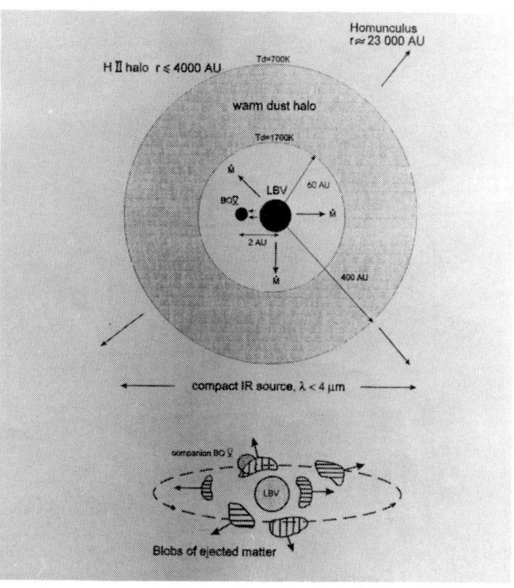

Fig. 4.   a: A schematic model of the η Car system; b: details of the central binary. Note that the radius of the Homunculus is $\approx$ 23000 AU, extending far beyond this picture.

4. Could η Car be a Binary?

A close inspection of the light curve shows ten minima of short duration (a few days?) which are most pronounced ($\sim 0.^m1$) in the UV and sometimes invisible at optical wavelengths. Analysis gives $JD_{min} = 2447631.39 + 52.^d38E$ with a stadard error in JD_{min} of $\simeq 6.^d7$.

There are, however, a few problems with the binary model: (i) relatively large errors in the O-C values of JD_{min} occur; (ii) the minima vary in depth from one to the other; (iii) some predicted minima are not observed; (iv) a monitoring campaign October, 1992, to July, 1993, did not show any of the predicted minima.

From the depth of the 'eclipses' in the various pass-bands one deduces that the eclipsed body must be a star with spectral type $\sim$ B0.5V; i.e. R=10 - 20 R$_\odot$, M $\sim$ 12M$_\odot$, log T$_{eff}$ $\simeq$ 4.45 and M$_{bol}$ $\simeq$ -6.2, at a separation of 1.4 AU.

5. Other Light Variations

In addition to the decrease in circumstellar extinction, we note other variations in brightness. η Car's total brightness variation shows the following differences with respect to the variation resulting from a decrease in CS extinction only: (i) there is excess radiation due to Balmer-line emission at $\lambda \sim$ 4000 A; (ii) there is excess radiation at $\lambda \sim 2\mu$m; (iii) there is a decrease in intensity at $\lambda > 4$ μm. This is clear from Fig. 3 which shows the total light amplitude (A) versus $1/\lambda$ for η Car during the period 1971-1992 (fat line, broken between the near-IR L and M pass-bands).

We identify the excess at $\lambda \sim 4000$ Awith either the H II halo, an accretion disk, or a hot spot. The excess at $\lambda \sim 2$ μm is from either a warm-dust halo, the H II halo, or an accretion disk. See Fig. 3 where the dotted curve represents the increase in amplitude caused by the central emission regions. The decrease of the intensity beyond 4 μm would be due to the expansion of the Homunculus; a 13% increase in the radius of the Homunculus between 1974 and 1992 causes a 27% decrease in the luminosity of the cold dust over the same period.

Reflected optical light also decreases due to Homunculus expansion. This is represented by the line labeled "reflection" in Fig. 3.

The sum of all these effects is capable of explaining the secular brightness increase of η Car.

Further particulars of this investigation will appear in A&A (van Genderen *et al.*, 1993).

References

Bath, G.T., 1979, Nature 282, 274
Feinstein, A., 1967, The Observatory 87, 287
Feinstein, A., Marraco, H.G., 1974, AA 30, 271
O'Connell, D.J.K., 1956, Vistas in Astron. 2, 1165
Maeder, A., Meynet, G., 1987, AA 182, 243
Manfroid, J. et al., 1991, ESO Scientific Report No. 8
Manfroid, J. et al., 1993, ESO Scientific Report No. 12
de Vaucouleurs, G, Eggen, O.J., 1952, PASP 64, 185
van Genderen, A.M., The, P.S., 1984, SpScRev 39, 31
van Genderen, A.M., de Groot, M., The, P.S., 1993, AA (in press)

THE GRADUAL ACCELERATION OF THE OUTFLOW OF VX-SAGITTARII: A STELLAR EVOLUTIONARY EFFECT?

NATHAN NETZER

ORT-Braude College, ISRAEL

Abstract. VX-Sagittarii is a red supergiant with a superwind which is observed in several maser lines. They provide an evidence that the outflow velocity keeps growing considerably at large distance from the star. It is argued that this phenomenon can be explained by stellar evolutionary effects.

As a rule, the outflow velocity for late type stars correlates with the mass loss rate and from that it is suggested that the mass loss rate was higher in the past and is decreasing now. The mass of VX Sagittarii can be estimated on this basis and is about 40–$50\,M_\odot$

1. Introduction

VX Sagittarii is a late type supergiant, located in an OB association and on this basis its distance is assumed to be $1700\,\mathrm{pc}$ (Humphreys *et al* 1972). From here one can infer a luminosity of about $4 \times 10^5\,L_\odot$. Assuming an effective temperature of 2500 K one gets a photospheric radius of $2.5 \times 10^{14}\,\mathrm{cm}$. The mass loss rate is estimated to be $2 \times 10^{-5}\,M_\odot\,\mathrm{yr}^{-1}$ (Knapp *et al* 1989, Netzer, 1989). It has been observed in four maser lines by Chapman and Cohen (1986), each suggesting a different outflow velocity. There is a SiO maser line whose width is about $2\,\mathrm{km\ s}^{-1}$, but it comes from a zone interior to the dust shell and therefore its association with acceleration processes is rather questionable. Further out, several stellar radii away, where dust has formed already, one observes the H_2O and mainline (1667 MHz) OH masers, suggesting an expansion velocity of 6—$11\,\mathrm{km\ s}^{-1}$. Considerably further out, at about 80 stellar radii ($2 \times 10^{16}\,\mathrm{cm}$), there is a 1612 MHz OH maser shell that expands at $19\,\mathrm{km\ s}^{-1}$. A more recent observation, of thermal CO lines, coming from the entire envelope but reflecting mainly its outer parts, suggests an outflow velocity of about $30\,\mathrm{km\ s}^{-1}$ (Knapp *et al*, 1989). From these observations it follows that either the outflowing matter keeps accelerating while moving away from the star, or that stellar evolutionary effects are involved, due to which recently lost material has left the central star with a smaller velocity than did material lost previously.

It might appear that the handicap of the stellar evolutionary scenario is that the crossing time of the envelope from the star to the OH 1612 MHz maser shell is of the order of 200 years, which is quite short on the evolutionary time scale. On the other hand, the outer zones, that dominate the CO emission, are far too dilute to explain the phenomenon by acceleration due to interaction of radiation and matter. Their crossing time, on the contrary, is of the order of several thousand years. Here stellar evolutionary effects can no longer be ruled out. The various possibilities are discussed below.

2. Radiative Transfer and Gradual Acceleration

If we assume that we are witnessing a gradual acceleration pattern in VX Sagittarii, we need first to discuss the equation of motion. It looks the following way:

$$v\frac{dv}{dr} = \frac{1}{c}\int_0^\infty \chi_\nu F_\nu\, d\nu - \frac{G\,M_\star}{r^2} = \frac{\chi_F\, L_\star}{4\pi r^2 c} - \frac{G\,M_\star}{r^2}, \tag{1}$$

where $M_\star$ is the stellar mass, $L_\star$ the bolometric luminosity, F_ν is the flux density (in units of energy per unit time per incident unit area per unit frequency), χ_ν is the dust opacity as a function of frequency and χ_F is the flux averaged opacity. This equation is very easy to solve in the case of a constant χ_F (cf. Goldreich and Scoville, 1976) yielding

$$v^2 = v_\circ^2 + \alpha\left(\frac{1}{r_\circ} - \frac{1}{r}\right), \tag{2}$$

where

$$\alpha = \frac{\chi_F\, L_\star}{2\pi c} - G\,M_\star$$

It might imply that the outflow velocity would get its terminal value at a distance comparable with the dust formation distance. As a matter of fact, however, χ_F is far from constant. At first sight, it might appear that it would decrease as we proceed outwards. Stellar radiation in the visible and near infrared is absorbed by the dust and then reemitted in longer wavelengths. The latter interacts more weakly with the dust and tends to escape. Two important considerations are, however, missing from this simplistic argument, namely (1) the dust to gas ratio is not constant throughout the envelope. Dust flows through the gas supersonically. The dust-to-gas ratio is reduced therefore by the ratio of v_g/v_d where v_g is the gas outflow velocity and v_d is the dust outflow velocity. The dust reaches its terminal velocity much faster than does the gas and hence in the inner parts of the envelope the opacity is most significantly reduced and it grows gradually outwards (Netzer and Elitzur, 1993). This drift phenomenon has been known, but its implication on the opacity has been disregarded by many authors. It is important for low mass loss rate stars, where below a certain mass loss rate interaction of radiation and dust can no longer overcome the stellar gravity. As the mass loss rate grows, dust and gas are better dynamically coupled and v_g/v_d is closer to 1. (2) The idea that the mutation of the spectral distribution of the radiation within the dust shell tends to decrease the flux averaged opacity, is not always correct. In oxygen stars (and VX Sagittarii falls within this category), the optical properties of the dust are dominated by silicates. The latter have a resonance in absorption at a wavelength of $9.7\,\mu$m. The secondary radiation is capable of interacting with the dust even more efficiently than does the incident stellar radiation and therefore χ_F may increase. Due to the combination of both effects we are likely to see acceleration at larger distances. A full quantitative calculation shows that even

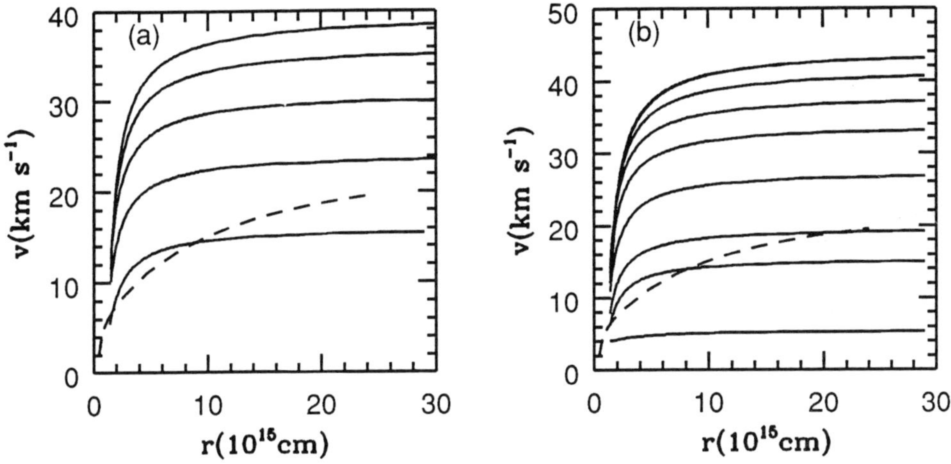

Fig. 1.   The velocity profiles for oxygen supergiants with a luminosity of $4 \times 10^5 L_\odot$ and (a) a mass of $30 M_\odot$ and mass loss rates (from the top to the bottom) in $M_\odot$ yr^{-1} of 3×10^{-4}, 1×10^{-4}, 3×10^{-5}, 1×10^{-5}, 5×10^{-6} (b) a mass of $10 M_\odot$ and mass loss rates in $M_\odot$ yr^{-1} of 3×10^{-4}, 1×10^{-4}, 3×10^{-5}, 1×10^{-5}, 3×10^{-6}, 1×10^{-6}, 6×10^{-7} and 3×10^{-7}. The dashed curve is a one that fits the observation of Chapman and Cohen.

though the rise in χ_F makes the acceleration more gradual than what eq. 2 would imply, it is short of explaining the phenomenon observed in VX-Sagittarii.

As an additional possibility, Chapman and Cohen (1986), suggested that dust grains keep growing as a results of accretion of material. Combined with the dust drift effect and radiative transfer in silicates, it might give rise to the desired acceleration pattern, provided that properties of the accreting species, such as density, abundance and condensation temperature have the appropriate values. Here lies also the handicap of such a model: a material that condenses at a relatively high temperature, will be depleted already at small distances, contributing virtually nothing to a gradual pattern of acceleration; a material that condenses at low temperature (such as ice) will do it so far away, that density would be low and kinetic considerations will slow its accretion and as a result there will be a very small contribution to acceleration. There must be some minimal abundance of the species and also a low solid density, so that the geometrical cross section of the grains will increase significantly as a result of accretion. The latter requirement remains valid even if we have in mind that cross sections are bigger because of deviation from sphericity. As we see in figure 2, we can match all those parameters to get gradual acceleration, but it is very hard to think of a species with the desired parameters. Moreover, if there were a one, the phe-

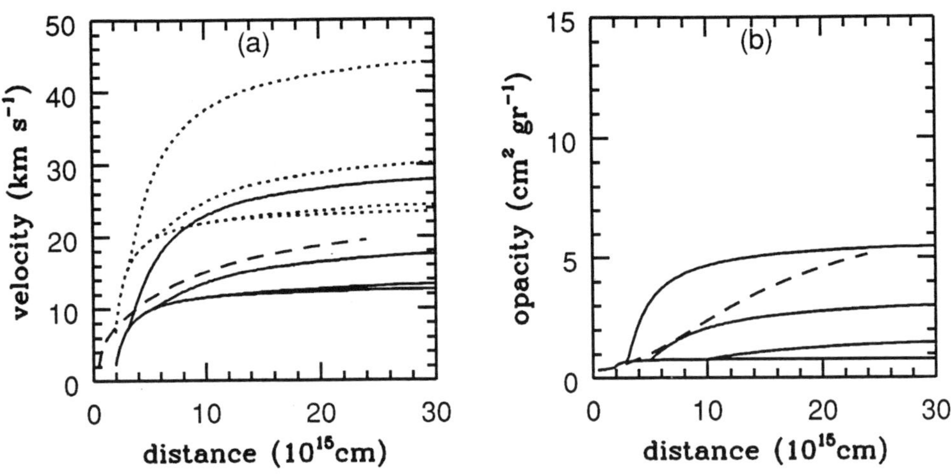

Fig. 2.   (a) the gas outflow velocity (solid lines) and the dust outflow velocity (dotted lines) for a series of models with where the initial grain size is 0.05 μm (the density of silicates is 3.3 g cm^{-3}). It is assumed that the grains start growing at a distance (from the top to the bottom) of 3×10^{15} cm, 5×10^{15} cm, 1×10^{16} cm, 3×10^{16} cm. The dashed line is an estimate by Chapman and Cohen (1986) for the velocity profile of VX Sagittarii. (b) The opacities according to the the same models and the opacity (dashed) required to produce the appropriate velocity profile for VX Sagittarii. The accreting material is assumed to have a density of 1 g cm^{-3} and a relative abundance by mass of 1:250

nomenon of gradual acceleration would be much more common in late type stars than we actually see. Observations done by Bowers and Johnston (1990) suggest that even though there is gradual acceleration in late type supergiants, it is not as gradual as in the case of VX Sagittarii, which is exceptional. It can usually be explained by radiative transfer alone. Therefore a different kind of a model is required.

3. The Stellar Evolutionary Model

As we have seen, no model based on radiative transfer could explain the difference observed in VX Sagittarii between the outflow velocity as observed in the H$_2$O maser line and the one observed in OH 1612 MHz. It is needless to say that no radiative transfer can explain the gap between the outflow velocity inferred from the latter maser line and the one inferred from CO lines which is about 30 km s^{-1} and is dominated by much further parts of the circumstellar envelope. Those parts of the envelope are far too optically thin to support such effects. The latter gap, however, relates to a zone whose crossing time is of the order of 10^3

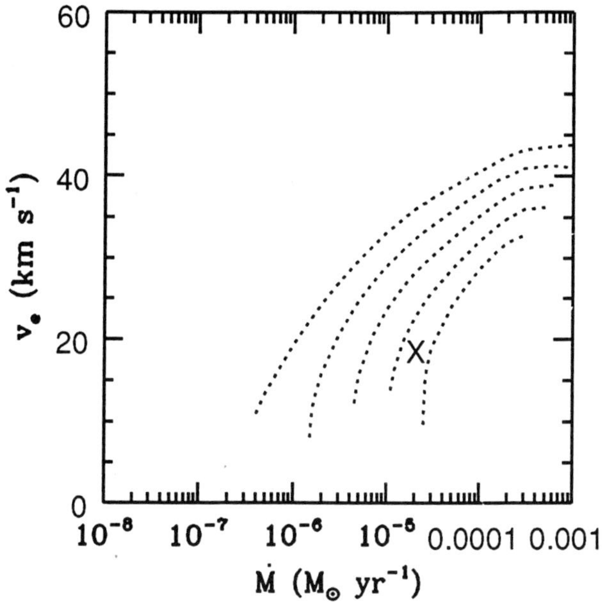

Fig. 3.   Terminal outflow velocity versus mass loss rate for calculated models stars. The effective temperature is 2500 K and the luminosity is $4 \times 10^5 L_\odot$. The grain size is assumed 0.05 μm. The curves correspond, from top to bottom, to supergiants with masses of 10, 20, 30, 40 and 50 $M_\odot$.

years and hence we cannot rule out the impact of stellar evolutionary effects. The discussion in section 2 may give us an important clue.

For a model star of given dust properties, the outflow velocity is determined chiefly by the luminosity and the mass loss rate. To the extent that the stellar mass is important, it tends to lower the outflow velocity. For small mass loss rates, when the coupling of dust and gas is weak and momentum transfer to the gas is not efficient, the gravitational force can set a lower limit on the mass loss rate driven by interaction of dust and gas.

As mentioned in section 1, the mass loss rate of VX Sagittarii is about $2 \times 10^{-5} M_\odot$ yr^{-1}. Ignoring the effect of gravity we would get for it an outflow velocity of about 40 km s^{-1}. In order to get the outflow velocity of 19 km s^{-1} observed in the OH maser shell (which is, as mentioned, at about 80 stellar radii), we must assume that the stellar mass should be about 40-50 $M_\odot$. If we assume that in the past the mass loss rate of VX Sagittarii was as high as $1 \times 10^{-4} M_\odot$ yr^{-1}, which is not unusual for late type supergiants, we get that due to the better coupling of dust to gas, the outflow velocity must have been about 30 km s^{-1}, which is what we observe in the CO lines. It seems therefore that the effect that we observe is a gradual decrease of the mass loss rate over the last several thousand years. Moreover: The current mass loss rate is not much above the minimum limit, as we see in figure 3. The outflow velocity is very strongly dependent upon the mass loss rate and hence, even on a short time scale like

200 years, which is the crossing time up to the OH maser shell, a small change in the mass loss rate may have resulted in a significant decrease of the outflow velocity. That might be the reason why the outflow velocity observed in H_2O and OH 1667 MHz masers is smaller. Thus it seems likely that VX Sagittarii is in transition towards a new stage in its evolution, as the current phase of mass loss is being terminated. Obviously, given the value of its mass, its evolution must lead eventually to a type II supernova explosion.

Acknowledgements

I wish to thank Gary Mamon for helpful comments, as well as the computer center of the Technion–Israel Institute of Technology and the Astronomy department, Indiana University for their technical help.

References

Bowers, P. F. and Johnston, K. J.: 1990, *Ap. J.* **354**, 676
Chapman, J. M. and Cohen, R. J.: 1986, *M.N.R.A.S.* **220**, 513
Goldreich, P. and Scoville, N.: 1976, *Ap. J.* **205**, 144
Knapp, G. R., Sutin, B. M., Phillips, T. G., Ellison, B. N., Keene, J. B., Leighton, R. B., Masson, C. R., Steiger, W., Veidt, B. and Young, K.: 1989, *Ap. J.* **336**, 822
Humphreys, R. W., Strecker, D. W. and Ney, E. P.: 1972, *Ap. J.* **172**, 75
Netzer, N.: 1989, *Ap. J.* **342**, 1068
Netzer, N. and Elitzur, M.: 1993, *Ap. J.* **410**, 701

THE COMPLEX Hα PROFILE OF THE HYPERGIANT HR 8752 DURING 1970 - 1992.

J. SMOLINSKI[1], J. L. CLIMENHAGA[2], Y. HUANG[3], SH. JIANG[3],
M. SCHMIDT[1] and O. STAHL[4]

[1]*N.Copernicus Astronomical Center, 87-100 Torun ul. Chopina 12/18, Poland*
[2]*Department of Physics and Astronomy, University of Victoria, Victoria B.C., Canada*
[3]*Beijing Astrophysical Observatory, Beijing, China*
[4]*Landessternwarte, Konigstuhl 12, D-6900 Heidelberg, Germany*

Abstract.

The profiles of Hα observed during the 1970 − 1992 period in the binary hypergiant HR 8752 (G0 Ia) are presented. We distinguish five typical Hα profiles designated as A, B, C, D and E types according to the number of emission and absorption features. The profiles of Hα are complex and contain several emission and absorption components, with: −130 km/s in emission or absorption, −84 km/s in absorption, −49 km/s in emission and about +6 km/s in emission. All of them are rather stable in radial velocities except of the main absorption component in the P Cygni profile with −84 km/s. The frequency of appearance and the periods of duration of the occurrence of the components is discussed. The duration times range between about 3 to 10 months for various components. The red emission component E_2 is particularly interesting. Possible explanations of its origin are discussed.

A long-term acceleration of the absorption component in the P Cygni profile is found; it can be interpreted as monotonous acceleration of the stellar wind.

1. Introduction

De Jager (1980), following a suggestion by Van Genderen, suggested the name hypergiants for the group of super-supergiants. Hypergiants form a well defined class of supergiants on the H-R diagram (Humphreys, 1992).

HR 8752 = HD 217476 (G0 Ia), the most luminous late-type star in our Galaxy is a hypergiant and lies near the Humphreys - Davidson − De Jager limit in the H-R diagram. It is a massive binary system (Stickland and Harmer, 1978), with period of about 620 days (Smolinski, 1986), contained in a common expanding envelope and surrounded by the HII region, the site for observed [NII] (Sargent, 1965) and radio emissions (Smolinski *et al.*, 1977; Higgs *et al.*, 1978). It has been intensively observed because of its conspicious location close to upper limit of stellar occurrence in the H-R diagram (Piters *et al.*, 1988) and its significance for understanding evolution of massive stars.

In 1969, one of us (J.S.) began a series of systematic spectroscopic observations of HR 8752 in his long−term observational program of supergiants of spectral type F0-K5, at the Dominion Astronomical Observatory, Victoria, Canada (Smolinski, 1971; Smolinski *et al.*, 1989). This star was also observed and analysed by Lambert and coworkers (see Lambert and Luck, 1978).

The investigations of HR 8752 confirmed the complexity of the circumstellar surrounding and led to the discovery of new phenomena in the star's extended

Space Science Reviews **66**: 231–236, 1994.

© 1994 *Kluwer Academic Publishers. Printed in Belgium.*

atmosphere and expanding envelope, not completely understood at present. One of them is the shape of the Hα profile, which consists of several emission and absorption components, varying in time. Imhoff (1977) proposed some explanations of it. It seems to us, that the detailed investigations of long-term observations of the Hα profile might shed some light on the nature of this system.

2. Spectroscopic observations and measurements

This paper is mainly based on observations obtained at the Dominion Astrophysical Observatory, Victoria, Canada, using the 48 inch (120 cm) tele-

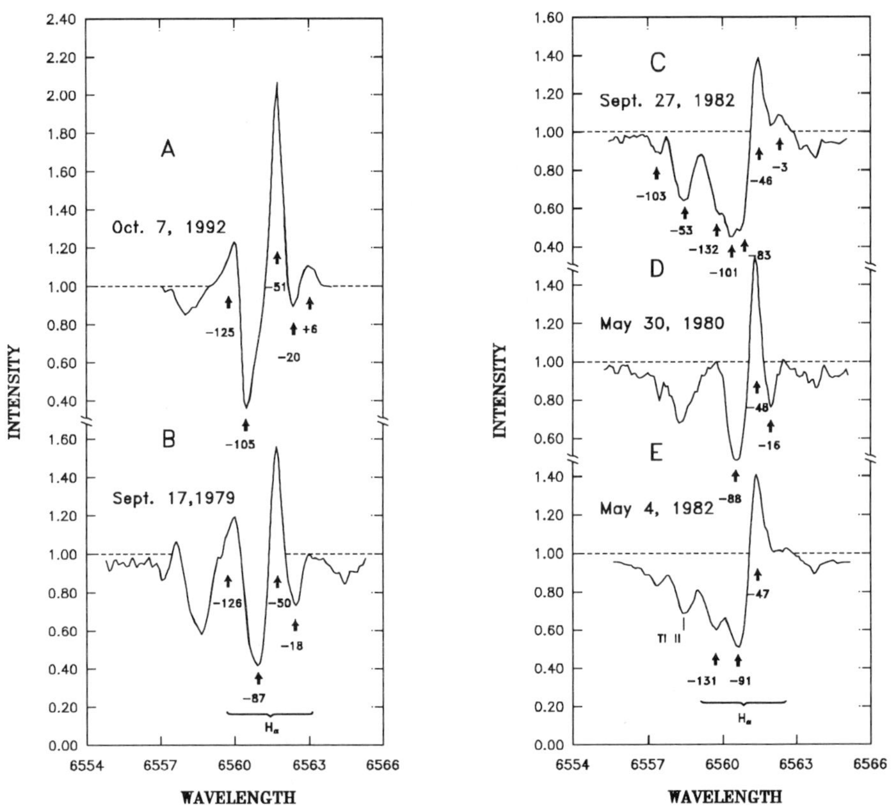

Fig. 1.   Typical profiles of Hα observed during an observing period of about 23 years. The horizontal broken lines show the continuum levels of the spectra. The numbers give the radial velocities of the components. The profiles are plotted according to the number of emission and absorption features.

scope in the coude focus (Smolinski, 1971). Besides, other observations have been made: in 1988 with the 2.2 m telescope at Calar Alto, Spain with a coude spectrograph, in 1991 with the 70cm telescope of the Landessternwarte Konigstuhl,

Heidelberg with a fiber-linked echelle spectrograph and in 1992 with the 2.16 m telescope of the Beijing Astronomical Observatory at its Xinglong Station with the French ISIS spectrograph (dispersion 2.7 A/mm).

We discuss in this paper over one hundred high dispersion spectra in the spectral region around Hα. For all spectra intensity tracings have been obtained and the radial velocities of all components of the Hα profile have been measured in the same way.

3. Results and discussion

On the basis of our material we distinguish five characteristic shapes of the Hα profile during the 23 years period of investigation. Some characteristic examples of these profiles are shown in Fig.1 in order of the number of emission and absorption components. The complexity of the Hα profile, which consists of several emission and absorption features is apparent from the Figure. We designated these typical profiles as A, B, C, D and E with the following characteristics.

TABLE I

The Estimated Duration Time of Typical Profiles.

Type of profile	Appearance number	Estimated duration time
A	5	3 months
B	6	10 months
C	2	4 months
D	5	8 months
E	2	3 months

Profile A: It consists of three emission components, i.e. E_0, E_1 and E_2 with radial veloctics (V_r) of about -130, -49 and $+6$ km/s respectively, and of two absorption components, namely A_2 and A_3 with V_r of about -100 and -84 km/s respectively.

Profile B: There are two emission components in this profile, i.e. E_0 and E_1 having the same V_r as in profile A. The E_0 component changes its intensity with time (being in emission or in absorption).

Profile C: It consists of two emission components E_1 and E_2. The component E_0 occurs only in absorption. There are three absorption components A_1, A_2 and A_3 with V_r of about -130, -100 and -84 km/s respectively.

Profile D: It is a classical P Cygni profile. There is the emission component E_1 and an absorption component which we designated as A_2+A_3 to indicate its possible connection with these two components.

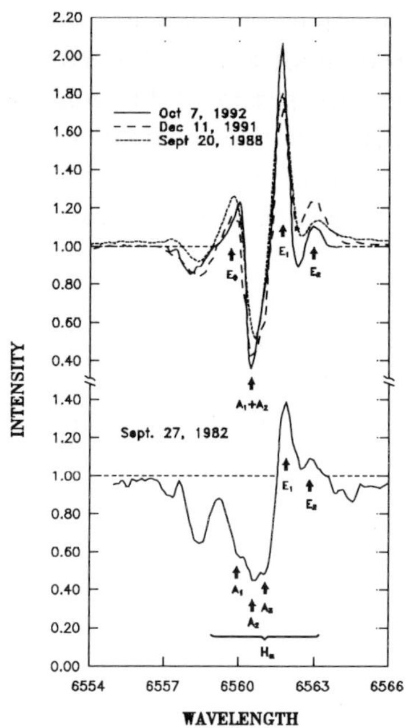

Fig. 2. Profiles of $H\alpha$ with the red emission component E_2. The radial velocities of the components are given in Table 2.

Profile E. It is also a classical P Cygni profile containing E_1 and A_2+A_3 but with an additional absorption component A_1 seen also in profile C.

It is worth noting that the $H\alpha$ profile has had the P Cygni shape during the full 23 years period of observation. However, the additional emissions, E_0 and E_2, and absorptions A_1 and the splitting of A_2+A_3 into two separate components have only appeared during some specific periods. The emissions,

TABLE II

Radial Velocities of Absorption and Emission Components of $H\alpha$ Profile Shown in Figure 2.

Date	Radial Velocity (km s^{-1})						
	A_1	A_2	A_3	A_2+A_3	E_0	E_1	E_2
Sept. 27, 1982	−132	−101	−83	−96	−	−46	−3
Sept. 20, 1988	−	−109	−81	−102	−143	−49	+8
Dec. 11, 1991	−	−	−	−100	−140	−51	+15
Oct. 7, 1992	−	−	−	−105	−125	−51	+6

E_0 and E_2 change in intensity, while E_1 is constant. The radial velocities of the emission and absorption components are rather stable but for the systematic changes of V_r of the (A_2+A_3) absorption component.

The estimated duration times and the number of appearances of the various particular profile types are given in Table 1. We have drawn the following conclusions on the basis of the results.

1. All the five classes of $H\alpha$ profile have repeatedly appeared at least twice during our period of observations.

2. Particular profile types have appeared during three up to ten months (see Table 1). We consider these times as minimal due to gaps in the observations. Accordingly, they are a result of some kind of long-term processes.

The profile changes will be discussed in details in the succesive papers.

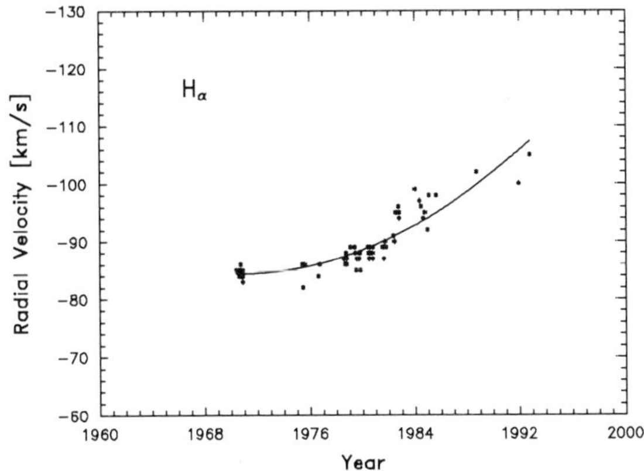

Fig. 3.   The changes of the radial velocities of the absorption component in P Cygni profile. The solid line is the second order fit. Note the acceleration of the stellar wind since 1970.

The emission component E_2 is the most enigmatic one (Smolinski *et al.*, 1982). Fortunately, we have observed it four times in the $H\alpha$ profile. It was also noticed by Arrelano Ferro (1985). Its intensity changed and it may appear as a broad emission. The four profiles of $H\alpha$ which contain E_2 are shown in Fig. 2 and its radial velocities are given in Table 2. Its time is at least three months. In three cases it appears together with the emission components E_0, although on Sept. 27, 1982 it was in absorption as A_1. The origin of E_2 and its interpretation is not clear. One may only suggest, that it can be connected with the binary nature and may be it is a result of the heating of the stellar wind of the primary by the secondary or it is related to a possible accretion disc around the secondary.

As was already mentioned the radial velocity of the absorption component of the P Cygni profile changes with time. There are time interwals when this absorption has splitted into two absorption components A_2 and A_3. Such a splitting is only seen in high resolution spectra. The origin of the splitting into A_2 and A_3, the time of its appearance and the duration times are not known. It seems, that the profile with this absorption component has some similarity to those observed in the long-period Cepheids χ Cyg and SU Vul mentioned by Luck and Lambert (1981), which might result from shock waves driven by the stellar pulsation.

The changes of the radial velocities of the absorption component A_2+A_3 in the P Cygni profile of Hα are shown in Fig.3. Very apparent is the increase of V_r since 1970. This phenomenon may be interpreted as a monotonous acceleration of the expanding envelope around HR 8752 system or as the acceleration of the stellar wind. It seems that the acceleration began in 1970, at the time when HR 8752 (Smolinski, 1971) showed enhanced mass loss, which may have resulted in shell ejection. Assuming this shell to be the seat of the Hα absorption component, its relative position with reference to the star over its subsequent years is found by integrating the velocity over the time since 1970. The importance of these observations is that it gives the opportunity to probe the velocity–distance profile of the stellar wind of this hypergiant. Since further observations are very desirable, we will continue the investigations of this acceleration.

Acknowledgements

This work was in part supported from the grant no. 2 1237 91 01 financed by the Polish Committee for Scientific Research. We appreciate the Referee's useful comments.

References

De Jager, C.: 1980, *The Brightest Stars*, D. Reidel Publ. Co., Dordrecht, Holland

Ferro, A. A.: 1985, *Rev. Mexicana Astron. Astrof.* **11**, 113.

Higgs, L. A., Feldman, P. A. and Smolinski, J.: 1978, *Astrophys. J.* **220**, L109.

Humphreys, R. M.: 1992, in C. De Jager and H. Nieuwenhuijzen (eds.), *Instabilities in evolved super- and hypergiants*, 13.

Imhoff, C. L.: 1977, *Astrophys. J.* **214**, 773.

Lambert, D. L. and Luck, R. E.: 1978, *Monthly Notices Roy. Astron. Soc.* **184**, 405.

Luck, R. E. and Lambert, D. L.: 1981, *Astrophys. J.* **245**, 1018.

Piters, A., De Jager, C., and Nieuwenhuijzen, H.: 1988, *Astron. Astrophys.* **196**, 115.

Sargent, W. L. W.: 1965, *Observatory* **85**, 33.

Smolinski, J.: 1971, in *Colloquium on Supergiant Stars*, ed. M. Hack, Osservatorio Astronomico di Trieste, Trieste, p.68

Smolinski, J., Feldman, P. A. and Higgs, L. A.: 1977, *Astron. Astrophys.* **60**, 277

Smolinski, J., Climenhaga, J. L., and Fletcher, J.M.: 1982, *Inf. Bull. Var. Stars*, No. 2229.

Smolinski, J., Climenhaga, J. L., and Fletcher, J.M.: 1986, *Symposium on Luminous Stars and Associations in Galaxies*, eds. De Loore, C.W.H., Willis, A.J., Laskarides, 89.

Smolinski, J., Climenhaga, J. L., and Fletcher, J.M.: 1989, *Physics of Luminous Blue Variables*, eds. Davidson, K., Moffat, A. F. J., Lamers, H. J. G. L. M., 131.

Stickland, D. J. and Harmer, D. L.: 1978, *Astron. Astrophys.* **70**, L53.

SPECTRAL ANALYSES OF WOLF-RAYET STARS: THEORY, RESULTS, CONCLUSIONS

WOLF-RAINER HAMANN*

Institut fur Theoretische Physik und Sternwarte der Universitat Kiel, D-24098 Kiel, Germany

Abstract. Stratified Non-LTE models for expanding atmospheres became available in the recent years. They are based on the idealizing assumptions of spherical symmetry, stationarity and radiative equilibrium. From a critical discussion we conclude that this "standard model" is basically adequate for describing real Wolf-Rayet atmospheres and hence can be applied for quantitative spectral analyses of their spectra.

By means of these models, the fundamental parameters have been determined meanwhile for the majority of the known Galactic WR stars. Most of them populate a vertical strip in the Hertzsprung-Russell diagram at effective temperatures of $\approx 35\,\mathrm{kK}$, the luminosities ranging from $10^{4.5}$ to $10^{5.9}\ L_\odot$. Only early-type WN stars with strong lines and WC stars are hotter. The chemical composition of WR atmospheres corresponds to nuclear-processed material (WN: hydrogen burning in the CNO cycle; WC: helium burning). Hydrogen is depleted but still detectable in the cooler part of the WN subclass.

Different scenarios for the evolutionary formation of the Wolf-Rayet stars are discussed in the light of the empirical data provided from the spectral analyses. Post-red-supergiant evolution can principally explain the basic observational properties, except the rather low luminosities of a considerable fraction of WN stars. Among the alternative scenarios, close-binary evolution can theoretically produce the least-luminous WN stars. However, final conclusions about the evolutionary formation of the WR stars are not yet possible.

Key words: stars: Wolf-Rayet – stars: expanding atmospheres – stars: mass-loss – stars: fundamental parameters – stars: element abundances

1. Theory

Wolf-Rayet spectra are characterized by bright and broad emission lines, in some cases accompanied by blue-shifted absorption features (P-Cygni profiles). Their spectral appearance is the consequence of a strong stellar wind, which provides a large line-emitting volume and broadens the lines by Doppler shift due to radial expansion. This explanation for WR spectra was suggested already by Scheiner (1890) and Beals (1929) and is nowadays widely accepted, while Underhill (e.g., 1991) still favors an alternative model.

1.1. MODEL CALCULATIONS

The spectral analysis of WR stars, i.e. the determination of their parameters and atmospheric composition, requires adequate models which account for the extreme non-LTE situation, the spherical symmetry and the radial expansion. A first approach to the quantitative modelling of WR spectra was made by Castor & Van Blerkom (1970) who treated the non-LTE radiation transfer for one "representative" point in the WR atmosphere by means of photon escape probabilities

* Present address: *Universitats-Sternwarte, Scheinerstr. 1, D-81679 Munchen, Germany*

Space Science Reviews **66**: 237–251, 1994.

© 1994 *Kluwer Academic Publishers. Printed in Belgium.*

("Sobolev approximation"). While they applied their method to the He II lines of two WN6 stars, it was used for various abundance determinations for H and CNO elements by, e.g., Willis & Wilson (1978), Smith & Willis (1982, 1983), Nussbaumer *et al.* (1979), Nugis (e.g. 1982a,b; 1991). Another class of one-point models applied nebula approximations (recombination theory) to WR atmospheres (Smith & Hummer 1988, Torres 1988). The results from these one-point model analyses remained in parts contradictory.

Today we know that in most WR atmospheres the ionization varies strongly with the radius in the sense that recombination proceeds towards outer layers. Thus different spectral lines are formed in different parts of the atmosphere, making one-point models obsolete.

Stratified non-LTE models for expanding atmospheres were developed during the last decade independently by Hillier and by Hamann *et al.* ("Kiel group"). Both codes can treat the line formation transfer in the comoving frame (i.e. not relying on Sobolev's approximation). Hillier (cf. 1990) solves the full set of equations by linearization, suppressing those terms coupling more than (optionally) one, three or five spatial points. The Kiel group has adapted the "iteration with approximate lambda operators" (Hamann 1985, 1986, 1987; Hamann & Wessolowski 1990; Koesterke *et al.* 1992) to the situation of spherical expansion. In the "local" version, both iteration methods are presumably equivalent. Many further details are handled differently in both codes. Nevertheless the results from both codes agree remarkably well.

The available model atmospheres for WR stars (both versions) are based on simplifying assumptions called the "standard model", which represents the most-idealized form of an expanding atmosphere. Spherical symmetry, together with stationarity and homogeneity, implies that the equation of continuity relates the velocity field $v(r)$ and the density structure $\rho(r)$ for a given mass-loss rate $\dot{M}$. No self-consistent hydrodynamic calculations for WR atmospheres are possible yet, as the acceleration mechanism is not identified. Instead, the velocity field must be pre-specified *ad hoc*, and one usually assumes the law

$$v(r) = v_\infty \left(1 - \frac{R_*}{r} \right)^\beta , \tag{1}$$

for its supersonic part, where the terminal velocity v_∞ and the exponent β are free parameters ($\beta = 1$ seems to be a good choice in most cases). Fortunately, for strong winds the results are not sensitive to details of the velocity law, as the lines are formed at large radii where the terminal velocity is almost reached. The electron temperature stratification follows from the assumption of radiative equilibrium.

Both codes (from Hillier and from Hamann's group) have achieved about the same degree of sophistication in their latest versions. Complex model atoms for, e.g., N or C can be treated with typically ≈ 100 energy levels and ≈ 1000 line transitions. Low-temperature dielectronic recombination (LTDR) is accounted

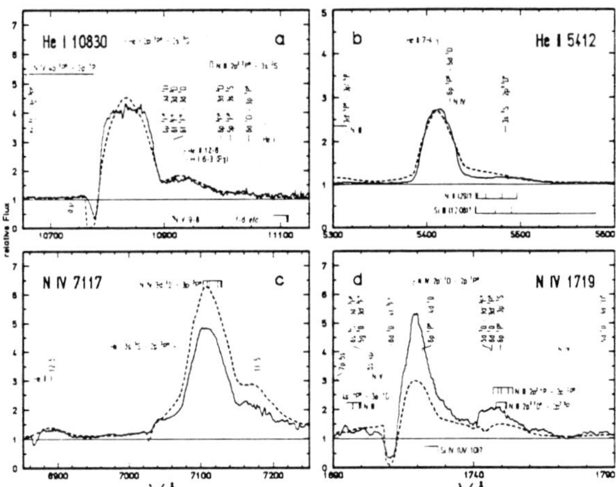

Fig. 1. Example fit to the observed WN6 spectrum of WR 136 (full-drawn) by a model (dashed; parameters: $T_* = 55\,\mathrm{kK}$, $L = 10^{5.5}\,\mathrm{M_\odot/yr}$, $v_\infty = 1700\,\mathrm{km/s}$, $\beta_N = 1.5\,\%$, $\beta_H = 12\,\%$). After Hamann *et al.* (1993b)

for in an approximate way. Frequency redistribution of line photons by electron scattering is included in the Formal Integral. The neglection of blanketing effects from "metal" lines is one deficiency of the present models.

1.2. THE VALIDITY OF THE "STANDARD MODEL"

The assumptions of the "standard model" can only be justified *a posteriori* from the comparison between its predictions and the observations. Fig. 1 shows, as a typical example, a comparison between synthetic and observed line profiles for the WN6 star WR 136. The four selected lines demonstrate that a satisfactory agreement of the main features is achieved. However, some minor discrepancies can also be recognized. The predicted absorption feature of the He I P-Cygni profile (Fig. 1a) is deeper than observed. Moreover, the flat-topped shape of the emission part is not fully reproduced. These discrepancies are still unexplained. The extended wings of the He II line (Fig. 1b) which are caused by electron-scattering frequency redistribution are weaker observed than calculated. This might indicate inhomogeneities in the wind (Hillier 1984), as the strength of the emission lines depends roughly on ρ^2 while the Thomson opacity scales linearly with the density. The observed nitrogen spectrum is much worse reproduced than the helium or hydrogen lines. Individual nitrogen lines can deviate by a factor of two (cf. Fig. 1c and d). These problems might be attributed to deficient atomic data, the neglection of metal line blanketing, or again to clumping effects.

A unique opportunity for testing the "standard model" is provided by the eclipsing WR+O binary system V444 Cygni. Hamann & Schwarz (1992) showed that both the continuum eclipse light curve and the helium emission lines in the composite spectrum can be consistently reproduced on the basis of the "standard

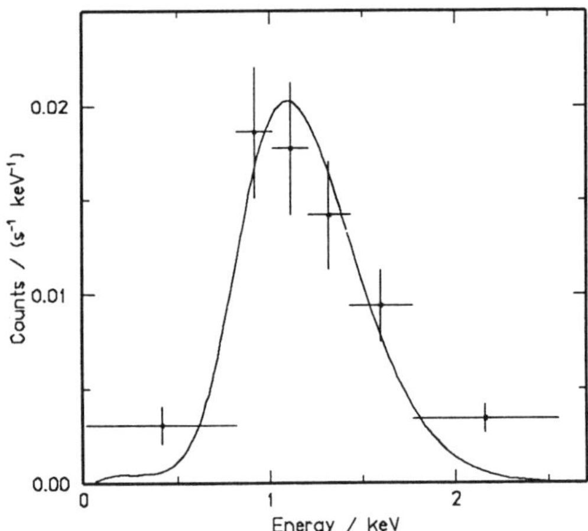

Fig. 2.   ROSAT observations of WR 110 (WN6), compared to a semi-empirical two-component model (cf. Baum *et al.* 1992). The hot component producing the X-rays has a temperature of $2\,10^6$ K and a filling factor of 5%, while the normal gas component is in radiative equilibrium and the stellar parameters are as obtained by Hamann *et al.* (1993a) from the analysis of the line spectrum. A column density of $n_\mathrm{H} = 5\,10^{21}\,\mathrm{cm}^{-2}$ is adopted for the interstellar absorption. From Wessolowski *et al.* (in preparation)

model".

X-ray emission has been observed from a couple of WR stars with EINSTEIN and ROSAT. Apart from possible binarity, this cannot be explained within the framework of the "standard model". Baum *et al.* (1992) constructed a semi-empirical two-component model, assuming that filaments of hot, shocked gas are homogeneously distributed throughout the WR atmosphere. The temperature T_X and the filling factor X_{fill} of the hot component are free parameters, while the ambient "normal" component is described by the "standard model" assumptions (especially, radiative equilibrium). The observed EINSTEIN count rates can be explained with that model for reasonable choices for T_X and X_{fill}. Recent, still preliminary calculations (cf. Fig. 2) reveal that even the spectral distribution of the X-rays as observed with ROSAT PSPC can be reproduced. These studies allow the important conclusion that the "standard model", although neglecting the X-ray emitting hot component, remains valid for describing the visual and UV spectrum. The number of ionizing X-ray photons is too small to compete with the high recombination rates present in the relatively dense WR winds.

Let us summarize the arguments supporting the "standard model". The main features of the observed spectra are reproduced. The light curve of V444 Cygni can be explained. The mass-loss rates derived from radio-emission measurements agree well with those from spectral line analyses (Schmutz *et al.* 1989), which is an argument against clumping being very effective. A hot component producing the observed amount of X-rays cannot significantly disturb the "standard model" atmosphere.

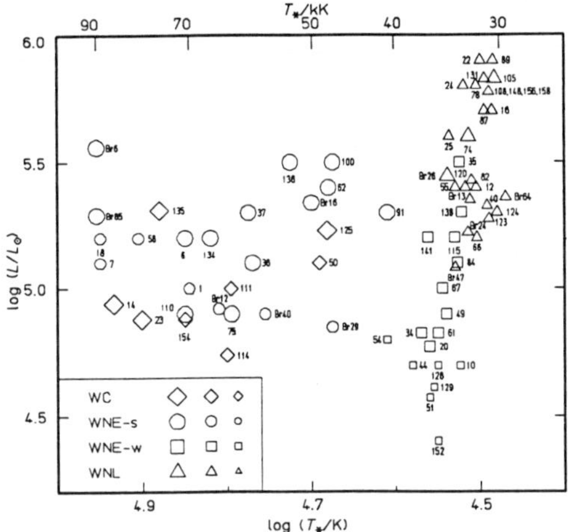

Fig. 3.   Empirical HRD of the WR stars. The Galactic stars are identified by their WR numbers (van der Hucht *et al.* 1981), while Br labels refer to LMC objects (Breysacher 1981). The spectral subgroups are distinguished by different symbols (see inlet), while their size indicates whether the mass-loss rate is large ($> 10^{-4.2}\ M_\odot$/yr), small ($< 10^{-4.8}\ M_\odot$/yr), or intermediate. Data taken from Hamann *et al.* (1993a) for Galactic WN, Koesterke *et al.* (1991) for LMC WN, and Koesterke *et al.* (in preparation) for WC stars.

On the other hand, the discrepancy between predicted and observed electron-scattering line wings seems to indicate some clumping effect. This would imply that spectral analyses tend to over-estimate the mass-loss rates by perhaps a factor of two.

Schmutz (1991) performed pilot calculations simulating metal line blanketing. His results suggest that the effective temperatures might be under-estimated by about 20% if blanketing is neglected. Consequently, the luminosities might be under-estimated by about 0.3 dex via a too small Bolometric Correction. Note that empirical luminosities can be derived only for those stars for which the distance is known from an open cluster or association membership.

2. Analyses and Results

Hillier's models have been applied so far mainly for pilot studies: He and H lines of the WN5 prototype WR 6 (Hillier 1987a,b), N and C lines in the same star (Hillier 1988), and carbon lines in the WC5 prototype WR 111 (Hillier 1989). Systematic spectral analyses with Hillier's code are presently performed at the University College London, and the results for 24 WN stars (accounting partly for He, H, C, N, O, Si) are presented for the first time at this conference (Crowther *et al.*, these proceedings).

The Kiel group started with pure-helium models, which were tested in a couple of pilot analyses (e.g. Hamann *et al.* 1988). A whole grid of models was published (Wessolowski *et al.* 1988) and applied for coarse analyses of 30 Galactic WR stars

(Schmutz *et al.* 1989). This study was extended recently to the vast majority (53) of the known Galactic single WN stars (Hamann *et al.* 1993a). 19 WN stars in the Large Magellanic Cloud (LMC) were analyzed by Koesterke *et al.* (1991). Detailed analyses of four WN stars with a determination of their hydrogen abundance were presented by Hamann *et al.* (1991). For a pilot study with nitrogen we selected the WN6 star WR 136 (Hamann *et al.* 1993b). The carbon models were tested in Kiel by application to the same star WR 111 (Hamann *et al.* 1992) Hillier had used. A whole grid of WC models for one fixed value of the carbon abundance (60% by mass) was published by Koesterke *et al.* (1992). A comprehensive analysis of 25 Galactic WC5-7 stars with individual He+C model calculations has been completed recently (Koesterke *et al.* in preparation).

The analyses of WN stars by means of pure-helium models performed by the Kiel group lead to the empirical Hertzsprung-Russell diagram (HRD) shown in Fig. 3. The WNL stars (*late* WN subtypes: WN7 ... WN9) are relatively cool ($T_* \approx 35\,\mathrm{kK}$)* and populate a strip between $\log L/L_\odot = 5.2 ... 5.9$. Stars with WNE-w spectra (*early* WN subtypes – WN2 ... WN6 – and *weak* lines) form an extension of that strip ($T_* \approx 35...40\,\mathrm{kK}$) to lower luminosities down to $\log L/L_\odot = 4.5$. The WNE-s stars (*early* subtypes, *strong* lines) are distinctively hotter ($T_* > 50\,\mathrm{kK}$), have the largest mass-loss rates, and scatter between $\log L/L_\odot = 4.9$ and 5.6. The WN stars of the LMC ("Br" labels) share their locations with their Galactic counterparts.

First results from recent analyses of Galactic single WC stars by means of He-C-models are also included in Fig. 3. Represented are eight stars of intermediate subtype (WC5-7) with known distances. These WC stars populate the same region in the HRD as the WNE-s stars.

The presence or absence of hydrogen in the helium-dominated WN atmospheres is not easy to assess, because all H I lines coincide with He II lines. Conti *et al.* (1983) inspected the emission line equivalent widths of the He II Pickering series, looking whether those series members blended with H I Balmer lines are systematically stronger than the unblended members. Hamann *et al.* (1993a) extended this study to their large sample. The occurrence of hydrogen is found to correlate strictly with the HRD position (cf. Fig. 4). Only the coolest WN stars (most WNL and part of the WNE-w) show hydrogen. But the discrimination does not follow the spectroscopic classification as WNL or WNE. WNE-s stars are hydrogen-free, except of WR 136 which has a very small hydrogen mass fraction (12% according to Hamann *et al.* 1993b) and is therefore drawn grey in Fig. 4.

The mass-loss rates of the Galactic WN stars are plotted versus luminosities in Fig. 5. Langer (1989b) expected from theoretical reasons that the mass-loss rates of hydrogen-free stars (corresponding to the open symbols in Fig. 5) should depend only on the current stellar mass as the only significant parameter. From the empirical data available at that time he suggested the relation

---

* T_* is the effective temperature referring to the radius of the hydrostatic core radius R_*. See Sect. 3 for discussion

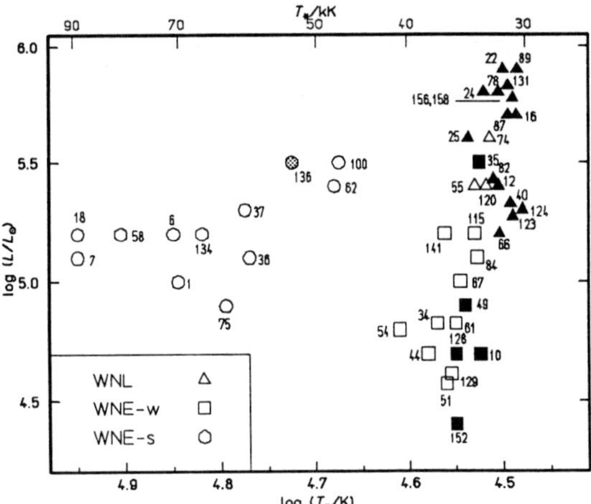

Fig. 4.   HRD of the subset of Galactic WN stars for which the presence (filled symbols) or absence (open symbols) of hydrogen has been established. The stars are again identified by their WR numbers, while the symbol shape refers to the spectral subclass (see inlet). From Hamann *et al.* (1993a)

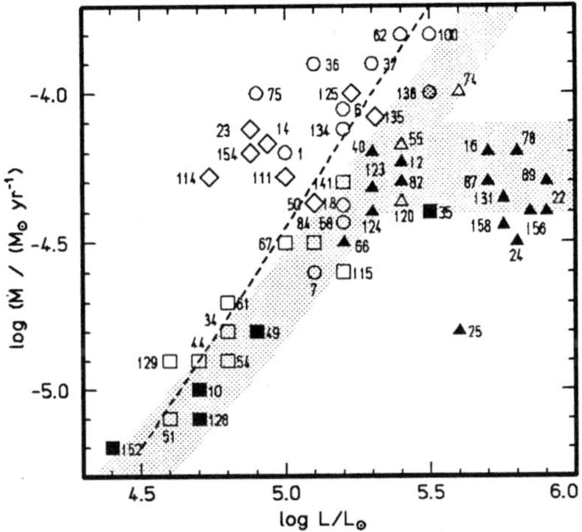

Fig. 5.   Mass-loss rates of Galactic WR stars versus their luminosities. Different symbol shapes distinguish between the spectral subclasses (cf. inlet of Fig. 3), while filled or open symbols indicate whether hydrogen is detectable or absent, respectively. The dashed line indicates a tentative relation for the hydrogen-free stars (Eq. 2). The shadowed regions represent the $\dot{M}$-L-relations adopted by Schaller *et al.* (1992) for their evolutionary calculations (horizontal strip: "WNL"; inclined strip: hydrogen-free stars).

$\log[\dot{M}/(M_\odot\mathrm{yr}^{-1})] = (0.6...1.0)\,10^{-7}(M/M_\odot)^{2.5}$. With the mass-luminosity relation for helium-burning stars (Langer 1989a) this yields the inclined, shaded strip in Fig. 5. (The upper limit shown is twice the lower limit, as applied by Schaller *et al.* (1992) for the second set of their evolutionary calculations.)

The results in Fig. 5 (still only considering the open symbols) indicate a steeper slope. Although there is a large scatter, the WN data may be roughly represented by a relation

$$\log \dot{M}/(M_\odot \mathrm{yr}^{-1}) = 1.5 \log L/L_\odot - 11.95 \,. \tag{2}$$

Interestingly, this line has the same slope (1.5) as found for OB stars, while it lies at higher $\dot{M}$ by about 2.0 dex. The four less-luminous WNE-w stars *with* hydrogen (filled squares) follow the just established $\dot{M}$-L-law for hydrogen-free WN stars. Most WC stars lie to the left of this relation.

The other WN stars with hydrogen (filled symbols, mainly WNL) have roughly constant $\dot{M}$, independent of L. The shaded horizontal strip in Fig. 5 indicates the range of $\log \dot{M}/(M_\odot \mathrm{yr}^{-1}) = -4.4 \dots -4.1$, as used by Schaller *et al.* (1992) for the "WNL" phase* of their tracks. The observed mass-loss rates are slightly lower (-4.5 ... -4.2).

3. Conclusions

3.1. THE EVOLUTIONARY STATUS OF THE WR STARS

More than a decade ago, Maeder (1982) discussed essentially four alternative scenarios for the formation of WR stars, all being designed to explain how nuclear-processed material can appear at the surface of massive stars.

Post-red-supergiant evolution: Continuous mass-loss peels down the hydrogen layers while the star evolves to a red supergiant (RSG), which makes the track then turning back to the blue side of the HRD.

Quasi-homogeneous evolution: Internal mixing, e.g. induced by rapid rotation, transports continuously the products of nuclear burning to the surface.

"Conti's scenario" O→Of→WN→WC: For very massive stars, the continuous mass-loss removes the hydrogen envelope before the star can evolve to the red.

Close-binary evolution: The hydrogen envelope is lost by mass transfer, preventing the star from evolving to the red side of the HRD. This might work also for less massive stars than the other scenarios.

The relevance of these scenarios can now be discussed in the light of the empirical data which became available in the meantime from the advanced spectral analyses described in the preceding section. A further source of empirical constraints, namely from stellar statistics, is not considered here.

The post-RSG scenario has been elaborated by the "Geneva group" (Maeder & coworkers) in a series of papers, and the most recent evolutionary tracks are displayed in Fig. 6. The empirical results from the spectral analyses are included in the same figure for comparison.

---

* In the context of evolutionary calculations, "WNL" or "WNE" means phases in which hydrogen is present or absent at the surface, respectively. The quotation marks shall distinguish these terms from the spectroscopic classification as WNL or WNE which refers to the relative strengths of N and He lines from different ionization stages.

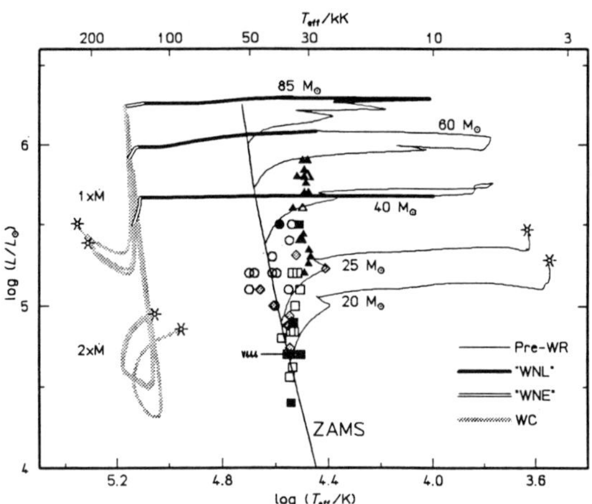

Fig. 6. Hertzsprung-Russell diagram. Discrete symbols indicate the locations of the Galactic WR stars as obtained from our analyses. Triangles, circles or squares refer to WNL, WNE-s or WNE-w stars, respectively, while filled or open style indicates whether hydrogen has been detected or not. WC stars are represented by shaded diamonds. The arrow points to the position of the WN component in the eclipsing binary system V444 Cygni (after Hamann & Schwarz 1992). The theoretical tracks are taken from Schaller *et al.* (1992) (metallicity $Z = 0.020$) for different initial masses (labels). The tracks are for the "standard" assumptions on the mass-loss rates ("$1 \times \dot{M}$"). Tracks for twice as strong mass-loss significantly differ only in the WC phase for which they are also shown ("$2 \times \dot{M}$"). The evolutionary stages, defined according to the surface composition, are indicated by different drawing styles (see inlet). Note that the subclass definitions "WNL" and "WNE" used here refer to the presence or absence of hydrogen at the surface and thus can be directly compared with the representation of the observed stars by filled or open symbols. The effective temperature used as abscissa refers to the hydrostatic core radius in case of evolutionary tracks (T_*), but to the visible radius (Rosseland optical depth of 2/3) in case of the analyzed stars ($T_{2/3}$). The problem of identifying temperatures is addressed in the text.

The luminosities are discrepant for the majority of WN stars. The lowest track producing WR stars starts with 40 $M_\odot$ at the zero-age main sequence (ZAMS). During the WC phase the luminosities decrease and pass through the observed range, if the high mass-loss tracks ("$2 \times \dot{M}$") are considered.

It is generally believed that the mass-loss during main-sequence and red-supergiant stages is driven by radiation pressure. The corresponding theory predicts a dependence of the mass-loss rate on the metallicity Z. In the calculations of post-RSG evolution (e.g. Schaller *et al.* 1992) therefore a scaling $\dot{M} \propto Z^{1/2}$ is adopted for these stages. As the mass-loss causes the evolution to proceed from the RSG to the WR stage before the supernova explosion can occur, WR progenitors must be more massive in low-metallicity environments, and hence the lower limit and the average of WN luminosities are expected to be higher. This prediction, however, is not corroborated by the empirical HRD (cf. Fig. 3) which reveals that the WN stars in the LMC are not more luminous than their Galactic counterparts. Thus, if the post-RSG scenario applies for the WR formation, the adopted scaling of $\dot{M}$ with Z seems to be not adequate.

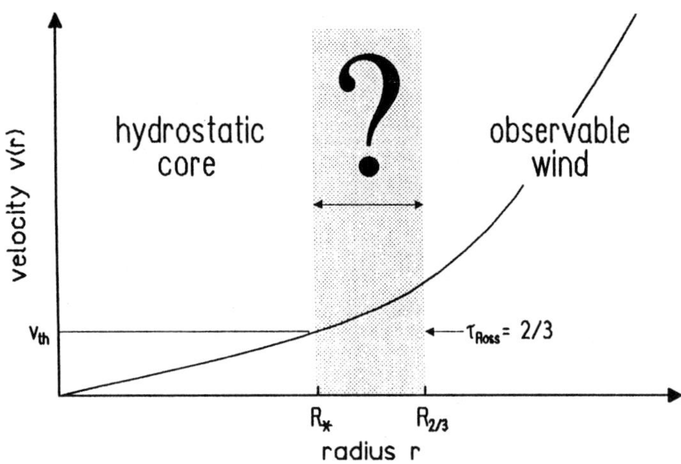

Fig. 7.   Lower part of the velocity field in a WR atmosphere (schematic representation). The radii of the sonic point (R_*) and of Rosseland optical depth 2/3 ($R_{2/3}$) are indicated. The range between R_* and $R_{2/3}$ is neither observable nor can it be described adequately by any available theory, and hence its extension is not known.

Discussing the effective temperatures of WR stars is not straightforward due to the spherical extension of their atmospheres. Contrary to the case of "normal" plane-parallel atmospheres, the definition of an effective temperature now depends on the choice of a reference radius ($L = 4\pi R_{\mathrm{eff}}^2 \, \sigma T_{\mathrm{eff}}^4$, with L being the well-defined quantity). The theoretical models for WR atmospheres are usually characterized by the parameter R_* which denotes the radius of the inner boundary located at some large optical depth (e.g., at $\tau_{\mathrm{Rosseland}} = 30$) or, practically almost identical, located at the base of the wind (e.g. at the sonic point). The effective temperature T_* which refers to this radius R_* has been used as abscissa of the empirical HR diagrams shown above (Figs. 3 and 4). However, R_* is not an observable quantity in a strict sense. Only the optically thin part of the atmosphere can be seen from the outside, and only for that part the adopted velocity law (cf. Eq. 1) is observationally confirmed. R_*, however, depends on the assumption that the same velocity law holds in the invisible, optically thick part as well, which is rather unlikely. The optical continuum radiation received by the observer emerges from a "photosphere" where the radial Rosseland optical depth reaches 2/3. Instead of R_*, this radius $R_{2/3}$ can be considered as an observable quantity which can be determined from a spectral analysis.

This situation is schematically sketched in Fig. 7. Typical WR winds have already reached 10% (WNL, WNE-w subtypes) to 70% (WNE-s, WC) of their terminal velocity before they become optically thin in the continuum* . Assuming

---

* This has, by the way, interesting implications for the acceleration mechanism of WR winds and the so-called momentum problem. Just in the strongest WR winds, the major part of acceleration happens already below the photosphere (whether by radiation pressure or another mechanism is an open question). Considering only that fraction of momentum transferred to the wind in its optically

the standard velocity law, the corresponding ratio $R_{2/3}/R_*$ amounts to 1.1 or 3, respectively (cf. Hamann *et al.* 1993a). In any case, $R_{2/3}$ lies at supersonic velocities, i.e. outside the hydrostatic domain.

Evolutionary models, on the other hand, are generally restricted to the hydrostatic domain and thus can only predict R_*. A hydrodynamical theory which could describe the moving subphotospheric layers is not available, as the driving mechanism is even not yet identified. Any theoretical "corrections" to obtain $R_{2/3}$ (cf. Maeder 1990, Schaller *et al.* 1992) are necessarily based on arbitrary assumptions, especially concerning the velocity law.

It must be emphasized that the region between R_* and $R_{2/3}$ ("?" in Fig. 7) is of unknown extension, as it is neither observable nor can it be described by an adequate theory. Consequently, a direct comparison between effective temperatures from evolutionary calculations and from empirical analyses is not possible. The only definite comparison can be made between T_* from evolutionary models and $T_{2/3}$ from analyses, expecting that $T_{2/3} < T_*$ because $R_{2/3} > R_*$, and attributing the difference to the extension of the supersonic, subphotospheric region.

This type of comparison was actually shown in Fig. 6. The empirical temperatures $T_{2/3}$ are smaller than the evolutionary predictions (except for the beginning of the "WNL" phase), and are therefore not in conflict with each other. If this type of tracks really applies, it seems that the photospheric radius stops shrinking when a photospheric temperature of $T_{2/3} \approx 30\,\mathrm{kK}$ is reached (at which, remarkably, helium becomes fully ionized and the WR mass-loss starts), while the contraction of the hydrostatic core proceeds further. In the WC phase the star appears about ten times larger than it would be in hydrostatic equilibrium.

The chemical composition of WN atmospheres, as far as analyzed yet, is obviously the result of (partial or complete) hydrogen burning in the CNO cycle. Helium is dominating, while hydrogen is depleted. Nitrogen contributes 1-2% to the mass, while carbon is underabundant (Hillier 1988; Hamann *et al.* 1993b; Crowther *et al.*, these proceedings). The latter reflect the timescales in the CNO reaction cycle: almost all CNO from the initial composition is quickly converted into N^{14}.

For 12 WN stars the hydrogen abundances have been determined quantitatively and can be compared in detail with the predictions of the post-RSG evolutionary calculations (cf. Fig. 8). For the group of WNL stars with high luminosity, the hydrogen mass fraction lies in the range predicted by the tracks with 40 or 60 $M_\odot$ initial mass. Striking is the lack of "WNE" stars (no hydrogen) at these high luminosities. The less luminous "WNL" stars have a small hydrogen abundance, contrary to the trend predicted by the evolutionary calculations (which, in their present state, fail anyhow in producing WN stars having such low luminosities).

---

thin part, the "single scattering limit" is never exceeded by more than a factor of ≈ 20 even for the subclasses with strongest winds (WNE-s, WC). This can probably be achieved by radiation pressure on lines, if multiple scattering and ionization stratification are properly taken into account (Lucy & Abbott 1993).

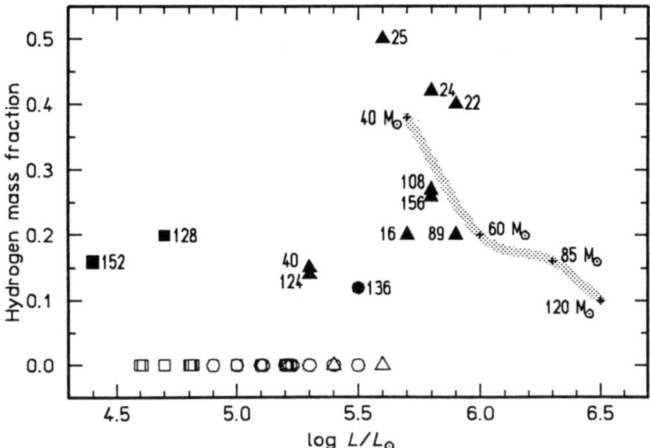

Fig. 8.   Hydrogen abundance versus luminosity for Galactic WN stars (labels: WR numbers). Most abundances are from the analyses by Crowther *et al.* (private communication, cf. also these proceedings), some others from Hamann *et al.* (1991: WR 22, 40, 128; 1993b: WR 136; unpublished: WR 89). The luminosities are throughout from Hamann *et al.* (1993a). Open symbols crowding at zero hydrogen abundance indicate the luminosities of those WN stars showing no hydrogen signature in their H I-Balmer/He II-Pickering decrement (after Hamann *et al.* 1993a). The shaded line gives the hydrogen abundance towards the end of the "WNL" phase as predicted by post-RSG evolutionary calculations (Schaller *et al.* 1992), the labels indicating the corresponding initial mass. The transition to the "WNE" phase (i.e. zero hydrogen) is predicted to happen almost instantaneously.

Remarkably, the existing "WNE" stars (no hydrogen detectable) populate exactly the same luminosity range as the latter group of "WNL" stars.

WC stars obviously present material which was partly processed by the 3α-reaction. The intermediate subtypes (WC5-7) analyzed so far by means of stratified models (Hillier 1989; Hamann *et al.* 1992; Koesterke *et al.*, in preparation) show carbon-to-helium ratios between 0.25 and 1.5 (by mass), which is in the range predicted by the post-RSG evolutionary models. While Koesterke *et al.* (in prep.) cannot find any correlations between the C/He ratio and other parameters, Smith & Hummer (1988) obtained a connection between abundances and subtype from their analyses with recombination theory. Smith & Maeder (1991) concluded that these results are confirming the post-RSG calculations. The determination of oxygen abundances in WC and WO stars should be subject to future work.

The above discussion about the relevance of the post-RSG evolution for the WR formation might be summarized as follows. Basic empirical properties of the WR stars are principally explained by the available evolutionary calculations. However, the scenario can only apply for the most luminous part of the Galactic WN sample. The agreement could be possibly improved by a suitable adjustment of the free parameters entering the models, e.g. the mass-loss during the course of the evolution. A hydrodynamic theory for the non-static, but sub-photospheric layers is required in order to explain the observed radii and related effective temperatures.

Quasi-homogeneous evolution leading to WR stars has been studied by Maeder (1987). Depending on the rotational velocity, a bifurcation occurs: while slowly rotating stars become red supergiants as usual, rapidly rotating stars are mixed in their interior and evolve at almost constant effective temperature. They first climb up and then descent in luminosity. By suitably adjusting the adopted mass-loss rate in the course of evolution, the observed range of luminosities can probably be reproduced. However, the expected chemical evolution is in conflict with the observation. Homogeneous evolution implies a continuous change of the surface abundances. When hydrogen is exhausted at the end of the "WNL" phase, a gradual carbon enrichment must become visible almost immediately after the onset of helium burning. I.e., "WNE" stars can probably not be produced within this scenario, and therefore it can not apply in general for the formation of the WR stars.

"Conti's scenario" producing WR stars by strong mass-loss works only for extremely massive stars, if radiation-driven winds on the main sequence are adopted as usual. Such luminous WR stars are not observed in our Galaxy. However, recent theoretical work on the pulsational stability of massive stars (Kiriakidis *et al.* 1993, Glatzel *et al.* 1993) predict so-called dynamical strange-mode instabilities being effective in massive stars. This might provide a mechanism driving violent mass-loss even from stars with less extreme masses and could possibly explain the formation of WR stars in the observed luminosity range without going through a RSG stage (cf. Langer, these proceedings). This scenario must be further elaborated and compared with all observational constraints.

Close-binary evolution can produce WR stars when the hydrogen envelope is lost by Roche-lobe overflow and a bare helium core is left (for an overview see, e.g., Vanbeveren 1991). In contrast to the other scenarios, this channel can work also for moderate stellar masses and is therefore predestinated for explaining those WR stars with lowest luminosities. Indeed, the WR component of the WN5+O6 binary V444 Cygni is located just in this part of the HRD, as indicated in Fig. 6 (after the analysis by Hamann & Schwarz 1992). The Galactic census by van der Hucht *et al.* (1988) comprises 12 WN+O, 12 WC+O and 2 WN+WC double-lined binary systems obviously containing two massive stars.

15 WN stars and, remarkably, only one (even uncertain) WC star are yet suspected as single-lined spectroscopic binaries (SB1). These stars may indeed have a low-mass companion, or the observed radial velocity variations might be due to pulsations or wind instationarities. However, pointing out the SB1 stars (most of them having fortunately independent luminosities because their distance is known from cluster/association membership) in the empirical HRD (Fig. 3) reveals no tendency towards lower luminosities, but rather the contrary. This does not corroborate the hypothesis that all WN stars of low luminosity were formed in close-binary systems. Why is their majority not known as SB1 systems? Can they escape from detection, or has the companion disappeared? Concluding, close-binary evolution is a nice theoretical possibility for the formation of WR

stars, but at present there is no empirical proof that a major subgroup of those WR stars which are apparently single or with a low-mass companion has evolved through this channel.

Present theories for the evolutionary formation of WR stars have been confronted with a bulk of empirical data obtained from recent spectral analyses of WR stars with advanced atmospheric models. It seems that different channels attribute to the WR formation. The most luminous WN stars might represent a post-RSG stage, but the quantitative agreement with corresponding evolutionary calculations is still poor. The less luminous part of the Population I WR stars evolved perhaps in close-binary systems. However, no definite conclusions can be drawn at present.

References

Baum E., Hamann W.-R., Koesterke L., Wessolowski U.: 1992, *Astron. Astrophys.* **266**, 402

Beals C.S.: 1929, *Monthly Notices Roy. Astron. Soc.* **90**, 202

Breysacher J.: 1981, *Astron. Astrophys. Suppl. Ser.* **43**, 203

Castor J.I., Van Blerkom D.: 1970, *Astrophys. J.* **161**, 485

Conti P.S., Leep E.M., Perry D.N.: 1983, *Astrophys. J.* **268**, 228

Glatzel W., Kiriakidis M., Fricke K.J.: 1993, *Monthly Notices Roy. Astron. Soc.* **262**, L7

Hamann W.-R.: 1985, *Astron. Astrophys.* **148**, 364

Hamann W.-R.: 1986, *Astron. Astrophys.* **160**, 347

Hamann W.-R.: 1987, in *Numerical Radiative Transfer*, W. Kalkofen (ed.), Cambridge University Press, p. 35

Hamann W.-R., Schwarz E.: 1992, *Astron. Astrophys.* **261**, 523

Hamann W.-R., Wessolowski U.: 1990, *Astron. Astrophys.* **227**, 171

Hamann W.-R., Schmutz W., Wessolowski U.: 1988, *Astron. Astrophys.* **194**, 190

Hamann W.-R., Dunnebeil G., Koesterke L., Schmutz W., Wessolowski U.: 1991, *Astron. Astrophys.* **249**, 443

Hamann W.-R., Leuenhagen U., Koesterke L., Wessolowski U.: 1992, *Astron. Astrophys.* **255**, 200

Hamann W.-R., Koesterke L., Wessolowski U.: 1993a, *Astron. Astrophys.* **274**, 397

Hamann W.-R., Koesterke L., Wessolowski U.: 1993b, *Astron. Astrophys.* (in press)

Hillier D.J.: 1984, *Astrophys. J.* **280**, 744

Hillier D.J.: 1987a, *Astrophys. J. Suppl.* **63**, 947

Hillier D.J.: 1987b, *Astrophys. J. Suppl.* **63**, 965

Hillier D.J.: 1988, *Astrophys. J.* **327**, 822

Hillier D.J.: 1989, *Astrophys. J.* **347**, 392

Hillier D.J.: 1990, *Astron. Astrophys.* **231**, 116

van der Hucht K.A., Conti P.S., Lundstrom I., Stenholm B.: 1981, *Space Sci. Rev.* **28**, 227

van der Hucht K.A., Hidayat B., Admiranto A.G., Supelli K.R., Doom C.: 1988, *Astron. Astrophys.* **199**, 217

Kiriakidis M., Fricke K.J., Glatzel W.: 1993, *Monthly Notices Roy. Astron. Soc.* **264**, 50

Koesterke L., Hamann W.-R., Schmutz W., Wessolowski U.: 1991, *Astron. Astrophys.* **248**, 166

Koesterke L., Hamann W.-R., Wessolowski U.: 1992, *Astron. Astrophys.* **261**, 535

Langer N.: 1989a, *Astron. Astrophys.* **210**, 93

Langer N.: 1989b, *Astron. Astrophys.* **220**, 135

Lucy L.B., Abbott D.C.: 1993, *Astrophys. J.* **405**, 738

Maeder A.: 1982, *Astron. Astrophys.* **105**, 149

Maeder A.: 1987, *Astron. Astrophys.* **178**, 159

Maeder A.: 1990, *Astron. Astrophys. Suppl. Ser.* **84**, 139

Nugis T.: 1982a,b, in *Wolf-Rayet stars: Observations, Physics, Evolution*, W.H. de Loore and A.J. Willis (eds.), IAU Symp. 99, p. 127, p. 131

Nugis T.: 1991, in *Wolf-Rayet Stars and Interrelations with Other Massive Stars in Galaxies*, Proc. IAU Symp. 143, K.A. van der Hucht and B. Hidayat (eds.), Kluwer, Dordrecht, p. 75

Nussbaumer H., Schmutz W., Smith L.J., Willis A.J., Wilson R.: 1979, in *The First Year of IUE*, A.J. Willis (ed.), UCL, London, p. 259

Schaller G., Schaerer D., Meynet G., Maeder A.: 1992, *Astron. Astrophys. Suppl. Ser.* **96**, 269

Scheiner J.: 1890, *Spectralanalyse der Gestirne*, Verlag von Wilhelm Engelmann, Leipzig

Schmutz W.: 1991, in *Wolf-Rayet Stars and Interrelations with Other Massive Stars in Galaxies*, Proc. IAU Symp. 143, K.A. van der Hucht and B. Hidayat (eds.), Kluwer, Dordrecht, p. 39

Schmutz W., Hamann W.-R., Wessolowski U.: 1989, *Astron. Astrophys.* **210**, 236

Smith L.F., Hummer D.G.: 1988, *Monthly Notices Roy. Astron. Soc.* **230**, 511

Smith L.F., Maeder A.: 1991, *Astron. Astrophys.* **241**, 77

Smith L.J., Willis A.J.: 1982, *Monthly Notices Roy. Astron. Soc.* **201**, 451

Smith L.J., Willis A.J.: 1983, *Astron. Astrophys. Suppl. Ser.* **54**, 226

Torres A.V.: 1988, *Astrophys. J.* **325**, 759

Underhill A.B.: 1991, *Astrophys. J.* **383**, 729

Vanbeveren D.: 1991, *Astron. Astrophys.* **252**, 159

Wessolowski U., Schmutz W., Hamann W.-R.: 1988, *Astron. Astrophys.* **194**, 160

Willis A.J., Wilson R.: 1978, *Monthly Notices Roy. Astron. Soc.* **182**, 559

LINE BLANKETED NON-LTE ATMOSPHERE MODELS FOR WOLF-RAYET STARS

WERNER SCHMUTZ

Institute of Astronomy, ETH Zentrum
CH-8092 Zurich, Switzerland

Abstract. A standard non-LTE Wolf-Rayet star atmosphere model is compared with an identical model but including line-blanketing. The structures of the two models are presented in detail and the implications of blanketed models for spectroscopic analyses are discussed.

Key words: Stars: Wolf-Rayet

1. Introduction

Todays non-LTE model calculations for Wolf-Rayet stars successfully reproduce the basic observed spectroscopic properties (Schmutz 1991a). The standard approach assumes spherical geometry, a monotonic velocity law, homogenity, and time independence (Hillier 1991). Most of the analyses were based on atmospheres composed of helium only (e.g. Hamann et al. 1989) but more sophisticated calculations including CNO model atoms have been published and will become standard in the future (e.g. Hillier 1988, Hamann et al. 1992). In the near future it will become possible to treat in non-LTE large model atoms of CNO and other important metals with thousands of spectral lines. However, in order to synthesize a line blanketed spectrum thousands of lines are still not sufficient. The blanketing effect is produced by tens of thousands of transitions with non-negligible optical depths (see Table I). Thus, even with the capabilities of present day computers it is not yet possible to compute line blanketed atmospheres where every significant line is treated individually. In order to build line blanketed models certain simplifications have to be made. Two completely different approaches have been proposed and successfully applied: One method is that of Anderson (1991), the other that of Schmutz (1991b). The present paper is based on the latter method which uses a Monte Carlo sampling technique to evaluate mean opacities that include the effects of line blanketing.

The method does not only allow to treat line blanketing but in the same computation, the radiative force is also evaluated. The numerous spectral lines whose opacities add up to the line blanketing effect are also those, that contribute to the radiative acceleration. Thus, the method allows to construct hydrodynamic line blanketed atmosphere models. Such "complete" models have been realized by Schaerer & Schmutz (1993, see also this volume) for O type stars. However, for Wolf-Rayet models it turned out, that the radiative force is insufficient to drive the mass loss rates needed to explain the observed emission line strengths. Therefore,

Space Science Reviews **66**: 253–261, 1994.

© 1994 *Kluwer Academic Publishers. Printed in Belgium.*

TABLE I

Number of transitions with a Sobolev optical depth larger than 0.1. The values are taken from the calculation presented in Section 3. Roman numbers denote the ionization stages.

Element	I	II	III	IV	V	VI	VII	VIII
He	94	105	0	0	0	0	0	0
C	0	0	88	85	0	0	0	0
N	0	0	134	314	79	0	0	0
O	0	2	97	388	285	26	0	0
Mg	0	0	30	61	20	158	0	0
Si	0	0	21	183	4	25	0	0
S	0	0	33	40	74	125	11	0
Fe	0	0	0	971	3851	14722	2442	240
Ni	0	0	0	280	343	2099	8409	409

it is not yet possible to construct hydrodynamic Wolf-Rayet models and only the line blanketing aspect can be treated with the present model calculations.*

In the next section an introduction to the computation method is given. In Sect. 3, the properties of one blanketed model is compared in detail with its non-blanketed counterpart. In the last section, the consequencies of using blanketed models for spectroscopic analyses are discussed.

2. Method

This description of the computation technique is only a brief outline. More details can be found in Schmutz (1991b; see also Schaerer & Schmutz these proceedings). A model is computed by an iteration between two distinct program parts: The solution of the non-LTE atmosphere problem and the evaluation of the line blanketing. The non-LTE atmosphere problem consists of the non-linearly coupled equation systems for the radiation transfer and for statistical equilibrium of the population processes. This part is solved with the approximative lambda operator method as formulated by Hamann (1987). The second part, the evaluation of the line blanketing, is a formal solution of the radiation transfer with a Monte Carlo simulation. This calculation is a modification and further development of

* In a recent publication Lucy & Abbott (1992) published a model of a Wolf-Rayet star with a radiation driven wind. A comparison of their stellar parameters and mass loss rate with observed Wolf-Rayet properties (cf. Hamann et al. 1989) reveals that they have not yet reproduced a real Wolf-Rayet star. Nevertheless, their paper is an important step forward to understand Wolf-Rayet winds: their calculation demonstrated the importance of the changing ionization condition in the wind (cf. Sect. 3.2) for transferring a maximum amount of momentum from the radiation field to the wind.

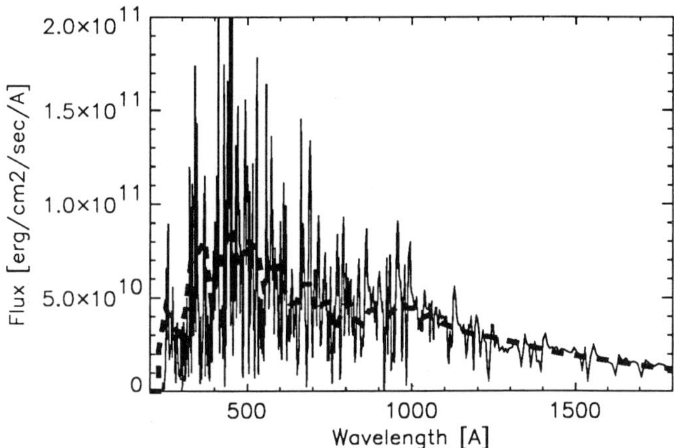

Fig. 1.   Emergent astrophysical flux of the line blanketed model discussed in Sect. 3. In the MC part 28,000 lines are included explicitly, whereas in the non-LTE atmosphere part (dashed curve, see also Fig. 6) the blanketing effect is approximated by averaged opacity coefficients. The statistical noise of the Monte Carlo simulation is about 2%, i.e. much smaller than the flux fluctuations. This structure is produced by the numerous spectral lines and their non-regular wavelength distribution.

the approach by Abbott and Lucy (1985). The Monte Carlo radiation transfer takes into account scattering, absorption, and re-emission processes. Since for the Monte Carlo radiation transfer the same non-LTE opacities and emissivities are adopted as in the non-LTE atmosphere, it yields the same emergent flux. Figure 1 displays the two emergent fluxes. The difference between the two solutions is that the Monte Carlo radiation transfer includes explicitly tens of thousands of lines (see Table I), whereas in the non-LTE atmosphere their blanketing effect is approximated by average scattering and absorptive opacity coefficients for a given wavelength band (in the present implementation 20 A). These average coefficients, or more exactly, the blanketing factors that modify the unblanketed coefficients, are evaluated in the Monte Carlo simulation. Thus, the two parts depend on each other in that the first part uses the blanketing factors that are evaluated in the second part, and the second part needs the non-LTE populations and temperature structure of the first for the continuum opacities and emissivities. Typically 3 or 4 iterations are needed to get consistency between the two parts.

In order to calculate the ionization and excitation of the metals in the Monte Carlo part, a radius dependent radiation temperature is used together with a modified Saha-Boltzmann equation (Eq. 2 of Schmutz 1991b). This ionization temperature is determined by the ionization ratio of helium for which the non-LTE populations are known. Corresponding to the three ionization stages of helium two such ionization temperatures $T_R(r, IP)$ result, that are used depending on the ionization potential of the ratio to be calculated. Figure 2 shows the radial dependence of these ionization temperatures of the model discussed in Sect. 3. At large optical depths LTE conditions are recovered, but departure from LTE

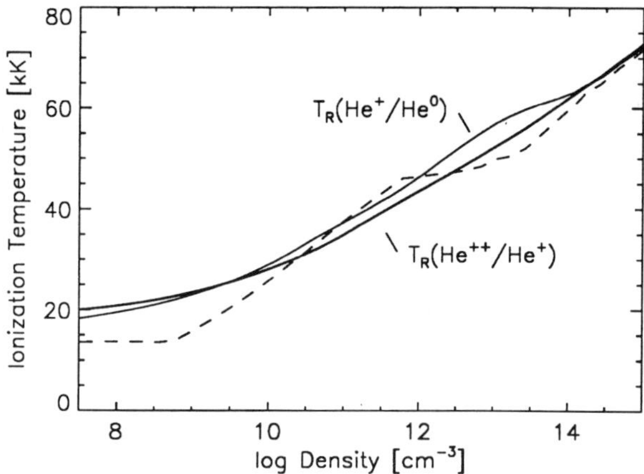

Fig. 2.   Ionization temperatures $T_R(r, IP)$ for IP = 24.6 and 54.4 eV, determined from the calculated helium non-LTE populations. These ionization temperatures are used to derive (with Eq. 2 of Schmutz 1991b) the ionization equilibrium of the metals (see Table II). The electron temperature is shown by the dashed line.

start already at about $\tau_R \approx 5$, due to the large contribution of scattering opacity. The inferred ionization structure of the metals are summarized in Table II.

3. Detailed Comparison of a Non-Blanketed with a Blanketed Atmosphere

In this section two models are compared that have identical stellar parameters and identical atmosphere structures but one model includes line blanketing and the other one is calculated without blanketing. The stellar parameters are $T_* = 45000$ K, $R_* = 5$ R$_\odot$, $\dot{M} = 10^{-4}$ M$_\odot$yr^{-1}, $V_\infty = 2500$ km s^{-1}. It is not intended that this parameter combination represents a particular Wolf-Rayet star but roughly, they are typical for an early WN type with strong emission lines.

3.1. THE ATMOSPHERE STRUCTURE

The atmosphere structure of the models are given in graphical form in Figs. 3 and 4. In all these graphs the ion density N_{ion} is used as space variable. Rosseland optical depth 1 is at $N_{ion} \approx 10^{12}$ cm^{-3}, where the radius is about twice the inner boundary radius, the expansion velocity is about half the terminal velocity, and the local electron temperature is roughly equal to the effective temperature. Figure 3 also shows that for optical depths larger than about 2 ($N_{ion} > 10^{13}$cm^{-3}) the atmosphere is not extended, i.e. there is no ambiguity in the definition of the model radius R_*.

The temperature structure of the blanketed model is only slightly higher than that of the unblanketed model (Fig. 4). The small temperature difference has only minor significance for the outcome of the calculated observables. The much more important difference between the two models can be found in the angular

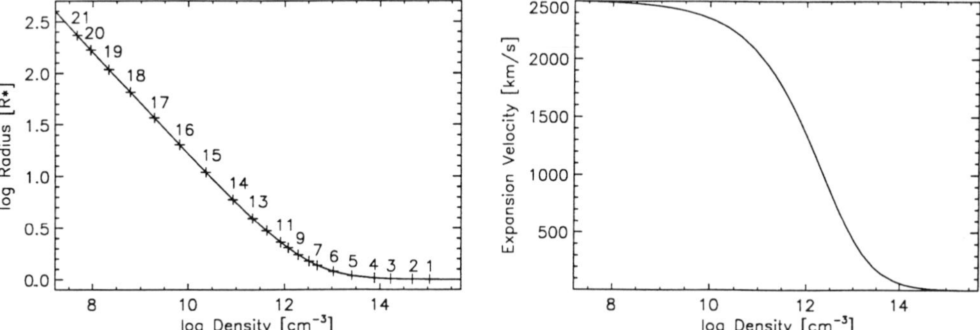

Fig. 3.   Left panel: Radial extension of the atmosphere in units of the inner boundary radius. The plus signs indicate the locations for which in Table II the ionization structure is listed. Right panel: Expansion velocity of the atmosphere.

TABLE II

Main ionization stages of the elements He through Cu at locations denoted by a depth point number. The geometrical locations of the depth points are indicated in Fig. 3.

depth point	He	C	N	O	Ne	Na	Mg	Al	Si	P	S	Cl	Ar	K	Ca	Sc	Ti	V	Cr	Mn	Fe	Co	Ni	Cu
1	III	V	VI	V	V	V	V	V	V	VI	VII	VII	VII	VII	VII	VI	VI	VI	VII	VII	VII	IV	VII	IV
2	III	V	VI	V	V	V	V	IV	V	VI	VII	VII	VII	VII	VI	VI	VI	VI	VII	VII	VI	IV	VII	IV
3	III	V	VI	V	V	V	IV	IV	V	VI	VII	VII	VII	VII	VI	VI	VI	VI	VII	VII	VI	IV	VI	IV
4	III	V	VI	V	V	V	IV	IV	V	VI	VII	VII	VII	VII	VI	VI	VI	VI	VII	VII	VI	IV	VI	IV
5	III	V	VI	V	V	V	IV	IV	V	VI	VII	VII	VII	VII	VI	VI	VI	VI	VII	VII	VI	IV	VI	IV
6	III	V	VI	V	V	V	IV	IV	V	VI	VII	VII	VII	VII	VI	VI	VI	VI	VII	VII	VI	IV	VI	IV
7	III	V	V	V	V	V	IV	IV	V	VI	VII	VII	VII	VI	VI	VI	VI	VI	VII	VII	VI	IV	VI	IV
8	III	V	V	V	V	IV	IV	IV	V	VI	VII	VII	VII	VI	VI	V	V	VI	VI	VI	VI	IV	VI	IV
9	III	V	V	V	IV	IV	IV	IV	V	VI	VII	VI	VI	VI	VI	V	V	VI	VI	VI	VI	IV	VI	IV
10	III	V	V	V	IV	IV	IV	IV	V	VI	VII	VI	VI	VI	VI	V	V	VI	VI	VI	VI	IV	VI	IV
11	III	V	V	V	IV	IV	IV	IV	V	VI	VI	VI	VI	VI	VI	V	V	VI	VI	VI	VI	IV	VI	IV
12	III	V	V	V	IV	IV	IV	IV	V	VI	VI	VI	VI	VI	VI	V	V	VI	VI	VI	VI	IV	VI	IV
13	III	V	V	IV	IV	IV	IV	IV	V	VI	VI	VI	VI	VI	V	V	V	VI	VI	VI	VI	IV	VI	IV
14	III	V	IV	IV	IV	IV	IV	IV	V	VI	VI	VI	V	V	V	V	V	VI	VI	VI	VI	IV	VI	IV
15	III	V	IV	IV	IV	III	III	IV	V	VI	V	V	V	V	V	IV	V	V	V	V	V	IV	V	IV
16	III	IV	IV	IV	III	III	III	IV	V	V	V	V	IV	V	IV	IV	V	V	V	V	V	IV	V	IV
17	III	IV	IV	IV	III	III	III	IV	V	V	V	IV	IV	IV	IV	IV	V	V	V	V	V	IV	V	IV
18	II	IV	IV	III	III	III	III	IV	V	IV	IV	IV	IV	IV	IV	IV	V	V	V	V	IV	IV	IV	IV
19	II	III	III	III	III	III	III	IV	IV	IV	IV	IV	IV	IV	IV	IV	V	V	IV	IV	IV	IV	IV	IV
20	II	III	III	III	III	III	III	IV	IV	IV	IV	IV	IV	IV	IV	IV	V	V	IV	IV	IV	IV	IV	IV
21	II	III	III	III	III	III	III	IV	IV	IV	IV	IV	IV	IV	III	IV	V	V	IV	IV	IV	IV	IV	IV

distribution of the radiation field.

In Fig. 5 the Eddington factors and the mean radius of the last interaction of a photon are displayed. These figures illustrate that the radiation field of the blanketed model is much more isotropic than that of the unblanketed model. In both models most of the escaped photons are created below $\tau_R \approx 1$. However,

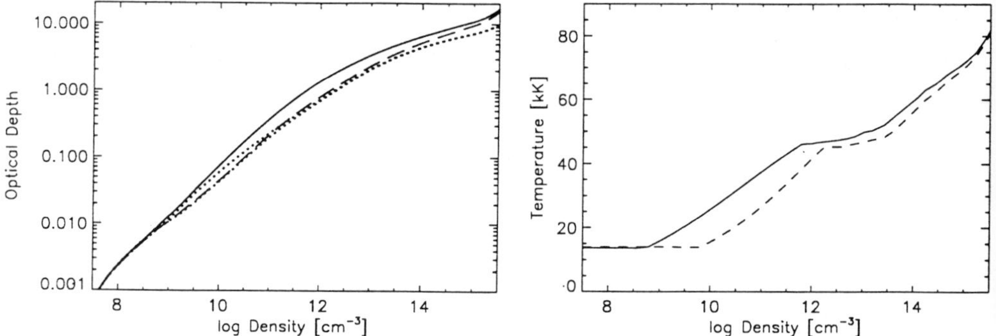

Fig. 4.   Left panel: Rosseland optical depth of the blanketed model (full drawn line) and unblanketed model (dashed line). The doted lines give the Thomson scattering optical depth. Right panel: Temperature structure of the blanketed model (full drawn line) and unblanketed model (dashed line).

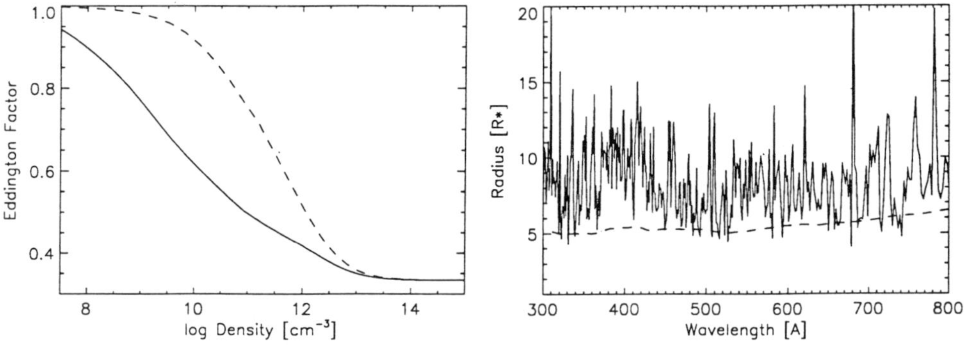

Fig. 5.   Left panel: Eddington factor $f = \int K_\nu \, d\nu / \int J_\nu \, d\nu$ of the blanketed model (full drawn line) and unblanketed model (dashed line). The Eddington factor is a measure of the isotropy of the radiation field. For $f = 1/3$ the radiation field is isotropic, if $f = 1$ all photons are streaming in the outward direction.
Right panel: Mean radii of the last interaction — in most cases a scattering event — of an escaped photon. This mean interaction location is significantly larger in the blanketed model (full drawn line) than in the unblanketed atmosphere (dashed line). The curve of the unblanketed model has been smoothed.

in the blanketed model, the scattering by the numerous spectral lines produces a diffuse radiation field out to large radii. This "trapped" radiation field has an important effect on the ionization equilibrium of the non-LTE populations. As illustrated in Fig. 6, in the unblanketed model He^{++} recombines at $N_{\mathrm{ion}} \approx 10^{10}\,\mathrm{cm}^{-3}$ (cf. Table II), whereas the recombination region of the blanketed model is at $N_{\mathrm{ion}} \approx 10^{9}\,\mathrm{cm}^{-3}$.

3.2. Observable Quantities

The most pronounced difference between the blanketed and the unblanketed model is that in the blanketed model the emergent flux in the far UV (228 – 911 A) is strongly modulated by the effect of the numerous spectral lines. The emergent flux has local minima where there is a high density of lines and there are maxima at wavelengths regions that contain only relatively few transitions (Figs. 1 and 7). Unfortunately, there is no line of sight to a Wolf-Rayet star where the absorption by the interstellar medium would allow the observation of this spectral region. Therefore, there are only indirect ways to investigate the radiation field at these wavelengths.

The strongest evidence for the blanketing action in the far UV are the FeV and FeVI emission lines in the 1200 – 1600 A region (Koenigsberger 1990). These lines have their upper levels in common with the strong transitions in the far UV. Since the far UV lines are optically thick, the observed emissions are a consequence of the branching ratios to the other possible transition channels. The model calculations do not (yet) predict the strengths of the observable lines, however, this is an interesting aspect for future improvements. A quantitative prediction of the observable iron line strengths is probably the best test of line blanketed atmospheres.

Another indirect test of the far UV radiation field is the ionization conditions in HII regions around Wolf-Rayet stars. Model calculations for nebulae indicate that by using the predicted energy distribution from blanketed models a better agreement with the observed nebular lines is obtained (Esteban et al. 1993). However, the ionization conditions depend on the integrated radiation field and as Fig. 7 shows, the mean energy distribution of the unblanketed model roughly agrees with the blanketed one.

The spectral parameters of Wolf-Rayet stars are determined by analyzing their helium emission lines. Table III lists the predicted equivalent widths of selected helium transitions of the two models. The differences reflect the shifted sizes of their emitting regions (Sect. 3.1). Obviously, the line strengths are strongly influenced by line blanketing. The changed strengths imply that different stellar parameters are deduced from blanketed models. However, nothing in favor or against blanketed models can be deduced from the helium line strengths. Schmutz

TABLE III

Predicted equivalent widths of helium emission lines.

model	HeI λ10830	HeI λ5876	HeII λ1640	HeII λ4686	HeII λ10124
unblanketed	-140	-13	-110	-230	-270
blanketed	-63	-7	-250	-530	-960

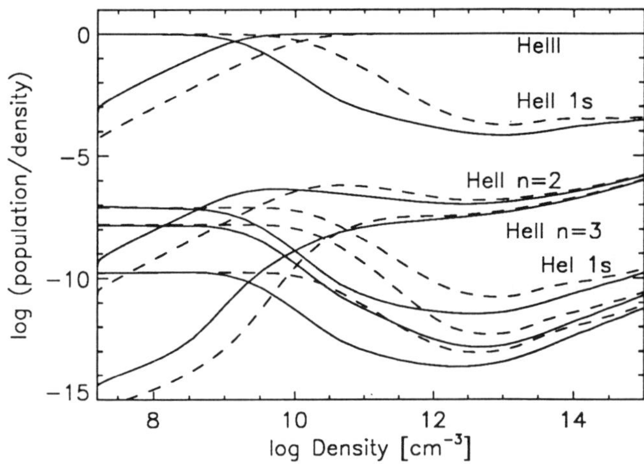

Fig. 6.   Relative populations of selected non-LTE levels. The populations of the blanketed model
are displaied with the full drawn lines, those of the unblanketed model with the dashed lines.

(1991a) compared the predicted ratios of emission lines with the observations and
concluded that the models basically reproduce the observations. Inspection of the
helium line rations reveals that the blanketed models yield the same ratios.

4. Discussion

¿From the first calculation that included line blanketing, Schmutz (1990) inferred
that smaller luminosities result from analyses with blanketed models. I have now
calculated a grid of line blanketed models and determined revised parameters for
a few early type WN stars. These results confirm this conclusion quantitatively.
E.g. the WN5 star HD50896 gets a luminosity that is smaller by a factor of
two. Similar reductions hold for other early type Wolf-Rayet stars. This result is
not very welcome. Howarth and Schmutz (1992) found that for WR binaries the
spectroscopically deduced luminosity is about a factor of two smaller than that
inferred from its mass. Now, by lowering the luminosities by another factor of
two this discrepancy grows to alarming size. This indicates that at some point
Wolf-Rayet stars are not yet really understood. Among others possibilities there
is one that has a good chance to be the correct one: By assuming the winds to be
homogeneous the average mass loss rates could be seriously overestimated if the
winds are significantly clumped (Hillier 1992). Correspondingly, the predicted
blanketing effect would also be overestimated since the blanketing effect is a
function of the average mass loss rate. This would help to reconcile the inferred
luminosities. On the other hand, assuming significantly lower mass loss rates
would bring difficulties for the theory of stellar evolution. It would then be
unclear how the wind could peel a WN star down to a WC star. Thus, it is
difficult to decide whether the strong influence of line blanketing on the deduced

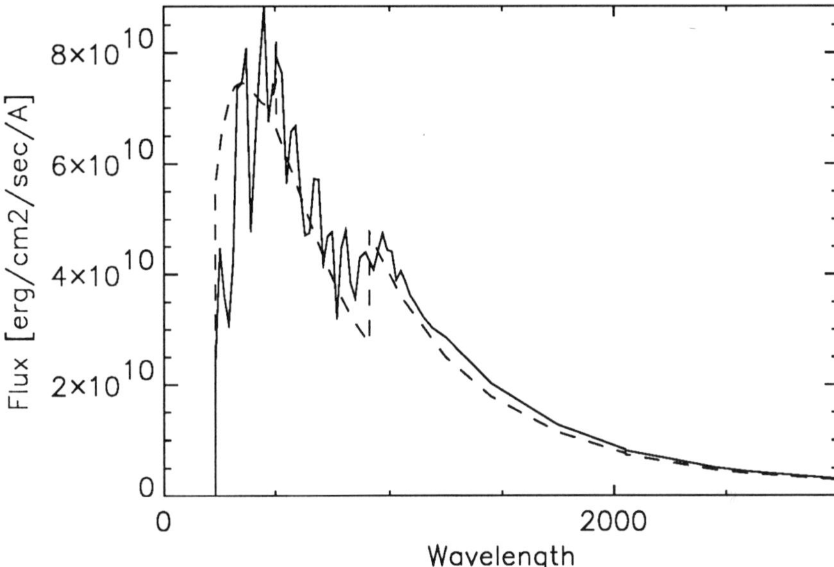

Fig. 7.   Emergent astrophysical flux of the blanketed model (full drawn line) and the unblanketed model (dashed line).

luminosities of Wolf-Rayet stars is realistic or whether it is an artifact of wrong assumptions.

References

Abbott D.C., Lucy L.: 1985, *ApJ* **288**, 679

Anderson L.: 1991, in L. Crivillary, I. Hubeny, D.G. Hummer, eds., *Stellar Atmospheres: Beyond Classical Models (NATO ASI Series C Vol 341)*, Kluwer: Dordrecht, p. 191

Esteban C., Smith L.J., Vilchez J.M., Clegg R.E.S.: 1993, *A&A* **272**, 299

Hamann W.-R.: 1987, in W. Kalkofen, ed., *Numerical Radiative Transfer*, Cambridge University Press: Cambridge, p. 35

Hamann W.-R. Leuenhagen U., Koesterke L., Wessolowski, U.: 1992, *A&A* **255**, 200

Hamann W.-R., Koesterke L., Wessolowski U.: 1993, *A&A* **247**, 397

Hillier D.J.: 1988, *ApJ* **327**, 822

Hillier D.J.: 1991, *IAU Symp.* **143**, p. 59

Hillier D.J.: 1992, *A&A* **247**, 455

Howarth I.D., Schmutz W.: 1992, *A&A* **261**, 503

Koenigsberger G.: 1990, *A&A* **235**, 282

Lucy L., Abbott D.C.: 1992, *ApJ* **327**, 822

Schaerer D., Schmutz W.: 1993, *A&A*, submitted

Schmutz W.: 1990, in C.D. Garmany, ed., *Properties of Hot Luminous Stars, ASPC,* **7**, Brigham Young University: Provo, p. 117

Schmutz W.: 1991a, *IAU Symp.* **143**, p. 39

Schmutz W.: 1991b, in L. Crivillary, I. Hubeny, D.G. Hummer, eds., *Stellar Atmospheres: Beyond Classical Models (NATO ASI Series C Vol 341)*, Kluwer: Dordrecht, p. 191

THE FUNDAMENTAL PARAMETERS OF THE CENTRAL STARS OF EIGHT WR RING NEBULAE

LINDA J. SMITH
Department of Physics and Astronomy
University College London
Gower Street, London WC1E 6BT, UK

and

CESAR ESTEBAN
Instituto de Astrofsica de Canarias
38200 – La Laguna, Tenerife, Spain

Abstract. We describe work that has recently been completed on deriving the fundamental parameters of eight WR stars through the photoionization modelling of their surrounding nebulae using non-LTE WR flux distributions. The resulting effective temperatures range from 57 000–71 000 K for the WN4–5 stars and $<$30 000–42 000 K for the WN6–8 stars. The derived stellar parameters are compared with those obtained from stellar emission line modelling. We find good agreement for the hot early WN stars, indicating that the non-LTE WR flux distributions have essentially the correct shape in the crucial far-UV region. We find lower temperatures for the four cooler late WN stars, particularly for the two WN6 stars. For the nebulae surrounding these stars, we find that the model flux distributions produce too much nebular ionization. We suggest that these discrepancies arise because of the lack of line-blanketing in the WR atmospheres. For the WO1 central star of G2.4 + 1.4, with strong nebular He II 4686 Å emission, we derive a temperature of 105 000 K, somewhat less than previous estimates. The positions of our eight WR stars on the H-R diagram are compared with the evolutionary tracks of Maeder (1990) for solar metallicity. In common with previous workers, we find that our derived luminosities are too low, giving an initial mass range of 25–40 $M_\odot$, below that expected for the majority of WR stars.

Key words: Stars – Wolf-Rayet – Fundamental parameters – Ring nebulae

1. Introduction

In this paper we describe work done in collaboration with J.M. Vlchez (IAC, Spain) and R.E.S. Clegg (RGO, UK) on the photoionization modelling of eight Wolf-Rayet (WR) ring nebulae. A complete description of this work can be found in Esteban et al. (1993).

Approximately 10% of Galactic WR stars are surrounded by ring nebulae. These nebulae are produced by two principal mechanisms (e.g. Chu 1991): the stellar wind sweeping up ambient interstellar material to form a shell; and/or the ejection of matter from the central star during an earlier evolutionary phase. Since these nebulae appear to be photoionized by their central stars, it is possible to obtain WR effective temperatures and luminosities by modelling the observed excitation of the nebulae.

The basic technique employed in photoionization modelling is to take a phys-

Space Science Reviews **66**: 263–270, 1994.

© 1994 *Kluwer Academic Publishers. Printed in Belgium.*

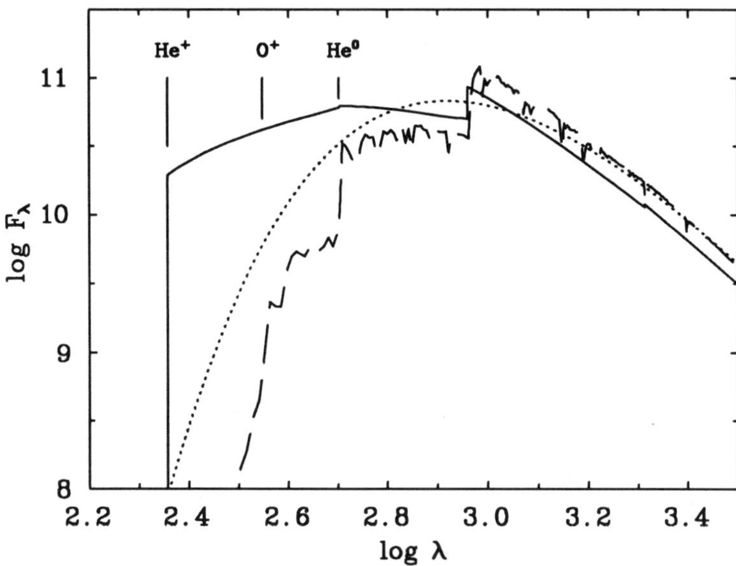

Fig. 1.   The emergent flux from a pure helium, unblanketed WR model atmosphere (solid line) compared with that of a Kurucz LTE model atmosphere (dashed line) and a blackbody (dotted line) with the same effective temperature of 35 000 K. The WR model has far more flux in the He I continuum.

ical model of the nebula (describing such parameters as radius, shell thickness, distance, electron density and abundances) and using a model stellar energy distribution, compute the theoretical nebular line strengths for comparison with observations. Further models are then generated using a grid of stellar energy distributions and/or nebular physical models until the theoretical and observed nebular line strengths agree to within the observational uncertainties. The best-fitting stellar energy distribution then gives the effective temperature and luminosity of the central star required to reproduce the observed level of excitation in the nebula.

In applying this technique to WR nebulae, it is essential that realistic non-LTE WR energy distributions are used. In Fig. 1, the flux distribution of a pure helium WR model atmosphere (Schmutz et al. 1994) is compared with that of a blackbody and a Kurucz LTE model atmosphere for the same effective temperature of 35 000 K. It is apparent that the WR model has far more flux shortward of the He^0 edge at 504 A. The observed levels of nebular excitation and ionization depend critically on the shape of this far-UV continuum. For example, the observed [O III]λ5007/[O II]λ3727 line ratio depends on the flux at the O$^+$ ionization edge at 353 A, and the presence or absence of nebular He II λ4686 is a sensitive indicator of the flux below 228 A. Thus by computing photoionization models for WR ring nebulae, we can test how realistic the model WR flux distributions are and determine whether the ring nebulae analyses give the same luminosities and effective temperatures as stellar emission line profile analyses.

2. Observations

TABLE I

Observed Sample of WR Ring Nebulae.

Nebula	Star		SpT	Telescope
S308	HD 50896	WR 6	WN5	INT+AAT
NGC 2359	HD 56925	WR 7	WN4	INT
NGC 3199	HD 89358	WR 18	WN5	AAT
RCW 104	HD 147419	WR 75	WN6	ESO
G2.4+1.4	Sand 4	WR 102	WO1	INT
NGC 6888	HD 192163	WR 136	WN6	INT
RCW 58	HD 96548	WR 40	WN8	Literature
M1-67	209 BAC	WR 124	WN8	INT

We have obtained spatially resolved spectroscopy of a large sample of WR ring nebulae which have been published in a series of papers (Esteban et al. 1990, 1991, 1992) and Esteban & Vlchez (1992). The details concerning the eight nebulae discussed in the present paper are given in Table I which lists the name of the nebula and central star plus spectral type, and the telescope used for the observations. The parameters for RCW 58 have been taken from Rosa (1987) and Kwitter (1984) who showed that the nebula is chemically enriched. The observed spectra have a resolution of 1–6 A and have been used to determine electron temperatures T_e, electron densities N_e, and abundances as input to the photoionization models. We find that three nebulae (S 308, NGC 6888 and M1–67) are chemically enriched and consist almost entirely of stellar ejecta. Two nebulae (G2.4 + 1.4 and RCW 104) show evidence for a small amount of chemical enrichment and the remaining nebulae have H II region abundances. A representative spectrum of the nebula S 308 is shown in Fig. 2. This nebula has the highest T_e in our sample and has an electron density N_e of $\leq 100\,\mathrm{cm}^{-3}$. The lowest excitation object is M1–67 (Esteban et al. 1991) which has $T_e = 6200\,\mathrm{K}$, a high N_e of 1050 cm^{-3} and no observable [O III] emission. Nebular He II is detected in only one object – G 2.4 + 1.4 which has an extreme WO1 central star.

3. Photoionization Models

As described in detail in Esteban et al. (1993), we have modelled the eight WR ring nebulae using the photoionization code described by Harrington et al. (1982)

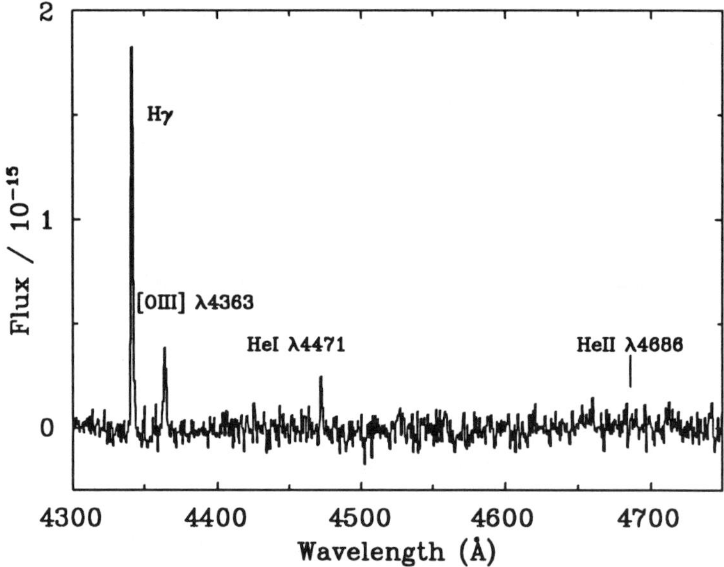

Fig. 2.   A portion of the spectrum of S 308 covering various emission lines used in the photoionization modelling. The strength of [O III] λ4363 relative to $\lambda\lambda$4959, 5007 gives an electron temperature of 14 400 K. The tight upper limit on the strength of He II λ4686 ($<$ 0.03 Hβ) is important in constraining the effective temperature of the central star WR 6 (WN5).

and Clegg et al. (1987). Briefly, we used a grid of forty pure helium unblanketed WR model flux distributions from Schmutz et al. (1994) with temperatures ranging from 30 000–90 000 K. We extended this basic grid to higher effective temperatures (74 000–166 000 K) by incorporating ten models from Schmutz et al. (1992) with flux below 228 A. For all these models, the shape of the emergent spectrum depends on two parameters T_* and R_t. T_* is defined as the effective temperature at the base of the wind with a radius R_*, and R_t is the "transformed radius" which is a measure of the wind density. For the lowest excitation nebula in our sample M1–67, we used two models constructed for the Ofpe/WN9 star R84 in the LMC (unblanketed – Schmutz et al. 1991; line-blanketed – Schmutz 1991) with effective temperatures of 28 500 and 28 000 K respectively. The models were normalized by adjusting the luminosities to equal the observed, de-reddened stellar flux at v (5160 A) for each central star.

We assumed that each nebula could be represented by a thin, hollow, spherical shell of uniform density, ionized solely by the central WR star. This physical model requires five input parameters in addition to the nebular abundances; the distance d, shell radius R, volume filling factor ϵ, electron density N_e, and shell thickness ΔR. The last three parameters are not well known: the nebulae appear clumpy implying $\epsilon \leq 1$; some of the nebulae (e.g. S 308) are at the low density limit ($N_e \leq 100\,\mathrm{cm}^{-3}$); and ΔR is poorly known since the shells are not well resolved. Fortunately, these three parameters can be constrained since the observed radio flux $S_\nu \propto N_e^2 \epsilon \Delta R$. In practice, because of the uncertainties, we constructed several different models for each nebula by exploring a wide parameter space. For example for S 308, we constructed an initial input model with our best estimates of $\Delta R = 0.08\,\mathrm{pc}$ and $N_e = 50\,\mathrm{cm}^{-3}$ which, with the observed 5 Ghz radio flux, gives $\epsilon = 0.21$. We then constructed a further four input models with extreme ϵ values of 0.01 and 1.0, and ΔR a factor of ten larger. Conversely for M1–67, ϵ, ΔR and N_e are well determined and the largest uncertainty is the He abundance. We therefore constructed two input models with $N(\mathrm{He})/N(\mathrm{H}) = 0.10$ and 0.22.

4. Results

For each physical input model and the grid of model atmospheres, we have computed the nebular line strengths, the electron temperature and the O^{++}/O^+ ratio for comparison with the observed values. We find that different combinations of input models and model atmospheres are able to reproduce the observed nebular line strengths and ratios within the observational uncertainties. In the case of S 308, we can discount the two input models with $\epsilon = 0.01$ and high values of N_e since the computed T_e and O^{++}/O^+ ratio are too low. For the three remaining input models, a minimum T_* of $55\,000\,\mathrm{K}$ is required to match the observed T_e and the O^{++}/O^+ ratio. The tight upper limit on nebular He II $\lambda 4686$ (Fig. 2) allows us to constrain T_* and R_t since models with $T_* = 90\,000\,\mathrm{K}$ and $R_t > 3\,\mathrm{R}_\odot$ produce observable He II. In total, we find that eight combinations of input model and model atmosphere are able to reproduce the observed S 308 spectrum within the observational uncertainties. To derive single representative values for the stellar and nebular parameters, we have calculated mean values weighted by a quantitative measure of the extent to which the observed spectrum, selected line ratios, and T_e are reproduced within the uncertainties. In this way, we derive $T_* = 71\,000 \pm 14\,000\,\mathrm{K}$ for WR 6, the central star of S 308.

For the low excitation nebula M1–67, the lack of [O III] provides a strong constraint on T_* since all models with $T_* \geq 35\,000\,\mathrm{K}$ produce $O^{++}/O^+ > 0$. On the other hand, models with $T_* = 30\,000\,\mathrm{K}$, while giving $O^{++}/O^+ = 0$, cannot reproduce either T_e or the He I line intensities, independently of the assumed He abundance. Of the two lower temperature models constructed for the Ofpe/WN9 star R84 (Sec. 3), we find that the line-blanketed model with $T_* = 28\,000\,\mathrm{K}$ produces the best fit, particularly for T_e and the observed S^{++}/S^+ ratio. We conclude that $T_* < 30\,000\,\mathrm{K}$ for the WN8 central star of M1–67.

We also encountered problems in obtaining acceptable fits for the nebulae NGC 6888, RCW 104 and RCW 58. In general, we find that the predicted level of nebular ionization is too high as indicated by the O^{++}/O^+ ratio and/or T_e. We experimented by introducing an approximate correction for line-blanketing using the R84 models, and obtained improved fits.

TABLE II

Model results: adopted stellar parameters.

Nebula	S 308	NGC 2359	NGC 3199	RCW 104
Central star	WR 6	WR 7	WR 18	WR 75
Spectral type	WN5	WN4	WN5	WN6
L ($10^3 L_\odot$)	414^{+190}_{-206}	121^{+93}_{-75}	506^{+172}_{-156}	141^{+113}_{-43}
T_* (10^3 K)	71 ± 14	67 ± 11	57 ± 14	≈35
R_* ($R_\odot$)	4 ± 2	2.6 ± 1.0	7 ± 2	10 ± 3
Q(H) (10^{48} s^{-1})	29^{+15}_{-16}	8^{+6}_{-5}	37 ± 12	6^{+5}_{-2}

Nebula	G 2.4+1.4	M1–67	NGC 6888	RCW 58
Central star	WR 102	WR 124	WR 136	WR 40
Spectral type	WO1	WN8	WN6	WN8
L ($10^3 L_\odot$)	120^{+93}_{-66}	125^{+124}_{-41}	279^{+93}_{-89}	136^{+170}_{-60}
T_* (10^3 K)	105 ± 21	≤30	42 ± 8	$30-35$
R_* ($R_\odot$)	1.1 ± 0.4	≥14	10 ± 4	12^{+6}_{-3}
Q(H) (10^{48} s^{-1})	8^{+6}_{-4}	4^{+4}_{-2}	15 ± 6	5^{+6}_{-2}

TABLE III

Comparison of Effective Temperatures.

WR	SpT	This work	Crowther, Smith & Hillier (1994)
WR 6	WN5	71000 ± 14000	74000
WR 7	WN4	67000 ± 11000	91000
WR 18	WN5	57000 ± 14000	72000
WR 75	WN6	≈35000	68000
WR 124	WN8	≤30000	33000
WR 136	WN6	42000 ± 8000	78000
WR 40	WN8	$30000-35000$	36000

The average temperatures T_*, luminosities L, stellar radii R_* and the number of ionizing photons s^{-1} $Q(H)$ are given in Table II for the eight WR central stars. The uncertainties reflect the dispersion about the mean value and the error on the distance for the distance dependent parameters. Overall, we find that all the nebulae, with the exception of RCW 104, are optically thin and are thus density- rather than radiation-bounded. From Table II, there is a clear difference in effective temperature between the early and late WN subtypes. The early WN stars have temperatures between 57 000–71 000 K and the late WN6–8 stars have lower temperatures of $\leq 30\,000$–42 000 K. The hottest star by far is the extreme WO1 star with $T_* = 105\,000$ K. This temperature is somewhat lower than the value of 150 000 K derived by Dopita et al. (1990) using a similar technique but blackbody energy distributions. We are able to reproduce the strong nebular He II emission at a lower temperature because the WR model flux distributions have significantly more far-UV flux than equivalent blackbodies.

5. Discussion and Conclusions

In Table III we compare our average derived temperatures with those of Crowther et al. (1994, these proceedings) which are based on line profile fits to the stellar emission lines using non-LTE model atmospheres including hydrogen, helium and metals. We find good agreement for the three early WN stars, particularly when it is considered that acceptable fits to the nebular spectra were obtained over the temperature range 55 000–90 000 K for each star. This agreement leads us to conclude that the model WR flux distributions have essentially the correct shape in the crucial far-UV region of the spectrum for $T_* > 55\,000$ K. For the four late WN stars, we find lower temperatures which is particularly striking for the two WN6 stars. These four stars are the central stars of the nebulae with which we had most difficulty in obtaining acceptable fits without introducing line-blanketing to reduce the level of nebular ionization. We therefore conclude that the currently available WR model atmospheres for late WN stars need to incorporate line-blanketing. Schmutz (1994, these proceedings) discusses the first available line-blanketed WR model atmospheres; it will clearly be worthwhile to repeat the photoionization modelling using these new models to discern the effects on both early and late WN central stars.

Finally, we have compared the positions of our eight WR stars on the HR diagram with the evolutionary tracks of Maeder (1990) for solar metallicity. We find, in common with previous workers (e.g. Hamann 1994, these proceedings), that the observed luminosities are too low. Specifically, our derived luminosities indicate an initial mass range of 25–40 $M_\odot$ which is below that expected for the majority of WR stars (see however D. Vanbeveren, these proceedings. This initial mass range is consistent with that found by comparing the observed abundances of the stellar ejecta nebulae (S 308, NGC 6888 and M1–67) with the predicted stellar surface abundances of Maeder (1990) (Esteban et al. 1992).

References

Chu, Y.-H.: 1991, in van der Hucht, K.A. & Hidayat, B. (eds.), IAU Symp. No. 143 *Wolf-Rayet Stars and Interrelations with other Massive Stars in Galaxies*, Kluwer, Dordrecht, p. 349.

Clegg, R.E.S., Harrington, J.P., Barlow, M.J. & Walsh, J.R.: 1987, *Ap. J.* **314**, 551.

Crowther, P.A., Smith, L.J. & Hillier, D.J.: 1994, these proceedings.

Dopita, M.A., Lozinskaya, T.A., McGregor, P.J. & Rawlings, S.J.: 1990, *Ap. J.* **351**, 563.

Esteban, C. & Vlchez, J.M.: 1992, *Ap. J.* **390**, 536.

Esteban, C., Vlchez, J.M., Manchado, A. & Edmunds, M.G.: 1990, *A&A* **227**, 515.

Esteban, C., Vlchez, J.M., Smith, L.J. & Manchado, A.: 1991, *A&A* **244**, 205.

Esteban, C., Vlchez, J.M., Smith, L.J. & Clegg, R.E.S.: 1992, *A&A* **259**, 629.

Esteban, C., Smith, L.J., Vlchez, J.M. & Clegg, R.E.S.: 1993, *A&A* **272**, 299.

Harrington, J.P, Seaton M.J., Adams, S. & Lutz, J.H.: 1982, *M.N.R.A.S.* **199**, 517.

Kwitter, K.B.: 1984, *Ap. J.* **287**, 840.

Maeder, A.: 1990, *A&A Suppl.* **84**, 139.

Rosa, M.R.: 1987, in Appenzeller I. & Jordan C. (eds.), IAU Symp. No. 122 *Circumstellar Matter*, Kluwer, Dordrecht, p. 457.

Schmutz, W.: 1991, in Crivellari L., Hubeny I. & Hummer D.G. (eds.), *Stellar Atmospheres: Beyond Classical Models*, NATO ASI Ser. C, Kluwer, **341**, p. 191.

Schmutz, W., Leitherer, C., Hubeny, I., Vogel, M., Hamann, W.-R. & Wessolowski, U.: 1991, *Ap. J.* **372**, 664.

Schmutz, W., Leitherer, C. & Gruenwald, R.: 1992, *P.A.S.P.* **104**, 1164.

Schmutz, W., Vogel, M., Hamann, W.-R. & Wessolowski, U.: 1994, in preparation.

TAILORED ANALYSES OF 24 GALACTIC WN STARS

PAUL A. CROWTHER and LINDA J. SMITH

Department of Physics and Astronomy, University College London
Gower Street, London, WC1E 6BT UK

and

D. JOHN HILLIER

Institute of Astronomy and Astrophysics, Scheinerstrasse 1
W-8000 Munich 80, Federal Republic of Germany

Abstract. The fundamental properties of 24 Galactic WN stars are determined from analyses of their optical, UV and IR spectra using sophisticated model atmosphere codes (Hillier, 1987, 1990). Terminal velocities, stellar luminosities, temperatures, mass loss rates and abundances of hydrogen, helium, carbon, nitrogen and oxygen are determined. Stellar parameters are derived using diagnostic lines and interstellar reddenings found from fitting theoretical continua to observed energy distributions.

Our results confirm that the parameters of WN stars span a large range in temperature ($T_* = 30$–90,000 K), luminosity ($\log L_*/L_\odot = 4.8$–5.9), mass loss ($\dot{M} = 0.9$–12×10^{-5} $M_\odot$ yr^{-1}) and terminal velocity ($v_\infty = 630$–3300 km s^{-1}). Hydrogen abundances are determined, and found to be low in WNEw and WNEs stars ($<15\%$ by mass) and considerable in most WNL stars (1–50%). Metal abundances are also determined with the nitrogen content found to lie in the range N/He = 1–5 $\times 10^{-3}$ (by number) for all subtypes, and C/N ~ 0.02 in broad agreement with the predictions of Maeder (1991). Enhanced O/N and O/C is found for HD 104994 (WN3p) suggesting a peculiar evolutionary history. Our results suggest that single WNL+abs stars may represent an evolutionary stage immediately after the Of phase. Since some WNE stars exist with non-negligible hydrogen contents (e.g. WR136) evolution may proceed directly from WNL+abs to WNE in some cases, circumventing the luminous blue variable (LBV) or red supergiant (RSG) stage.

Key words: stars:atmospheres – stars:Wolf-Rayet – stars:abundances – stars:winds – stars:mass-loss

1. Introduction

Previous tailored analyses including the effects of metals have been performed for only two WN stars (Hillier, 1988; Hamann *et al*, 1993). The reason for the sparsity of such models to date is computational time. Grids are both less expensive and time consuming than sequences of model calculations for individual stars. Using model grids (e.g. Schmutz *et al*, 1989) no verification is made that the derived parameters lead to the correctly shaped line profiles, that adopted interstellar reddenings are appropriate, or that the effects of hydrogen and metals are indeed negligible. A complete investigation into the current state of the Standard Model requires analyses, such as those presented here, from which deficiencies of the Standard Model can hopefully be identified and corresponding improvements made.

Space Science Reviews **66**: 271–275, 1994.

© 1994 *Kluwer Academic Publishers. Printed in Belgium.*

2. Observations and analysis technique

Our analyses are primarily based on new high S/N optical observations of Galactic WN stars (WN2-9) obtained with the INT and AAT telescopes covering $\lambda\lambda 3800$–7300 at resolutions of 0.1–3A, with additional ESO observations provided by Dr W. Schmutz. He I $\lambda 10830$ profiles have been obtained from Howarth & Schmutz (1992) and Schmutz *et al* (1989). High resolution UV spectra are taken from St-Louis (1990).

Detailed analyses are performed using the W-R Standard Model of Hillier (1987, 1990). This sophisticated W-R atmospheric code simultaneously solves the statistical equilibrium equations and the radiative transfer equations for hydrogen, helium and metals in the comoving frame under the constraint of radiative equilibrium. Stellar parameters are derived using diagnostic hydrogen (Hβ), He I ($\lambda 5876$)* and He II ($\lambda 5411$) lines, and an absolute continuum flux (M$_v$). Metal (CNO, Si) abundances are estimated from fits to UV and optical profiles. C/N is estimated from UV collisional lines and N/He determined from optical recombination lines.

3. Results

From their spectral appearance and derived stellar parameters, WN stars readily fall into one of four broad categories, as described below. In general, N/He and C/N ratios are broadly consistent with evolutionary predictions (Maeder, 1991). Mass loss rates derived from spectroscopic analyses are typically ~ 0.1 dex higher than those found from radio measurements. No significant dependence of mass loss rate on stellar luminosity is found, in contrast to that found for OB stars (Garmany & Conti, 1984). The correlation of temperature with surface flux ($\dot{M}/4\pi R_*^2$) found by Howarth and Schmutz (1992) is broadly confirmed except for the peculiar WN3 star WR46 (HD 104994). Average stellar parameters of the WN subtypes are listed in Table I.

3.1. WNE*s* STARS

High stellar temperatures (~ 70kK) and mass loss rates ($\sim 10^{-4}$ M$_\odot$ yr^{-1}) are found for the WNE*s* stars. The presence of hydrogen is detected for WR136 (WN6) alone (H/He=0.3). Predicted electron scattering wings are weaker than observed, suggesting clumping may be important (Hillier, 1991). In general, N IV and C IV line strengths can simultaneously be matched to better than a factor of two, although in general N III and N V line strengths are not well reproduced. With standard CNO equilibrium values very strong C IV profiles result, necessitating the adoption of considerably lower C/He ratios (factor of ~ 5).

---

* $\lambda 10830$ is adopted as the He I diagnostic if $\lambda 5876$ is weak or absent

TABLE I

Theoretical comparison of WN subtypes

Subtype	WN2-3w no He I single	WN2w no He I binary	WN3-4w He I	WN3-4s	WN5-6s
Sample	2	1	2	1	7
T_* (kK)	$\sim$90	$\sim$90	50	90	70$\pm$10
log ($L_*/L_\odot$)	5.3$\pm$0.1	5.3	4.8–5.0	5.1	5.2$\pm$0.2
$R_*/R_\odot$	2	2	4	1.5	3$\pm$1
$\dot{M}_{-5}$	$\sim$1	4	1.0	3.0	9$\pm$3
$\bar{v}_\infty$ (km s^{-1})	2850	3300	1900	1550	1900
$\dot{M}v_\infty/(L_*/c)$	3–10	30	10	18	80$\pm$45
$(b-v)_0$	$-$0.31$\pm$0.01	$-$0.14	$-$0.30$\pm$0.01	$-$0.09	$-$0.07$\pm$0.02
H/He (#)	0	0	$\sim$1	0.0	0–0.3
X_N (%)	1.4		1.5	1.4	1.4
C/N (#)	0.03		0.02	0.03	0.01
M_v	$-$2.9$\pm$0.1	$-$3.0?	$-$3.2$\pm$0.4	$-$3.5	$-$4.4$\pm$0.6
B.C.	$-$5.6$\pm$0.3	$-$5.3	$-$4.1$\pm$0.1	$-$4.6	$-$3.8$\pm$0.3
T_N (kK)	75:	75:	60-75	90:	60-80

Subtype	WN5-6w single	WN5-6w binary	WN7+abs	WN7–8	WN9+abs
Sample	2	2	3	6	1
T_* (kK)	35–40	55	32-33	32-36	30
log ($L_*/L_\odot$)	5.4	5–5.3	5.4–5.9	5.4-5.9	5.85
$R_*/R_\odot$	11–13	3–5	17–29	13-27	32
$\dot{M}_{-5}$	5$\pm$2	4–11	1.5–6	4-10	4
$\bar{v}_\infty$ (km s^{-1})	1550	1550	2150	850	1170
$\dot{M}v_\infty/(L_*/c)$	10	30–40	7–10	5–15	3.6
$(b-v)_0$	$-$0.24$\pm$0.04	$-$0.09$\pm$0.02	$-$0.29$\pm$0.01	$-$0.22$\pm$0.04	$-$0.28
H/He (#)	0–0.3	0–0.1	3–4	0.01–1.5	1.5
X_N (%)	1.4	1.2	0.8	1.2	1.5
C/N (#)	0.01	0.01	0.03	0.04	0.08
M_v	$-$5.4	$-$4.6$\pm$0.6	$-$6.4$\pm$0.6	$-$6$\pm$1.0	$-$7.1
B.C.	$-$3.3$\pm$0.1	$-$3.6$\pm$0.1	$-$3.1$\pm$0.1	$-$3.0$\pm$0.1	$-$2.8
T_N (kK)	40-60	40-60	35-40	35-40	$\sim$30

W-R stars analysed are WR1, 2, 6, 7, 16, 18, 22, 24, 25, 40, 46, 78, 108, 110, 123, 124, 128, 134, 136, 138, 141, 152, 156. Following Koesterke *et al* (1991) weak and strong lined WNE stars are defined such that W_λ(5411 He II) is less than or greater than 40A.

Parameters of WR108 (WN9+abs) determined using the distance (5-6kpc) derived from an investigation of IS Na I D lines (Crowther, 1993).

T_* for WR46 is determined from relative strengths of O V-VI lines.

3.2. WNEw STARS

Two distinct groups of weak lined WNE stars are found. The first has weak or absent He I emission and non-Gaussian He II profiles (WN3-4w), and the second has strong He I emission with Gaussian He II profiles (WN5-6w). We determine both low luminosities (log L/L$_\odot$ < 5) and mass loss rates ($\sim 10^{-5}$ M$_\odot$ yr^{-1}) for the former group, with stellar temperatures of $\sim$50kK (or 70kK) resulting from helium (or metal) analyses. The second group consists of known binaries from Smith & Maeder (1989). Using estimated WR:O light ratios for these stars, resulting stellar parameters are typical of WNEs stars. WR46 (WN3p), without detectable He I emission, is found to have a very high stellar temperature ($\sim$90kK) from metal profile fits and an enhanced oxygen content (factor of 10) although this star may have a close white dwarf companion (van Genderen $et~al$, 1991). WR2 (WN2) appears to be unique since He II profiles are Gaussian and He I is absent. It requires a binary solution for reasonable line profile fits.

3.3. WNL STARS

Low temperatures, low wind velocities and high luminosities are found for WNL stars. Hydrogen contents are found to be moderate (H/He$\sim$1), except for WR123 (WN8) for which H/He<0.1. N/He and C/N ratios are found to be typical of CN equilibrium values with a high C/N ratio (0.1) found for WR156 (WN8). Line strengths of N III are reasonably well predicted for WN7–8 stars within a factor of two, with N IV generally weak.

3.4. WNL+ABS STARS

The WNL+abs stars* are found to have lower mass-loss rates ($\sim$2) and higher terminal velocities ($\sim$3) than WNL stars and have significant hydrogen contents (H/He$\sim$3–4). Excellent fits to N III-IV profiles are found using stellar temperatures $\sim$15% higher for WN7 stars. A low N/He ratio (0.002) is preferred for WR25 (WN7+abs) and a high C/N ratio ($\sim$0.1) is found for WR108, newly classified as WN9+abs (Crowther, 1993).

4. Evolutionary status of WN stars

Agreement with the theoretical single star evolutionary tracks of Schaller $et~al.$ (1992) for the WN stars is poor (however see Vanbeveren, this volume). The low luminosities obtained for most WN stars suggest that stars of lower initial mass may reach the WR phase than is currently predicted ($\sim$25 M$_\odot$). The very high stellar temperatures predicted for pure helium WNE stars is not confirmed by observation.

Normal WNL and WNE subtypes are concluded to be post LBV (or RSG) stars, with evolution proceeding from WNL to WNE with decreasing hydrogen

---

* '+abs' classification is given to stars with intrinsic absorption lines.

content and decreasing stellar core. Derived stellar parameters of the WNL+abs stars (H/He, v_∞, log L_*) are very similar to some Of stars (e.g. HD 152408, ζ Pup), suggesting an intimate evolutionary connection. Hence, not all massive W-R stars are necessarily post-LBV or RSG objects. The major distinction between extreme Of stars and the WNL+abs stars is the higher mass-loss rates of the Wolf-Rayet stars. Very similar chemical abundances are found for massive Of stars, such as ζ Pup, for which Bohannan *et al* (1990) find H/He=4 and Pauldrach (these proc) determine C/N=0.12. Lower hydrogen contents are found for LBVs, typically H/He=2–3 (Crowther & Willis, these proc), suggesting an evolutionary sequence:

O $\rightarrow$ Of $\rightarrow$ WNL+abs ($\rightarrow$ LBV) $\rightarrow$ WNL $\rightarrow$ WNE*s* ($\rightarrow$ WC $\rightarrow$ SN)

for high initial mass stars. Indeed, since some WNE stars exist with non-negligible hydrogen contents (e.g. WR136) evolution may proceed directly from WNL+abs to WNE in some cases. For lower initial mass stars, the WNL+abs phase may be absent due to insufficient mass-loss. Contradictory to these findings, however, is the observed presence of WN6–7 stars in clusters and absence of WN8–9 stars (Moffat, 1989).

Acknowledgements

We are particularly grateful to Ian Howarth, Werner Schmutz and Nicole St-Louis for providing us with UV, optical and near-infrared spectra. LJS and PAC acknowledge financial support from the SERC.

References

Bohannan, B., Voels, S.A., Hummer, D.G. & Abbott, D.C., 1990. *Astrophys. J.*, **365**, 729.
Crowther, P.A. PhD thesis, University of London, 1993.
Garmany, C.D. & Conti, P.S., 1984. *Astrophys. J.*, **284**, 705.
Hamann, W-R., Wessolowski, U. & Koesterke, L., 1993. *Astr. Astrophys.* submitted.
Hillier, D.J., 1987. *Astrophys. J. Suppl.*, **63**, 947.
Hillier, D.J., 1988. *Astrophys. J.*, **327**, 822.
Hillier, D.J., 1990. *Astr. Astrophys.*, **231**, 116.
Hillier, D.J., 1991. *Astr. Astrophys.*, **247**, 455.
Howarth, I.D. & Schmutz, W., 1992. *Astr. Astrophys.*, **261**, 503.
Koesterke, L., Hamann, W-R., Schmutz, W. & Wessolowski, U., 1991. *Astr. Astrophys.*, **248**, 166.
Maeder, A., 1991. In: *Evolution of Stars: The Photospheric Abundance Connection IAU Symposium 145*, eds. Michaud, G. & Tutukov, A., p. 221, Kluwer, Dordrecht.
Moffat, A.F.J, 1989. *Astrophys. J.*, **347**, 373.
Schaller, G., Schaerer, D., Meynet, G. & Maeder, A., 1992. *Astr. Astrophys. Suppl.*, **96**, 269.
Schmutz, W., Hamann, W-R. & Wessolowski, U., 1989. *Astr. Astrophys.*, **210**, 236.
Smith, L.F. & Maeder, A., 1989. *Astr. Astrophys.*, **211**, 71.
St-Louis, N. PhD thesis, University of London, 1990.
van Genderen, A.M. & et al., , 1991. In: *Wolf-Rayet Stars and Interrelations with other Massive Stars in Galaxies, IAU Symposium 143*, eds. van der Hucht, K.A. & Hidayat, B., p. 129, Kluwer, Dordrecht.

OXYGEN AND CARBON ABUNDANCES FOR THE WO STARS

ROBIN L. KINGSBURGH
Universidad Nacional Autonoma de Mexico

and

M.J. BARLOW and P.J. STOREY
University College London

Abstract. We present relative carbon and oxygen abundances derived via an optically thin recombination line analysis for five WO stars, and compare the derived abundances to recent evolutionary models. New recombination coefficients for O^{4+}, O^{5+} and O^{6+} ions have allowed total oxygen abundances to be derived. The final C/He values range between 0.4 and 0.8 by number, consistent with C/He ratios previously derived for WC stars. O/He values range between $0.1 - 0.4$, with C/O ratios between $2.1 - 4.8$.

A comparison of the derived abundances with the evolutionary models of Maeder (1990) and Schaller et al. (1992) shows promising agreement. We find reasonably tight agreement between the abundances derived for the WO stars. The degree of enhancement for the oxygen abundances in regions of low metallicity predicted by Maeder (1990) is not corroborated by our results.

Additionally we present a revised, quantified classification scheme for WO subtypes. We extend the class to lower excitation, WO5, and place MS 4 (=WR 30a) in this class. Equivalent widths of the strongest lines of MS 4 are also presented. Finally, we present new observations of DR 1, a WO3 star located in the dwarf irregular galaxy IC 1613.

Key words: Stars: Wolf Rayet – Stars: Abundances

1. Introduction

The WO stars are a rare group of highly evolved Wolf-Rayet stars whose spectra are dominated by lines of highly ionized stages of oxygen. Initially massive stars which have presumably evolved through the WC stage, the WO stars are thought to be immediate precursors of supernovae. Spectra of WO stars are characterized by strong lines of OVI, OV, CIV and HeII. The increased strength of the oxygen lines is believed to reflect a physical enhancement of oxygen over the WC stars, due to α-particle capture by carbon nuclei. The case to physically distinguish the strong oxygen line WR stars from the WC stars was first made by Barlow & Hummer (1982; BH82), who introduced the WO spectral class.

A spectrophotometric analysis of the population I WO Wolf-Rayet stars is presented, based on UV observations obtained with the IUE and optical observations obtained at the AAT with the RGO spectrograph and IPCS as detector. New AAT observations of MS 4 are also presented, along with new observations of DR 1 obtained at the WHT with the ISIS spectrograph, and two EEV CCDs as detectors.

Space Science Reviews **66**: 277–280, 1994.

© 1994 *Kluwer Academic Publishers. Printed in Belgium.*

2. Revised Classification Scheme

BH82 present the original WO classification scheme. We have added quantitative criteria relating the OV 5592 A and OVI 3811,34 A relative line strengths, which can be used alone to define the WO subclass, and also to discriminate the WC4 stars from the WOs. Typically for a WC4 star EW(OVI)/EW(OV)=0.2. For Sand 2, a WO4 star, this ratio is 2.6. The revised classification criteria are presented in Table 1. We have also extended the WO class to lower excitation, and now include the WO5 subclass.

2.1. MS 4: a WO5 Star

The equivalent widths for MS 4(=WR 30a) are presented in Table 2. MS 4 has previously been classified as WC4+O4 (Moffat & Seggewiss 1984) and WO4+O4 (Smith et al. 1990). In MS 4, the EW(OIV 3400) > EW(OVI 3811,34), hence this star would be classified as WO4 in the original scheme. However, for MS 4, EW(OVI)/EW(OV)~0.6 versus ~3 for the WO4 stars and EW(OVI)/EW(CIV) is also less for MS 4 than for the WO4 stars. The excitation of MS 4 is therefore clearly lower than that of the WO4 stars, and this led us to define a new WO5 class, in which we place MS 4.

2.2. DR 1: a WO3 Star in IC 1613

DR 1, the WO star in the dwarf irregular galaxy IC 1613, is surrounded by an H II region with unusually strong nebular He II 4686 in emission. D'Odorico & Rosa (1982) first noted that the WR star in this H II region had peculiar characteristics and classified this star as WCpec or WC+WN. Davidson & Kinman (1982) subsequently analyzed the star and surrounding HII region. They observed strong O VI 3811,34 A in the central star and suggested it may be a member of the WO class. Garnett et al. (1991) have more recently suggested that the star may be a WO4 star, based on observations in the 3660–4940 A range. Our spectral coverage extends from 3200–8600A; a full analysis of the star and surrounding nebula will be presented elsewhere.

Here we present the equivalent widths for the stellar features (Table 2) and abundances for the central star (Tables 3 and 4). The nebular continuum was calculated and subtracted, before measuring the equivalent widths. Using these values and the classification criteria of Table 1, we classify DR 1 as a WO3 star.

3. Abundances

A case B optically thin recombination analysis of radiative and dielectronic recombination lines at optical and UV wavelengths has been completed for lines of C, He and O, following the method of Hummer, Barlow & Storey (1982). New recombination coefficients for O^{4+}, O^{5+} and O^{6+} ions have allowed total oxygen abundances to be derived for the first time. $T=5\times10^4$ K and $n_e=10^{11}$ cm^{-3} were assumed for all cases. We generally see good agreement between the abundances derived from different lines for the same ion. We have omitted lines which we suspect are thick or partially thick.

For DR 1, no estimate of the O^{4+} abundance could be made (O IV 3400 can be affected by optical depth effects, and no UV spectra were available); the O^{6+} abundance was derived from O VI 5290 and the O^{5+} abundance was derived from O V 5590. In order to estimate the total oxygen abundance, we adopted the mean O^{4+}/O^{6+} ratio for Sand 1 and Sand 2 (two WO4 stars) multiplied by 0.5 (the ratio of the O IV and O VI equivalent width ratios).

Table 3 presents the abundances by number and Table 4 presents the abundances by mass together with a comparison with recent evolutionary model results of Schaller et al. (1992).

3.1. COMPARISON WITH THEORY

A comparison of the derived abundances with the $40M_\odot$ and $60M_\odot$ evolutionary models of Maeder (1990) and Schaller et al. (1992) shows good agreement with model stages 36 and 37 at initial Z=0.02. Maeder (1990) predicted a greater enhancement of oxygen abundances in his lower metallicity models. We find, however, resonably tight agreement between the abundances derived for the WO stars, independent of metallicity. Maeder (1990) predicted C/O ratios of ~0.5 for initial Z=0.005 and C/O~2 for Z=0.02 (see also Smith & Maeder, 1990). For Sand 1 and Sand 2, located in the Magellanic Clouds, we find C/O ratios of $2-3$, versus C/O ratios of $3-5$ for Sand 4 and Sand 5, located in the Milky Way. Hence the predicted extent of processing of carbon into oxygen in massive stars in low metallicity galaxies is not corroborated quantitatively, although a trend of decreasing C/O with decreasing initial metallicity may be present. Additionally, we do not find the total C+O/He ratios (Table 3) to be in accord with those given by Smith & Maeder (1990), who predict C+O/He>1 for the WO stars; here we find C+O/He ranges from 0.44 to 1.2 by number.

TABLE I

Revised WO Classification Scheme.

class	$\log\frac{EW(OVI)}{EW(CIV)}$	$\log\frac{EW(OVI)}{EW(OV)}$	other criteria	members
WO1	>0.7	>0.9	no OIV;OV≥CIV	Sand 4
WO2	0.2→0.7	0.65→0.9	no OIV;OV<CIV	Sand 5
WO3	−0.3→0.2	0.45→0.65	OIV<OVI;OV<<CIV	DR 1
WO4	−0.8→−0.3	0.25→0.45	OIV≃OVI;OV<<CIV	Sand1,Sand2
WO5	<−0.8	0.0→0.25	OIV>OVI;OV<<CIV	MS 4

TABLE II
MS 4 and DR 1 Equivalent Widths.

	OIV 3400	OVI 3811,34	CIV+HeII 4658,86	OVI 5290	OV 5590	CIV 5801,12
MS 4 (WO5)	14	10	62	3.5::	18	170
DR 1 (WO3)	204[a]	360	330	54:	120:	690

TABLE III
WO Abundances By Number.

	$\frac{C^{4+}}{He^{2+}}$	$\frac{O^{4+}}{He^{2+}}$	$\frac{O^{5+}}{He^{2+}}$	$\frac{O^{6+}}{He^{2+}}$	C/He	O/He	C/O	$\frac{C+O}{He}$
Sand 1	0.81	0.11	0.19	0.078	0.81	0.38	2.1	1.20
Sand 2	0.50	0.045	0.094	0.032	0.50	0.17	2.9	0.67
Sand 4	0.53	0	0.042	0.12	0.53	0.16	3.3	0.69
Sand 5	0.36	0	0.024	0.052	0.36	0.076	4.8	0.44
DR 1	0.71	(0.049)	0.16	0.067	0.71	0.28	2.5	0.99

TABLE IV
WO Abundances By Mass and Comparison to Models.

	Sand 1 WO4	Sand 2 WO4	Sand 4 WO1	Sand 5 WO2	DR 1 WO3	Model Stages 37	36	37
Z_{init}	0.002	0.005	0.02	0.02	0.001	0.02	0.02	0.02
X_{He}	0.20	0.32	0.31	0.42	0.24	0.29	0.33	0.30
X_C	0.49	0.47	0.49	0.45	0.50	0.44	0.42	0.43
X_O	0.31	0.21	0.20	0.13	0.26	0.25	0.22	0.25

References

Barlow, M.J. & Hummer, D.G.: 1982, '' in C.W.H. de Loore & A.J. Willis, ed(s)., *IAU Symp. No. 99, Wolf–Rayet Stars*, Reidel:Holland, 387

Davidson, K. & Kinman, T.D.: 1982, *PASP* **94**, 634

D'Odorico, S. & Rosa, M.: 1982, *A&A* **105**, 410

Garnett, D.R., Kennicutt, R.C., Chu, Y. & Skillman, E.D.: 1991, *ApJ* **373**, 458

Hummer, D.G., Barlow, M.J. & Storey, P.J.: 1982, '' in C.W.H. de Loore & A.J. Willis, ed(s)., *IAU Symp. No. 99, Wolf–Rayet Stars*, Reidel:Holland, 79

Maeder, A.: 1990, *A&AS* **84**, 139

Moffat, A.F.J. & Seggewiss, W.: 1984, *A&AS* **58**, 117

Schaller, G., Schaerer, D., Meynet, G. & Maeder, A.: 1992, *A&A* **96**, 269

Smith, L.F. & Maeder, A.: 1990, *A&A* **241**, 77

Smith, L.F., Shara, M.M. & Moffat, A.F.J.: 1990, *ApJ* **348**, 471

HD 190918: ARBITER BETWEEN MODELS FOR WOLF-RAYET STARS

ANNE B. UNDERHILL

*Department of Geophysics and Astronomy, University of British Columbia, Vancouver,
B.C., V6T 1Z4, Canada*

and

GRANT M. HILL

*Department of Astronomy, University of Western Ontario, London, Ont., N6A 3K7,
Canada*

Abstract. New orbital elements for the 112.4 day WN4.5+O9.5Ia spectroscopic binary HD 190918 are presented. Solutions consistent with the O star being a main sequence star or a supergiant are obtained when $i = 25°$ or $20°$, respectively. We predict the X-ray flux to be expected from this system and note how an observation of the X-ray flux from HD 190918 would help one to choose the most probable value of the inclination.

1. Introduction

The spectral type of HD 190918 is WN4.5+O9Ia, however there is no certainty that the luminosity class of the O star is Ia rather than V. Twenty-five yellow-green spectrograms at 30 Å mm^{-1} have been measured independently by each of us for radial velocity. A solution for orbital elements was made for the O star starting from the orbital elements of Fraquelli, Bolton, & Horn (1983); then a solution was made for the WR star using the radial velocities from the broad emission line He II $\lambda5411$ and keeping P, e, and ω fixed at the values found for the O star. The results are given in Table 1.

The masses of the stars of HD 190918 and the dimensions of the orbit are given in Table 2 as functions of i. The velocity changes of the sharp C III $\lambda5696$ emission line were compared with the predicted orbital velocities of the O star. The C III emission line follows the orbital velocity of the O star rather well except that at all phases it is displaced shortward by an additional 14 km s^{-1}.

Some WR binary stars, including V444 Cygni and CQ Cephei, are known as X-ray sources (Pollock 1987). However HD 190918 was not reported by Pollock because the Einstein satellite did not observe its field (Harnden 1993 private communication). The equations of Usov (1992) describe the X-ray flux generated by colliding winds. We used these equations to estimate the X-ray flux expected from HD 190918. We examined four cases: Cases I and II in which the rate of mass loss from the WR star is $5 \times 10^{-5} M_\odot$ yr^{-1} and the O star is a main sequence star or a supergiant respectively, and Cases III

Space Science Reviews **66**: 281–284, 1994.

© 1994 *Kluwer Academic Publishers. Printed in Belgium.*

and IV in which the rate of mass loss from the WR star is $5 \times 10^{-6} M_\odot$ yr^{-1} and the O star is a main sequence or a supergiant star respectively. The separation of the centers of the stars, calculated from the orbital elements, runs between 1.46×10^{13} and 1.99×10^{13} cm when the O star is type O9V and between 1.81×10^{13} and 2.46×10^{13} cm when the O star is a supergiant.

When a ring-like disk, Model I, is used to represent the line emitting region of the WR star of HD 190918, an inner radius for the ring of about 3×10^{13} cm (cf. Bhatia and Underhill 1988) leaves ample room for the O star to move around the WR star pretty well in the clear. If one adopts a model, Model II, in which the WR star has a dense spherical wind with $\dot{M}_{WR} = 5 \times 10^{-5} M_\odot$ yr^{-1}, the companion moves through the wind at the above distances from the center of the WR star.

In order to estimate L_x we assume that the radius of a main sequence companion is 12 $R_\odot$ or 8.35×10^{11} cm and that of a supergiant companion is 35 $R_\odot$ or 2.44×10^{12} cm. These are typical radii for hydrogen-burning stars which have the masses of the stars given in Table 2 for the cases $i = 25°$ and $i = 20°$. The predicted X-ray fluxes are given in Table 3 for four sets of parameters. When r_{OB}/R_{OB} is less than unity we use equation (81) of Usov (1992); when this ratio is approximately equal to or greater than unity we use equations (89) and (95). We assume that $V_{WR}^\infty = V_{Ostar}^\infty = 1600$ km s^{-1}.

2.   Discussion

The orbital elements of HD 190918 do not allow us to distinguish between the O star being a main sequence or a supergiant star. The sharp C III line is formed in plasma which appears to move with the O star but which shows a component of motion toward the observer of 14 km s^{-1} relative to the motion of the center of mass of the system.

The plasma forming the strong He II λ5411 line of the WR star is displaced by 91 km s^{-1} longward from the velocity shown by the O star to be the line-of-sight velocity of the center of mass of the system. This substantial positive displacement (known for more than 40 years) cannot be explained using Model II; in terms of Model I it suggests infall of plasma from the disk to the surface of the WR star. Similar negative displacements are known for the emission lines from the C and N ions in WR spectra. The negative displacements suggest outflow of hot plasma contained in filaments and projected against the face of the star.

Four conclusions can be drawn from the predicted X-ray fluxes for HD 190918:

1. If the observed L_x is of the order of or less than 10^{31} ergs s^{-1}, Case III is reasonable; it implies that the O star is a mainsequence star and that $\dot{M}_{WR} = 5 \times 10^{-6} M_\odot$ yr^{-1}.

2. If the observed L_x is of the order of 10^{32} ergs s^{-1}, Case 1 is reasonable;

TABLE I

Orbital Elements for HD 190918

Element (Units)	O Star[a]	WR Star[b]
Period (days)	112.4 ± 0.2	112.4^c
Eccentricity	0.391 ± 0.067	0.391^c
ω (degrees)	198.9 ± 10.7	18.9^c
T_{peri} (BJD-2440000.0)	7420.5 ± 3.6	7410.6 ± 3.2
γ (km s^{-1})	-20.9 ± 0.7	70.2 ± 4.6
K (km s^{-1})	16.9 ± 2.1	34.4 ± 7.4

[a]From three absorption lines: He II $\lambda 5411.52$, O III $\lambda 5592.37$, and He I $\lambda 5875.68$.
[b]From one broad emission line: He II $\lambda 5411.52$.
[c]Value adopted from the O-star solution.

TABLE II

The Masses and Dimensions of HD 190918 as Functions of the Inclination of the Orbit

Item (Units)	$i = 30°$	$i = 25°$	$i = 20°$	$i = 15°$
$M_{O\,Star}$ ($M_\odot$)	6.6	10.9	20.6	47.5
$M_{WR\,Star}$ ($M_\odot$)	3.2	5.4	10.1	23.3
$a_{O\,Star}$ ($R_\odot$)	69.1	1.7	101.1	133.5
$a_{WR\,Star}$ ($R_\odot$)	140.6	166.4	205.6	271.7

TABLE III

Predicted X-ray Fluxes from HD 190918

Case	$\dot{M}_{WR}$ ($M_\odot$ yr^{-1})	$\dot{M}_{OB}$ ($M_\odot$ yr^{-1})	r_{OB}/R_{OB} Range[a]	L_x (ergs s^{-1})
I	5×10^{-5}	10^{-8}	0.24–0.33	A few $\times 10^{32}$
II	5×10^{-5}	10^{-6}	0.92–1.25	A few $\times 10^{35}$ with extinction
III	5×10^{-6}	10^{-8}	0.75–1.02	A few $\times 10^{31}$ with extinction
IV	5×10^{-6}	10^{-6}	2.29–3.12	A few $\times 10^{34}$

[a]The quantity r_{OB} is the distance from the center of the OB star to the shocked region; R_{OB} is the radius of the OB star. When $r_{OB}/R_{OB} \approx 1.0$, the dense plasma of the shocked region may extinguish most of the X rays. The range gives the value of r_{OB}/R_{OB} from periastron to apastron.

it implies that the O star is a mainsequence star and that $\dot{M}_{WR} = 5 \times 10^{-5} M_\odot \ \mathrm{yr}^{-1}$.

3. If the observed L_x is of the order of 10^{34} ergs s^{-1}, Case IV is reasonable; it implies that the companion is a supergiant and that $\dot{M}_{WR} = 5 \times 10^{-6} \ M_\odot \ \mathrm{yr}^{-1}$.

4. If the observed L_x is of the order of 10^{35} ergs s^{-1}, Case II is reasonable; the O star is a supergiant and $\dot{M}_{WR} = 5 \times 10^{-5} M_\odot \ \mathrm{yr}^{-1}$.

To resolve which case describes HD 190918 best, it is necessary to obtain an X-ray observation of HD 190918. The radio flux gives no sure information about the rate of mass loss in the wind from the WR star because, as Underhill (1986) pointed out, the radio flux may originate in a disk-driven wind as well as in a spherical wind from the WR star.

3. Conclusions

We prefer Model I to Model II for representing the line-emitting regions of WR stars like HD 190918 because it describes a situation in which an explanation can be found for the large systematic positive displacements of the He II lines and the smaller systematic negative displacements of the lines from the C and N ions (not presented here). It also provides a more or less clear space in which the O star can move around the WR star without seriously disturbing the flow of the wind. Underhill (1991) has presented additional reasons why Model I is to be preferred over Model II.

The implication of adopting Model I is that WR stars are young massive stellar objects with a geometric structure like that of a T Tauri star or a Herbig Ae/Be object. Wolf-Rayet stars may be formed when molecular clouds condense to form massive stars in the presence of larger than normal interstellar magnetic fields (Bhatia and Underhill 1988). The presence of WR stars among groups of O and B stars implies that in some of the birthing regions for massive stars the interstellar magnetic fields are larger than normal. The ring-like disks of WR stars may be expected to dissipate fairly rapidly leaving normal hydrogen-burning stars having masses of 10–20 $M_\odot$ or so.

References

Bhatia, A. K. and Underhill, A. B.: 1988, *Astrophys. J. Suppl. Ser.* **67**, 187
Fraquelli, D. A., Bolton, C. T., and Horn, J.: 1983, unpublished preprint
Pollock, A. M. T.: 1987, *Astrophys. J.* **320**, 283
Underhill, A. B.: 1986, *Publ. Astron. Soc. Pacific* **98**, 897
Underhill, A. B.: 1991, *Astrophys. J.* **383**, 729
Usov, V. V.: 1992, *Astrophys. J.* **389**, 635

MIXING PROCESSES AND STELLAR EVOLUTION

J.-P. ZAHN
Observatoire de Paris, Section de Meudon, 92195 Meudon, France
Observatoire Midi-Pyrenees, 14 ave. E. Belin, 31400 Toulouse, France
Astronomy Department, Columbia University, New York, NY 10027, U.S.A.

Abstract. We review some of the mixing processes that may influence the evolution of massive stars, such as penetrative convection, and put our emphasis on those which occur in the radiative envelope. There the main transport mechanisms are a thermally driven meridian circulation, which departs significantly from the classical Eddington-Sweet description, together with turbulent motions generated by the differential rotation. This rotation-induced mixing will surround the convective core of such stars with a region of decreasing helium content, which may prevent semi-convection from ever appearing; the extent of this region depends sensitively on the rotation rate. The results are compared with those obtained earlier by Mestel (1953). Work is in progress to verify whether stars which rotate sufficiently fast may be thoroughly mixed.

Key words: Stellar evolution, rotation, abundances

1. Introduction

In its standard form, the theory of stellar structure and evolution has been extremely successful in explaining most of the observed properties: the existence of a main sequence, the mass-luminosity-radius relations, the evolution towards the red giant branch, the pulsation of variable stars, the evolution of close binaries, to quote only the major ones. However, there are some disturbing discrepancies, which are reviewed by Andre Maeder in these proceedings. In most instances, they may be ascribed to an extra mixing which occurs in regions where the Schwarzschild citerion predicts a stable stratification, and this is why the organizers of this meeting have asked me to discuss the mixing processes that may influence stellar evolution.

Many of such processes have been identified (cf. Zahn 1983); we shall restrict our attention only on those that are relevant to massive stars. The most efficient are the mechanisms which extend the convective core beyond the limit assigned by the Schwarzschild criterion: penetrative convection and semi-convection. For the first, we have still no satisfactory treatment, as we shall see; this is not too surprising, since convection as such is far from being mastered, in spite of all efforts that have been spent. With semi-convection the situation is even worse, but it is not clear whether it would come into play if other types of mixing, which will be discussed below, were properly taken in account.

After recalling our present understanding of penetrative convection, I shall focus mainly on the mixing processes which are likely to occur within the radiation zone. The problem is far from new: it dates back to Von Zeipel (1924),

Space Science Reviews **66**: 285–297, 1994.
© 1994 *Kluwer Academic Publishers. Printed in Belgium.*

Eddington (1925) and Vogt (1925). But progress has been slow, mainly because turbulence is omnipresent in stars: even when the motions are very slow they do not remain laminar. This presents a major difficulty, because in most instances turbulence is handled with recipes which can be tested experimentally; unfortunately, we lack of such experiments with stars, and we can only test the consequences of our assumptions and approximations by comparing the theoretical predictions with observed properties. We have practiced this for 40 years with the mixing-length treatment of stellar convection, and there is such a need that we should proceed likewise to elaborate a coherent description of the mixing in radiative interiors. The major part of this paper will be devoted to it.

2. Convective penetration

Penetration is observed in all circumstances where the convectively unstable domain is not constrained within rigid boundaries: in the laboratory, the Earth atmosphere, the oceans, etc. The question is thus not whether penetration does occur, or not, but how far the convective motions penetrate into the adjacent stable domain.

The estimates which have been given so far for the extent of penetration suffer from the neglect of viscous dissipation. This may look as an excellent approximation, since the viscosity of stellar interiors is extremely small. However convection is highly turbulent, and therefore the effects of viscosity are tremendously enhanced: it is as if the viscosity ν were replaced by the turbulent viscosity $\nu_t \approx v\ell$, with ℓ and v being the characteristic size and velocity of the largest turbulent eddies. The ratio ν_t/ν is the Reynolds number of the turbulent flow: in stellar convection zones, it can easily reach 10^{11} or 10^{12}.

Therefore such inviscid estimates must be considered as upper limits for the penetration. The one given by Roxburgh (1978, 1989) has the great merit of being parameter-free. It states that convection from a stellar core penetrates to a radius r_{pen} given by the following integral:

$$\int_0^{r_{\text{pen}}} (L_{\text{total}} - L_{\text{rad}})\, d\left(\frac{1}{T}\right) = 0 \,, \tag{1}$$

with T being the temperature, L_{total} the total luminosity generated by the nuclear reactions, and L_{rad} the total luminosity carried by radiation. In fig. 1 they are sketched versus the inverse of the temperature: in the unstable core, L_{total} first increases rapidly, and then gradually levels off, because the nuclear sources are extremely sensitive to the temperature. The radiative luminosity L_{rad} cannot exceed that associated with an adiabatic temperature gradient; in the core of massive stars, it is unable to match L_{total}, and the excess flux is transported by convection. But L_{rad} increases steadily, and it catches up with L_{total} at r_{KS}, where Karl Schwarzschild's criterion would locate the boundary of the convective core: from there on, radiation suffices to carry the whole energy flux, and there is no

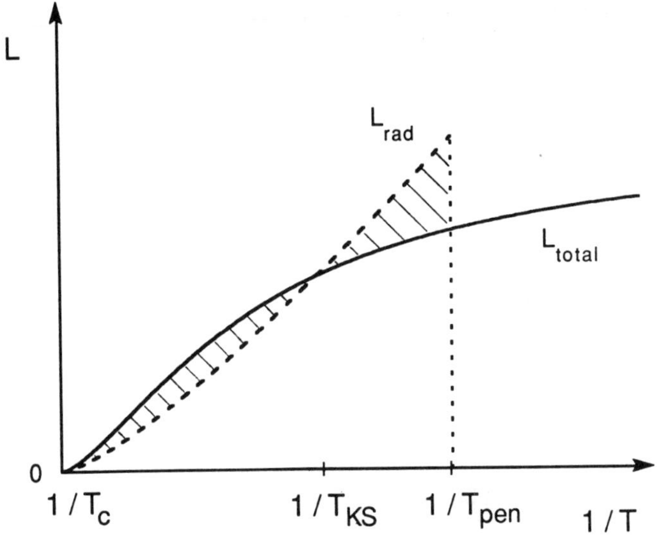

Fig. 1.   Illustration of the Roxburgh criterion for convective penetration; in abscissae the inverse temperature, which stars at its central value $1/T_c$. When one applies the Schwarzschild criterion, the core extends to T_{KS}; with Roxburgh's prescription, it reaches T_{pen}, for which the two hatched areas are equal.

need anymore for convective transport. However, as Roxburgh has shown, in the inviscid limit the convective motions penetrate to the level where the two areas hatched in fig. 1 equal each other.

Recently, with several collaborators, Roxburgh has tried to refine his integral constraint, and to assess the effect of viscous dissipation; he reports himself on their results in these proceedings. The difficulty remains to estimate the turbulent viscosity, which is treated as a free parameter in their numerical simulations.

Similar calculations, also in two dimensions, have been performed earlier by Hurlburt *et al.* (1986); in a more recent work, Hurlburt *et al.* (1993) find that the penetration scales as predicted by Zahn (1991), but again it depends on the viscous dissipation, which can only be guessed. Direct three-dimensional simulations are still in a preliminary state (Nordlund *et al.* 1992; Brandenburg *et al.* 1993; Cattaneo & Malagoli, private communication). In all calculations performed so far, the radiative diffusivity is artificially enhanced to accommodate the insufficient spatial resolution, and this yields often a penetration depth which is unrealistically large.

To wrap up, convective penetrations poses no longer a theoretical problem, but the result, namely the extent of penetration, depends entirely on a parameter which we are not able, as yet, to determine from first principles. Hence it is difficult to do much better than parametrizing directly the penetration depth, as is common practice among the colleagues who build stellar models. The only advice one would like to give them is to chose the core radius as the characteristic length, rather than the local pressure scale height, for reasons that were explained in Zahn (1991) and Roxburgh (1992).

3. Rotation-induced turbulence

It was Schatzman (1969) who the first suggested that stellar radiation zones should be the seat of some mild turbulence, in order to explain the relative homogeneity of the surface material: mild enough so that it would not interfer with the vertical transport of heat, but efficient enough so that it would prevent gravitational settling. The best candidate to produce such a turbulence is the differential rotation of the star, which is liable to many instabilities, the strongest of these being the familiar shear instability (Spiegel & Zahn 1970; Zahn 1974, 1975): it grows on a dynamical time which is of the order of a rotation period. Other instabilities may also intervene, but they are probably of lesser importance, because they proceed on much longer timescales. Let us recall the main properties of the shear instabilities that arise in stellar radiation zones.

3.1. TURBULENCE DUE TO THE VERTICAL SHEAR

A shear flow becomes unstable whenever its Reynolds number exceeds some critical value of order 1000, provided there is no force which can oppose the turnover of the turbulent eddies. The buoyancy is one of such forces, and it prevents a vertical shear from becoming unstable when, according to the Richardson criterion,

$$\frac{N^2}{(dV/dz)^2} \geq R_c , \tag{2}$$

where $V(z)$ is the horizontal velocity, which depends only on the vertical coordinate z, and N the buoyancy frequency. The latter is given by

$$N^2 = \frac{g}{H_P}(\nabla_{\mathrm{ad}} - \nabla) \tag{3}$$

with g being the gravity, H_P the pressure scale height, and $\nabla = \partial \ln T / \partial \ln P$ the usual logarithmic gradients of stellar structure theory. The critical Richardson number R_c depends somewhat on the flow profile, but it is always lower than 1/4.

The stability condition above applies in the limit of adiabatic perturbations. In a real star, radiative damping acts to smooth out these perturbations, and the criterion becomes less stringent: turbulence will be sustained as long as (Townsend 1958; Dudis 1974; Zahn 1974)

$$\frac{N^2}{(dV/dz)^2}\frac{v\ell}{K} \leq R_c . \tag{4}$$

Here K is the radiative diffusivity, and ℓ and v are again the size and the velocity of the turbulent eddies, with $v\ell \ll K$. The largest eddies allowed are those which satisfy marginally this inequality (4); they produce a turbulent viscosity

$$\nu_t = \frac{1}{3}\, v\ell = \frac{1}{3} R_c\, K \frac{(dV/dz)^2}{N^2}\,. \tag{5}$$

When translated into spherical coordinates, and after horizontal averaging, this expression becomes (see Zahn 1992)

$$\nu_t = \frac{8}{45} R_c\, K \left(\frac{\Omega}{N}\right)^2 \left|\frac{d\ln\Omega}{d\ln r}\right|^2 \tag{6}$$

Now the nuclear reactions build up a gradient of molecular weight which stiffens the stratification; in that case, one retrieves the original Richardson criterion, with $N^2 = (g/H_P)\, d\ln\mu/d\ln P$. In spherical coordinates, the stability criterion reads

$$\left|\frac{d\ln\mu}{d\ln r}\right| > \frac{8}{15} R_c \left(\frac{\Omega^2 r}{g}\right) \left|\frac{d\ln\Omega}{d\ln r}\right|^2 ; \tag{7}$$

this is the condition a μ-gradient has to satisfy in order to inhibit the turbulence generated by the vertical shear, in the presence of radiative damping.

3.2. TURBULENCE DUE TO THE HORIZONTAL SHEAR

In contrast, nothing can prevent a horizontal shear from becoming unstable, not even the Coriolis force, although it plays to some extent the role of a restoring force. Thus the differential rotation in latitude will generate turbulence as soon as its Reynolds number exceeds its critical value of about 1000.

Not much is known about this turbulence, except that it probably resembles the geostrophic turbulence observed on Earth, in the oceans and in the atmosphere. Therefore, it should be highly anisotropic, with much more efficient transport in the horizontal directions than in the vertical. The effect of such a turbulence is to reduce the differential rotation which generates it, much as we see in thermal convection, which tends to establish an adiabatic stratification.

What proof do we have that such turbulence occurs in stars? Below the solar convection zone, thanks to helioseismology, we know that the rotation becomes quickly uniform with depth, and this can only be achieved by a turbulent viscosity much stronger in the horizontal than in the vertical direction (Spiegel & Zahn 1992). We shall thus assume that this anisotropic turbulence exists throughout the radiative interior of rotating stars, and that it enforces a rotation which depends mainly on depth, and very little on latitude.

3.3. TURBULENT EROSION OF THE ADVECTIVE TRANSPORT

Such an anisotropic turbulence induces negligible mixing in the vertical direction – even worse: it acts to inhibit the transport of matter by a large scale flow. As we shall see below, the star is the seat of a meridian circulation, whose vertical component is given by

$$u(r, \theta) = U(r)P_2(\cos \theta) \, . \tag{8}$$

The continuity equation yields the amplitude of the horizontal component:

$$V = \frac{1}{6\rho r} \frac{d}{dr} [\rho r^2 U] \, . \tag{9}$$

The effect of this turbulence is to limit the inhomogeneities on horizontal surfaces, with the result that the vertical transport of a chemical behaves as a diffusion. The effective diffusivity is given by

$$D_{\text{eff}} = \frac{|rU(r)|^2}{30 \, D_h} \, , \tag{10}$$

with D_h being the turbulent diffusivity in the horizontal direction. This theoretical prediction made by Chaboyer and Zahn (1992) has been verified by Charbonneau (1992) through two-dimensional numerical simulations. The outcome is that the vertical transport of a chemical is much less efficient than the transport of angular momentum, which is little affected by this turbulent erosion: its large-scale advection by the meridian flow would subsist even if Ω were rigorously constant on horizontal surfaces.

Unfortunately, it is not possible to determine *ab initio* the horizontal diffusivity D_h, and this is admittedly the weak point of our approach. However, don't we face the same problem when fitting a convective envelope to an atmosphere with the mixing-length treatment? In Zahn (1992) we suggested a plausible parametrization for this quantity D_h, which is

$$D_{\text{eff}} = \frac{C_h}{30} \, |rU| \left| \frac{U}{2V - \alpha U} \right| \, , \tag{11}$$

in terms of the components of the meridian flow and of the angular momentum profile

$$\alpha = \frac{1}{2} \frac{d \ln(r^2 \, \Omega)}{d \ln r} \, . \tag{12}$$

The parameter C_h is strictly smaller than 1, and it needs to be calibrated against the observations.

4. Meridian circulation

Since Eddington (1925) and Vogt (1925) we know that a rotating star cannot, in general, achieve radiative equilibrium, and that it is therefore the seat of a large scale circulation. The way this circulation was treated until now, since the pioneering work of Sweet (1950), was to take some rotation law – usually just uniform rotation, for sake of simplicity – and to calculate the corresponding meridian flow. However such a flow advects angular momentum: thus it induces differential rotation and keeps modifying the rotation profile. Although one was well aware of this feed-back, it was rarely taken into account, mainly because one was unable to handle the turbulence generated by the differential rotation. The only attempt was by Sakurai (1986, 1991), who tackled the full two-dimensional problem, assuming that the flow remained laminar.

However, with our assumption of an anisotropic turbulence the problem reduces to one dimension in space, since the rotation rate depends only on the radial coordinate. Then it is possible to derive an explicit expression for the meridian velocity, which we write schematically as

$$U(r) = U_\Omega\{\Omega^2\} + U_{\text{bar}}\left\{\frac{d^3\Omega^2}{dr^3}, \frac{d^2\Omega^2}{dr^2}, \frac{d\Omega^2}{dr}\right\} + U_\mu\left\{\frac{d^2\tilde{\mu}}{dr^2}, \frac{d\tilde{\mu}}{dr}, \tilde{\mu}\right\}. \quad (13)$$

The first term represents the classical Eddington-Sweet velocity, whereas the last describes the effect of chemical inhomogeneities, which are produced mainly by the nuclear reactions ($\tilde{\mu}$ is the amplitude of the horizontal variation of the molecular weight).

This effect was first described by Mestel (1953): in his treatment, the inhomogeneities are advected by the meridian flow until the "μ-currents" are strong enough to cancel the meridian velocity $U(r)$ at some depth. Thereafter, the star is divided in two homogeneous regions, separated by a "μ-barrier" of increasing strength, which can only be prevented if the rotation approaches the break-up speed, i.e. if the centrifugal force is close to the local gravity.

In our approach the situation is quite different: the chemical inhomogeneities are eroded by the anisotropic turbulence, and they saturate at the level

$$\frac{\tilde{\mu}}{\mu} = -\frac{r^2}{6D_h}U\frac{d\ln\mu}{dr}. \quad (14)$$

The main feedback is accomplished here by the baroclinic term U_{bar} in (13), which is ignored in the classical treatment. This term involves the derivatives of the rotation rate Ω, and it thus arises whenever the star is in differential rotation. As may be expected, it plays an important role in the evolution of the rotation profile.

The flux of angular momentum within the star is the sum of the advective flux carried by the meridian flow and of the viscous flux, which is due to the turbulence generated by the vertical shear (cf. §3.1):

$$\mathcal{F} = -\frac{4\pi}{5}r^4\rho\,\Omega\,U - 4\pi r^4\rho\nu_t\frac{\partial\Omega}{\partial r}. \quad (15)$$

During their main-sequence evolution, low-mass stars are spun down very efficiently because they are the seat of a wind which is forced to rotate with

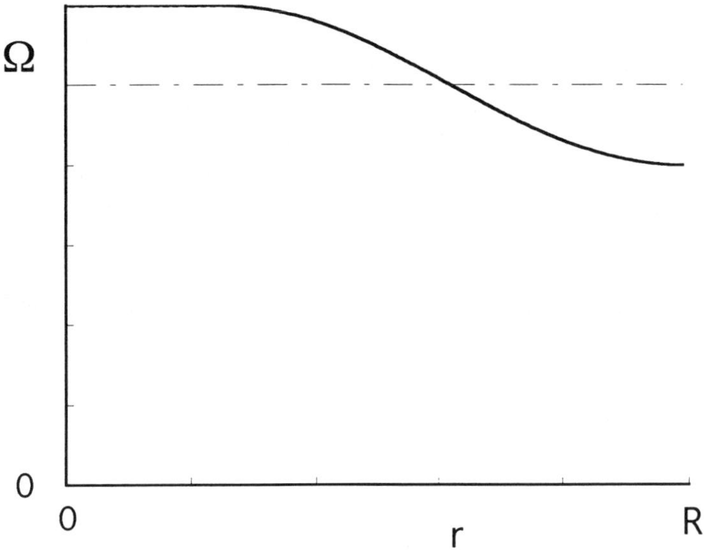

Fig. 2. The asymptotic rotation profile in a homogeneous massive star. The rotation is assumed to be uniform in the convective core, which occupies here 1/4 of the radius.

an extended magnetosphere, as was shown by Schatzman (1962). The angular momentum which is lost has to be carried through the radiative interior, and this is achieved mainly by the meridian flow. In other words, it is the loss of angular momentum through the wind which causes of the meridian circulation.

On the contrary, massive stars lose little angular momentum during their main-sequence phase, because their wind takes off near the surface: for this reason, they remain fast rotators as they evolve. Therefore the net flux of angular momentum within these stars is negligibly small, with the advective and viscous fluxes cancelling each other. Hence such a star reaches an asymptotic regime, after a Kelvin-Helmholtz time, in which

$$\frac{1}{5}U = -\nu_t \frac{d\ln\Omega}{dr};\tag{16}$$

such a rotation state has been anticipated by Randers (1941). Note that the meridian flow vanishes completely in the inviscid limit: this property was established by Busse (1981, 1982), but his claim was met by sharp criticism, on the ground that he had neglected the classical Eddington-Sweet term in the expression (13) of the meridian velocity.

As a matter of fact, the flow predicted by (16) differs substantially from the Eddington-Sweet circulation. In that classical treatment, the vertical velocity tends to a non-zero limit near the surface in the case of uniform rotation (Sweet 1950), and for more general rotation laws $\rho U \rightarrow cst$ (Baker & Kippenhahn 1959). According to (16) and to (5), the vertical velocity behaves here as

$$U \approx \frac{R}{t_{\mathrm{KH}}} \left(\frac{\Omega^2 R^3}{GM} \right)^3 \left(\frac{\rho_m}{\rho} z^4 \right), \tag{17}$$

with the usual notations; $z = (R - r)/R$ is the normalized depth, $\rho_m(r)$ the mean density and $t_{\mathrm{KH}} = GM^2/RL$ the Kelvin-Helmholtz time. Thus the flow is weaker than in the classical case, where $U \approx (R/t_{\mathrm{KH}})(\Omega^2 R^3/GM)$, and its vertical component vanishes near the surface. But let us stress that this comparison is only valid for the massive stars, which conserve their angular momentum.

Using (16) and the full expression (13) for $U(r)$ (see Zahn 1992), on can derive the asymptotic rotation profile: it is depicted in fig. 2 for a homogeneous star. The core rotates somewhat faster than the surface, but the departure from uniform rotation is rather modest:

$$\frac{\Delta\Omega}{\Omega} \approx \frac{\Omega^2 R^3}{GM}. \tag{18}$$

This profile depends little on the turbulent viscosity: it satisfies essentially the zero circulation condition $U_\Omega(r) + U_{\mathrm{bar}}(r) \equiv 0$.

5. Rotation-induced mixing in massive stars

In order to describe the mixing that occurs in the radiative envelope of a massive star, one has to solve simultaneously an equation for the transport of angular momentum

$$4\pi r^2 \frac{\partial\Omega}{\partial t} = -\frac{1}{r^2} \frac{\partial r^2 \mathcal{F}}{\partial r}, \tag{19}$$

with the flux $\mathcal{F}$ given by (15) and another equation for the transport of the mean molecular weight

$$\frac{\partial\mu}{\partial t} = \frac{1}{r^2} \frac{\partial}{\partial r} \left[r^2 \rho (D_t + D_{\mathrm{eff}}) \frac{\partial\mu}{\partial r} \right] + \dot\mu. \tag{20}$$

The flux of angular momentum is defined in (15), $D_t \approx \nu_t$ is the turbulent diffusivity associated with the vertical shear instability, D_{eff} is the effective diffusivity arising from the turbulent erosion of the meridian flow, which has been given in (10), and $\dot\mu$ is the increase of molecular weight produced by the nuclear reactions. In writing (20) we have assumed that the molecular weight is determined essentially by the abundance of one species, such as helium on the main-sequence. Otherwise one would have to follow separately the evolution of all relevant constituants, and to sum up their contributions to μ.

Endal and Sofia (1978, 1981) and more recently Pinsonneault *et al.* (1989, 1990) have used the same set of equations to model very successfully the spin-down and the lithium depletion of low-mass stars. We differ from them mainly on two points: we include the baroclinic terms in the expression (13) of the meridian

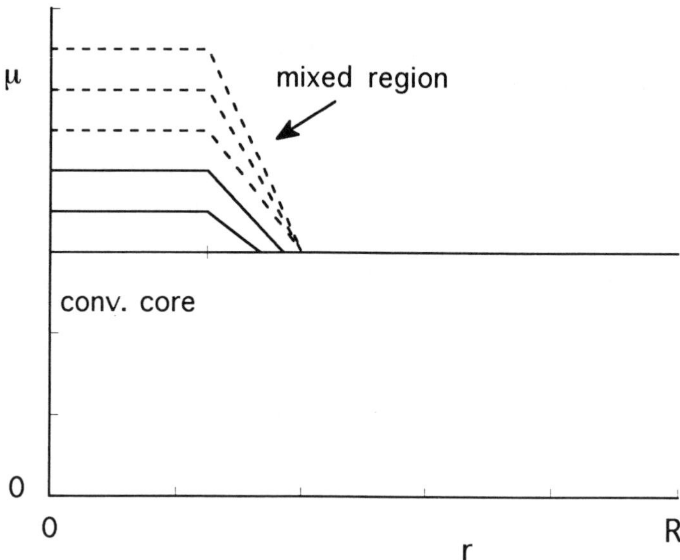

Fig. 3.   Sketch of the molecular weight profile, as the massive star evolves on the main-sequence. First the mixed zone expands into the radiation zone (continuous lines), until it is halted by the (steepening) μ-gradient (dashed lines).

velocity, and we treat the transport by the meridian flow as an advection, instead of a diffusion as they do.

Work is in progress to interface the governing differential equations with a stellar evolution code, and to build their numerical solutions. In the meanwhile, we cannot resist the temptation of making some crude predictions on how mixing proceeds in such a star – fully aware this cannot substitute for the actual calculations.

As the star evolves, the concentration of helium increases in the convective core, and helium-rich material diffuses into the radiative envelope as described by (20). At the beginning, the μ-gradient is too weak to prevent the vertical shear instability, which then dominates the transport, and we may neglect D_{eff} as compared to D_t. Referring back to (6) and to (18), and assuming that all quantities are constant, including the gradients, we get the following rough estimate for the turbulent diffusivity

$$D_t \approx \nu_t = \frac{8}{45} R_c \, K \left(\frac{\Omega}{N}\right)^2 \left|\frac{d\ln\Omega}{d\ln r}\right|^2 \approx \frac{R^2}{t_{\mathrm{KH}}} \left(\frac{\Omega^2 R^3}{GM}\right)^3 , \tag{21}$$

having dropped all numerical coefficients which are of order unity. Thus the helium-rich core expands approximately as

$$d^2 \approx D_t \, t , \tag{22}$$

with d being its extent into the radiation zone (see fig.3).

Since the molecular weight increases roughly as t/t_{MS}, with t_{MS} being the main-sequence lifetime, the μ-gradient in the radiation zone varies as

$$\left| \frac{d \ln \mu}{d \ln r} \right| \approx \frac{t}{t_{\text{MS}}} \frac{R}{d} = \frac{R\, d}{D_t\, t_{\text{MS}}} \, . \tag{23}$$

But we have seen that the vertical shear instability will be suppressed when the gradient of μ becomes strong enough. According to (7) this occurs when

$$\left| \frac{d \ln \mu}{d \ln r} \right| > \frac{8}{15} R_c \left(\frac{\Omega^2 r}{g} \right) \left| \frac{d \ln \Omega}{d \ln r} \right|^2 \approx \left(\frac{\Omega^2 R^3}{GM} \right)^3 , \tag{24}$$

i.e. when the helium-rich core extends to

$$\frac{d}{R} \approx \frac{t_{\text{MS}}}{t_{\text{KH}}} \left(\frac{\Omega^2 R^3}{GM} \right)^6 . \tag{25}$$

Thereafter, the transport of helium is not entirely suppressed, but it proceeds on a much slower pace, because it relies then only on the meridian advection, which is intrinsically of lesser efficiency (cf. §3.3), and which is reduced moreover because the profile of Ω keeps adjusting itself such as to cancel the meridian flow.

Needless to state again that (25) is a very crude estimate: we ignored all numerical factors of order unity, and the μ-profile was approximated by straight lines. But we may draw the conclusion, at least in qualitative terms, that the convective core of a massive star is surrounded by a region of decreasing helium content, whose extent depends sensitively on the rotation speed (fig.3).

At first sight, our result ressembles that obtained by Mestel (1953), since we both get a "μ-barrier" which prevents the star from being completely mixed, unless the rotation speed is rather high. But there is an important difference: in Mestel's treatment, the extent d of the mixed region varies stepwise with $\eta = \Omega^2 R^3/GM$, and there is a threshold, at $\eta = 0.96$ in the model he uses, below which there is no mixed zone at all: the Ω-currents are checked as soon as they emerge from the core, in his own words, and a barrier is built which prevents any further exchange of matter between core and envelope. (To accommodate such an extreme value of η, Mestel assumes that Ω increases with depth.) In contrast, in our solution d varies continuously as η^6, which allows for some extra mixing outside the convective core even when the rotation rate is moderate.

6. Conclusions

Let us summarize. Today, we still lack of a satisfactory treatment for convective penetration. The best we can do is to parametrize the extent of penetration in our stellar structure codes: in massive stars it is preferable to prescribe it as a fraction of the radius of the convective core.

In the radiative envelope, some mild mixing is achieved through the combined action of a thermally driven meridian circulation and of turbulent motions generated by differential rotation. This circulation is weaker than in the classical Eddington-Sweet theory, and most of the transport of elements such as helium is mediated by the turbulence. Work is in progress to describe the mixing through numerical simulations. In the meanwhile, some qualitative predictions can be derived from the transport equations, but they need to be validated by the actual calculations. They apply only to the main-sequence phase, during which the structural changes are rather modest.

The helium-rich material produced in the convective core diffuses into the radiation zone, until the gradient of molecular weight has become strong enough to suppress the turbulence. The mixed region extends up to a level which is determined by the rotation speed. In all likelihood, it is this rotational mixing which determines the gradient of molecular weight outside the convective core, and not the semi-convection which is presently implemented in most stellar structure codes. It also appears that a massive star which rotates sufficiently fast may be completely mixed, up to its surface, as suggested by some observations.

References

Baker, N., Kippenhahn, R.: 1959, *Zeitschrift fur Astrophysik* **48**, 140

Brandenburg, A., Jennings, R.L., Nordlund, A, Rieutord, M., Stein, R.F., Tuominen, I.: 1993, *Journal of Fluid Mechanics*, (in press)

Busse, F.H.: 1981, *Geophysical and Astrophysical Fluid Dynamics* **17**, 215

Busse, F.H.: 1982, *Astrophysical Journal* **259**, 759

Charbonneau, P.: 1992, *Astronomy and Astrophysics* **259**, 134

Chaboyer, B., Zahn, J.-P.: 1992, *Astronomy and Astrophysics* **253**, 173

Dudis, J.J.: 1974, *Journal of Fluid Mechanics* **64**, 65

Eddington, A.S.: 1925, *Observatory* **48**, 78

Endal, A.S., Sofia, S.: 1978, *Astrophysical Journal* **220**, 279

Endal, A.S., Sofia, S.: 1981, *Astrophysical Journal* **243**, 625

Hurlburt, N.E., Toomre, J., Massaguer, J.: 1986, *Astrophysical Journal* **311**, 563

Hurlburt, N.E., Toomre, J., Massaguer, J., Zahn, J.-P.: 1993, *Astrophysical Journal*, (in press)

Mestel, L.: 1953, *Monthly Notices of the RAS* **113**, 716

Nordlund, A, Brandenburg, A., Jennings, R.L., Rieutord, M., Ruokalainen, J., Stein, R.F., Tuominen, I.: 1992, *Astrophysical Journal* **392**, 647

Pinsonneault, M.H., Kawaler, S.D., Demarque, P.: 1990, *Astrophysical Journal, Supplement Series* **74**, 501

Pinsonneault, M.H., Kawaler, S.D., Sofia, S., Demarque, P.: 1989, *Astrophysical Journal* **338**, 424

Randers, G.: 1941, *Astrophysical Journal* **94**, 109

Roxburgh, I.W.: 1978, *Astronomy and Astrophysics* **65**, 281

Roxburgh, I.W.: 1989, *Astronomy and Astrophysics* **211**, 361

Roxburgh, I.W.: 1992, *Astronomy and Astrophysics* **266**, 291

Sakurai, T.: 1986, *Geophysical and Astrophysical Fluid Dynamics* **36**, 257

Sakurai, T.: 1991, *Monthly Notices of the RAS* **248**, 457

Schatzman, E.: 1962, *Annales d'Astrophysique* **25**, 18

Schatzman, E.: 1969, *Astronomy and Astrophysics* **3**, 331

Spiegel, E.A., Zahn, J.-P.: 1970, *Comments on Astrophysics and Space Physics* **2**, 178

Spiegel, E.A., Zahn, J.-P.: 1992, *Astronomy and Astrophysics* **265**, 106

Sweet, P.A.: 1950, *Monthly Notices of the RAS* **110**, 548

Townsend, A.A.: 1958, *Journal of Fluid Mechanics* **4**, 361

Vogt, H.: 1925, *Astronomische Nachrichten* **223**, 229

Von Zeipel, H.: 1924, *Monthly Notices of the RAS* **84**, 665

Zahn, J.-P.: 1974, *Stellar Instability and Evolution*, (ed. P. Ledoux, A. Noels and R.W. Rogers; Reidel, Dordrecht), 185

Zahn, J.-P.: 1975, *Memoires Soc. Royale des Sciences Liege*, 6e serie 8, 31

Zahn, J.-P.: 1983, *Astrophysical Processes in Upper Main Sequence Stars*, (ed. B. Hauck and A. Maeder; Observatoire de Geneve), 253

Zahn, J.-P.: 1991, *Astronomy and Astrophysics* **252**, 179

Zahn, J.-P.: 1992, *Astronomy and Astrophysics* **265**, 115

OVERSHOOTING FROM CONVECTIVE CORES:THEORY AND NUMERICAL SIMULATION

IAN W. ROXBURGH

Astronomy Unit, Queen Mary and Westfield College, University of London, Mile End Rd., London E1 4NS, United Kingdom.

Abstract. Convective overshooting increases the fraction of the star which is effectively mixed, thus altering models of stellar evolution. If the feed back of overshooting on the structure of the star is neglected the estimated extent of overshooting is very small. If the feed back is included in these estimates then the adiabatic core is extended by a distance comparable to a substantial fraction of the radius of the unstable region. An upper limit on convective overshooting is given by the integral constraint (Roxburgh 1978,1989) with viscous dissipation neglected. If this constraint is applied to small convective cores then the maximum extent of the penetration region is shown to be at most about 0.18 times the radius of the core independent of the details of energy generation and opacity. The ratio of the maximum penetration distance to the scale height at the edge of the "classical boundary" varies very strongly with core size, and modelling overshooting by taking the penetration distance as a multiple of the scale height is likely to give misleading results. Numerical simulations of two-dimensional compressible convection in a fluid where the central regions are naturally convectively unstable, and the surrounding layers are stable, have been undertaken for different values of the Prandtl number. The results indicate that for low Prandtl numbers viscous dissipation is of decreasing importance and the simple integral condition gives a reasonable estimate of the extent of overshooting. Stellar seismology offers the possibility of detecting the location of the core - envelope interface through a periodic variation of the small frequency separation with frequency.

Key words: Stars – Convection

1. Convective Penetration: Mixing-length and unstable mode-calculations

The penetration of convective motion from an unstable region into the surrounding stable region increases the fraction of the star which is effectively mixed, thus affecting the subsequent evolution. This overshooting is now recognised as important in stellar evolution and is increasingly being incorporated into evolutionary calculations of stellar models; reliable estimates of the extent of convective penetration are however difficult to determine.

Early attempts to quantify the extent of convective penetration from convective cores by Saslaw and Schwarzschild (1965) and Roxburgh (1965) were incorrect in that they did not include the feed back of the overshooting on the structure of the star. In both calculations the structure of the star was taken as given by a standard Cowling model with no overshooting where the convective core is slightly superadiabatic $\Delta\nabla \geq 0$, the boundary is where $F_{rad} = F_{total}$ and $\Delta\nabla = 0$, the envelope is stable and in radiative equilibrium. The variation of $\Delta\nabla = (\nabla - \nabla_{ad})$, where $\nabla = \mathrm{dlogT/dlogP}$, is shown in Figure 1. Inside the convective core the superadiabatic gradient is given by the mixing length theory of convection with $F_{conv} = F_{total} - F_{rad}$. Typically $\Delta\nabla \cong 10^{-8}$ in the convective core and rises

Space Science Reviews **66**: 299–308, 1994.
© 1994 *Kluwer Academic Publishers. Printed in Belgium.*

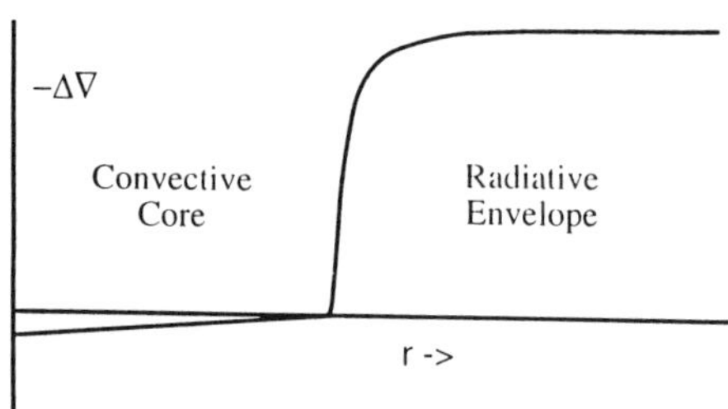

Fig. 1.   Variation with radius of the superadiabatic gradient $\Delta\nabla$ in a Cowling model without overshooting. $\Delta\nabla$ is small and positive in the unstable region and rapidly changes to large negative values in the stable envelope.

very steeply to a value of order 10^{-1} in the radiative envelope.

Saslaw and Schwarzschild used a Cowling model as described above and calculated the eigenfunction of the lowest order ($\ell = 1$) unstable adiabatic convective mode, and found that the penetration distance of the radial velocity was very small. Roxburgh likewise took a standard Cowling model but determined the penetration depth by a mixing length argument determining the distance a convective "eddy" rising from deep inside the convective core penetrated into the stable radiative zone before coming to rest. Both authors obtained a very small penetration distance claiming that overshooting was unimportant.

The feed-back of the convective overshooting on the structure of the star is however important; in the overshoot region "eddies" with even very small convective velocities carry a large amount of energy unless the stratification is very close to adiabatic; the overshoot region therefore adjusts to be very slightly subadiabatic until the convective velocities have fallen to sufficiently small values so that the energy carried by convection is small. Within the overshoot region the energy carried by convection goes negative so that the energy carried by radiation exceeds the total flux, the temperature gradient then changing to that required for the energy to be carried by radiation in a very narrow "boundary layer". This is illustrated in Figures 2 and 3.

This feed-back was incorporated into the "eddy" estimate by Shaviv and Salpeter (1973) using a non-local mixing length model, these authors found a significant extension of the almost adiabatic core, a result that been subsequently reproduced by several authors. The problem however is in quantifying the actual extent of the overshoot region; whereas the extent of the unstable almost adiabatic region is insensitive to the detailed mixing length model, the extent of the overshoot region does depend on the details of the model. When seeking to incorporate the effects of convective penetration into evolutionary calculations of

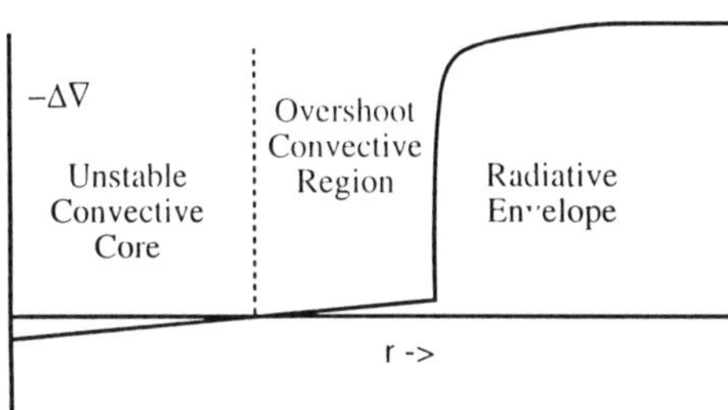

Fig. 2.   Variation with radius of the superadiabatic gradient $\Delta\nabla$ in a Cowling model with overshooting. In the overshoot region $\Delta\nabla$ remains small and negative before adjusting to large negative values.

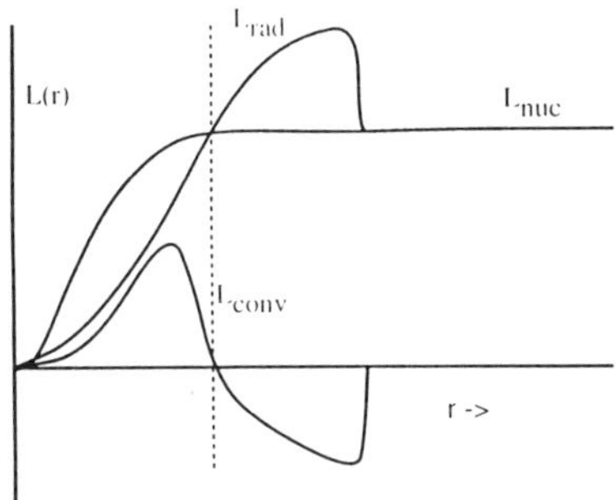

Fig. 3.   Variation with radius of the luminosities L_{rad}, L_{conv}, and the total luminosity $L_{nuc}.L_{conv}$ goes negative in the overshoot region so that L_{rad} exceeds the total luminosity L_{nuc}.

stellar models, it is quite common to seek to parametrise this uncertainty using a simple ad-hoc recipe in which the adiabatic core extends a distance d $= \alpha\,H_p$ where H_p is the pressure scale height at the edge of the "classical core" where:

$$L_{rad} = L_{nuc}\ [H_p = -P/\mathrm{dlog}P/\mathrm{dr}].$$

This can be quite misleading since $H_p \longrightarrow \infty$ as r $\longrightarrow$ 0, the smaller the size of the "classical" core the larger the size of the core including convective overshooting; a most unlikely situation!

As far as I am aware the inclusion of feed back into the most lowest order eigenmode approach has not appeared in the literature, although the present author did undertake such a calculation for the $\ell = 1$ mode quite some years ago (cf Roxburgh 1989). In this calculation the energy carried by convection was determined from the eigenfunctions for the radial velocity v_1 and temperature perturbation T_1 averaged over a sphere,

$$F_c = \lambda \, c_p \, \rho \, v_1 \, T_1,$$

where λ is an amplitude factor determined by the condition that

$$F_c + F_{rad} = F_{total}.$$

In this calculation the super adiabatic gradient cannot be imposed from other considerations (eg by using the mixing length model) but has to be determined self consistently, that is the departure from the adiabatic gradient $(\nabla \text{-} \nabla_{ad})$ has to be such that the eigenfunctions v_1 and T_1 give just the correct variation of convective flux so that $F_c + F_{rad} = F_{total}$ at all points in the convective region. Again this calculation demonstrated significant convective penetration.

This type of calculation could be generalised by taking a spectrum of unstable modes with relative amplitudes given by some other considerations, eg that the energy distribution amongst modes followed a Kolmogorov law, or was proportional to the growth rates of the modes, but there are so many unjustified assumptions in this approach that little confidence could be placed in the solution.

2. The Integral Constraint

An alternative approach, integrating the heat equation over the whole of the convecting region (unstable plus penetration regions) yields an inequality that gives an estimate of the maximum extent of the overshoot region (Roxburgh 1976, 78, 89). The heat equation is

$$\frac{\partial}{\partial t}(\rho S) + \frac{\partial}{\partial x^i}(\rho u_i S) = -\frac{1}{T}\frac{\partial F_i}{\partial x^i} + \frac{\Phi}{T} + \frac{\epsilon}{T} \tag{1}$$

where ρ is the density, T the temperature, u_i the velocity, S the entropy, F_i the flux carried by radiation and Φ the viscous dissipation. Φ is positive definite and it is this property that gives an inequality that provides an upper limit on the extent of convective penetration. Integrating this equation over a sphere V containing the convective core, such that $u_i = 0$ outside V, taking turbulent averages and a assuming a statistically stationary state gives

$$\int_V \overline{\frac{1}{T}\left(\frac{\partial F_i}{\partial x^i} - \epsilon\right)}\, dV = \int_V \overline{\left(\frac{\Phi}{T}\right)}\, dV > 0 \tag{2}$$

where the over bar denotes the averaging operator. Integrating by parts and using Gauss's theorem gives the integral condition in the form

$$\int_V \overline{(F_i - \Gamma_i)\frac{1}{T^2}\frac{\partial T}{\partial x^i}}\, dV = \int_V \overline{\left(\frac{\Phi}{T}\right)}\, dV > 0 \tag{3}$$

where Γ_i is the energy flux due to nuclear reactions and is defined as $\partial \Gamma_i / \partial x^i = \epsilon$. As shown in Roxburgh (1989) if the convective core is almost adiabatic the average of the fluctuating quantities on the left hand side of this inequality can be well approximated by the mean field values so that the inequality reduces to

$$\int_0^{r_c} (L_{rad} - L_{nuc}) \frac{1}{T^2} \frac{dT}{dr} \, dr = \int_0^{r_c} \frac{\Phi}{T} 4\pi r^2 \, dr > 0 \tag{4}$$

It is the inequality in this form that will be used to set an upper limit on convective penetration by neglecting viscous dissipation, that is setting $\Phi = 0$ in equation (4).

If we confine our attention to "Cowling type" models where the opacity and energy generation per unit mass are $\kappa = \kappa_0 \rho^\alpha T^{-\beta}$, and $\epsilon = \epsilon_0 \rho T^\eta$ and the equation of state is that of an ideal gas then the calculations are quite straight forward. To a good approximation the stratification in the core and penetration region is adiabatic with $P \propto T^{5/2}$ so the core is described by a polytrope of index 3/2, where $T = T_c \, \theta$ and ξ is a scaled value of the radius. For given values of η and $\varpi = (3/2 + \beta - 3\alpha/2)$ the maximum radius of convective core including penetration ξ_c is then determined in terms of the classical radius ξ_0 by the solution of the equation (Roxburgh 1992)

$$\int_0^{\xi_c} (\theta^{\overline{\omega}} \xi^2 \frac{d\theta}{d\xi} + \frac{B}{A} \int_0^{\xi} \theta^{\eta+3} \xi^2 \, d\xi) \, d\xi = 0 \tag{5}$$

where A and B are determined by the central values of pressure and density. If we compare models with the same central conditions then B/A is given in terms of ξ_0, the radius of the core without overshooting by

$$\frac{B}{A} = \frac{(-\theta^{\overline{\omega}} \xi^2 \frac{d\theta}{d\xi})_{\xi = \xi_0}}{\int_0^{\xi_0} \theta^{\eta+3} \xi^2 \, d\xi} \tag{6}$$

Numerical solutions for $(\overline{\omega}, \eta) = (1.5, 16)$, corresponding to electron scattering are given in Table 1.

If this criterion is applied to convective cores where the central conditions remain unchanged, the penetration distance ranges from 0.25 of the radius of the classical core (for small cores) up to 0.35 of the classical core radius (for large cores). It would therefore be more reasonable to model the convective penetration by $d = \alpha \, r_c$ where r_c is the radius of the classical convective core (where $L_{rad} < L_{nuc}$) and to take $\alpha = 0.3$ or to take the mass of the core as given by $M_c/M_o = 1.85$. It should be noted that the results derived here compare the core with penetration to the classical core with the same central values of pressure and temperature since this is how convective penetration would be included in the construction of a stellar model. However the convective core has to be matched onto a radiative envelope, the change in core mass from the

TABLE I

Solution for $\kappa = \kappa_0$, $\epsilon = \epsilon_0 \, \rho T^{16}$

ξ_0	ξ_c	ξ_c/ξ_0	d/H_p	M_c/M_0
1.000	1.251	1.251	0.213	1.802
1.200	1.539	1.283	0.350	1.845
1.400	1.851	1.322	0.550	1.873
1.500	2.023	1.349	0.689	1.887
1.600	2.224	1.390	0.886	1.918

inclusion of overshooting produces a change in central values so that one cannot simply read of the changes in a model from the above tables. It is necessary to make a complete stellar model with and without overshooting to make a meaningful comparison (cf Roxburgh 1978).

3. Numerical Simulation

The phenomenon of convective penetration can also be studied by numerical simulation, although it is not of course possible to simulate the conditions in stellar interiors. Such numerical calculations have been undertaken by several authors, usually with somewhat artificial physics in which the initial model has an unstable layer with polytropic index n [n + 1 = dlogP/dlog T] less than 1.5, surrounded by one or two stable layers where the polytropic index is large (typically n = 3); the conductivity K(T) is a discontinuous function of temperature in such models. Recently however Roxburgh and Simmons (1993) numerically studied penetration in a plane parallel layer where the conductivity K(T) was a continuous function of temperature

$$K(T) = K_0[(T/T_0)^3 + 0.6(T_0/T)^5].$$

The conductivity therefore has a minimum at T = T_0 and the central regions are unstable whereas the surrounding layers are stable; the initial distribution of polytropic index n [n+1 = dlogP/dlogT] are shown in Figure 4. The result of these calculations demonstrated that at least in the parameter regime that could be studied, viscous dissipation was of decreasing importance for decreasing Prandtl numbers, and that the so called "Roxburgh criterion" gave a reasonable estimate of the extent of convective penetration. Of course these calculations were only for moderately unstable regions (Rayleigh numbers in the centre of the layer ($R_a \simeq 5 \; 10^6$), and for relatively large Prandtl numbers ($\nu/K = \sigma \geq 0.01$), and are therefore a long way from the conditions of turbulent convection in stellar cores.

In the plane parallel case the integral condition reduces to

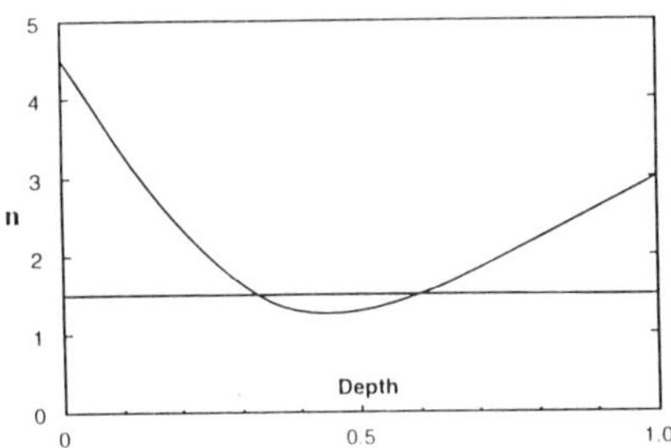

Fig. 4.   Variation of the polytropic index n where n+1= dlogP/dlogT with depth in the initial unstable radiative model.

$$\int_0^h \left\langle (F_i - \Gamma_i)\frac{1}{T^2}\frac{\partial T}{\partial x^i} \right\rangle \, dz = \int_0^h \left\langle \frac{\Phi}{T} \right\rangle \, dz \tag{7}$$

We define

$$I_0(z) = \left\langle (F_i - \Gamma_i)\frac{1}{T^2}\frac{\partial T}{\partial x^i} \right\rangle \;, V(z) = \left\langle \frac{\Phi}{T} \right\rangle \tag{8}$$

$$I_u = \int_{F \leq \Gamma} I_0(z)\, dV, I_p = \int_{F \geq \Gamma} I_0(z)\, dV \tag{9}$$

$$V_u = \int_{F \leq \Gamma} \left\langle \frac{\Phi}{T} \right\rangle \, dV, V_u = \int_{F \geq \Gamma} \left\langle \frac{\Phi}{T} \right\rangle \, dV \tag{10}$$

where I_u is the flux integral over the unstable region ($F \leq \Gamma$), I_p is the flux integral over the stable region ($F \geq \Gamma$), V_u the viscous dissipation over the unstable region and V_p that from the stable region. The variations of I_p, I_u, V_p, V_u are shown in Figure 5; as can be seen from this diagram the contribution to the integral constraint from viscous dissipation decreases with decreasing Prandtl number. In Figure 6 we show the variation of the flux integrand $I_0(z)$ and viscous dissipation $V(z)$ for a particular set of parameters, and in Figure 7 the various contributions to the energy flux. Figure 8 compares the variation of the (time and horizontal average) of the superadiabatic gradient for the numerical solution with that from the traditional astrophysical solution with no overshooting. As expected the superadiabatic gradient is small and negative in the overshoot region.

4. Stellar Oscillations: Signature of the boundary of convective cores

Finally it is worth noting that in principle information on the location of the boundary of convective cores can be obtained from stellar seismology. The discontinuity in the derivatives of the sound speed at the boundary of a convective

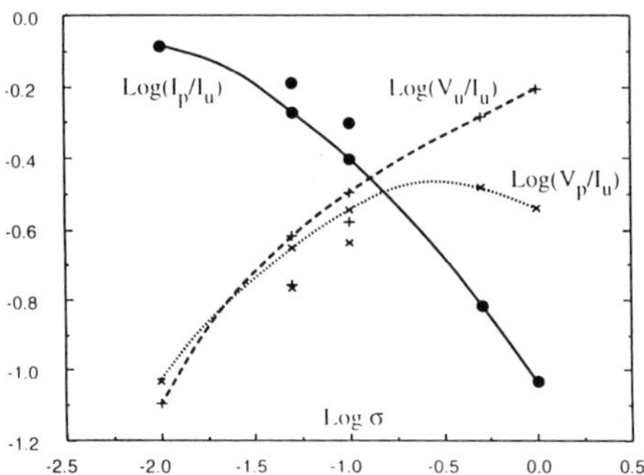

Fig. 5.   Variation with Prandtl number σ of the contributions to the Integral Constraint from the flux integral in the penetrative region I_p/I_u, and from viscous dissipation V_u/I_u and V_p/I_u. The lines connect points with the same Rayleigh number $R_a = R_o = 4.78\ 10^6$.

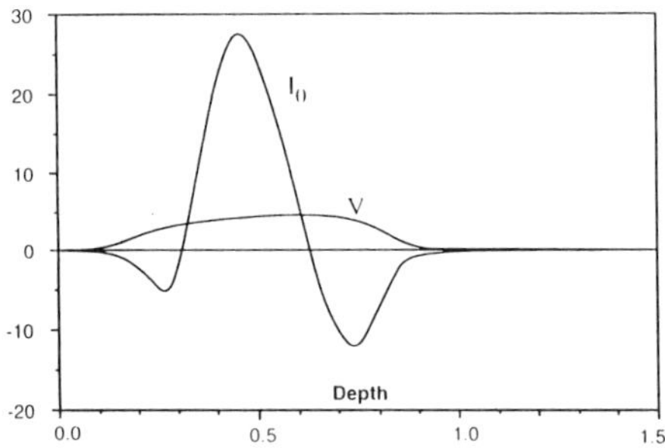

Fig. 6.   Variation with depth of the Integrands I_0 and V for a model with Prandtl number $\sigma = 0.05$, Rayleigh number at the minimum of the conductivity $R_a = 4.78\ 10^6$. The temperature at z = 1 is twice the value at the top of the layer.

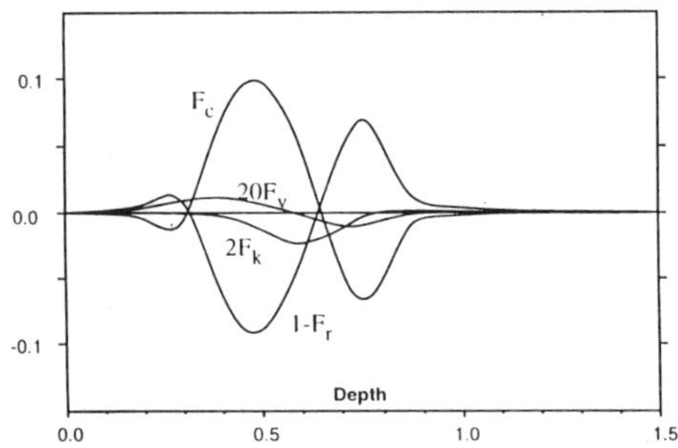

Fig. 7.   Variation with depth of the convective flux F_c, the radiative flux F_r, the kinetic energy flux F_k, and the viscous flux F_v, for a model with Prandtl number $\sigma = 0.05$, Rayleigh number at the minimum of the conductivity $R_a = 4.78\ 10^6$.

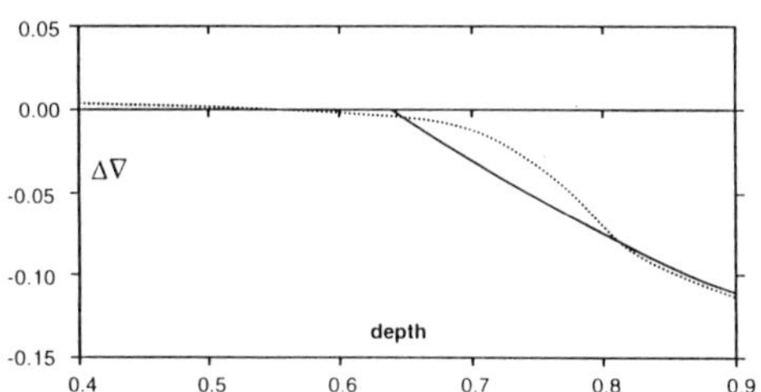

Fig. 8.   Variation with depth of the superadiabatic gradient $\Delta\nabla$ below the base of the unstable region for the traditional astrophysical solution (solid line), and for a numerical simulation with $R_a = 4.78\ 10^6$, $\sigma = 0.1$ (dotted line).

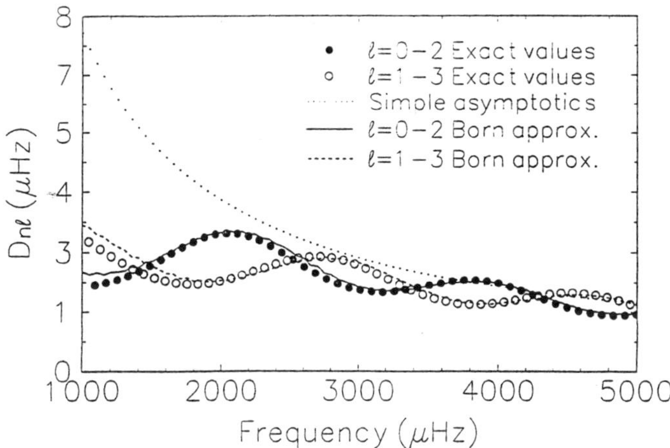

Fig. 9.   Small frequency separations for an evolved main sequence star with a convective core. The central hydrogen abundance is 0.5 whereas the initial hydrogen abundance was 0.7. The solid and open points are the exact values calculated from the stellar model whereas the continuous curves are the theoretical predictions obtained using the Born approximation.

core gives rise to a variation with frequency $\nu_{n,\ell}$ of the small separation in frequencies $D\nu_{n,\ell} = [\nu_{n,\ell} - \nu_{n-1,\ell+2}]/(2\ell+3)$. Figure 9 shows the variation of $D\nu_{n,\ell}$ for $\ell = 0$ and 1 for a simple evolved Cowling model where the hydrogen abundance in the convective core is 0.5 whereas the initial hydrogen abundance was 0.7. The "Born approximation" of Roxburgh and Vorontsov (1993) reproduces the exact values quite accurately.

This signature of the boundary between the convective core and radiative envelope offers the possibility of testing stellar models by observing their oscillation frequencies using a satellite mission such as PRISMA (Appourchaux et al 1993) or STARS (Jones et al 1993).

5. Acknowledgements

The author gratefully acknowledges helpful discussions with S.M. Chitre, D.Narisimha, J. Simmons and S.V. Vorontsov. This work was supported in part by the UK SERC under grant GR/F 52033.

References

Appourchaux,T. et al:1993, PRISMA: Probing Rotation and Interior Structure of Stars. Report of the Phase A study. European Space Agency, Paris, SCI (93) 3.

Jones,A. et al:1993, STARS: Seismic Telescope for Astrophysical Research from Space. European Space Agency, Paris.

Roxburgh,I.W.: 1965, *Mon. Not. Roy. Ast.Soc.* **130**, 223.

Roxburgh,I.W.:1976, The Internal Structure of the Sun and Solar Type Stars;in Basic Mechanisms of Solar Activity; ed. Bumba,V. and Kleczeck,J.; Reidel,p.451.

Roxburgh,I.W.: 1978, *A&A* **65**, 281.

Roxburgh,I.W.: 1989, *A&A* **211**, 361.

Roxburgh,I.W.: 1992, *A&A* **266**, 291.

Roxburgh,I.W. and Simmons,J.: 1992, *A&A* **277**, 102.

Roxburgh,I.W. and Vorontsov, S. V.: 1993, *Mon. Not. Roy. Ast.Soc.* **in press.**,

Saslaw,W.C. and Schwarzschild,M.: 1965, *Ap.J.* **142.**, 1468.

Shaviv,G. and Salpeter,E.: 1973, *Ap.J.* **184**, 91.

MASSIVE CLOSE BINARIES: OBSERVATIONAL CHARACTERISTICS

E.P.J. VAN DEN HEUVEL

*Astronomical Institute 'Anton Pannekoek' and Center for High Energy Astrophysics University of
Amsterdam, The Netherlands*

Abstract. Theoretically predicted evolutionary phases of massive close binaries are compared with
the observations. For the evolution up to the High-Mass X-ray Binary (HMXB) phase there is fair
agreement between theory and observation. Beyond the HMXB phase there is much uncertainty.
Notably it is puzzling why we observe so few systems consisting of a helium star and a neutron
star (Cygnus X-3 is the only one found so far), and why the incidence of double neutron stars
is so low. A better understanding of Common Envelope evolution is required in order to answer
these questions. The role of velocity kicks imparted to neutron stars during supernova collapse is
discussed. Such kicks might cause many runaway OB stars to be single.

1. Introduction

Figure 1 depicts the subsequent evolutionary stages through which a massive
close binary system with a not too small initial mass ratio ($q = M_2/M_1 \geq 0.4$)
is, theoretically, expected to pass (see for example Van den Heuvel 1974, 1978,
1983 and 1993 and references therein). For such systems the first phase of
mass transfer proceeds more or less "conservatively", i.e.: total mass and orbital
angular momentum of the system are expected to be conserved.

The different evolutionary phases and their expected durations are described
in the figure and its caption, to which we refer the reader for details.

In this paper we discuss in section 2 the possible identifications with observed
types of binary systems of the various evolutionary phases up till the X-ray binary
phase. We also discuss in this section the reasons for the differences between the
two observed classes of High Mass X-Ray Binaries (HMXB): the 'Standard'
HMXBs and the Be/X-Ray Binaries.

In section 3 as an intermezzo we discuss the evidence of kicks imparted to
neutron stars at birth, and the possible implications of these kicks for the origins
of HMXBs.

In section 4 we discuss the formation of the low-mass X-ray Binaries, since
these also are the products of the evolution of high-mass binary systems. This is
because, in order to produce a neutron star or a black hole, a star should have an
initial mass of $\geq 8 M_\odot$ and $\geq 30 - 40 M_\odot$ respectively (cf. Van den Heuvel and
Habets 1984; De Kool et al. 1987). Hence, the low-mass X-ray binaries, which
consist of a neutron star or a black hole together with a low-mass star ($\leq M_\odot$),
must also have started out as massive binary systems - in this case with a very
small mass ratio (≤ 0.15).

In section 5 we discuss the final evolution of high-mass X-ray binaries and the
identification of the expected final evolutionary stages with observed binary systems. Finally, we discuss the loose ends and gaps that exist in our understanding

Space Science Reviews **66**: 309–322, 1994.

© 1994 *Kluwer Academic Publishers. Printed in Belgium.*

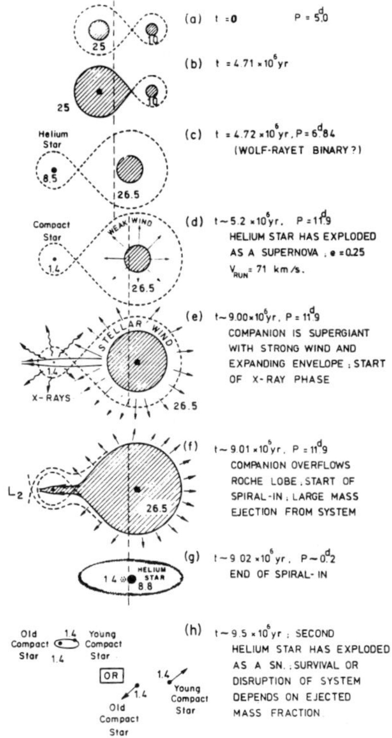

Fig. 1.   Conservative evolution of a close binary system with initial masses of 25 and 10 $M_\odot$. Each stage is labelled with the approximate age of the system and the orbital period in days. The numbers inside the representation of the stars indicate mass ($M_\odot$). The system becomes a wind-powered 'standard' massive X-ray binary for some 10^4 yr in stage e (see text). For the subsequent evolutionary stages f, g and h: see text.

of the evolution of massive binary systems.

2. Evolutionary phases up till the HMXB phase

2.1. A difference in evolution between the intermediate-mass and high-mass systems, and the production of Be-X-ray binaries

The Be/X-ray binaries have systematically lower masses (companion masses $\simeq 8 - 15 M_\odot$) and wider orbits ($P \sim 15^d$ to several years) than the "standard" HMXBs. This difference is simply explained by the lower relative mass fraction contained in the helium core produced in lower mass stars as compared to massive stars ($M_{He} \sim 0.1 M^{1.4}$, where M is the initial main-sequence mass of the star), in combination with the conservation of mass and orbital angular momentum during the first phase of mass transfer, as can be easily verified (see van den Heuvel 1983). Figure 2 depicts as an example the evolution of an intermediate-

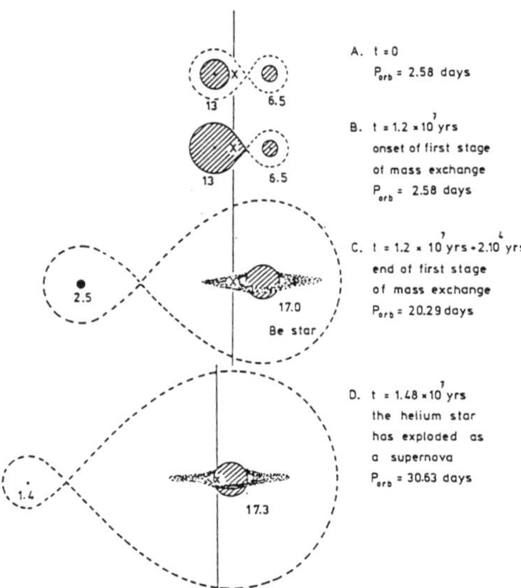

Fig. 2. Conservative scenario for the formation of a Be/X-ray binary out of a close pair of early B stars with masses of 13.0 $M_\odot$ and 6.5 $M_\odot$. The numbers indicate mass (in units of $M_\odot$). After the end of the mass transfer the Be star presumably has a circular disk or shell associated with its rapid rotation (induced by the previous accretion of matter with high angular momentum; from Habets (1985)).

mass close binary (component masses 13+6.5 $M_\odot$) into a Be/X-ray binary. One notices that during the first phase of mass transfer, its orbital period increases by a much larger factor (i.e.: ~ 8) than for the system of figure 1, also if that system would have had the same initial mass ratio 0.5 (in that case its period would have increased by only a factor 2.0, as one can easily verify). The rapid rotation of the mass-receiving original secondary in the Be system is due to the large amount of mass with angular momentum which this star recieved (by accretion through a disk). Due to the long orbital periods of the Be systems, tidal forces are insufficient to synchronize the rotation of the stars with their orbital periods. (Binary stars with early type main-sequence components and $P \geq 10^d$ are never synchronized (cf. van den Heuvel 1970)).

This explains why the Be stars in Be/X-ray binaries remained in rapid rotation whereas most of the standard HMXBs (which, with one exception have $P \leq 10^d$), have been tidally synchronized.

2.2. STATISTICAL PROPERTIES OF UNEVOLVED OB-BINARIES

2.2.1. distribution of mass ratios $\phi(q)$

These are subject to large selection effects, favouring the detection of systems with high mass ratios. For a recent discussion we refer to Hogeveen (1991; see also van den Heuvel 1993). It appears likely, from the work of Garmany et al. (1980) that the mass-ratio distribution of O-type binaries can be reasonably well approximated by a flat distribution, whereas the works of Abt & Levy (1978) and Trimble (1990) yield distributions $\phi(q)$ ranging from approximately flat to $\phi(q) \sim q^{-1}$ (Trimble). Hogeveen (1991) even obtained a distribution as steep as $\phi(q) \sim q^{-2}$ down to $q = 0.3$ to 0.4 and flat below $q = 0.3$.

Within the observational uncertainties the distributions of Trimble and Hogeveen cannot be distinguished form the distribution

$$\phi(q) = 2/(1 + q)^2 \tag{1}$$

suggested by Kuiper (1935). This distribution has the advantage that it does not become singular at $q = 0$. Studies of the evolution of synthetic stellar populations

TABLE I

Orbital periods (P^f), eccentricities (e), and system runaway velocities (v_g) that known Wolf-Rayet Binaries will obtain after the supernova explosion of their WR star, assuming a 1.4 $M_\odot$ neutron star to be left. It is assumed that the mass of the WR star at the time of the explosion is 10 $M_\odot$ (or present WR mass, if this is less than 10 $M_\odot$), and that the mass ejection is intantaneous an symmetric. For short-period systems also the period after the tidal circularization of the orbit is given, in parentheses. The last column lists the radial velocity amplitude of the O-star induced by the presence of the neutron star remnant of the WR star.

Name	$P(d)$	$M_{wr} + M_{OB}$ $M_\odot$	$P_f(d)$	e	$V_g(km/s)$	$K(km/s)$
AB 6	6.861	6.4+37	8.99	0.13	43.4	12
AB 8	16.664	13.5 +50	23.66	0.17	45.2	8
γ^2 Vel	78.5	25+35	130.8	0.24	33	5.4
HD94305	18.82	16+35	31.33	0.24	52	9
HD97152	7.886	20+35	13.1	0.24	70	11.6
HDE311884	6.34	50+60	8.5 (8.2)	0.14	57	9.6
HD152270	8.893	20+60	11.9	0.14	51	8.4
HDE320102	8.83	11+35	14.7	0.24	67	11.2
HD168206	29.707	13+25	61.8	0.33	53	8.6
HD186943	9.555	13+25	19.9	0.33	77	12.4
HD190918	112.8	9 +35	176.2	0.21	26	4.9
V444Cyg	4.212	11+30	6.2 (5.9)	0.19	77	13.1
GPCep	6.69	10+45.5	9.9 (9.6)	0.18	65	10.9
CQCep	1.64	23+19.3	4.3 (3.2)	0.41	155	27
CXCep	2.217	11+25.6	4.3 (3.7)	0.32	126	21.6

that contain a certain fraction of close binaries show that the results do not differ dramatically between a flat input distribution for q and that given by eq. (1)

(e.g. see Meurs and van den Heuvel 1989; Tutukov and Yungelson 1993, and references therein).

2.2.2. *the initial distribution of orbital periods*
For O-stars and early B-stars the distributions peak approximately around $P \sim 3^d$ (see the review by Vanbeveren 1991, and references therein). For wider unevolved systems, from $P \sim 10^d$ on, the distribution of log P (and $log\ a$) is approximately flat (cf. Abt 1983) implying $f(P) \propto P^{-1}$ and $g(a) \propto a^{-1}$, where f and g denote the distributions of P and a, respectively.

2.3. SYSTEMS AFTER THE FIRST PHASE OF MASS TRANSFER

The computations predict that these systems consist of a helium star and a massive OB main-sequence star. Systems of this type are well-known only among the highest mass binaries ($M_1 \geq 20 M_\odot$): the Wolf-Rayet binaries, as first suggested by Paczynski (1967), see van der Hucht (this volume) and Vanbeveren (this volume). Since no WR-stars with masses smaller than about 5 $M_\odot$ are known, apparently helium stars with masses $\leq 5 M_\odot$ do not produce the very strong winds and large mass loss which are so characteristic for the WR-stars, and make them observable at optical wavelengths. Due to the absence of such winds, helium stars less massive than about $5 M_\odot$ go unnoticed (their energy distribution is expected to peak in the extreme UV). A few Be-binaries such as ϕ Per have been suggested to harbour such lower mass helium stars (cf. Habets 1985).

Table 1 lists a number of representative WR binaries (after Abbott and Conti 1987). The table also lists the orbital periods P^f, eccentricities e and runaway velociteis V_g (of the centre of the gravity of the system) which would result if their WR-stars would explode in a spherically symmetric supernova, and leave an 1.4 $M_\odot$ star remnant (assuming that the mass of the WR star at the time of the explosion was either $10 M_\odot$, or its present mass, if that is smaller). The systems do not become unbound by the explosions, as less than half the system mass is ejected, cf. Blaauw (1961).

2.4. SYSTEMS WITH A YOUNG RAPIDLY SPINNING PULSAR: PSR 1259-63 AND LSI 61° 303

Already from the time of the discovery of the first accreting binary X-ray source Cen X-3, the question came up: why do we not observe young radio pulsars orbiting massive stars (cf. Graham-Smith, 1974)? The answers mostly given were: the wind of the massive companion disperses the pulsar signal beyond recognition, or: the neutron star's spin is braked so rapidly by the accretion of wind matter, that there is very little chance to observe the young pulsar phase. Neither may, however, be true, since it now seems clear that massive binary systems containing young neutron stars have been found, namely PSR 1259-63 and LSI 61^0 303 (a third one may be the Small Magellanic Cloud pulsar which describes a very elliptic orbit with $P \sim 50^d$ around a B-type star, Manchester, 1993).

PSR 1259-63 is a 47 ms radio pulsar in a 3.5 year very eccentric ($e = 0.87$) orbit around an B-type star. It has a magnetic field strength in the normal range for young pulsars (3.10^{11} G), and is eclipsed by the wind of the Be star for several months near periastron. Its spindown age is $\sim 3.10^5$ yr, suggesting that it is a normal young radio pulsar (Johnston et al. 1993).

LSI 61° 303 is one of the most peculiar Be/X-ray binaries known: it exhibits strong synchrotron radio outbursts with a period of 26.5^d, which is also the radial velocity and photometric period of its peculiar Be-type companion. It was noticed long ago by Maraschi and Treves (1981) that it is in the error box of a COS-B gamma-ray source, and these authors therefore suggested that we are dealing here with a young Crab-like pulsar moving in an eccentric orbit around the Be star. The radio outbursts in this case would occur near apastron, when the relativistic pulsar wind is able to push aside the wind matter of the Be-star.

Recently, GRO-Comptel observations have provided strong confirmation for this model: it was discovered that the gamma-ray flux from this system varies with a 26.5 day period, in phase with the radio outbursts (Hermsen, 1993). Thus here we have probably a very young pulsar orbiting a Be star.

2.5. SYSTEMS AFTER THE FIRST SN-EXPLOSION: RELATION TO RUNAWAY OB STARS

One notices from table 1 that for a considerable number of systems the runaway velocities expected after the SN-explosion of the WR-stars, are in the 50-170 km/s range i.e. of the order of the highest runaway velocities observed for OB-runaway stars.

Blaauw (1993) in an important recent paper showed that 6 out of 7 well-studied runaway OB stars are rapid rotators and have a higher than normal He-abundance, characteristics just as expected for the post-mass transfer mass-gaining components of massive close binaries (the cause for the rapid rotation of the mass-gainer is that the mass-transfer proceeds through a disk, feeding high angular momentum to the star- just like for Be stars in Be/X-ray systems depicted in figure 2). For the runaway stars ζ Oph, AE Aur, μ Col and 53 Ari the parent association is known (the Sco-Cen association for the case of ζ Oph, and the I Ori association for the other three). Blaauw showed that all these four runaway stars are "blue stragglers" in the HR-diagrams of these associations (they are located to the left of the -already evolved- upper main sequence of these associations). This is exactly what one would expect for the post mass-transfer components of the close binaries, as these stars have been rejuvenated by the large amount of fresh hydrogen which they received during the mass transfer.

All this evidence taken together strongly suggests that 6 out of these 7 "classical" runaway OB stars have been produced by mass-transfer in massive close binaries, in which subsequently the original primary star exploded as a SN.

If the SN-mass-ejection was spherically symmetric, one would expect them to still be accompanied by a neutron star or black hole in a bound orbit.

Searches for X-ray emission (other than coronal emission) from runaway OB stars have always been unsuccessful. However, it should be kept in mind that most of the Be/X-ray binaries are transient X-ray sources which often exhibit no X-ray emission at all between their transient outbursts. These X-ray quiet periods can last up to decades.

If the accretion rate is low, the Alfven radius R_A is located far from the neutron star, such that the centrifugal barrier produced by the neutron star's rotation is too great to allow matter at the Alfven surface to enter the magnetosphere and to accrete (cf. Stella, White and Rosner 1986). Therefore no accretion will take place during such phases, and one does not necessarily expect X-ray emission from neutron star companions of runaway OB stars (since, on the main-sequence, the winds of these stars are not very strong, and the accretion rate is expected to be low). Only during the brief period prior to overflowing its Roche lobe such a star may produce an X-ray binary. Thus here, like in so many cases, the absence of evidence is *not* the evidence of absence!

It should furthermore be noticed that the HMXBs are (in most cases: mild) runaway OB stars: as shown by van Oyen (1989) their galactic z-distribution is about twice as wide as that of other unevolved OB stars, and they do not occur in ("avoid") OB-associations, and a few of them (e.g. QV Nor= 1538-52; X Persei; SS433) are genuine extreme runaway OB stars with velocities of order $10^2 \, km/s$.

Thus the runaway OB stars have the charateristics of the mass-gaining components of massive close binaries- and the HMXBs are mostly (mild) runaway OBs. This all suggests that it is likely that many of the runaway OBs have resulted from mass-exchange binaries in which the original primary star has exploded as a supernova. Many of them might well have an unseen compact companion star (see especially also Stone, 1991). Alternatively if large asymmetries in the SN mass ejection occur (see next section), it is also possible that a sizeable fraction of the systems was disrupted in the first SN explosion.

3. The Evolutionary History of Low-Mass X-ray Binaries

Contrary to the formation of High-Mass X-ray Binaries, which represent a normal phase in the evolution of massive close binaries, the formation of LMXBs is an extremely rare event that requires very special initial conditions. This follows immediately from the fact that these systems are very long-lived: their mass transfer phase lasts $\gtrsim 10^8 \, yrs$, while there are only some 10^2 such systems in the Galaxy. Thus their formation rate in the Galaxy is only of order $\sim 10^{-6} \, yr^{-1}$, some 10^4 times lower than the supernova rate. Thus, not more than one out of every 10^4 supernovae produces a progenitor system of an LMXB (cf van den Heuvel 1983).

Since thus the probability for a neutron star (or a black hole) to end up with a low-mass companion star is extremely low, the formation of an LMXB must require a very extraordinary combination of initial system parameters. Space does not allow us to deal with the various possible models for forming LMXBs (with a neutron star as well as a black hole component). We refer to the excellent review by Webbink (1992); for models for the formation of the LMXBs that contain black holes, of which now definitely four have been identified, we refer to de Kool et al. (1987) and Romani (1992).

The importance of the discovery of these systems is that, despite the extremely low probability of forming them, already four of them have been found in our neighbourhood in the Galaxy (within a few kpc). This implies that there must be enormous numbers of stellar black holes in the Galaxy of order 10^8 or more (van den Heuvel 1992b). These must all be remnants of massive stars. We thus are only just beginning to observe the tip of this iceberg of massive-star remnants in the Galaxy.

4. Final Evolution of HMXBs, Cygnus X-3 and the Formation of Double Neutron Stars and Neutron-Star Plus Black Hole Binaries

When we first calculated the final evolution of a HMXB (van den Heuvel and De Loore 1973) we found that due to the great difference in mass between the neutron star and its massive companion the mass transfer by Roche-lobe overflow would make the orbital separation of the system shrink dramatically. This orbital shrinking occurs independently of whether the neutron star can accept all the matter that is transferred to it, or expells it from the system due to radiation pressure. We therefore concluded that the final result will be a helium star with a compact companion in a very narrow orbit, with a period of only a few hours, as depicted in figure 1.

We suggested in 1973 that the very peculiar 4.8 hours orbital period X-ray binary Cygnus X-3 (which is one of the intrinsically brightest X-ray sources in the Galaxy, at a distance of about 10 kpc, in the galactic plane; it shows gigantic radio outbursts and has a 4.8 hour infrared period and an X-ray lightcurve suggesting the presence of a strong stellar wind from its companion) is to be identified with such a close helium-star binary.

This suggestion was confirmed in 1991 by the discovery by van Kerkwijk et al (1991) that the infrared spectrum of Cyg X-3 shows very strong emission lines of helium and is identical to that of a nitrogen-type Wolf-Rayet star, i.e.: a massive helium star ($M \geq 5 M_\odot$). [Optical observations of Cyg X-3 are impossible as the visual extinction in its direction is $> 20^m$].

Refinements of the calculations of the final evolution of HMXBs were introduced by Taam et al. (1978) using the concept of Common-Envelope (CE) evolution (see also Taam and Bodenheimer 1992). These authors argued that a HMXB system will only survive as a binary if the initial orbit is sufficiently wide. In close systems, (i.e. for $P < 100^d$ if $M_2 = 20 M_\odot$) the neutron star will, according to this model, spiral completely into the core of its massive companion star and form a so-called Thorne-Zytkov star: a red super-giant with a neutron star in its center (Thorne and Zytkov 1977; Biehle 1991, Cannon et al. 1992).

If the concept of CE-evolution is correct, this will happen to all "standard" HMXBs and to more than half (and possibly as much as over $\frac{2}{3}$) of all Be/X-ray binaries. Thus, Cyg X-3 should then have resulted from a wide system like one of the Be/X-ray binaries. We have some doubts about this, however, since, judging from its WN6-7 spectrum the helium star in Cyg X-3 appears to be quite massive, more as what one would expect from a standard HMXB, thus: from a system with a short initial orbital period!

We wish to point out here that if the companion in a HMXB has an initial mass $\geq 30 M_\odot$ (as is probably the case in 4U 1700-37, Cygnus X-1 and 4U 1223-62) this massive star may never evolve into a red super-giant. Instead, because of the existence of the Humphreys-Davidson limit, it will lose most of its envelope mass as a blue super-giant and then become a WR-star (the LBV scenario for the most massive close binaries, Vanbeveren, 1991). Hence, in this case spiral-in is not required to expel the hydrogen-rich envelope, as the star will do this anyhow on its own accord. In such a situation no complete spiral-in needs to occur and even a close HMXB may still survive as a (very) close binary, as suggested in our 1973 model.

It thus seems quite likely to us that Cyg X-3 has resulted from the evolution of a standard HMXB like 4U 1700-37 (O6f-companion, $P = 3.4\ days$). It will be clear that a system like Cyg X-3 is an almost ideal progenitor for close double neutron star systems like the Hulse-Taylor pulsar PSR 1913+16 ($P = 7^h45^m$, $e = 0.615$) and PSR 1534+12, as was pointed out long ago (Flannery and van den Heuvel 1975; De Loore et al. 1975). If the initial system resembles Cygnus X-1, the binary after spiral-in will consist of a helium star and a black hole. Upon exploding, the helium star may leave a neutron star or a black hole. In the first case, an eccentric close binary will result, consisting of a (young) pulsar and a black hole.

Such systems must certainly exist in the Galaxy. We expect them, however, to be difficult to observe as the neutron star is a newborn one, most probably with a strong magnetic field and, therefore, with a short spin-down lifetime (at most a few million years). This is contrary to the PSR 1913+16 and 1534+12 systems where we observe the old recycled neutron star of the two, with a weak magnetic field ($\sim 10^{10}\ G$) and, therefore, a long spin-down lifetime of several times $10^8\ years$. This long lifetime is in fact the reason why we are able to observe such systems, despite their rather low birthrate in the Galaxy which is $< 10^{-5}\ yr^{-1}$ (Phinney 1991; Narayan et al. 1991; van den Heuvel 1992a). The neutron star plus black hole binaries will have an even lower birthrate but are not in this favourable situation of a long lifetime of their pulsars. (The birthrate of systems consisting of a black hole and a recycled neutron star -i.e.: in which the black hole is the second-born compact object in the system- is expected to be extremely low, according to our calculations).

5. Pulsar velocities and their possible origins

5.1. OBSERVED PULSAR VELOCITIES

It is long known that pulsars tend to be runaway objects: this is evidenced by their broad galactic z-distribution (Gunn and Ostriker 1970) and by the measurements of individual proper motions (Lyne et al. 1982). Recently, many new pulsar proper motions were published (Harrison et al. 1993) which brings their number of interferometrically measured proper motions to over 55. Furthermore, several individual young pulsars in SN-remnants were discovered that have excessively large space velocities. The most extreme case is that of PSR 1757-24 in the "Bird-like" SNR G5.4-1.2, which has most likely a transverse velocity of 2300 km/s, and certainly $> 1000\,km/s$ (Frail and Kulkarni 1991). Another similar case is that of PSR 2224+65 which has a space velocity $> 800\,km/s$, directed along the galactic plane (Cordes et al. 1993).

Correcting for selection effects, the root-mean-square tangential velocity in the sample of Harrison et al. (1993) is 248 km/s, implying a mean excess space velocity (with respect to the local rest frame) of $\sim 300\,km/s$, which is of the order of the local escape velocity from the Galaxy. In Harrison et al.'s sample of pulsars, 5 (or: $\sim 10\%$) have tangential velocities ranging from 550 km/s to 2300 km/s.

5.2. POSSIBLE ORIGINS OF PULSAR VELOCITIES

5.2.1. a binary disruption

Can these very high velocities have been made by binary disruption (i.e.: without asymmetry-kicks)? The answer is: in theory velocities up to $\sim 1400\,km/s$ can be produced by the explosion of massive helium stars ($\sim 10 M_\odot$) in very close binaries with a neutron star companion (the shortest orbital period around a $10 M_\odot$ helium star without making this star overflow its Roche lobe is $1 - 2^h$). Although this is a (extreme) possibility for obtaining pulsars with high space velocities, it seems unlikely that as much as 10 percent of all pulsars would result from such systems. This is the more so, since the bulk of the helium star plus neutron star systems will contain helium stars with masses between $\sim 2.5 M_\odot$ and 4 $M_\odot$. Such systems will not be disrupted in a symmetric SN-explosion of the helium star. Also, such systems will undergo a second spiral-in, in which their mass is reduced further, before exploding (Phinney 1991; Van den Heuvel 1992a) such that they will finish in very narrow orbits which will never disrupt in a symmetric second supernova explosion either (cf. Van den Heuvel 1992a).

So, these systems might produce binary radio pulsars with extremely short orbital periods, which coalesce by gravitational radiation losses very soon after their formation (van den Heuvel 1992a; Tutukov and Yungelson 1993), which might explain why we have not observed such systems. Or, alternatively, these systems are disrupted in the second supernova explosion, due to an asymmetry-kick that is imparted to the newborn neutron star in the explosion. We will argue, however, that this is quite unlikely. Dewey and Cordes (1987) have argued that asymmetry-kicks are absolutely required in order to avoid producing too many double neutron star systems like PSR1913+16. However, they did not include the occurrence of the second spiral-in of systems with $M_{He} < 4 M_\odot$ in their calculation, which will drastically alter their conclusions for the following reasons.

Due to the second spiral-in the orbital velocities of these systems will be $> 500\,km/s$. So, a randomly directed kick of $\sim 300\,km/s$ (the average observed in Harrison at al.'s sample) will disrupt less than half these systems, since statistically half of the kicks will be in retrograde directions with respect to the orbital motion. Thus, one still would expect more than half of the systems to remain bound after the second supernova explosion: we thus keep producing large numbers of binary pulsars.

5.2.2. arguments for kicks intrinsic to the SN-collapse

These come from three independent sources:

1. Independently of whether or not kicks are required to avoid overproducing double neutron stars, the observed orbital eccentricities of the Be/X-ray binaries (on average: $e \sim 0.5$) are far larger than expected in the case that the helium-star progenitors of their neutron stars had undergone a symmetric supernova explosion (see also figure 2). Average kick velocities of 50-100 km/s (and in some cases $> 250\,km/s$) are required to explain these orbital eccentricities (Verbunt and van den Heuvel 1994; van den Heuvel 1993).

2. As outlined above, it is highly unlikely that of order ten percent of all pulsars obtained their space velocities $> 550\sqrt{2}\,km/s$ through binary disruption in symmetric SN explosions. The most likely alternative seems therefore to us to be that indeed the large observed space velocities of pulsars are due to "kicks", i.e. due to asymmetries in the mass ejection during the supernova explosion itself.

3. The misalignment of the spinaxis of the binary pulsar PSR1534+12 with the normal to the orbital plane (Wolszczan 1992) is clear evidence for a kick imparted to the last-born neutron star in the system; this is because the observed pulsar is the one which was spun-up by accretion during an X-ray binary phase (cf. Wolszczan 1992), so its spinaxis is expected to be normal to the orbital plane. The misalignment with this normal can only be explained by a kick imparted to this plane in the second supernova.

We thus conclude from the work of Harrison et al. and from the other presented evidence (see in particular also Bailes 1989) that it is most likely that pulsars receive in their supernova explosions asymmetry-kicks that on average are of order 300 km/s, and in 10 percent of the cases are between 550 and 1000 km/s.

5.3. IMPLICATIONS OF KICKS FOR THE FORMATION OF HMXBs, AND THE BINARY ORIGIN OF RUN AWAY OB STARS

An average kick of 300 km/s will, independently of its direction, unbind most high-mass binaries with orbital periods $> 30^d$ during the first supernova explosion; the relative orbital velocity in a system with a total mass of 20 $M_\odot$ and $P = 30^d$ is about 180 km/s. So, a random kick of 300 km/s will unbind over 2/3 of these systems (only kicks with velocity vectors in a rather narrow cone around the direction opposite to the orbital velocity will leave bound systems). Similarly, random kicks of 200 km/s will unbind most systems with orbital periods $> 150\, days$.

If kicks of this magnitude are included it may well be that more than half of all runaway OB stars are single despite them having originated from supernova explosions in binaries. It will, however, be clear that the highest runaway velocities will come from the closest systems, which are unlikely to be disrupted even by kicks of 300 km/s.

Attempts to calculate the evolution of stellar populations that contain a realistic fraction of binaries produce many helium star plus neutron star systems (of order 10^4 in the Galaxy at any one time, cf. Lipunov 1992, 1993; Tutukov and Yungelson 1993). We see, however, only one Cygnus X-3 system in the Galaxy. Clearly, either the results of the population synthesis models cannot be trusted, or most of the helium star plus neutron star systems do not resemble Cygnus X-3. Indeed, helium stars with $M \lesssim 5 M_\odot$ probably don't have the strong WR-like wind mass loss that the helium stars $> 5 M_\odot$ exhibit. As the bulk of the predicted helium star binaries will have $M_{He} < 5 M_\odot$, this might be a reason why we don't observe them. An alternative explanation might be that most of the progenitor systems spiral-in completely and produce Thorne-Zytkov objects.

Also, the observed birthrate of double neutron star systems like PSR 1913+16 is much smaller than predicted by the above mentioned population synthesis models. These predict a galactic formation rate of double neutron stars of $\gtrsim 10^{-4} yr^{-1}$ (Tutukov and Yungelson 1993). Here the same explanation may apply as for the absence of a large number of helium star plus neutron star systems. Alternatively, if indeed $\sim 10^4$ non-X-ray emitting helium star plus neutron star systems are present in the Galaxy, it may be that most double neutron stars are formed at such short orbital periods (due to the occurrence of the second spiral in phase) that they coalesce quickly (i.e.: within a few million years after they were born), such that they have a very low probablity for being observed (Van den Heuvel 1992a; Tutukov and Yungelson 1993).

It will be clear that we will only be able to answer these questions if we have a better understanding of the physics of Common Envelope evolution, a still rather poorly understood process.

References

Abbott, D.C. and Conti, P.S. 1987, Ann. Rev. Astron. Astrophys. *25*, 113-150

Abt, H.A. 1983, Annual Rev. Astron. Astrophys. *21*, 343-372

Abt, H.A. and Levy, S.G. 1978, Astrophys. J. Suppl. *36*, 241

Bailes, M. 1989, Astrophys.J. *342*, 917

Biehle, G.T. 1991, Astrophys. J. *380*, 167

Blaauw, A. 1961, Bull. Astron. Inst. Netherlands, *15*, 265-291

Blaauw, A. 1992, in: "Massive Stars: Their Lives in The Interstellar Medium", ASP Conf. Proceedings # 35 (eds. J. Cassinelli and E. Churchwell), Astron. Soc. of the Pacific, San Francisco, p. 207.

Cannon, R., Eggleton, P.P. Zytkow, A.N. and Podsiadlowski, P. 1992, Astrophys. J. *386*, 206

Cordes, J.M., Romani, R.W. and Lundgren, S.C. 1993, Nature *362*, 133

De Kool, M., van den Heuvel, E.P.J. and Pylyser, E. 1987, Astron. Astrophys. *183*, 47

De Loore, C., De Greve, J.P. and De Cuyper, J.P. 1975, Astrophys. Sp.Sci. *36*, 219

Dewey, R.J. and Cordes, J.M. 1987, Astrophys.J. *321*, 780

Flannery, B.P. and van den Heuvel, E.P.J. 1975, Astron. Astrophys. *39*, 61

Frail, D. and Kulkarni, S.R. 1991, Nature, *352*, 785

Garmany, C.D., Conti, P.S. and Massey, P. 1980, Astrophys.J. *242*, 1063

Graham-Smith, F. 1974, Proc. 16th Solvay Conf. on Physics, Univ. of Brussels Press, 133

Gunn, J.E. and Ostriker, J.P. 1970, Astrophys. J. *160*, 979

Habets, G.M.H.J. 1985, "Advanced Evolution of Helium Stars and Massive Close Binaries", Ph.D. Thesis, University of Amsterdam

Harrison, P.A., Lyne, A.G. and Anderson, B. 1992, in: "X-ray Binaries and Recycled Pulsars" (eds. E.P.J. van der Heuvel and S.A. Rappaport), Kluwer Acad. Publ., Dordrecht, pp. 155-160

Hermsen, W. 1993, (priv. com.)

Hogeveen, S. 1991, "The Mass Ratio Distribution of Binary Stars", Ph.D. Thesis, University of Amsterdam (132 pp)

Hogeveen, S.J. 1992, Astrophys. Sp. Sci., *196*, 299

Johnston, J. Manchester, R.N. Lyne, and D'Amico, N. 1993, IAU Circ. # 5794 (18th May)

Kuiper, G.P. 1935, Publ. Astron. Soc. Pacific. *47*, 15

Lipunov, V.M. 1992, "Astrophysics of Neutron Stars", Springer Verlag, Heidelberg, 322 pp.

Lipunov, V.M. 1993, paper presented at the conference "Evolutionary Relations in the Zoo of Interacting Binaries", Monte Porzio (Rome), June 1993

Lyne, A.G., Anderson, B. and Salter, M.J. 1982, Mon. Not. R.A.S. *201*, 503

Manchester, R.N. 1993 (priv. com.)

Maraschi, L. and Treves, A. 1981, Mon.Not.R.A.S. *201*, 503

Meurs, E.J. and van den Heuvel, E.P.J. 1989, Astron. Astrophys. *226*, 88

Narayan, R., Piran, T. and Shemi, A. 1991, Astrophys. J. *379*, L17

Paczynski, B. 1967, Acta Astronomica *17*, 355

Phinney E.S. 1991, Astrophys. J. *380*, L17-22

Romani, R.W. 1992, Astrophys. J., *399*, 621

Stella, L., White, N.E. and Rosner, R. 1986, Astrohys.J. *308*, 669

Stone, R.C., 1991. Astron. J. *102*, 333-349

Taam, R.E. and Bodenheimer, P. 1992, in: "X-ray Binaries and Recycled Pulsars" (eds. E.P.J. van den Heuvel and S.A. Rappaport), Kluwer Acad. Publ., Dordrecht, p. 281

Tutukov, A.V. and Yungelson, L.R. 1993, Mon.Not.R.A.S. *260*, 675

Vanbeveren, D. 1991, Space Sc. Rev. *56*, 249-311

Vanbeveren, D. 1991, Astron. Astrophys. *252*, 159

van den Heuvel, E.P.J. 1970a, in: "Stellar Rotation" (ed. A. Slettebak), New York, Gordon & Breach, pp. 78-86

van den Heuvel, E.P.J. 1974, in: Proc. 16th Solvay Conference on Physics (Univ. of Brussels Press, Brussels), p.119

van den Heuvel, E.P.J. 1978, in: "Physics and Astrophysics of Neutron Stars and Black Holes" (eds. R. Giacconi and R. Ruffini), North Holl. Pub. Comp., Amsterdam, pp. 828-871

van den Heuvel 1992a, in: "X-ray Binaries and Recycled Pulsars" (E.P.J. van den Heuvel and S.A.

Rappaport, eds.), Kluwer Acad. Publ. Dordrecht, pp. 233-256

van den Heuvel, E.P.J. 1992b, in: "Proc. Internat. Space Year Confer. (Satellite Symp. No. 3), ESA ISY-3 (July 1992), pp. 29-36

van den Heuvel, E.P.J. and De Loore, C. 1973, Astron. Astrophys. *25*, 387

van den Heuvel, E.P.J. and Habets, G.M.H.J. 1984, Nature *309*, 698

van Kerkwijk, M.H., Charles, P.A., Geballe, T.R., King, D.L., Miley, G.K., Molnar, L.A. and van den Heuvel, E.P.J. 1992, Nature, *355*, 703-705

van Oyen, J. 1989, Astron. Astrophys., *217*, 115

Verbunt and van den Heuvel, E.P.J. 1993 in: "X-ray Binaries" (eds. W.H.G. Lewin, J.A. van Paradijs and E.P.J. van den Heuvel), Cambridge University Press (in press)

, Webbink, R.F. 1992, in: "X-ray Binaries and Recycled Pulsars (eds.E.P.J. van den Heuvel & S.A Rappaport) Kluwer Acad Publ., Dordrecht, pp. 269-280

Wolszczan, A. 1992, in: "X-ray Binaries and Recycled Pulsars" (E.P.J. van den Heuvel and S.A. Rappaport eds.) Kluwer Acad. Publ. Dordrecht, p. 93

DOPPLER TOMOGRAPHY OF O-TYPE BINARIES: THE PHYSICAL PROPERTIES OF SEVEN SYSTEMS

L. R. PENNY, W. G. BAGNUOLO and D. R. GIES*
CHARA, Department of Physics and Astronomy, Georgia State Univ., Atlanta, GA 30303 USA

Abstract. We have analyzed UV photospheric lines of seven O-type binaries, by means of cross-correlation and Doppler tomographic methods, with the goal of estimating the physical properties of the individual stars. These systems are HD 1337 (AO Cas), HD 47129 (Plaskett's star), HD 57060 (29 UW CMa), HD 37043 (Iota Ori), HD 215835 (DH Cep), HD 152218, and HD 152248. Mass ratios have been obtained primarily from a cross-correlation technique, but also by several other techniques. The tomographic techniques allow us to separate the spectra of the components. We then can estimate the individual spectral types and luminosity classes of the stars (and hence T_{eff} and $\log g$, respectively), the luminosity ratio, and projected rotational velocities.

We discuss the physical properties of these O-type binaries. These are some of the early results of a large scale project involving 36 O-type double-lined binary systems (from the catalog of Batten *et al.* 1989) which we will study using IUE and complementary ground-based data.

1. IUE Data Extraction and Processing

The basic steps of our analysis are data extraction and processing, determining the radial velocity orbits, separating the spectra, and analyzing the spectra. The IUE spectra (SWP, high dispersion) were obtained through the NASA/GSFC IUE Regional Data Analysis Facility. We produced a matrix of pseudo-rectified spectra sampled with a common wavelength grid. Routines were written in IDL (Interactive Data Language) to: (1) remove data spikes, (2) bin the spectra on a $\log \lambda$ scale (bin size = 10 km s^{-1}), (3) smooth the data with a Gaussian transfer function (FWHM = 40 km s^{-1}, (4) align each spectrum with a global average using cross-correlation shifts from the vicinity of strong interstellar lines (after which the interstellar lines are removed by interpolation), and (5) rectify each spectrum by a spline fit to a series of pseudo-continuum zones.

2. Radial Velocities and Orbits

We determined radial velocities for these spectra by cross-correlating each spectrum with a standard, single star spectrum (we used HD 54662 [O6.5 V] for DH Cep, HD 34078 [O9.5 V] for HD 152218, and HD 46149 [O8.5 V] for HD 152248). We used essentially the full wavelength range of the spectra (1200 - 1900 A) but we set to unity regions surrounding the strong P Cygni lines of N V $\lambda1240$, Si IV $\lambda1400$, and C IV $\lambda1550$, since these features reflect wind and not center-of-mass motion. The relative velocities were then determined by fitting two Gaussians to the composite cross-correlation functions. The cross-correlation functions for DH Cep are shown as a function of orbital phase in Figure 1.

* Guest Observer with the International Ultraviolet Explorer Satellite

Space Science Reviews **66**: 323–326, 1994.
© 1994 *Kluwer Academic Publishers. Printed in Belgium.*

TABLE I

Orbital Elements (Primary/Secondary)

HD (P/S)	P (days)	K (km s^{-1})	$m \sin^3 i$ ($M_\odot$)	$a \sin i$ (10^6 km)	Std. Dev. of Fit (km s^{-1})	UV Flux Ratio
215835 (P)	2.110953(8)	224(2)	13.3(9)	6.5(1)	10.8	0.80
215835 (S)	2.11092(1)	260(6)	11.5(8)	7.5(2)	16.6	
152218 (P)	5.58101(6)	147(6)	13(2)	11.3(5)	21.5	0.50
152218 (S)	5.58075(8)	198(10)	10(2)	15.2(8)	44.6	
152248 (P)	5.55333(8)	192(6)	20(2)	14.7(5)	31.8	0.95
152248 (S)	5.55336(2)	212(4)	18(2)	16.2(1)	27.3	

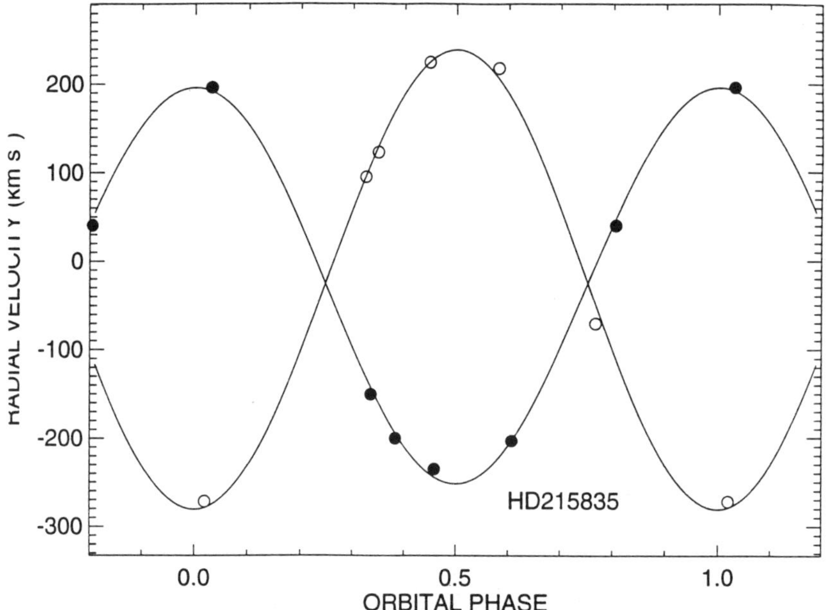

Fig. 1.   Radial velocities of 6 IUE spectra of DH Cep plotted against orbital phase. The filled (open) circles represent observed velocities found by cross-correlation of the primary (secondary). Solid lines are drawn for both the primary and secondary radial velocity solutions.

We used the velocity data (together with ground-based data from Hill *et al.* 1974 for HD 152218 and HD 152248) to search for possible orbital periods using the Fourier transform method of Roberts *et al.* (1987); because of the large gaps between observations, many candidate periods exist. We then used these periods with the program of Morbey & Brosterhus (1974) to determine orbital elements. Table 1 shows the orbital parameters obtained for these three stars. Numbers in

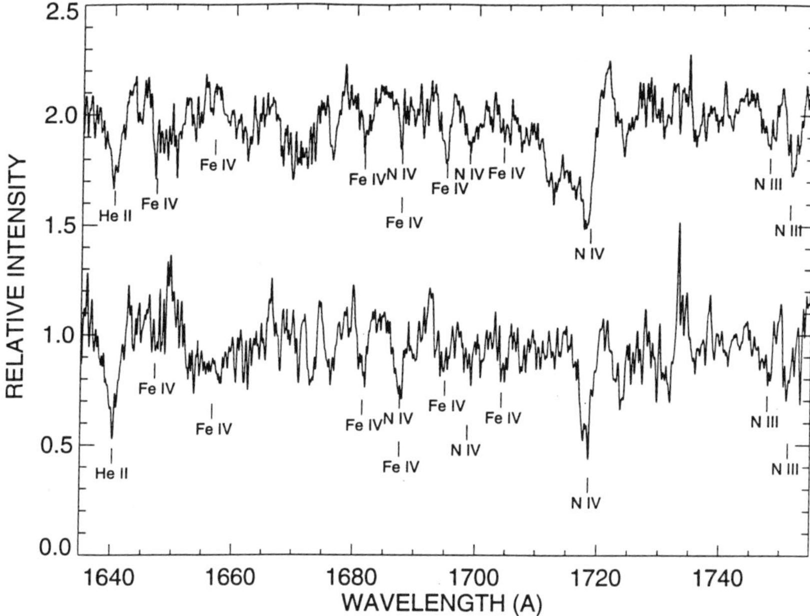

Fig. 2.   Spectra of the primary (*top*) and secondary (*below*) components of DH Cep, from tomography of 6 *IUE* spectra, in the 1635-1755 A region. Note the He II λ1640, Fe IV, and N IV λ1718 lines in secondary which indicate a later type.

parentheses indicate the errors in the last digit quoted. Circular orbits have been assumed in each case.

3. Spectral Analysis and Physical Properties

We used the etimated velocites and flux ratios (from cross-correlation analysis and Petrie's method) to reconstruct the individual spectra using the tomography algorithm (Bagnuolo & Gies 1991; Bagnuolo *et al.* 1992). Separated spectra of DH Cep are shown on Figure 2. An IDL routine allowed a comparison of each recovered spectrum with comparison spectra in the O-Star Atlas (Walborn *et al.* 1985). Spectral typing was done using 15 temperature and 8 luminosity sensitive features (Bagnuolo *et al.* 1994). The spectral types were translated into effective temperatures using the calibrations of Howarth & Prinja (1989). Table 2 lists the physical properties of the seven stars analyzed to date. (Properties not yet determined indicated by dots.) Note that masses indicated by an asterisk are $m \sin^3 i$. When data as in Table 2 is available for a larger sample of stars, we will take a 'census' of the evolutionary status of these systems.

This work is supported through NASA Astrophysics Data Program grant NAG 5-1994.

TABLE II

Physical Properties of Seven Binaries

Star (P/S)	$M/M_\odot$	Interpolated Spectral Type	T_{eff} (K)	$\log L/L_\odot$	$R/R_\odot$	$\log g$	$V \sin i$ km s^{-1}
AO Cas (P)	12.4	O9.5 I	32,000	5.06	10.9	3.29	120
(S)	18.3	O8.1 V	36,000	4.82	6.3	4.00	130
Plaskett's (P)	42.5	O7.3 I	35,100	5.80	21.5	3.40	75
(S)	51.0	O6.2 I	38,400	5.75	17.4	3.70	310
29 CMa (P)	16.0	O7.5-8 Iabf	33,750	5.30	13.0	3.20	135
(S)	19.0	O9.7 Ib	29,000	4.80	10.0	3.70	100
Iota Ori (P)	39.0	O9 III	32,000	5.40	16.0	3.60	69
(S)	19.0	B1 III	21,000	4.40	11.0	3.65	46
DH Cep (P)	13.3*	O5.3 V	44,600	...	...	...	...
(S)	11.5*	O6.3 V	41,100	...	...	...	...
HD152218 (P)	13.4*	O9 III	32,000	...	...	...	...
(S)	9.9*	O9.8 III	29,000	...	...	...	...
HD152248 (P)	14.7*	O8.2 Ib	32,800	...	...	...	...
(S)	16.2*	O7.7 Ib	34,100	...	...	...	...

References

Bagnuolo, W.G., & Gies, D.R.: 1991, *Astrophysical Journal* **376**, 266

Bagnuolo, W.G., Gies, D.R. & Wiggs, M.S.: 1992, *Astrophysical Journal* **385**, 708

Bagnuolo, W.G., Gies, D.R., Hahula, M.E., Wiemker, R., & Wiggs, M.S.: 1994, *Astrophysical Journal*, **423**, in press

Batten, A.H., Fletcher, J.M., & MacCarthy, D.G.: 1989, *Pub. Dom. Astrophys. Obs. Victoria* **17**,

Hill, G., Crawford, D.L., & Barnes, J.V.: 1974, *Astronomical Journal* **79**, 1271

Howarth, I.D., & Prinja, R.K.: 1989, *Astrophysical Journal, Supplement Series* **69**, 527

Morbey, C.L., & Brosterhus, E.B.: 1974, *Publications of the ASP* **86**, 455

Roberts, D.H., Lehar, J., & Dreher, J.W.: 1987, *Astronomical Journal* **93**, 968

Walborn, N.R., Nichols-Bohlin, J., & Panek, R.J.: 1985, ,
    NASA Report RP-1155, Washington, DC

EVOLUTION OF MASSIVE CLOSE BINARIES.

MCB's: a Disappearing Curiosity or an Important Missing Link in the Mind of a Lot of Massive Star Astrophysicists.

D. VANBEVEREN

Dept. of Physics, V.U.B., Pleinlaan 2, 1050 Brussels, Belgium.

Abstract. This paper briefly reviews the competition between massive single star and massive close binary evolution the last two decades. The status of the binary evolutionary model is summarized, the assumptions and simplifications are critically discussed. Using all computations performed since 1970, general conclusions are drawn and a comparison with massive single star evolution is presented. Special attention is given at the assumptions behind the commonly accepted model for the mass gainer and a new accretion model is proposed. The binary results in combinarion with single star evolution are compared with observations of massive stars with emphasis on the HR diagram, star number counts, WR stars, SN 1987A, OBN and OBC stars.

Key words: Evolution – Binaries

1. Previous Reviews on Massive Close Binaries and Abreviations Used Here.

The last two decades only five comprehensive and more or less complete reviews appeared on massive close binary evolution i.e. Paczynski (1971), Thomas (1977), van den Heuvel (1978), de Loore (1980) and Vanbeveren (1991). The numerous binary reviews in proceedings of conferences are not quoted. I do not want to discuss their reliability but due to the limited number of pages such a review is usually far from being complete. In this review I will try to be as complete as possible as far as the work after 1991 is concerned. For the studies before 1991 I invite the reader to consider the reviews given above.

The following abreviations will be used: CHB = core hydrogen burning, CHeB = core helium burning, SW = stellar wind, RLOF = Roche lobe overflow, SN = supernova explosion, CC = compact companion, MCB = massive close binary, WR = Wolf–Rayet star, RSG = red supergiant.

2. Massive Single Stars Versus Massive Close Binaries The Last Two Decades.

In this section I will scematically present my personal view on the history of massive star evolution and the competition between single stars and massive binaries.

From a statistical study Kuhi (1973) concluded that 73 % of all WR stars are binaries with an OB type companion. Accounting for possible undetected components, it was concluded that all WR stars were formed by binary evolution. The

Space Science Reviews **66**: 327–347, 1994.

© 1994 *Kluwer Academic Publishers. Printed in Belgium.*

orbital masses of WR components of close binaries allowed to conclude that WR stars form a subset of massive stars whereas the observations and interpretation of early type X–ray binaries also placed these objects in the massive star range.

This has lead to extended studies of MCB's and their evolution from the late sixties up to late seventies, early eighties in Europe, i.e. Warsaw (Paczynski, Ziolkowski, ...), Gottingen (Kippenhahn, Weigert, Thomas,...), Amsterdam (van den Heuvel, Heise, Savonije,...), Moscow (Tutukov, Yungelson, Massevitch,...), Brussels (de Loore, De Greve, Vanbeveren, Packet, Hellings,...). We can safely state that 1970 - 1980 is a MCB evolutionary period.

What happened ? In 1980 it was shown that the statement of Kuhi (1973) is statistically biased (Vanbeveren and Conti,1980) and that the real WR+OB binary frequency was more like 40 % – 50 %. From a binary point of view this lower percentage was no problem because at that moment, based on the qualitative binary evolutionary scenario of van den Heuvel and Heise (1972) it was expected that the number of WR+CC = the number of WR+OB binaries. The statement 'number of WR+CC = number of WR+OB' was strenghtened by observations of Moffat, Seggewiss, Isserstedt,... (late 70's, early 80's) who reported periodic variations in a large number of single lined WR stars; these variations were interpreted in terms of the presence of compact companions.

The lack of hard X–rays in these suspected WR+CC binaries was a problem. van den Heuvel (1976) suggested that the SW in WR stars is so large that it absorbs these hard X–rays. However detailed computations where the process of hard X–ray formation was combined with the details of absorption of these X–rays in the SW of the WR star (Vanbeveren et al., 1982) revealed the contrary. In the latter paper we critically discussed the assumptions behind the statement WR+CC = WR+OB. The predicted number of WR+CC binaries depends obviously on the survival probability of the binary during the first SN. We argued that this survival probability could be much lower in MCB's than commonly accepted at that time. This small probability now becomes very plausible in view of the recent study of Harrison et al.(1992) where the authors rediscuss the runaway velocity distribution of pulsars and conclude that on the average the velocities are much higher than thought in the past. If these velocities reflect the kick velocities the compact star gets during the SN, i.e. reflects the asymmetry of the SN ejecta, it follows indeed that the probability for the system to remain bound after the first SN may be very low, thus the theoretically expected number of WR+CC systems could be very low as well. Since then the interpretation of the periodic variations in single lined WR stars has changed also i.e. they are due to intrinsic variability of the SW of the WR star.

It is clear from the discussion above that from 1980 on the importance of the MCB channel in order to produce WR stars decreased drastically. A single star scenario was proposed by Conti and this started the flourishing period of massive single star evolution with Maeder, Meynet, Chiosi, Bertelli, Bressan, Allongi, Nasi, Screenivasan, Woosley, Weaver, Doom, Langer, El Eid,... We can safely conclude that 1980 – 1992 is a massive single star evolutionary period.

The flourishing period of single stars corresponds to the period of increasing computer power. The computation of large data sets became feasible without too large effort but ...this was done only for single stars until recently. The reason why binaries were not considered is simply that major conclusions were reached and a general MCB evolutionary scenario was obtained already before say 1982. Realizing that including a bit of overshooting or a different SW mass loss rate formalism will only give (small) qualitative differences is not directly a stimulae in order to spend your time producing such a MCB data set. Even more, a binary data set not only has to consider different primary masses but also different mass ratios and different periods.

The publication of large data sets but for single stars only caused in my opinion the formation of a general single star mind among massive star astrophysicists, understandable from a psychological point of view but very unfortunate as I will demonstrate in sections 5, 6, 7, 8 and 9.

3. The Massive Binary Evolutionary Model: Present State.

Our binary code follows the evolution of both components simultaneously, new opacity tables (Rogers and Iglesias, 1992) were implemented, overshooting and stellar wind mass loss are included. We use the SH criterion for semi-convection and the RLOF model for mass exchange, i.e. when the radius of the primary star becomes larger than a critical radius, mass transfer starts; the mass loss rate is then computed by imposing the condition that the stellar radius equals the critical radius; this mass transfer continues as long as the star has the tendency to expand. The variation of the period and the mass ratio during RLOF is calculated using the model of Vanbeveren et al. (1979). When sufficient mass has been removed (i.e. when almost all hydrogen rich layers are removed), a star stops expanding and the RLOF process ends. The further evolution depends on whether the star resembles a WR star or not, i.e. on whether large SW mass loss occurs or not.

The structure of the stars is computed assuming spherical symmetry. What about rotation ? Let us assume that primaries of non-evolved MCB's rotate synchroneously with the orbit. It then follows that the equatorial velocity v = $2\pi R/P$ (R = stellar radius, P = system period). Using an average orbital period of observed non-evolved MCB's = 10 days it can easily be calculated that on the average a CHB O type primary then has a v.sini around 60 - 70 km/sec. It remains to be demonstrated whether such low rotational velocities affect the evolution of massive primaries. Anyhow from the observed v.sini of CHB single stars (Howart and Prinja, 1989) it looks as if the effect of rotation on evolution is not larger for massive primary components than for single stars.

What about gravitational distortion ? From the shape of the equipotential surfaces of close binary components it is clear that except for a minority of case A systems the deviation from spherical symmetry is very small during the CHB phase of the primary. During hydrogen shell burning when the primary fills or nearly fills its Roche lobe, the star is centrally condensed and only the outermost mass layers are affected by gravitational distortion. Since the subsequent evolutionary phases are very short and since the layers which have eventually been affected by this distortion are removed due to RLOF on a 10000 yrs timescale, it is understandeble that this effect has never been explicitely included in order to study the evolution of MCB's.

The behaviour of the mass gainer during RLOF or how does a star react when it accretes matter at very high rates? The classical way in order to study this effect dates from the early seventies (Benson, 1970a,b) and relies on the 'critical' assumption that the accretion process does not destroy the radiative equilibrium condition in the outer layers. This model treats thermohaline mixing as an instanteneous process (Ulrich, 1972; Kippenhahn et al., 1980; Packet, 1988) but neglects the effect of rotation.

In the present review I want to propose an alternative assumption in order to study the gainer's behaviour. It is a well known fact that accretion in a binary may cause a rapid spinning up of the outer layers of the gainer (up to velocities close to the break up velocity) and eventually of the whole star (the Be star phenomenon?). Rapid rotation may drive efficient turbulence and/or convection (Zahn, the present proceedings). Just like in a protostellar phase it is not unthinkable that rapid accretion is capable to make the whole star convective, it is the star is homogenized. I will discuss the consequences of this model in section 4.2.2. It has been used succesfully in order to explain the overluminosity and helium enrichment of the optical star of the standard massive X–ray binary Vela X–1 in a paper published in the present proceedings (Vanbeveren, Herrero, Kunze and van Kerkwijk).

The first SN: when the mass loser explodes, what is the probability that the system remains bound ? As argued already in the previous section this probability may be quite low.

The evolution after the first SN: when the system was disrupted, the further evolution is the evolution of a single star BUT with a structure which was modified by accretion and which may deviate significantly from the structure of a normal single star. When the system remained bound a massive X–ray binary will be formed. Its further evolution is described by the spiral–in process. The only study treating this process in a more or less realistic way is the two dimentional treatment of Bodenheimer and Taam (1984). Unfortunately very soon after the onset of the spiral–in their computations failed to converge so that stating that after the spiral–in process the system consists of a hydrogen deficient CHeB star + CC (WR+CC) is still speculation. However the existence of double pulsars indicates that there should be systems which can either survive a spiral–in phase or avoid it. We will consider the last possibility in sections 5.3 and 5.4.

A large set of binary evolutionary calculations for the Magellanic Clouds including updated physics and the new OPAL opacities will appear in Astron.

Astrophys. Suppl.(de Loore and Vanbeveren,1993). We made these computations using the same binary code as the one used by de Loore and De Greve (1992) and De Greve and de Loore (1992) who performed a massive star evolutionary data set for the Galaxy (lateron De Greve, 1993,used again the same code in order to compute a data set for the intermediate mass case B binaries also for the Galaxy). However while adapting the code for the Magellanic Clouds we found a computer bug which makes the Galaxy computations very unreliable, i.e. the computer program computed the structure of the stellar interior adopting a metallicity Z=0.03 and using Cox–Stewart opacities while the outermost layers were computed with Z=0.02 and Los Alamos opacities. In this way a Z=0.03 Cox-stewart interior was fitted to a Z=0.02 Los Alamos exterior. This is one of the reasons why we decided to recompute the Galaxy binary set over again (de Loore and Vanbeveren, 1994a) but now with updated physics and OPAL opacities. Together with the MC computations they form a complete and corresponding set of case B massive close binary evolutionary tracks with various metallicities. A full analysis will be presented in de Loore and Vanbeveren (1994b).

4. The Computations.

The last two decades evolutionary computations of MCB's were performed for a wide range of system parameters, i.e. $9 \leq M_1/M_\odot \leq 100$, $3 \leq P \leq 50$ days, $0.3 \leq q \leq 1$, $0.002 \leq Z(\text{metallicity}) \leq 0.03$, covering the majority of the OBSERVED non-evolved massive close binaries.

4.1. THE BEHAVIOUR OF THE MASS LOSER IN EARLY CASE B BINARIES.

4.1.1. General conclusions.

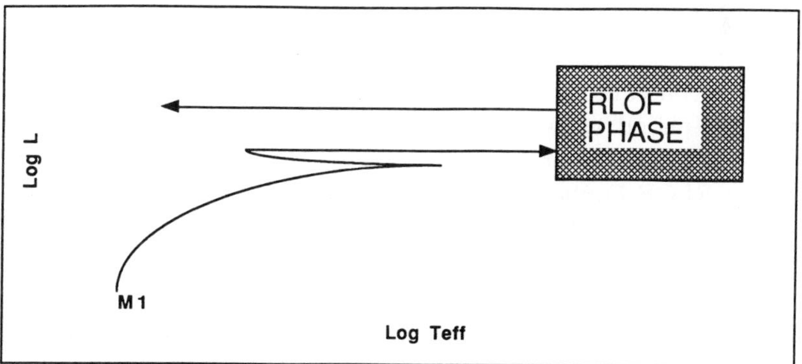

Fig. 1.   Evolutionary track of the primary of an arbitrary case B MCB.

Figure 1 shows the general evolutionary behaviour of a massive mass loser in a case B MCB. Prior the the RLOF the star evolves as a normal single star. As the star expands its radius reaches a critical value and RLOF starts. The RLOF phase lasts as long as the star has the tendency to expand. This tendency stops when almost all hydrogen rich layers are removed and this does not depend on the detailed physics of the RLOF process itself i.e. the structure of a star at the end of RLOF is largely independent from the assumptions during RLOF, from the period, from the mass ratio.*The knowledge that there exists a radius from where large mass loss occurs is much more important than the knowledge how to determine this radius or how to treat this mass loss process.* This is the reason why in fig. 1 the RLOF is shown as a black box; given the input the output does not or only marginally depend on the details of what really happens in this box.

Semi-convection is unimportant for the evolution of the mass loser, overshooting implies only quantitative differences whereas MCB evolution in the Galaxy, LMC, SMC is very similar (Hellings and Vanbeveren, 1981; de Loore and Vanbeveren, 1994b).

The larger the period the larger are the M_{dot} values encountered during RLOF and thus the more violent will be the effect of accretion. Suggestion: could it be that the homogenization model of the mass gainer as discussed in section 3 applies particulary in late case B/C systems?

The influence of SW: we have to separate SW mass loss in O type stars from the SW mass loss in LBV's.

a. The M_{dot} in O type stars: when current semi–empirical M_{dot} relations are used in evolutionary codes we conclude that SW mass loss in O type stars affects massive star evolution only if $M \geq 40\ M_{\odot}$ (resp. 55 $M_{\odot}$ and 80 $M_{\odot}$) for the Galaxy (resp. the LMC and the SMC).

b. The M_{dot} in LBV's: we first realise that the M_{dot} in LBV's is very similar to the M_{dot} encountered during a typical RLOF in MCB's. This leads to the formulation of the LBV scenario for MCB's (Vanbeveren, 1991), i.e. *when a star enters a LBV phase prior to the RLOF, the RLOF will never occur.* This has two important consequences: since LBV mass loss is more or less spherically symmetric, the period variation follows a Jeans mode, i.e. systems who went through the LBV scenario are expected to have larger periods than systems who experienced a RLOF. Furthermore, no accretion effects on the companion star are expected, i.e. no rejuvenation and no alterred atmospheric abundances (see also section 4.2).

The further evolution of the mass loser after RLOF/LBV depends on whether the remnant is a WR star or not i.e. on whether the CHeB star loses mass by SW at WR rates or not.. We use a non-Anne Underhill definition of a WR star, i.e. a WR binary component is a core helium burning star with mass larger than 5 $M_{\odot}$ in the blue part of the HR diagram (this minimum mass value is the minimum value of observed WR components in binaries). The definition implies that all WR binary components have atmopherical hydrogen abundances $X_{atm} \leq$

0.4–0.5. The results of detailed computations with different SW mass loss rate formalisms (remind that already in 1979 we used WR mass loss rates depending on the luminosity) can be summarized as follows:

a. for WN stars we expect a very unique nitrogen to carbon number ratio i.e. $N/C = 100 \ (N/C)_\odot$ (Vanbeveren and Doom, 1980),

b. hydrogen less WR binary components satisfy a unique M-L relation (Vanbeveren and Packet, 1979; Vanbeveren, 1991) i.e.

$$\log L/L_\odot = 3.4 + 1.786 \log M/M_\odot$$

c. this M-L relation does not depend on Z (Hellings and Vanbeveren, 1981; de Loore and Vanbeveren, 1994b)

d. the properties above do not depend on the details of the evolution of the progenitors, the SW mass loss rate formalism during the WR phase, the treatment of the outer layers of the stars, the opacities, the way how semi-convection was treated, the uncertain CO reaction rate during the CHeB phase.

4.1.2. Comparison with Single Star Computations.

TABLE I

Helium burning lifetimes (in 10^5 years) for Z=0.02/Z=0.002.

Initial mass	primary	single1	single2
9	29.5/28.9		31/27
12	18/16		16/15
15	12.1/10.6	22/24	12/11
20	8.3/7.2	14/16	7.8/8.1
40	5.9/4.7	6.6/8.9	4.7/4.4
60	4.3/3.9	7.0/6.1	3.8/3.7

I first compare the helium burning lifetimes of our binary computations with the single star results of Maeder and Meynet (1989) and Maeder (1991) (table 1, the colums primary and single1). As can be noticed the helium burning lifetimes predicted for massive single stars resulting from the latter papers were almost a factor 2 larger than the binary values. This situation was very uncomfortable since from pure physical grounds one expects that the single star values should be smaller than or equal to the binary values. Fortunately there was Cesare Chiosi at the stellar population conference in Brazil (1991) who showed using physical arguements that the single star helium burning lifetimes of the Geneva group had to be erronomous. The Geneva group found the computer bug and new tracks were computed (Schaller et al., 1993). In table 1 I show the new helium burning timescales as well (colum single2) and as can be noticed the binary and single star values correpond much better now. However a warning is appropriate here: *all results and conclusions which are based on the Geneva evolutionary results which are published before 1992 have to be treated with caution.* This is a fortiori the case for all theoretical predictions of star numbers, especially the WR star numbers as well as for all population synthesis models with special emphasis on the influence of WR stars and Red Supergiants where these tracks were used.

The M–L relation of single WNE and WC stars (Maeder, 1983) has a very similar shape as the one for WNE/WC binary components, HOWEVER the minimum mass of a single WNE star resulting from standard single star evolution (Schaller et al., 1993) is much larger than for WNE binary components. This is particularly the case for the Magellanic Clouds. We will return to this point in section 5.5.

4.2. THE BEHAVIOUR OF THE MASS GAINER IN EARLY CASE B BINARIES.

As was discussed in section 3, the behaviour of the gainer can be studied either by assuming that accretion does not destroy the radiative equilibrium in the outer layers, either by assuming that accretion makes the whole star convective.

4.2.1. *During Accretion The Radiative Equilibrium in The Outer Layers is Not Destroyed and Thermohaline Mixing is Included.*

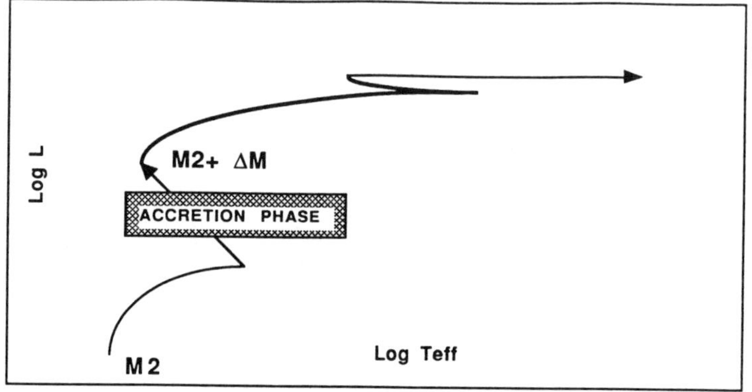

Fig. 2.　Evolutionary track of the secondary of an arbitrary MCB.

As for the mass loser also for the gainer it is possible to present a general evolutionary behaviour (fig. 2). The following effects are noticed:

a. rejuvenation: although a MCB with a WR component may be 6.10^6 years old, the mass gainer may look like a star with a lifetime 2.10^6 yrs,

b. the gainer may have alterred CNO abundances but the helium to hydrogen ratio is almost normal,

c. in the HR diagram, the accretion stars can hardly be distinguished from normal single stars (de Loore and Vanbeveren, 1994b),

d. the general evolutionary behaviour of mass gainers is largely independent from the metallicity (de Loore and Vanbeveren, 1994b),

e. when the moment the accretion starts and the total amount of accreted mass are fixed, how the star looks like after the accretion phase is largely independent from the details of the accretion phase itself. This is the reason why in fig. 2 the evolutionary track in between the start and the end point is simply replaced by the box 'accretion phase'.

4.2.2. *During Accretion The Star Becomes Completely Convective.*

For typical evolutionary tracks I refer to the paper Vanbeveren et al. these proceedings where this model was introduced for the first time and applied to the X–ray binary Vela X-1. We deduce the following properties:

a. a more pronounced rejuvenation compared to the model discussed in section 4.2.1. During the accretion phase of a late case B/ case C system it is possible that an over–contact phase is avoided and the RLOF could proceed in a more or less conservative way.

b. The gainer has alterred CNO abundances and the helium to hydrogen ratio is significantly increased ($\epsilon = N_{He}/(N_H+N_{He}) = 0.3$ is not extraordinary). Could this explain the helium discrepancy in a number of massive blue supergiants?

c. In the HR diagram, these accretion stars are distinctly different from normal single stars; they may be significantly overluminous. An extended study of the evolutionary behaviour of accretion stars is presented by Vanbeveren and de Loore (1994).

4.3. The Evolution of Massive Interacting Binaries with Large Periods: late case B, case C.

Let us first note:

a. there are very few massive binaries observed with large or very large periods, corresponding to late case B or case C binaries;

b. quantitative simultaneous evolutionary computations for late case B or case C systems up to the end of the RLOF have never been performed yet, i.e. the expected evolutionary behaviour of these systems relies on qualitative properties only.

The fact that very few late case B, case C systems are known may however be due to observational selection. It takes much more telescope time in order to determine the system parameters of a large period binary; an observer will therefore prefer an early case B, case A. Accounting for such human effects and trying to quantify them, it follows that late case B, case C binaries may be at least as important as early case B, case A. Since 35 – 40 % of the massive stars are double with a period appropriate for early case B, it may very well be that the real interacting massive binary frequency is more like 70 – 80 % leaving only 20 – 30 % for real single stars.

How do late case B, case C systems evolve ? The answer to this question can to some extend be predicted accounting for the following general evolutionary properties:

a. the larger the period the more violent is the RLOF phase, i.e. the larger is the primary mass loss,

b. as for early case B, also for late case B, case C the RLOF should stop when the primary stops expanding.

A yellow or red supergiant primary stops expanding and returns to the blue part of the HR diagram as a consequence of mass loss due to RLOF when almost all hydrogen rich layers are removed. It follows then that for the majority of late case B, case C systems the total primary mass loss due to RLOF should be very similar to the total mass loss of a massive primary in an early case B binary (i.e. the final remnant mass of a primary in an interacting binary is to a large extend independent from the initial period of the system).

Is the RLOF phase in late case B, case C massive binaries conservative or non – conservative? Here we can adopt the suggestion of van den Heuvel (1983) which is based on a comparison of the RLOF timescale of the primary and the accretion timescale of the secondary (let me repeat however that no realistic simultaneous binary evolutionary computations exist in order to quantify this suggestion): systems with initial mass ratio larger than 0.4 evolve conservative, i.e. the mass transfer is stable and the secondary is able to accrete all the mass lost by the primary. Systems with initial mass ratio lower than 0.4 enter a common envelope phase very soon after the onset of the RLOF and no significant mass transfer occurs, i.e. most of the mass lost by the primary has to leave the system taking with a large part of the orbital angular momentum.

Accepting this model it follows then that post–RLOF late case B/case C systems with initial mass ratio larger than 0.4 should have very large periods and mass ratios larger than 2; a post–RLOF late case B/case Csystem with initial q $\leq$ 0.4 should be a system with very short period (perhaps smaller than a day) and mass ratio $\leq$ 1.

The suggestion that late case B, case C systems enter a common envelope phase is based on the fact that a star should swell enormously when the accretion rate is large. However this swelling is a direct consequence of the assumption that the outer layers of the star remain in radiative equilibrium during accretion. When on the contrary large accretion makes the whole star convective (section 3), this rapid increase of the stellar radius could be avoided and also here the mass transfer could proceed in a conservative way. The final product should then be a binary with very large period as well and mass ratio $\geq$ 1.

5. Comparison With Observations.

Remark: the WNE, WNL notation was first introduced in 1980 (Vanbeveren and Conti, 1980) i.e. WNL = WN with hydrogen (this corresponded at that time meanly to WN7,8,9) and WNE = WN without hydrogen (WN$\leq$6). The main reason for this separation was an evolutionary one i.e. WNL stars are core helium burning and hydrogen shell burning stars (explaining their larger luminosity) whereas the luminosity in WNE stars is produced by the 3α–process only. However the original definitions have been changed somehow by Conti and Massey (1989) and this causes some confusion in literature.

In the following I will use the original hydrogen definition but I think it is necessary to make up our minds what to use in the future.

5.1. THE MASS-LUMINOSITY RELATION OF WR STARS.

a. Already in 1979 a unique ML relation was proposed for hydrogen less WR binary components, thus for WNE and WC type stars (Vanbeveren and Packet, 1979) i.e.

$$\log L/L_\odot = 3.4 + 1.79 \log M/M_\odot$$

The relation holds for WNE and WC binary components with mass larger than 5 $M_\odot$. Later revisions gave an almost identical result (Vanbeveren, 1991; de Loore and Vanbeveren, 1994b). Already in the 1979 paper we argued that the ML relation of single WNE/WC stars has to be similar to the relation given above provided that single stars lose enough mass during say the RSG phase.

b. The ML relation of WNE/WC binary components in the Magellanic Clouds is very similar to the Galactic one (de Loore and Vanbeveren, 1994b).

5.2. THE WR BINARIES.

Detailed modelling of all known WR binaries in the Galaxy reveals the following results:

a. if we can believe the spectral types (and in a second place the luminosity class) of the OB companions in WR binaries it can be concluded that *the RLOF process was essential in the formation of a large number (the majority?) of WR binaries* (Vanbeveren, 1987, 1989, 1991),

b. only the WR binaries with a giant or supergiant component (γ^2Velorum as an example) may have had a LBV past, i.e. RLOF has not played a fundamental role in their formation (the LBV scenario, see section 4.1.1),

c. the majority of WR components of close binaries have had progenitors with initial mass between 20 - 40 $M_\odot$ (moderate overshooting, Vanbeveren, 1988), i.e. the statement 'the bulk of WR stars have had initial masses larger than 40 $M_\odot$' is not valid. This statement was proposed by Conti et al. (1983) based on the Galactic distribution of O and WR stars. However comparison between both distribution was done by the eye in the latter paper and evolutionary computations were used without overshooting and a SW mass loss rate formalism during CHB which predicts too high values when compared to the formalism of de Jager et al. (1988). I have redone the same excercise but using statistics and standard evolutionary models (section 5.5) with overshooting and the de Jager et al. SW formalism (Vanbeveren, 1988a) and concluded that the majority of WR stars should have had initial masses larger than 20 $M_\odot$, corresponding to the binary conclusion.

d. From evolutionary point of view we do not expect any difference between the WR binary population in the MC and in the Galaxy (de Loore and Vanbeveren, 1994b).

e. WR + OB binaries with a late case B, case C history (section 4.3) should either have very large periods and q $\geq$ 1–2, either very small periods and mass ratio $\leq$ 1. Possible candidates for the large period scenario are HD 192641, HD 193077 and HD 193693.

5.3. THE WR+CC SYSTEMS.

As was discussed in section 2 the number of WR+CC could be very small. However if there exists such systems we have to find formation mechanisms. The spiral–in phase of an OB+CC binary (standard X–ray binary) is a possibility but the physics are still very poorly understood; it is expected that after spiral–in the system has a period of a few houres; this model may find it difficult in order to explain one of the best known WR+CC candidate HD 50896 with a period of 3.7 days (Firmani et al. 1980).

The LBV scenario (section 4.1.1) offers the possibility to form WR+CC and thus eventually double pulsars without the spiral–in phase. If in a OB+CC system the OB star is massive enough so that it will experience a LBV phase, the spiral–in phase can be avoided and a WR+CC can be formed with relatively large period (days rather than a few houres as expected from the spiral–in process).

5.4. THE FORMATION OF DOUBLE PULSARS.

Once a WR+CC binary (or more generally a massive CHeB star+CC binary) has been formed, the formation of a double pulsar obviously only depends on the subsequent SN explosion of the CHeB star. The big uncertainty in this scenario affecting the theoretically predicted number frequency of double pulsars is the formation of the CHeB+CC system, i.e. either through an LBV phase (previous section) or through the spiral–in phase.

There may be a class of binaries where double pulsars can be formed directly from the OB+CC phase without encountering a WR+CC phase and avoiding a spiral–in phase. As will be discussed in more detail in section 7, if binaries with primary masses between 10 $M_\odot$ and 20 $M_\odot$ and with mass ratio very close to one evolve in a conservative way, the OB type mass gainers remain in the blue part of the HR diagram during their entire live (a possible explanation for the blue progenitor of SN 1987A). Since they do not expand a spiral–in phase does not need to occur and the system experiences a second SN explosion as a OB+CC system, thus possibly forming a double pulsar without a previous spiral–in phase.

If the high runaway velocities in pulsars (van den Heuvel, these proceedings) really reflect the asymmetry of the ejecta of a previous supernova explosion, the survival probability of a massive binary during its first and eventual second SN explosion may be much lower than thought until now, in this way predicting a much lower frequency of double pulsars and solving the question why so few double pulsars have been found until now.

5.5. THE STANDARD MODEL OF MASSIVE STAR EVOLUTION OR HOW WRONG CAN CONCLUSIONS BE IF IN A THEORETICAL PREDICTION BINARIES ARE OMITTED ?

Let us recall the meaning of the standard model of massive single star evolution (Schaller et al., 1993): M_{dot} of de Jager et al. (1988), overshooting $\alpha = 0.2$, during the core hydrogen burning and the red supergiant phase $M_{dot} \propto Z^{0.5}$, during the WR phase $M_{dot} \propto M^{2.5}$ independent of Z.

The corresponding standard model of massive close binary evolution is then obviously, M_{dot} of de Jager et al., overshooting $\alpha = 0.2$, during core hydrogen burning $M_{dot} \propto Z^{0.5}$, during the WR phase $M_{dot} \propto M^{2.5}$ independent of Z.

5.5.1. The HR Diagram of WR Stars.

In figure 3a, we compare the observed (see also the review of W.R. Hamann, these proceedings) and the theoretically predicted HR diagram positions of single WN stars in the Galaxy. Figure 3b is similar to figure 3a except that the theoretical binary predictions are considered. It is clear that standard single star tracks do not cover the observed WN region consistently. Binary tracks however do. The foregoing discussion is true a fortiori for the LMC. This means that either we have to accept that the majority of WR stars in the Galaxy and certainly in the MC's are binaries (a majority with until now undetected companions) either we have to conclude that there is something wrong with the assumptions of single star evolution. We will reconsider the latter possibility later on.

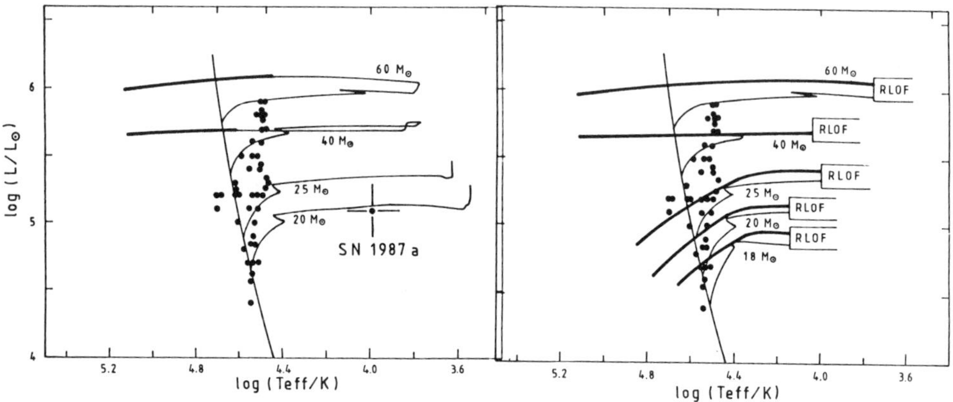

Fig. 3.   The observed WN location in the HRD compared to predictions of standard single star evolution (fig. 3a) and of primaries of MCB's (fig. b) (bold part of the tracks).

5.5.2. The WR/O Number Ratio.

TABLE II

The observed and theoretically predicted WR/O number ratio.

	theory(no bin)	theory(with bin)	observations
Galaxy	0.032	0.08	0.12
LMC	0.009	0.028	0.04
SMC	0.002	0.012	0.015

Warning: if one wants to compare the observed and theoretically predicted WR/O number ratio, one implicitly assumes that the observed O star sample is homogeneously distributed over the core hydrogen burning band. For the Galactic O type stars within 2.5 kpc from the sun this is not necessarily the case (Vanbeveren, 1988a).

Keeping the foregoing remark in mind I have repeated the study of Maeder (1991), but now with the new Geneva tracks with improved He burning timescales (Table 2). The theoretical ratio is computed first without binaries (as was done by Maeder, 1991) (the column theory(no bin)), secondly by using a 35 % O type binary frequency and thus using also the binary evolutionary results (column theory(with bin)). We notice that the theoretical results without the binaries do not correspond to the observations; including the binaries brings theory closer to observations however I think that it is still a matter of faith concluding that the comparison is satisfactory.

5.6. A SOLUTION FOR THE HRD PROBLEM (AND FOR THE WR/O PROBLEM?).

TABLE III

The total mass lost during the RSG phase of a 20 $M_\odot$ and 25 $M_\odot$ star predicted by the de Jager formula and the value needed in order to make the star return to the blue.

Initial mass	ΔM_{dJ}	ΔM_{needed}
20	3–4	10
25	8	14

Proposition (see also Vanbeveren, 1991):

the total mass lost during the Red Supergiant Phase of a massive star with initial mass $\leq 40\,M_\odot$ is considerably larger than predicted by the standard single star evolutionary computations,

i.e. the M_{dot} formula of de Jager et al. which is a mathematical formula describing the mass loss rate of a set of stars mainly composed by CHB stars, can not be applied during the RSG phase. Closer inspection of the HR diagram of the WR stars reveals that if we do not want to accept the conclusion that the majority of WR stars are close binary components, even a 20 $M_\odot$ single star (or even lower) must lose sufficient mass during the RSG in order to evolve into a WR star. As an illustration table 3 shows the mass lost by a 20 $M_\odot$ and 25 $M_\odot$ star when the de Jager formula is applied (from Schaller et. al, 1993) and the total mass loss which is needed in order to make the star return to the blue. It is interesting to remark that increasing the mass loss rate during the RSG, thus lowering the theoretically predicted minimum initial mass of single WR stars, will not only solve the HRD problem but it will also increase the predicted WR/O number ratio.

If the HR diagram of WR stars in the LMC is not significantly different from the Galactic one as was suggested by Hamann (these proceedings) the total mass lost during the RSG should only marginally depend on the metallicity, i.e. also in the LMC a 20 $M_\odot$ should evolve into a WR star. Although we will return to SN 1987A later on it is interesting to remark here that stating that the blue early B supergiant progenitor of SN 1987A was a single star with an initial mass of 20 $M_\odot$ is hard to reconcile with the HR position of LMC WR stars (see fig. 3a).

5.7. THE WC/WN NUMBER RATIO.

The WC/WN number ratio difference between the Galaxy and the MC's was discussed by Maeder (1991). Using single star evolution only he concluded that this difference could be explained by adopting a metallicity dependence of the SW mass loss rate during the RSG phase as predicted by the radiatively driven wind theory. Due to the fact that binary evolution was not included I fear that this conclusion is very uncertain. This uncertainty is even larger if the arguements of the previous subsection are considered.

An explanation for the Z–dependence of the WC/WN number ratio was proposed already 13 years ago (Vanbeveren and Conti, 1980) in a binary paper, i.e. the M_{dot} during the WR phase is Z dependent. If M_{dot} during the WR phase is lower in the MC's than in the Galaxy, the total CHeB phase is shorter whereas it takes longer before the nitrogen rich layers are removed from the star and carbon enhanced matter appears at the surface. Both effects decrease the theoretically expected WC/WN number ratio.

6. The Influence of The RLOF on The Distribution of Stars all over The HR Diagram.

About 100000-1 million years after the end of the RLOF in a massive binary, the mass loser explodes. Independent from whether the SN explosion disrupts the system or not, the remaining mass gainer looks like a single star. In star counts such a star will be counted as a normal star however as a consequence of accretion the actual position of the star in the HR diagram may have nothing to do with its position appropriate for the mass of this star at the moment of its formation. This question has been studied in two papers (Vanbeveren, 1988; Meurs and van den Heuvel, 1989). It can be concluded that if the mass transfer is conservative in close binaries with mass ratio larger than say 0.7, then about 40% of the massive stars which are observed as single stars, have not been formed as single stars but were originally the less massive components of close binaries. If this is true we may wonder what is the meaning of actual IMF's determined by means of star counts or by means of luminosity functions.

7. SN 1987A.

Let me first remind you that the blue progenitor of SN 1987A can be explained by single star evolutionary computations of a 20 $M_\odot$ star, but fine tuning of semi-convection is needed (parametrized diffusion ?), little overshooting, little mass loss during the RSG phase in the LMC. HOWEVER as stated already in section 5.6 if the observed location of LMC WR stars in the HR diagram is correct, it is hard to sustain the single star nature of the progenitor of SN 1987A (see fig. 3a).

Alternative binary suggestions:

a. late case B, case C evolution of a binary with mass ratio very close to one (for this model see the paper of P. Podsiadlowski, these proceedings); let us remark however that detailed simultaneous binary evolutionary computations for this model were never performed yet. Especially the behaviour of the gainer in these systems which may be subjected to extremely high mass accretion rates, remains speculation. As was discussed in section 4.2, they may very well become completely homogeneous. Whether in this case a star will explode as a blue supergiant is very unlikely,

b. early case B evolution ($P \leq 50$ days) of a binary with mass ratio very close to one assuming conservative mass transfer (de Loore and Vanbeveren, 1992). Detailed simultaneous evolutionary computations were performed for the systems 15+14.98 $M_\odot$, 12+11.98 $M_\odot$, 9+8.98 $M_\odot$, $P = 8 - 12 - 25$ days.

Also for the latter suggestions it is essential that during the accretion phase the outer layers of the mass gainer remain in radiative equilibrium. The conclusions do not hold any longer if instead the mass gainer would become completely convective during the accretion phase. Our 1992 computations were performed with old opacity tables. For the present review I have repeated the early case B computations but with the new Rogers-Iglesias opacities. Figure 4 shows a typical evolutionary track for the mass loser and for the mass gainer. The evolution of the mass gainer was followed until the end of its own core helium burning phase. As can be noticed the star remains in the blue part of the HR diagram i.e. the conclusions of our 1992 paper do not depend on the opacity tables used in the evolutionary code and thus a blue supergiant progenitor of a type II SN explosion is a natural artifact of the evolution of a massive binary with mass ratio very close to one and conservative RLOF. This conclusion does not depend on the metallicity or on the criterium which is used in order to treat semi–convection.

A criticism which is frequently heared is the exceptional mass ratio needed in order for this model to work and thus the expected low frequency of such events. However let us first remind that there is only one such an event known so that arguing in terms of probabilities is not very meaningfull. Even then, one does not have to ask how many binaries have a mass ratio so close to one but one has to ask 'among all SN explosions resulting from a 20 $M_\odot$ star, how many have had a say 12+11.98 $M_\odot$ history' and here of course enters again the IMF argument. Suppose that only 1 % of all 12 $M_\odot$ stars have a 11.98 $M_\odot$ companion. Using an IMF $\propto M^{-2.3}$ and assumig a conservative RLOF for binaries with mass ratio larger than 0.7 it follows that 6 % of all exploding 20 $M_\odot$ stars will explode in the blue as a consequence of the binary scenario discussed above.

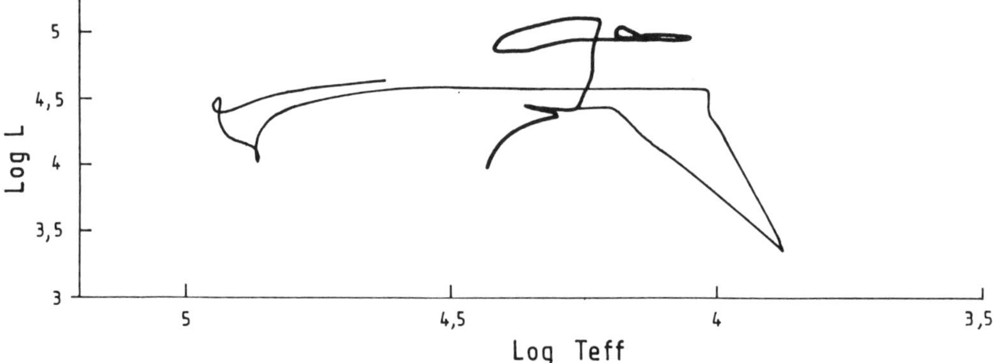

Fig. 4.   Evolutionary track of the primary and the secondary of a 12+11.98 $M_\odot$ close binary.

8. The OBN Stars and Binary Evolution

Although different models for OBN stars have been proposed in the past, I think we may not forget that at least 50 % of them are confirmed binaries. Binaries provide two natural mechanisms in order to form an OBN star. Firstly as a consequence of mass loss by RLOF nitrogen enhanced (and carbon/oxygen depleted) layers appear at the surface of all massive primaries. Secondly when nitrogen enhanced layes appear at the surface of the primary and when mass transfer goes on, these layers can be accreted by the secondary and also the secondary becomes nitrogen enhanced (and carbon/oxygen depleted) although due to the mixing with the unaffected outer layers of the secondary the effect may be less pronounced. Typical values for this enhancement have been computed by Vanbeveren, 1989 (see also Vrancken et al. 1992). In this framework the binary HD 163181 is very interesting: the primary and the secondary are OBN stars although the secondary is only slightly N enhanced (Hutchings, 1975). The primary however is a factor 2 less massive than the secondary but is the most luminous component. The only way to explain this by means of binary evolutionary models is admitting that the primary is a core helium burning star at the end of RLOF whereas the secondary has accreted sufficient nitrogen enhanced matter from the primary. The mass of the primary is 14 $M_\odot$ corresponding to the mass of a WR star. As a consequence this primary will probably evolve into a WR star in the next future.

We can finally remark that since the optical stars of massive X–ray binaries are former accretion stars, they may be nitrogen enhanced, slightly carbon/oxygen depleted. If the helium enhancement in Vela X–1 (see Vanbeveren et. al, these proceeding) is real then with the accretion model that we propose for the optical star of the system we expect a nitrogen enhancement of at least a factor four to five.

9. The OBC Stars and Binary Evolution.

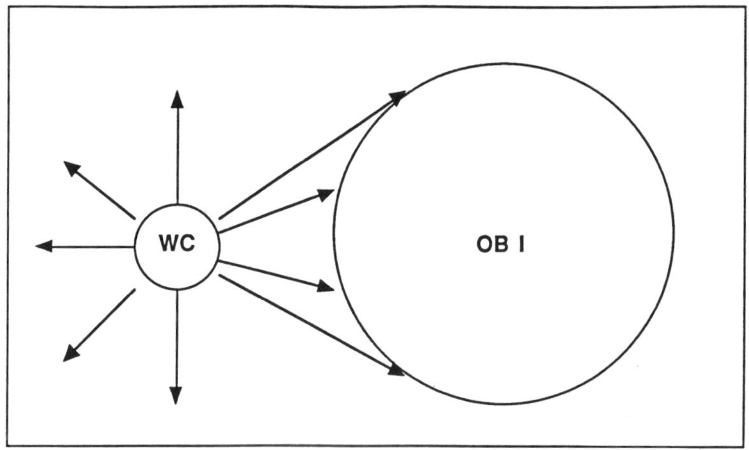

Fig. 5.   Wind accretion in a WC+OB I close binary .

If the OBC phenomenon is a real abundance effect, C being enhanced, binaries may give the solution (and is the only one until now). Suppose we have a WC+OBI close binary (see fig. 5). The WC star is losing mass at high rates in a more or less spherically symmetric way. If the OB companion is a supergiant and if the binary separation is not too large a significant fraction of the stellar wind can be accreted. Since the material coming from the WC star is composed maily of carbon (and oxygen), even a small amount of accreted matter implies a significant carbon enhancement in the outer layers. I have computed several cases (the formulation of the problem and the computations are very straightforward) and even when it is assumed that the carbon enhanced matter coming from the WC star is mixed with the whole unaffected outer layers of the OB supergiant carbon enhancement factors between two and four are not extraordinary. When the WC star ends its live and explodes, a 'single' OBC supergiant will be observed. The O supergiant companion of γ^2Velorum may evolve into such a OC supergiant.

10. Conclusions.

In this review I have summarized the present state of the art of massive close binary evolutionary computations. The assumptions and the physical ingredients have been discussed and from extended computations made the last two decades general conclusions were proposed. The binary computations are compared to single star evolutionary calculations. Then by combining binary and single star evolution I have critically compared the computations and observations with special emphasis on the theoretical predictions of number frequencies of special types of stars in large stellar groups. It was demonstrated that a theoretical statistical study based on evolution of large groups of stars may be quite meaningless if binaries are omitted. Furthermore a warning is appropriate: single star tracks of the Geneva group published before 1992 are wrong as far as the core helium burning phase is concerned, also for the massive star range. The core helium burning timescales may be a factor 2 too large. This means that all studies where these tracks were used have to be reinvestigated in order to see whether conclusions are critically affected. This should certainly be the case for all the WR number statistics. Some of them were repeated for the present review using the new Geneva tracks and the differences with earlier studies are quite large.

An inspection of massive star literature the last two decades revealed quite a strange scientific behaviour as far as a correct use of binary/single star evolutionary results is concerned; it occurs to me as if carefull reading of papers is out of fashion. A literature search usually does not go further than say 5 years and even this is an optimistic number (referees are not completely innocent here). In this way it happened that people reinvented results which were published some 10 years ago without referencing the original papers; these original papers very often are binary papers since many very interesting close binary studies were published at least one decade ago. I therefore want to end this review with two advices:

a. although the word BINARY may appear in the titel of a massive star paper or although the authors of a massive star paper may be known as BINARY SCIENTISTS, it may still be interesting also for massive single star scientists to read these papers carefully and then to use references properly,

b. do not use the word binary only there where the most complicated and fined tunned single star scenario fails.

References

Benson, R.S.: 1970a, *Ph. D. Thesis*, Univ. of California: Berkeley.
Benson, R.S.: 1970b, *Bull. Am. Astron. Soc.* **2**, 295.
Bodenheimer, P. and Taam, R.E.: 1984, *ApJ.* **280**, 771.
Conti, P.S. and Massey, P.: 1989, *ApJ.* **337**, 251.

Conti, P.S., Garmany, C.D., de Loore, C. and Vanbeveren, D.: 1983, *ApJ.* **274**, 302.

De Greve, J.P.: 1993, *Astron. Astrophys. Suppl.* **97**, 527.

De Greve, J.P. and de Loore, C.: 1992, *Astron. Astrophys. Suppl.* **96**, 653.

de Jager, C., Nieuwenhuyzen, H. and van der Hucht, K.A.: 1988, *Astron. Astrophys. Suppl.* **72**, 259

de Loore, C. and De Greve, J.P.: 1992, *Astron. Astrophys. Suppl.* **94**, 453.

de Loore, C. and Vanbeveren, D: 1992, *Astron. Astrophys.* **260**, 273.

de Loore, C. and Vanbeveren, D: 1993, *Astron. Astrophys. Suppl.* (in press),

de Loore, C. and Vanbeveren, D: 1994a, *Astron. Astrophys. Suppl.* (in press),

de Loore, C. and Vanbeveren, D: 1994b, *Astron. Astrophys.* (in press).,

de Loore, C.: 1980, *Space Sci. Rev.* **26**, 113.

Firmani, C., Koenigsberger, G., Bisiacchi, G.F., Moffat, A.F.J. and Isserstedt, J.: 1980, *ApJ.* **239**, 607.

Garmany, C.D., Conti, P.S. and Massey, P.: 1980, *ApJ.* **242**, 1063.

Harrison, P.A., Lyne, A.G. and Anderson, B.: 1992, '' in E.P.J. van den Heuvel and S.A. Rappaport, ed(s)., *X–ray binaries and recycled pulsars*, Kluwer Acad. Publ.: Dordrecht, 155.

Hellings, P. and Vanbeveren, D.: 1981, *Astron. Astrophys.* **95**, 14.

Howart,I.H. and Prinja, R.K.: 1989, *Astrophys. J. Suppl.* **69**, 527.

Hutchings, J.B.: 1975, *ApJ.* **200**, 122.

Kippenhahn, R., Ruschenplatt, G. and Thomas, H.C.: 1980, *Astron. Astrophys.* **91**, 175.

Kuhi, L.: 1973, *Wolf-Rayet and High Temperature stars*, M.K.V. Bappu and J. Sahade (eds): Reidel: Dordrecht, 205.

Maeder, A. and Meynet, G.: 1989, *Astron. Astrophys.* **210**, 155.

Maeder, A.: 1991, *Astron. Astrophys.* **242**, 93.

Meurs, E.J.A. and van den Heuvel, E.P.J.: 1989, *Astron. Astrophys.* **226**, 88.

Packet, W.: 1988, *Ph. D. Thesis*, Vrije Universiteit Brussel: Brussels.

Paczynski, B. : 1971, *Ann. Rev. Astron. Astrophys* **9**, 183.

Rogers, F. J. and Iglesias, C.A.: 1992, *ApJ* **79**, 507.

Schaller, G., Schaerer, D., Meynet, G. and Maeder, A.: 1993, *Astron. Astrophys.* **96**, 269

Thomas, H.C.: 1977, *Ann. Rev. Astron. Astrophys* **15**, 127.

Ulrich, R.K.: 1972, *ApJ.* **172**, 165.

van den Heuvel, E.P.J.: 1983, *Accretion Driven Stellar X-Ray Sources*, W.H.G. Lewin and E.P.J. van den Heuvel (eds): Cambridge Univ. Press : Cambridge

van den Heuvel, E.P.J. and Heise, J.G.: 1972, *Nature. Phys. Sci* **239**, 67.

van den Heuvel, E.P.J.: 1976, *I.A.U. Symp.* **73**, Reidel: Dordrecht

van den Heuvel, E.P.J.: 1978, *Physics and Astrophysics of Neutron Stars and Black Holes*, R. Giannoni and R. Ruffini (eds), North-Holland Publ. Co.: Amsterdam.

Vanbeveren, D. and Conti, P.S.: 1980, *Astron. Astrophys.* **88**, 230.

Vanbeveren, D. and Doom, C.: 1980, *Astron. Astrophys.* **87**, 77.

Vanbeveren, D. and Packet, W.: 1979, *Astron. Astrophys.* **80**, 242.

Vanbeveren, D., DeGreve, J.P., Van Dessel, E.L. and de Loore, C.: 1979, *Astron. Astrophys.* **73**,, 219.

Vanbeveren, D., Van Rensbergen W. and de Loore, C.: 1982, *Astron. Astrophys.* **115**, 69.

Vanbeveren, D. and de Loore, C.: 1994, *Astron. Astrophys.* (submitted),

Vanbeveren, D.: 1987, *Astron. Astrophys.* **182**, 207.

Vanbeveren, D.: 1988a, *Astron. Astrophys.* **189**, 109.

Vanbeveren, D.: 1988b, *Astrophys. Space Sci.* **149**, 1.

Vanbeveren, D.: 1989, *Astron. Astrophys.* **224**, 93.

Vanbeveren, D.: 1991, *Astron. Astrophys.* **252**, 159.

Vanbeveren, D.: 1991, *Space Sci. Rev.* **56**, 249.

Vrancken, M, De Greve, J.P., Yungelson, L. and Tutukov, A.: 1991, *Astron. Astrophys.* **249**, 411.

MASSIVE SINGLE STAR EVOLUTION: A COMPARISON WITH OBSERVATIONS

A. MAEDER

Geneva Observatory, CH-1290 Sauverny, Switzerland

Abstract. We firstly examine the critical model assumptions for massive stars, in particular regarding mixing, mass loss and metallicity. The comparisons of models and observations for main sequence stars reveal some interesting problems, such as the lack of O-stars close to the zero-age sequence, the so-called helium and mass discrepancies. We emphasize that this last discrepancy was probably due to the unsafe atmosphere modelling used by spectroscopists. The comparisons for supergiants enlighten a number of most interesting problems: the He and CNO abundances in blue supergiants, the distribution of supergiants in the HR diagram and above all the variations of the blue to red number ratios with metallicity. Then, we examine the properties and chemistry of WR stars and the observations and interpretations concerning the great changes of WR numbers in galaxies of different metallicities. Finally, we emphasize the main WR filiations.

Key words: Stars: evolution of – supergiant – Wolf-Rayet – Galaxies: stellar content of

1. Introduction

Massive stars are at the crossroads of many most important astrophysical problems. Due to their high luminosities, massive stars are visible at large distances in the Galaxy and also in the Universe thanks to their distinct features in the integrated spectrum of galaxies. They offer thus a powerful tool for studying stellar populations and star formation in highly redshifted galaxies. Also, they are essential for our understanding of nucleosynthesis and of the precursors of supernovae.

In this review, we examine, in full agreement with the title of this meeting, the most critical model assumptions and the way the models compare with the observations. This leads us to notice the many points of agreement, but also to enlighten the points of disagreement which need further development and represent the germs of future progresses in our understanding of massive star evolution.

2. Short Review of Some Critical Model Assumptions for Massive Stars

Some basic physical model ingredients, like opacities and nuclear reactions rates, still show some embarassing uncertainties. Particularly, the molecular opacities for red supergiants are insufficient, and the uncertainties on some nuclear rates such as $^{12}C(\alpha, \gamma)^{16}O$ remain rather large. The mass loss rates are of overwhelming importance in massive star evolution, since all model outputs are heavily influenced by mass loss. The rates generally used are those by de Jager et al. (1988); they are probably not too uncertain close to the main sequence, but for

Space Science Reviews 66: 349–363, 1994.

© 1994 *Kluwer Academic Publishers. Printed in Belgium.*

the supergiants of all colours the uncertainties can be quite large, in particular for the red supergiants.

The treatment of convection and mixing in stellar interiors is still a major problem for massive stars, more especially as supergiants are often so close to a neutral state between a blue and a red location, that any changes in the inner structure of the models may have produced a shift from blue to red and vice versa. We may identify the following different assumptions regarding convection and mixing in massive star models:
- Schwarzschild's criterion
- Schwarzschild's criterion and core overshooting
- Overshooting below the convective envelope
- Ledoux criterion
- Semiconvection or semiconvective diffusion
- Turbulent diffusion or other forms of rotational mixing

All these kinds of models are claimed by their authors to well fit the observations and the debate has been very vivid over recent years (Chiosi and Maeder 1986; Maeder and Meynet 1989; Brocato et al. 1990; Stothers 1991ab; Stothers and Chin 1990,1991, 1992ab). There is at present no definite theoretical or observational proof in favour of any model. However a few useful indications on the limits and possibilities of the various models must be mentioned.

Overshooting has been very much discussed in recent years and many claims have been made in favour of a large overshooting from convective cores with respect to what is predicted by Schwarzschild's criterion. However it seems now clear that the overshooting distance is limited to about (0.2–0.4) Hp, where Hp is the pressure scale height (Maeder and Meynet 1989; Stothers and Chin 1991; Napiwotzki et al. 1993; Meynet et al. 1993b). Some recent comparisons with observations favour Ledoux' rather than Schwarz schild's criterion for convection (Stothers and Chin 1992ab). The result may depend on the adopted mass loss rates, also recent theoretical works (Grossman et al. 1993) show that Ledoux' criterion has no bearing at all in a stratified stellar medium. Thus, both at the theoretical and observational levels, the situation still remains uncertain about convective criteria.

It is not always so clear whether the various authors speaking about semiconvection consider exactly the same physical process. Semiconvection occurs in zones which are convectively unstable according to Schwarzschild's criterion, but would not be so according to Ledoux'. Semiconvection may produce some mixing in zones with a gradient of the mean molecular weight. Various treatments of the problem have been made (Chiosi and Maeder 1986; Langer et al. 1989; Arnett 1991; Chiosi et al. 1992). Langer et al. (1985) propose a diffusion treatment which is equivalent to Schwarzschild's criterion when the mixing timescale is short compared to the evolutionary timescale and to Ledoux' criterion in the opposite case. These models have a great interest regarding the discussion about the blue progenitor of SN 1987 A (Langer 1991ac) as well as about the evolutionary status of blue supergiants.

Rotational mixing may have large consequences for massive star evolution. The basic reason is that the radiative viscosity is so large that dissipative processes (which are related to radiative viscosity) may have a timescale comparable to the evolutionary timescales of massive stars (Maeder 1987b). Mixing can produce homogeneous or nearly homogeneous evolution on the main sequence and thus directly lead to the formation of WR stars. As emphasized by Langer (1993), rotational mixing may also solve several problems: the existence of the WN+WC stars (Langer 1991b), the origin of N enhancement in OB supergiants and the alleged mass discrepancy for OB main sequence stars, and the blue progenitor of SN 1987 A. Interestingly enough, the claims in favour of semiconvection mean less mixing in convective zones with varying mean molecular weight, while the claims in favour of turbulent mixing mean more mixing in radiative zones. Whether both these claims are true is a major question. In this context, one must notice that both semiconvection and rotational mixing depend mainly on the radiative diffusivity. Thus, rather than advocating two different processes, one for explaining SN 1987 A and another one for the properties of OB stars, it seems advisable to examine whether a single physical process can account for the various constraints. It is likely that in future, rotation will be an unavoidable property in star models. The same is probably true for binarity, which can induce tidal mixing and mass transfer.

How does massive star evolution depend on metallicity Z is a major question. Indeed, Z can enter evolution through four possible doors. 1) The nuclear production: metallicity may influence the nuclear rates, a good example is the CNO cycle. In general, a very slight contraction or expansion to a new equilibrium state may strongly compensate the changed rate. 2) Opacity effects: in the interiors of massive stars, electron scattering is the main opacity source, which is independent on metallicity. Thus, contrarily to the case of low and intermediate mass stars, metallicity through opacity has no great direct effect on the inner structure of massive stars. 3) Stellar winds: only in the very external layers, metallicity Z may strongly influence the opacity and thus the atmospheres and winds. Wind models for O–stars by Abbott (1982) suggested a Z–dependence of the mass loss rates $\dot{M}$ of the form $\dot{M} \sim Z^{\alpha}$, with $\alpha = 1.0$. Models by Kudritzki et al. (1987) and by Leitherer and Langer (1991) gave a value α between 0.5 and 0.7. It is likely that this is the main effect by which metallicity may influence massive star evolution (Maeder, 1991). For yellow and red supergiants, there is however no model giving reliable enough mass loss rates and $\dot{M}$ vs. Z relation. This is a major uncertainty in post–MS evolution. 4) There is a ratio $\Delta Y / \Delta Z$ between the relative enrichments in helium and heavy elements of the order of 5 as established from low Z HII regions (Pagel et al. 1992). Thus, changes in Z imply large changes in Y, which have a direct effect on the models.

Binary mass transfer by Roche lobe overflow (RLOF), which is an extreme case of tidal interaction, drastically modifies the course of stellar evolution. It may thus significantly influence the comparison between observations and models (Vanbeveren 1991; Vanbeveren and de Loore 1993; see also Vanbeveren in this meeting). Estimating the exact importance of that effect is of major importance for most comparisons between models and observations. The fraction of all stars (single and binaries) undergoing RLOF is estimated to be 20 to 40% (Podladiowski et al. 1992). In section 6 below, we also discuss the value of this fraction. We may remark that this percentage does not account for the fact that some close binaries could be mixed by tidal interactions and would thus evolve homogeneously, without large increase of their radius and thus without RLOF. Indeed, the true importance of RLOF in many problems is still uncertain and complete simulations of star populations with binaries remain to be done.

3. Comparisons between Models and Observations on the Main Sequence

Several grids of models taking into account some of these effects have been made over recent years (Brunish and Truran 1982ab; Maeder 1990; Schaller et al. 1992; Alongi et al. 1993; Schaerer et al. 1993; Charbonnel et al. 1993; Meynet et al. 1993a; de Loore et Vanbeveren 1993). A general comparison of cluster sequences and evolutionary tracks has been made (Meynet et al. 1993b), which shows a very good general agreement. However, in the case of massive stars some specific problems are known to exist.

There is an apparent lack of O–stars close to the zero–age sequence (Garmany et al 1982). The point is also quite clear from recent gravity and T_{eff} determinations by Herrero et al. (1992). Chiosi et al. (1992) suggest that this is due to the fact that about 20% of the O–stars are still embedded in molecular clouds. Alternatively there may be no true ZAMS for O–stars, because nuclear reactions ignite early during the contraction phase (Appenzeller 1980) and may thus make stars inhomogeneous before the end of the contraction phase. Also for massive stars, the accretion timescale of the protostellar cloud is longer than the Kelvin–Helmoltz timescale. The consequence is that no pre–MS star should be visible (Palla et al. 1993), a fact which could contribute to hidden stars close to the ZAMS.

Another possible problem is the so–called mass discrepancy for O–stars. Spectroscopic masses derived from gravity and terminal velocity determinations were claimed to be smaller than those predicted by stellar models (Herrero et al. 1992 and ref. therein; Kudritzki et al. 1992). In other words, spectrocopy suggests that O–stars are overluminous for their masses and the discrepancy amounts to about 50%. An interpretation by Langer (1993) is that the overluminosity of O–stars is the sign of some rotational or tidal mixing enlarging the helium core. The reality of the mass discrepancy has been questioned recently by Lamers and Leitherer (1993). They show that there exist large discrepancies between theoretical and observed mass loss rates, the same being true for the terminal velocities; they also show that the discrepancies cannot be solved by adopting smaller masses for

O–stars. According to Schaerer and Schmutz (1993), the use of plane–parallel models for O–stars may lead to significant errors for spectroscopic gravities, masses and He abundances. It is likely that the claims about the mass discrepancy are due to previous unsafe modelling of stellar atmospheres or to mixing in the interiors. A recent study by Pauldrach (this meeting) tends to support this view.

The surface abundances in He and CNO elements offer a powerful test of stellar evolution. Evidences of CN processing are provided by He and N enhancements together with C depletion, while O–depletion only occurs for advanced stages of processing. The abundances may cover the range from solar values (C/N = 4, O/N = 10) to CNO equilibrium values in the extreme case which is reached in WN stars (C/N = 0.02, O/N = 0.1; Maeder 1983, 1987a). Models with mass loss but no extra–mixing predict He and N–enrichment in MS stars only for initial masses larger than about 50 $M_\odot$ depending on the mass loss rates. Models with rotational mixing may lead to an early appearance of the products of the CNO cycle (cf. Maeder 1987b). The observations of 25 OB stars by Herrero et al. (1992) show that most MS stars have normal abundances. Fast rotators are an exception and they present He end N enhancements. Among exceptions, the group of ON stars, i.e. O–stars with N–enrichments (Walborn 1976, 1988) contains at least 50% of short–period binaries (Bolton and Rogers 1978). An analysis of the association Per OB1 suggests (Maeder 1987b) that there is a bifurcation in stellar evolution: while most stars follow the tracks of inhomogeneous evolution, a fraction of about 15%, mainly composed of fast rotators and binaries, may evolve homogeneously and become ON blue stragglers. Herrero et al. (1992) spoke about an He discrepancy in connexion with the alleged mass discrepancy of O–stars. Clearly there is not so much of an He discrepancy on the main sequence, except for the case of fast rotators or close binaries. The He discrepancy however seems to be generally present in supergiants.

4. Comparisons Between Models and Observations for Supergiants

Many problems and controversies remain about supergiants. These stars are often close to a neutral state between a blue and a red location in the HR diagram (Tuchmann and Wheeler 1989, 1990). There, even minor changes in convection and mixing processes may greatly affect the evolution. Let us examine some critical problems.

4.1. HE AND CNO ABUNDANCES IN BLUE SUPERGIANTS

A lot of new observations have been made in recent years, which well support Walborn's hypothesis (1976, 1988) that ordinary OB supergiants are enriched in helium and nitrogen and depleted in carbon, as a result of CNO processed elements present at the stellar surface. According to Walborn, the particular case may be the small group of OBC supergiants, which just have cosmic abundances. This hypothesis is supported in particular by the results of Herrero et al. (1992) who showed that supergiants of type Ia, Iab and Of stars present He–enhancements. Like for all rules, there are exceptions: some B–supergiants do not exhibit He and N excesses. These authors also show that fast rotators of all luminosities present evidences of CNO processing. Similar enhancements of nitrogen and helium abundances have also been found for several post–MS B–type stars by Gies and Lambert (1992). Abundances determinations have also been made for B–supergiants in the LMC and SMC, which is particularly interesting in relation with studies on the status of the progenitor of SN 1987 A (Kudritzki 1990; Lennon et al. 1991). From these results it is clear that B–type supergiants in the Galaxy, the LMC and the SMC generally show evidences of CNO processing.

The above observations place severe constraints on stellar models. Most of them do not predict He and N–enrichments in blue supergiants at solar composition. This may be called an He discrepancy (Herrero et al. 1992). Blue loops, with the associated He and N–enrichments (as a result of dredge–up in red supergiants) only occur for $M < 20\ M_\odot$. This is the case for models with Schwarzschild's criterion and overshooting (Schaller et al. 1992), and with Ledoux criterion (Stothers and Chin 1992ab; Brocato and Castellani 1993). Models with semiconvection (Arnett 1991) have the same difficulty: at solar composition, the evolution goes straight to the red supergiants and there are no enriched blue supergiants. At lower metallicity like that of the LMC, most of the models have some blue loops and are thus in better position to explain the observed enrichments. Only complete comparisons in clusters at various Z will show the exact points of agreement or disagreement.

4.2. DISTRIBUTIONS OF THE SUPERGIANTS IN THE HR DIAGRAM

There seem to be relatively more stars out of the formal MS band than is predicted (see Meylan and Maeder 1982). The problem appears particularly serious at low metallicities as in the SMC and LMC. Instead of 10% of stars out of the MS, this fraction seems to be about 40%. In the Milky Way, excesses of A–type supergiants have also been suggested (Stothers and Chin 1977; Chiosi et al. 1978). The observed and theoretical numbers could be brought into an agreement if the MS would also include the B and A–type supergiants. This discrepancy is related to the so–called blue Hertzsprung gap (BHG), which is predicted by most models to occur at the end of the MS and which is not observed; instead, the true star distribution appears to be continuous from the MS to the A–type supergiants (Nasi and Forieri 1990; Fitzpatrick and Garmany 1990; Chiosi et al. 1992).

Various explanations have been proposed for the lack of BHG. Opacity effects may produce a paunch on the MS as shown above, but with present opacities (Iglesias et al. 1992) and mass loss rates, the paunch is located at too high luminosities to account for the observations (Schaller et al. 1992). Extended atmospheres and/or binaries (Tuchmann and Wheeler 1989, 1990) have been advocated. Mixing was shown to reduce but not to suppress the gap (Langer

1991c). The temperature scale may also be a problem. Indeed, a gap between T_{eff} = 35'000 K and 20'000 K corresponds only to a difference of 0.04 in (B–V) colour, which is quite small and may be blurred by several effects. Only the adjustment of individual isochrones on star clusters, together with a mapping of He and CNO abundances, will properly inform us about the reality of the gap problem and on the exact status of blue supergiants.

We also have to consider the red side of the distribution of the blue and yellow supergiants in the HR diagram. A net decrease appears in the distribution of LMC supergiants in the HR diagram to the left of an oblique line between log T_{eff} = 4.2 and 3.9 (Fitzpatrick and Garmany 1990). This feature is called the ledge. Two kinds of models are able to produce a high number of blue supergiants and give rise to a ledge: a) models with low mass loss, b) models with blue loops.

Clearly, models with low mass loss (Brunish and Truran 1982ab; Schaller et al. 1992) predict that most of the He–phase is spent in the blue supergiant phase directly after the MS. This may give a ledge. However such models do not make red supergiants, which is in disagreement with observations in the LMC and SMC. Models with blue loops also enhance the number of blue supergiants, and are simultaneously able to account for the He and N–enrichments in blue supergiants. This is the case of most models with overshooting, Ledoux criterion or semiconvection and masses smaller than about 20 $M_{\odot}$. For example, models with overshooting at $Z \leq 0.008$ have well developed blue phases at all masses (Schaerer et al. 1993). As shown by Meynet (1993), the number ratio B/R of blue to red supergiants for young clusters in the and Galaxy are well accounted for by these models. It is however not certain wether the loops are correctly predicted for stars with initial masses larger than 20 $M_{\odot}$.

4.3. THE STATISTICS OF BLUE AND RED SUPERGIANTS AND ITS VARIATION WITH METALLICITY

The number ratio B/R of blue to red supergiants was among the first stellar properties shown to vary through the Galaxy and to exhibit differences between the Galaxy and the Magellanic Clouds (van den Bergh 1968; Meylan and Maeder 1983; Humphreys and McElroy 1984; Brunish et al. 1986). The main trend is that B/R increases steeply with metallicity: for M_{bol} between –7.5 and –8.5, B/R is up to 40 or more in inner galactic regions and only about 4 in the SMC (Humphreys and McElroy 1984). A difference by an order of magnitude between the Galaxy and the SMC was also found on the basis of well selected clusters (Meylan and Maeder 1982).

The various star models are generally able to account for the occurence of blue supergiants, with B/R ratios in relatively good agreement with the observations (Brunish et al. 1986). One of the reasons is of course the flexibility offered by the uncertain mass loss rates. Indeed, B/R may change from infinity in case of no mass loss to about 0 for very high mass loss rates. Thus the real difficulty is not to account for an average observed B/R ratio, but to account also for the change of

B/R with metallicity. The models with Schwarzschild's criterion and overshooting (Schaller et al. 1992; Alongi et al. 1993), the models with Ledoux criterion (Brocato and Castellani 1993) and the models with semiconvection (Arnett 1991), even if they are able to fit the average B/R ratio in the LMC or in the Galaxy, appear all to predict higher B/R ratios at lower Z in contradiction with the observations. This is a major problem, which is not solved in any grids of models. We do not yet know what models with turbulent diffusion or rotational mixing are predicting at various Z, since grids of such models are not yet available.

5. Properties and Chemistry of Wolf–Rayet Stars

WR stars are nowadays considered as bare cores resulting mainly from the peeling–off by stellar winds of stars initially more massive than about 25 to 40 $M_\odot$. The main arguments are (Abbott and Conti 1987; Conti 1988; Lamers et al. 1991):

1. The low H/He ratio in WR stars, rarely larger than 3 in late WN stars and mostly 0 in other subtypes.
2. The CNO ratios show equilibrium values in WN stars.
3. The continuity of the abundances in the sequence of types

    O, Of, WNL, WNE, WCL, WCE and WO

    well corresponds to a progression in the peeling off the stars.
4. The observed mass loss rates in O–stars and in supergiants are high enough to remove the stellar envelopes during the stellar lifetimes, as well as the average mass loss rates in WR stars are sufficient to accomplish the further peeling off.
5. The low average masses between 5 and 10 $M_\odot$ of WR stars and their fitting to a mass–luminosity relation (Smith and Maeder 1991).
6. The presence of WR stars in young clusters and associations with ages smaller than $6 \cdot 10^6$ yr (Humphreys and McElroy, 1984).
7. The existence of transition objects Of/WN between Of and WR stars and between LBV and WN stars.
8. The presence of He and N–rich shells around some WR stars (Esteban and Vilchez 1992).
9. The WR/O and WN/WC number ratios are consistent with theoretical expectations and with their predicted variations according to the metallicity of galaxies (Maeder 1991).
10. The consistency of the T_{eff} of WR stars with that of He–rich bare cores when optically thick winds are accounted for (cf. Schaller et al. 1992; Schaerer and Maeder 1992).

There are two main groups of Wolf–Rayet stars, those of types WN and those of types WC. WN stars show the products of the CNO cycle. The late WN of types WN6–WN9, noted shortly WNL, generally still contain hydrogen with H/He ratios between 0 and 3 (Conti et al. 1983; Hamann et al. 1993; Willis 1991).

The early WN stars of types WN2–WN6, noted WNE, generally show no hydrogen in their spectra, but there are a few cases with H/He ratios between 0 and 0.5. A correlation of the H–content with the T_{eff} of WN stars, rather than with their subtypes, has been found by Hamann et al. (1993): the coolest WN stars show hydrogen and the hottest ones have none. One may remark however that the coolest group is mainly formed by WNL stars while the hottest group consists of WNE stars. Thus, both pictures are essentially consistent. We may conjecture that the presence or absence of hydrogen is a key physical factor strongly influencing the structure of the outer layers. The above observed abundance ratios well correspond to the equilibrium values of the CNO cycle (cf. Maeder 1983, 1987a), i.e. the C/N and O/N ratios are two orders of a magnitude smaller with respect to solar abundances. Interestingly enough, such abundances are rather independent on the various model assumptions and they mainly depend on the nuclear cross–sections. The good agreement with model predictions is very important; it indicates the general correctness of our understanding of the CNO cycle and of the relevant nuclear data.

WC stars contain no hydrogen and are mainly He, C and O–stars (Smith and Hummer 1988; Willis 1991; Nugis 1991). A most interesting finding is that by Smith and Hummer, who showed that the C/He ratio is increasing for earlier WC subtypes. Smith and Maeder (1991) emphasized that the (C+O)/He ratio is to be preferred to the C/He ratio which goes up and down during the He processing and to the O/C ratio which may give confusing results in galaxies of different metallicities. They propose the following calibration (in number ratios):

	(C+O)/He
WC9	0.03–0.06
WC8	0.1
WC7	0.2
WC6	0.3
WC5	0.55
WC4	0.7-1.0
WO	> 1

The sequence WC9 to WO appears to be a progression in the exposition of the products of He–burning and the rare WO types introduced by Barlow and Hummer (1982) just appear as the most extreme type in this sequence. Comparisons of observations and model predictions show a generally good agreement (Willis 1991, 1993). The above connection between WC subtypes and the (C+O)/He ratios is the key for understanding the differences in the distribution of WC stars in galaxies of different metallicities.

Evolutionary models also predict M–L–$\dot{M}$–R–T_{eff} relations for WR stars without hydrogen (WNE,WC). Such He–C–O cores have a rather simple internal structure with little composition difference between the center and the surface, so that the above relations are mostly independent on the way the WR stars have been formed. Models (Maeder 1991; Schaerer and Maeder 1992) show that the differences due to the metallicity of the parent galaxy concern mainly the lifetimes and the domain of the M–L–$\dot{M}$–R–T_{eff} relations occupied by WR stars. The basic connection is the mass–luminosity relation (Vanbeveren & Packet 1979; Maeder 1983; Langer 1989a; Schaerer and Maeder 1992):

$$\log L/L_{\odot} = 3.03 + 2.695 \log M/M_{\odot} - 0.461 \, (\log M/M_{\odot})^2$$

For $M > 10 \, M_{\odot}$, a linear relation may well fit. On the observational side, a mass–luminosity relation has been demonstrated (Smith and Maeder 1989; Smith et al. 1993); it implies a constant bolometric correction of –4.5.

6. The Relative Number Frequencies of WR Stars in Galaxies

WR stars are observed in several nearby galaxies, which gives statistical data on their relative frequencies at various metallicities. Data for nearby galaxies have been collected by Smith (1988) and Maeder (1991), who showed a strong increase of the number ratios WR/O and WC/WN with the metallicity of the parent galaxies.

The origin of the observed variations of the relative number of WR stars in the Galaxy, the LMC and SMC was a very debated subject. Originally, these variations were attributed to metallicity by Smith (1973) and Maeder et al. (1980), who suggested that high Z favours mass loss, which in turn favours the formation of WR stars. The dependence on metallicity was criticized by several authors, who attributed the differences in the WR populations mainly to changes in the IMF and SFR (Bertelli and Chiosi 1982; Garmany et al. 1982; Massey and Armandroff 1991). However, these claims are unlikely because no systematic difference of the IMF slope has been found between the Galaxy, the LMC and SMC (Humphreys and MacElroy 1984; Mateo 1988; Massey et al. 1989; Parker et al. 1992). Also, the galactic gradient of the WR number is much steeper than that of O–stars (Meylan and Maeder 1982; van der Hucht et al. 1988). Thus, IMF and SFR differences do not seem to be the key factor here, while it may be very important in starburst regions. Thirdly, a problem of the explanation based on metallicity was that regions of similar Z, as the LMC and the outer galactic regions, have differences in their WR populations (Massey and Armandroff 1991). The explanation is likely that in a given ring in the Galaxy there is a scatter in Z and that the average WR population is heavily weighted towards properties typical of higher Z (Smith and Maeder 1991).

In order to clarify the debate, let us make the following considerations. WR properties and statistics depend on many parameters: metallicity Z, star formation rate SFR, initial mass function IMF, age in bursts, duration of the bursts, binary frequency, etc. Thus, it is essential to distinguish: a) Regions in galaxies where the assumption of an average constant star formation rate over, say, the last $2 \cdot 10^7$ yr is valid, b) Regions or galaxies where a recent burst has occured so that the assumption of a constant SFR does not apply. The first case concerns for example large rings in the Milky Way and galaxies, where the SFR and ages play no role. There the metallicity effects on stellar evolution seem to be a dominant factor. This case is examined in this review, its proper understanding is a prerequisite for the studies of the second group. Indeed, case b) involves all possible parameters and concerns HII regions, blue compact galaxies, WR galaxies and starbursts. Some model properties for this case b) are discussed by Meynet (this meeting).

The WR/O, WC/WR (and WC/WN) ratios strongly increase with the metallicity of the parent galaxy (cf. Maeder et al. 1980; Smith 1988,1991). The general trend of growing WR/O ratios is also confirmed by the studies of the integrated properties of HII galaxies (Arnault et al. 1989; Conti 1991; Smith 1991; Vacca and Conti 1992; Mas Hesse and Kunth 1991; Mas Hesse 1992). In stellar models a growth of these ratios with Z is predicted (Maeder 1991; Maeder and Meynet 1994). The growth of WR/O results from the lowering of the minimum initial mass for forming WR stars and from the increase of the lifetimes with increasing Z (and mass loss). With Salpeter's IMF and standard mass loss rates, the models predict a WR/O ratio equal to 0.08 at $Z = 0.04$ and 0.28 for mass loss rates twice as large. This illustrates well the great influence of the mass loss rates on the values of the number ratios. At $Z = 0.002$, the WR/O ratios are in both cases 0.005, which is negligible. This is in agreement with the low fraction of WR stars in metal deficient galaxies. Also the study of the integrated spectrum and HeII 4686 feature in dwarf galaxies shows a total absence of WR contribution at Z lower than about 0.002 (Arnault et al. 1989; Smith 1991). The models show growth of WC/WN which results from the higher mass loss rates which lead to an earlier visibility of the products of He–burning. Interestingly enough, for increased mass loss rates, the WC/WN ratios go down again for $Z \geq 0.02$ instead of further increasing as expected. This results from the fact that the WN phase of the most massive stars may already be entered during the main sequence phase and is therefore much longer. Whether this is the right explanation for the apparently larger number of WN stars in starbursts (Vacca and Conti 1992) is still unknown.

Indeed, close comparisons between models and observations must account for the various channels of WR formation and in particular the binary RLOF, which seems to improve the comparisons (Vanbeveren 1991; Vanbeveren and de Loore 1993). These analyses, together with the new ones we have performed (Maeder and Meynet 1994), show that the best fit of the observed WR/O and WC/WN ratios is obtained with models having high mass loss rates, but that it is also necessary to assume that a certain fraction of the O–stars becomes WR stars as a result of binary mass transfer. This fraction was estimated to be 35% by Vanbeveren and de Loore (1993), which is in rather good agreement with the results by Podsiadlowski et al. (1992). The new comparisons we make also show that a better agreement is achieved if one takes into account the possibility that some WR stars result from RLOF in binaries. These new comparisons indicate that the fraction of WR stars which truly owe their existence to RLOF is highly variable with the metallicity of the galaxy, being nearly 100% at low Z like in the SMC and going down to values as low as about 10% in the inner regions of the Milky Way. It is important to emphasize that these fractions are not necessarily identical to the fractions of WR binaries.

7. The Distribution of WC Subtypes

The distributions of WC and WO stars in galaxies present a number of very distinct properties:

–1. It is well known that WC stars are more numerous in inner galactic regions (cf. van der Hucht 1988). More specifically, the later a WC subclass is, the inner is the limit in galactocentric radius beyond which no star of that subclass is found, (Smith and Maeder, 1991). Indeed, WC9 and WC8 stars are only found in inner galactic regions with higher Z. For metallicity Z as in the LMC, WC stars are mostly of types WC6–WC4. In dwarf galaxies, only WC4 and WO stars are found. Also, the above authors emphasize that at low Z values (lower than for the LMC) the majority of stars in the WC–WO group are WO stars.

–2. In the Milky Way, the luminosity of earlier WC subtypes is lower than for later WC subtypes (Lundstrom and Stenholm 1984; van der Hucht et al. 1988; Conti 1988).

–3. Stars of a given WC subtype seem brigther at lower Z as suggested by Smith and Maeder (1991), who point out that surprisingly LMC WC4 stars are even brighter than galactic WC5–WC6 stars.

The above facts are well explained on the basis of the relation between WC subtypes and the (C+O)/He ratio as shown by Smith and Maeder (1991). The entry points and lifetimes in the WC9 to WO sequence are highly dependent on mass and Z. At high Z, due to high mass loss the WC stage is entered early during the He–burning phase, so that the surface (C+O)/He ratio is low and thus the WC type is generally later than at lower Z. As evolution goes on, mass and luminosity decline and the ratio (C+O)/He decreases, and thus the sequence of earlier WC types is described. The entry point in the above sequence occurs at lower L and thus earlier WC types for lower stellar masses.

For lower metallicities, the entry in the WC phase occurs with higher (C+O)/He ratios, i.e. earlier WC types, and the evolutionary sequence up to type WO is shorter. This behaviour, illustrated by Figures 10 to 13 in Maeder (1991), explains the above properties (points 1 to 3). We emphasize that the fact that the luminosity for a WC subtype depends on the initial Z may have some consequence for the interpretation of the WC lines in the integrated spectra of galaxies, a fact which should be accounted for in populations synthesis with WR stars.

¿From the observations of WR stars in clusters and in galaxies, some preferential filiations between WR subtypes can be enlightened (van der Hucht, 1988; see also Moffat et al. 1986):

R < 8.5 kpc	WNL	$\rightarrow$	WCL		
R > 6.5 kpc	WNL	$\rightarrow$	WCE	$\rightarrow$	WO
			WNE	$\rightarrow$	no WC stars

R is the galactocentric radius. These filiations are very well supported by the model results (Maeder 1991). The first connection is typical of high M and Z: a long WNL phase, followed by a negligible WNE phase, emerges on the late and luminous part of the WC sequence. The second connection is typical of large masses with solar Z or lower, while the third one corresponds to lower masses. Thus, we see that the observations and understanding of WR stars in nearby galaxies has brought about the clarification of many problems. This allows us to apply these results for the interpretation of the integrated spectrum of galaxies.

On the whole, we see that in the domain of massive stars there are presently many encouraging points of agreement, but observations also show some difficulties which need to be explored and will require further improvements of the stellar models.

References

Abbott D.C.: 1982, *ApJ* **259**, 282

Abbott D.C., Conti P.S.: 1987, *ARAA* **25**

Alongi M., Bertelli G., Bressan A., Chiosi C., Fagotto F., Greggio L., Nasi E.: 1993, *A&AS* **97**, 851

Appenzeller I.: 1980, "Star Formation", 10th Saas–Fee Course, Ed. A. Maeder, L. Martinet, Geneva Obs., p. 3

Arnault P., Kunth D., Schild H.: 1989, *A&A* **224**, 73

Arnett D.: 1991, *ApJ* **383**, 295

Barlow M.J., Hummer D.G.: 1982, in *Wolf–Rayet Stars: Observation, Physics, Evolution*, IAU Symp. 99, Eds. C.W.H. de Loore, A.J. Willis, Reidel Publ., p. 387

Bertelli G., Chiosi C.: 1982, in *Wolf–Rayet Stars: Observation, Physics, Evolution*, IAU Symp. 99, Eds. C.W.H. de Loore, A.J. Willis, Reidel Publ., p. 359

Bolton C.T., Rogers G.L: 1978, *ApJ* **222**, 234

Brocato E., Buonanno R., Castellani V., Walker A.R.: 1990, *ApJS* **71**, 25

Brunish W.M., Truran J.W.: 1982a, *ApJ* **256**, 247

Brunish W.M., Truran J.W.: 1982b, *ApJS* **49**, 447

Brunish W.M., Gallagher J.S., Truran J.W.: 1986, *AJ* **91**, 598

Charbonnel C., Meynet G., Maeder A., Schaller G., Schaerer D.: 1993 *A&AS*, in press

Chiosi C., Bertelli G., Bressan A.: 1992, *ARAA* **30**, 235

Chiosi C., Nasi E., Sreenivasan S.R.: 1978, *A&A* **63**, 103

Chiosi C., Maeder A.: 1986, *ARAA* **24**, 329

Conti P.S.: 1988, *O–stars and WR stars*, NASA SP–497, Eds. P.S. Conti and A.B. Underhill, p. 81

Conti P.S.: 1991, *ApJ* **377**, 115

Conti P.S., Leep E.M., Perry D.N.: 1983, *ApJ* **268**, 228

Esteban C., Vilchez J.M.: 1991, in *Wolf–Rayet stars and interrelations with other massive stars in galaxies*, IAU Symp. 143, Ed. K.A. van der Hucht and B. Hidayat, Kluwer Acad. Publ. p. 422
Fitzpatrick E.L., Garmany C.D.: 1990, *ApJ* **363**, 119
Garmany C.D., Conti P.S., Chiosi C.: 1982, *ApJ* **263**, 777
Gies D.R., Lambert D.L.: 1992, *ApJ* **387**, 673
Grossman S.A., Narayan R., Arnett D.: 1993, *ApJ* **407**, 284
Hamann W.R., Koesterke L, Wessolowski U.: 1993, *A&A* **274**, 397
Herrero A., Kudritzki R.P., Vilchez J.M., Kunze D., Butler K., Haser S.: 1992, *A&A* **261**, 209
Humphreys R.M., Mc Elroy D.B.: 1984, *ApJ* **284**, 565
Iglesias C.A., Rogers F.J., Wilson B.G.: 1992, *ApJ* **397**, 717
de Jager C., Nieuwenhuijzen H., van der Hucht K.A.: 1988, *A&AS* **72**, 259
Kudritzki R.P., Gabler R., Kunze D., Pauldrach A.W.A., Puls J.: 1990, in *Massive Stars in Starbursts*, Ed. C. Leitherer et al., Cambridge Univ. Press, p. 59
Kudritzki R.P., Pauldrach A., Puls J.: 1987, *A&A* **173**, 293
Kudritzki R.P., Hummer D.G., Pauldrach A.W.A., Puls J., Najarro F., Imhoff J., 1992: *A&A* **257**, 655
Lamers H. J.G.L.M., Leitherer C.: 1993, *ApJ*, in press
Lamers H. J.G.L.M., Maeder A., Schmutz W., Cassinelli J.P.: 1991, *ApJ* **368**, 538
Langer N.: 1991a, *A&A* **342**, 155
Langer N.: 1991b, *A&A* **248**, 531
Langer N.: 1991c, *A&A* **252**, 669
Langer N.: 1993, *A&A* **265**, L17
Langer N., El Eid M.F., Baraffe I.: 1989, *A&A* **224**, L17
Langer N., El Eid M.F., Fricke K.J.: 1985, *A&A* **145**, 179
Leitherer C., Langer N.: 1991, in IAU Symp. No. 148 *The Magellanic Clouds*, Eds. R.F. Hanes, D.K. Milne, Kluwer Acad. Publ., p. 480
Lennon D.J., Kudritzki R.P., Becker S.T., Butler K., Eber F., Groth H.G., Kunze D.: 1991, *A&A* **252**, 498
Lundstrom I., Stenholm B.: 1984, *A&AS* **58**, 163
Maeder A.: 1983, *A&A* **120**, 113
Maeder A.: 1987a, *A&A* **173**, 247
Maeder A.: 1987b, *A&A* **178**, 159
Maeder A.: 1990, *A&AS* **84**, 139
Maeder A.: 1991, *A&A* **242** 93
Maeder A., Lequeux J., Azzopardi M.: 1980, *A&A* **90**, L17
Maeder A., Meynet G.: 1989, *A&A* **210**, 155
Maeder A., Meynet G.: 1994, *A&A* in press
Mas–Hesse J.M.: 1992, *A&A* **253**, 49
Mas–Hesse J.M., Kunth D.: 1991, *A&AS* **88**, 317
Massey P.: 1981, *ApJ* **246**, 153
Massey P., Armandroff T.E.: 1991, in *Wolf–Rayet stars and interrelations with other massive stars in galaxies*, IAU Symp. 143, Ed. K.A. van der Hucht and B. Hidayat, Kluwer Acad. Publ., p. 575
Massey P., Gamrany C.D., Silkey M., De Gioia–Eastwood: 1989, *AJ* **97**, 107
Mateo M.: 1988, *AJ* **331**, 261
Meylan G., Maeder A.: 1982, *A&A* **108**, 148
Meylan G., Maeder A.: 1983, *A&A* **124**, 84
Meynet G.: 1993, in *The feedback of chemical evolution on stellar population*, Eds. D. Alloin and G. Stasinska, Observ. Paris, p. 40
Meynet G., Maeder A., Schaller G., Schaerer D., Charbonnel C.: 1993a, *A&A*, in press
Meynet G., Mermilliod J.C., Maeder A.: 1993b, *A&AS* **98**, 477
Moffat A.F.J., Vogt N., Paquin G., Lamontagne R., Berrera L.: 1986, *AJ* **91**, 1386
Napiwotzki R., Rieschick A., Blocker T., Schonberner D., Wenske V.: 1993, in *Inside the Stars*, IAU Coll. No. 137, Eds. W. Weiss and A. Baglin, ASP Conf. Ser., vol. **40**, p. 461
Nasi E., Forieri C.: 1990, *Astrophys. Space Sci.* **166**, 229

Nugis T.: 1991, in *Wolf–Rayet stars and interrelations with other massive stars in galaxies*, IAU Symp. 143, Ed. K.A. van der Hucht and B. Hidayat, Kluwer Acad. Publ., p. 75

Pagel B.E.J., Simonson E.A., Terlevich R.J., Edmunds M.G.: 1992, *MNRAS* **255**, 325

Palla F., Stahler S.W., Parigi G.: 1993, in *Inside the Stars*, IAU Coll. No. 137, Eds. W. Weiss and A. Baglin, ASP Conf. Ser., vol. **40**, p. 437

Parker J.W., Garmany C.D., Massey P., Walborn N.R.: 1992, *AJ* **103**, 1205

Podsiadlowski Ph., Joss P.C., Hsu J.J.L.: 1992, *ApJ* **391**, 246

Schaerer D., Maeder A.: 1992, *A&A* **263**, 129

Schaerer D., Meynet G., Maeder A., Schaller G.: 1993, *A&AS* **98**, 523

Schaerer D., Schmutz W.: 1993, *A&A*, in press

Schaller G., Schaerer D., Meynet G., Maeder A.: 1992, *A&AS* **96**, 269

Smith L.F.: 1973, IAU Symp. 49, p. 15

Smith L.F.: 1988, *ApJ* **327**, 128

Smith L.F.: 1991, in *Wolf–Rayet stars and interrelations with other massive stars in galaxies*, IAU Symp. 143, Ed. K.A. van der Hucht and B. Hidayat, Kluwer Acad. Publ., p. 601

Smith L.F., Hummer D.G.: 1988, *MNRAS* **230**, 511

Smith L.F., Maeder A.: 1991, *A&A* **241**, 77

Smith L.F., Meynet G., Mermilliod J.C.: 1993, *A&A*, submitted

Stothers R.B.: 1991a, *ApJ* **381**, L67

Stothers R.B.: 1991b, *ApJ* **383**, 820

Stothers R., Chin C.W.: 1977, *ApJ* **211**, 189

Stothers R., Chin C.W.: 1992a, *ApJ* **390**, L33

Stothers R., Chin C.W.: 1992b, *ApJ* **390**, 136

Stothers R.B., Chin C.W.: 1990, *ApJ* **348**, L21

Stothers R.B., Chin C.W.: 1991, *ApJ* **374**, 288

Tuchman J., Wheeler J.C.: 1989, *ApJ* **344**, 835

Tuchman J., Wheeler J.C.: 1990, *ApJ* **363**, 255

Vacca W.D., Conti P.S.: 1992, *ApJ* **401**, 543

Vanbeveren D.: 1991, *A&A* **252**, 159

Vanbeveren D., de Loore C.: 1993, ASP Conf. Ser. **35**, 257

Vanbeveren D., Packet W.: 1979, *A&A*, **80**, 242

van den Bergh: 1968, *J.R.A.S. Canada* **62**, No. 4

van der Hucht K.A., Hidayat B., Admiranto A.G., Supelli K.P., Doom C.: 1988, *A&A* **199** 217

Walborn N.: 1976, *ApJ* **205**, 419

Walborn N.: 1988, in *Atmospheric Diagnostics of Stellar Evolution*, IAU Coll. 108, Ed. K. Nomoto, Springer Verlag, p. 70

Willis A.J.: 1991, in *Evolution of stars: the photospheric abundance connection*, IAU Symp. 145, Ed. G. Michaud and A. Tutukov, Kluwer Acad. Publ., p. 195

PRESUPERNOVA EVOLUTION OF THE MOST MASSIVE STARS

N. LANGER

MPI fur Astrophysik, D-85740 Garching, F.R.G.

Abstract. Taking as example a $60\,M_\odot$ star of solar metallicity, the state of the art of model calculations for very massive, from the main sequence to the supernova stage, is reviewed. It is argued that — due to the simple internal structure of Wolf-Rayet stars — the post main sequence evolutionary phases are currently those which are better understood. A brief discussion of the supernova outcome from very massive stars is given.

Then, the more uncertain main sequence evolution is discussed. A first attempt to incorporate results about pulsational instabilities of very massive stars in stellar evolutionary calculations is performed. On its basis, a new type of evolutionary sequence for very massive stars is obtained, namely O-star $\rightarrow$ Of-star $\rightarrow$ H-rich WNL $\rightarrow$ LBV $\rightarrow$ H-poor WNL $\rightarrow$ WNE $\rightarrow$ WC $\rightarrow$ SN. This scenario is shown to correspond better to many observed properties of very massive stars than the standard one. It includes a model for the prototype LBV P Cygni.

Key words: stellar evolution – massive stars – mass loss – supernovae

1. Introduction

Our knowledge about the evolution of massive stars becomes increasingly uncertain for larger considered initial masses. The main reason for this is that the evolution with time of the most basic stellar property — namely the stellar mass — appears to be less well predictable for more massive stars. For moderate masses (i.e. stars in the ZAMS mass range $8 - 20\,M_\odot$), mass loss rates, especially for the post main sequence evolution, are also largely uncertain; however, their total mass remains — to first order — unchanged during their whole life (cf. Maeder 1992). The basic theoretical problems for this mass range are internal mixing processes (cf. Langer 1994). Those are of less importance for the very massive stars (VMS; i.e. $M_{ZAMS} \gtrsim 40\,M_\odot$), which, however, provide a nice diagnostic tool for internal mixing processes (cf. Langer 1991).

The very massive stars do loose a significant, possibly a major fraction of their initial mass prior to core collapse. E.g., according to recent stellar evolution models (Schaller et al. 1992, Woosley et al. 1993), a $60\,M_\odot$ star terminates its life with only a few solar masses material left. The observational evidence in favour of the scenario that VMS do loose the major fraction of their initial mass before they explode as supernova is the following: the luminosity of a VMS is basically fixed by its actual mass. This is well known for main sequence stars, but it holds even much better for Wolf-Rayet (WR) stars, which obey a very tight mass-luminosity relation (Vanbeveren and Packet, 1979, Maeder 1983, Langer 1989). We know that in particular the most massive stars develop into WR stars, since on one side they have the largest mass loss rates, and on the other side they have the largest convective cores; i.e. for them it is most easy to remove the unprocessed part of the stellar envelope and to uncover material which did belong to the convective core formerly. The key point is that the most evolved, hydrogen less WR stars have rather small luminosities throughout (cf. e.g. Hamann, this volume), i.e. they have small masses. Consequently, the VMS have either to keep part of their hydrogen rich envelope and remain very luminous WNL stars until core collapse, or — once they have lost their hydrogen envelope — they have to loose mass very rapidly in order not to remain very luminous hydrogen less WR stars for a long time. While the first possibility appears to be excluded by the large observed WNL mass loss rates (Hamann et al. 1993), the second — together with the theory of the internal structure of WR stars (cf. Sect. 2) — leads to the concept of mass dependent mass loss rates for hydrogen less WR stars (Langer 1989a), which is actually well confirmed by the continuously increased data base and accuracy of known luminosities and mass loss rates of galactic WR stars (cf. Fig. 5 of Hamann, this volume).

In this paper we concentrate on two topics. The first is to outline the consequences of the concept of mass dependent WR mass loss for the resulting possible supernova explosion. Especially, the questions of the homogeneity of the resulting supernova class, and the possibility to discriminate supernovae of binary WR stars from those of single stars are addressed (Section 2). Then, in Section 3, by applying concepts which fit well to the hydrogen less stars to very massive main sequence stars, we discuss how VMS may get rid of their hydrogen envelope. Final remarks are given in Sect. 4.

2. VMS Evolution Beyond the Loss of their Hydrogen Envelope

When a massive star has lost its hydrogen containing envelope completely, its internal structure is rather simple and can be accurately prescribed as a function of just one parameter, e.g. the remaining stellar mass (Langer 1989, 1989a). Consequently, the thermal and mechanical structure of the star does *not* depend on the internal chemical profiles, and so it does *not* depend on the previous evolutionary phases (which is rather fortunate since they are very uncertain; cf. Section 3). This circumstance, together with the mass convergence as a consequence of mass dependent WR mass loss (Langer 1989a; cf. also Sect. 1), i.e. the fact that the final mass is almost independent of the initial mass for a wide range of ZAMS masses (Langer 1989a, Maeder 1992), makes the remnants of VMS look all very much alike, at the stage of core helium exhaustion. Recently, Woosley et al. (1993) investigated whether this causes a homogeneous class of presupernova events. They found that beyond core helium exhaustion the chemical structure, which to some extent still contains the information about the initial mass, becomes important again, due to the formation of shell energy sources, which makes the thermal and mechanical structure different for stars of different initial masses.

As an example, Woolsey et al. compare the outcome of a $60\,M_\odot$ evolution, which yields a $4.25\,M_\odot$ WC star as presupernova configuration, with a presupernova model resulting from the evolution of a $4.25\,M_\odot$ He-star without mass loss. Note that the convective core mass of both 4.25 $M_\odot$ stars at core helium exhaustion (the amount of mass lost through stellar winds is negligible afterwards) is just equal ($\sim 2\,M_\odot$), due to the fact that their thermal structure is fixed by the actual stellar mass at that time. Note that this gives the initial mass of the C/O-core at core He-exhaustion. However, at the time of core collapse, the C/O-core of the once very massive star has grown to more than 3 $M_\odot$, while that of the evolved He-star has still only 2 $M_\odot$ at the same stage. I.e., shell helium burning proceeded very differently in both stars.

However, note that the above example is an extreme case, since in reality a He-star (which might be born in a close binary system) will also loose mass during core helium burning; i.e. a 4.25 $M_\odot$ He-star at core collapse should have been somewhat more massive from the beginning. Moreover, despite the relatively large differences in the pre-SN structures of both stars, one may still talk of some kind of convergence: note that the initial 60 $M_\odot$ star, would it not have been evolved with mass dependent WR mass loss, would have ended up with a C/O-core mass of roughly 20 $M_\odot$ (cf. e.g. Maeder 1981) and thus probably collapse into a black hole. Compared to this case, both pre-SN objects of the above example are very similar.

TABLE I

Supernova ejecta in $M_\odot$.

$M_{initial}$	M_{final}	M_{He}	M_O	M_{Si}	$M_{^{56}Ni}$	SN Type
7 He	3.20	0.40	0.44	0.07	0.008 - 0.15	Ib ??
60 H	4.25	0.21	1.38	0.17	(0.24)	Ic ??
1.4 C/O	-	0	0.2	0.25	0.7	Ia

Nevertheless, one can work out the systematic differences of pre-SN models from single VMS compared to those originating from He-star models (i.e. close binaries). This has been done by Woosley et al. (1994), assuming that (sufficiently massive) He-stars are in fact WR stars and obey the same mass loss law as single WR stars. As a typical case of the investigated sample, we compare the amounts of various elements ejected in the final supernova explosion of a 7 $M_\odot$ He-star (which ends up as a 3.2 $M_\odot$ WC star) with the case of the 60 $M_\odot$ evolution mentioned above in Table I. Typical numbers for a Type Ia supernova are also given, for comparison. If the numbers of Table 1 are related with the spectral classification scheme for SNe Ia, Ib, and Ic (cf. Wheeler and Harkness 1990), one is tempted to conclude that the binary models fit better to Ib SNe, while the initially very massive stars may be associated with Ic's. However, only a comparison of theoretical spectra with observed ones is an appropriate mean to relate observed supernovae with stellar models, i.e. the above conclusion should be regarded as very preliminary (cf. Wheeler, this volume).

3. How do VMS Loose their Hydrogen Envelope?

In Sect. 1 it has been argued that very massive stars (at solar metallicity) very probably do loose their hydrogen envelope completely. However, it is not quite clear how they do it. For a $60\,M_\odot$ star, which we will analyze as an example in more detail in this Section, radiation pressure driven O-star winds are not sufficient to turn the star into a WR star (i.e. to drop the surface hydrogen mass fraction below ~ 0.40) during the core hydrogen burning phase. This statement holds for calculations using theoretical wind mass loss rates (Kudritzki and Hummer 1990), cf. Woolsey et al. 1993, as well as those using the on average somewhat larger observed rates (Lamers and Leitherer 1993; Leitherer, this volume), cf. Schaller et al. (1992). Even an increase of the main sequence mass loss rate by a factor of 2 — which may be at the limit of compatibility with observed rates — turns a $60\,M_\odot$ star into a WR star only at the very end of the main sequence phase (Meynet et al. 1993). Consequently, a Luminous Blue Variable (LBV) phase with very high mass loss rate is invoked in the standard evolutionary picture for such stars (cf. e.g. Schaller et al. 1992), which prevents the star from becoming a red supergiant and turns it into a WR star within a short time.

This standard picture has several major shortcomings, many of which could be removed if the main sequence mass loss of VMS were larger (cf. Meynet et al. 1993, and see below). In this context it is interesting to have again a look at the hydrogen less WR stars. Glatzel et al. (1993) have shown in a recent linear stability analysis, that those objects are violently unstable with respect to radial pulsations for WR masses in excess of $\sim 4\,M_\odot$. The growth rates of these instabilities are extremely large, and they are likely to induce a dynamical mass loss. If one estimates the induced mass loss rate as $\dot{M} \simeq \Delta M \sigma_i / P$, with ΔM the fraction of the envelope mass involved in the pulsations, σ_i the imaginary part of the eigenfrequency, and P the pulsational period, some $10^{-5}\,M_\odot\,yr^{-1}$ are the result, which is of the order of magnitude of the observed WR mass loss rates. Thus one may speculate that these pulsations provide the initiation mechanism for the WR mass loss. Kiriakidis et al. (1993) have shown that the same type of violently unstable pulsation may occur in very massive main sequence stars, provided that their metallicity is large enough and their effective temperature low enough. They have investigated a $60\,M_\odot$ star at 2% metallicity in detail and found it to be unstable for $4.63 \gtrsim \log T_{eff} \gtrsim 4.43$, with a maximum instability at about $T_{eff} = 35\,000\,K$. Note that this temperature corresponds exactly to the surface temperature of the still hydrogen rich WNL stars (cf. Hamann, this volume). Therefore, we have calculated a $60\,M_\odot$ evolutionary sequence (using OPAL opacities; cf. Iglesias et al. 1992) with the assumption that the mass loss rate for the pulsationally unstable models is proportional to the growth rate σ_i of the instability as computed by Kiriakidis et al. The proportionality factor k

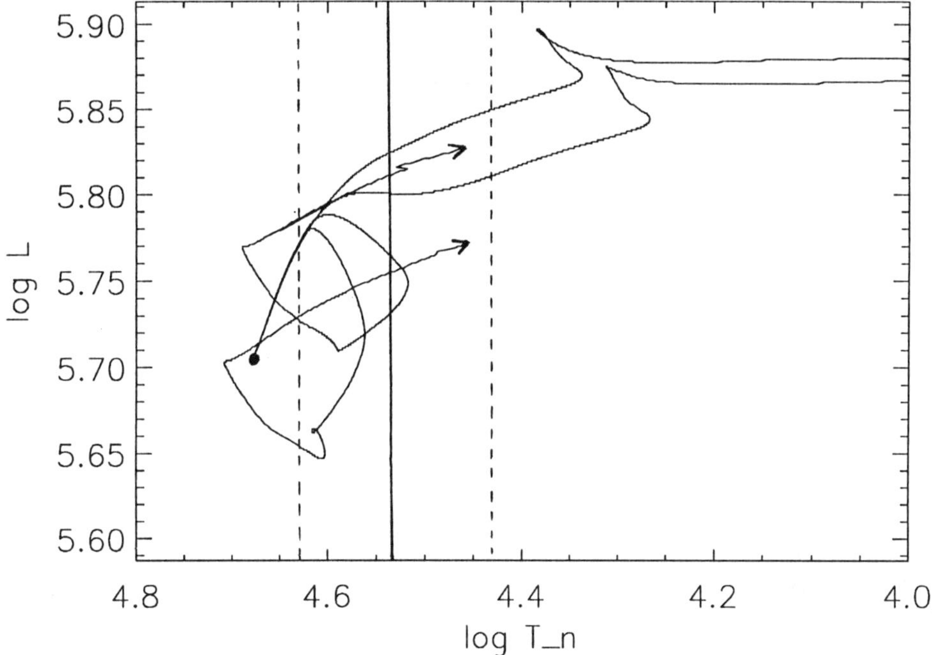

Fig. 1.   Evolutionary tracks of $60\,M_\odot$ stars with different assumptions on the mass loss rates for the pulsationally unstable models, during the core hydrogen burning phase, and the beginning of shell hydrogen burning. The upper track corresponds to the standard case (i.e. pulsations are ignored). The other 3 tracks correspond to different values of the proportionality coefficient k (see text). The heavy dot marks the zero age main sequence model. The effective temperature range for which the models are unstable is indicated (dashed lines), with the vertical full line denoting the maximum of the instability.

has been chosen such that a maximum mass loss rate of some $10^{-5}\,M_\odot\,yr^{-1}$ is achieved.

Fig. 1 shows the resulting evolution in the HR diagram for the main sequence phase for three values of k, and for the standard case. It is remarkable, that for k larger than some critical value the instability strip is no longer crossed by the evolutionary track during core hydrogen burning. Instead, the star turns into a WNL-type object, i.e. it maintains a mass loss rate of some $10^{-5}\,M_\odot\,yr^{-1}$ and soon shows large amounts of hydrogen burning products at the stellar surface (cf. also Fig. 2).

Note that this evolutionary scenario certainly remains somewhat speculative until self-consistent nonlinear calculations of WR-type outflows are available (work on this is in progress with P. Hoflich). Let us, however, summarize the advantage of this picture over the standard one — besides the fact that it puts together bits and pieces of what we know about VMS:

• A WNL phase during core hydrogen burning is obtained, which is consequently

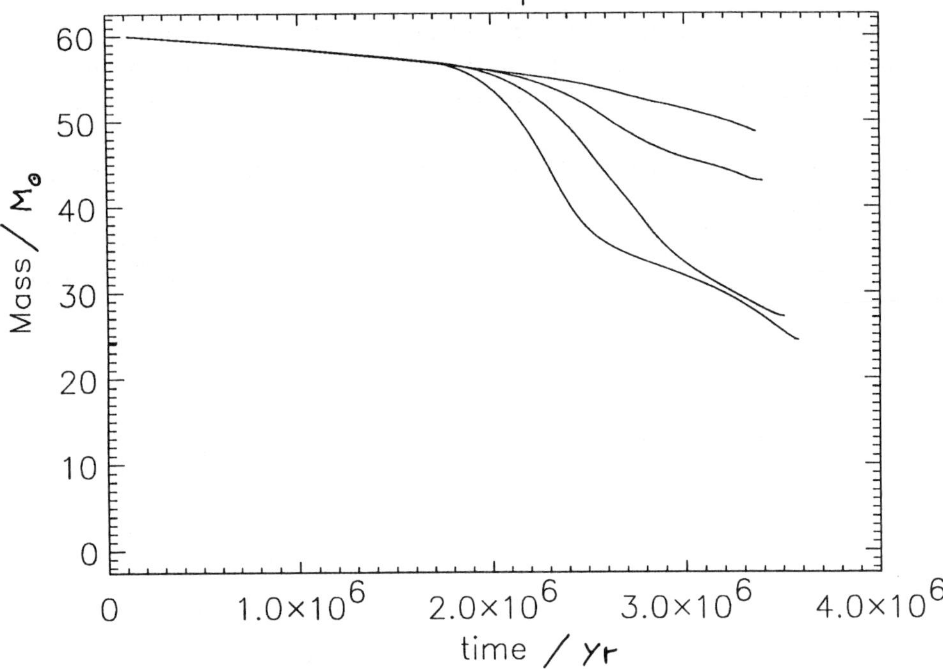

Fig. 2.   Evolution of the total mass with time for the 60 $M_\odot$ sequences shown in Fig. 1, until core hydrogen exhaustion.

long-lasting ($\sim 10^6\, yr$). This fits to the rather larger number of very luminous WNL objects (cf. Hamann, this volume).

- In contrast to the standard scenario, a WNL stage prior to the LBV phase is obtained. This seems actually required regarding the very small hydrogen abundances recently measured in P Cygni type LBVs (Smith et al. 1993, Najarro, priv. comm.), which appear to be *smaller* that many WNL hydrogen abundances.

- The stellar models which correspond to the LBV phase do show many properties of P Cygni: besides the HRD position, the direction and speed of the redward evolution correspond well to the observations (de Groot and Lamers 1992), which is difficult to obtain with standard calculations (El Eid and Hartmann 1993; cf., however, Maeder 1991). Also the obtained surface composition of $X_s \simeq 0.3$ (Deacon and Barlow 1991; Najarro, priv. comm.) and the actual mass of roughly 23 $M_\odot$ (Pauldrach and Puls 1990) are in good agreement (cf. Fig. 2).

- A second, post-LBV WNL-stage with a small (but positive) hydrogen surface abundance of about $X_s \simeq 0.10$ is obtained. Note that WNL models with small surface hydrogen mass fractions are not obtained by the standard scenario. It may correspond to stars like WR 136 (cf. Hamann, this volume).

- The luminosity of the WNE phase is considerably reduced, due to a smaller He-core mass resulting from core hydrogen burning. Note that the WNE luminosities obtained within the standard scenario are too large in order to correspond to the observed WNE stars (see again: Hamann, this volume).

Certainly, the evolutionary scenario for VMS involving pulsational mass loss, which leads to the sequence O-star $\rightarrow$ Of-star $\rightarrow$ H-rich WNL $\rightarrow$ LBV $\rightarrow$ H-poor WNL $\rightarrow$ WNE $\rightarrow$ WC $\rightarrow$ SN for the case of a 60 $M_\odot$ star of solar metallicity, deserves further investigation. Work on this line is in progress with W.R. Hamann, F. Najarro, A. Pauldrach and J. Puls.

4. Final remarks

The evolution of very massive stars is largely determined by their huge mass loss rates. However, for the evolutionary stages where the largest mass loss rates are achieved, i.e. the WR and LBV phases, we still do not have a quantitative mass loss theory. Consequently, even the main sequence evolution of VMS can not yet be accurately predicted by means of stellar evolution calculations. However, in Sect. 3 we have shown that incorporation of many observational and theoretical evidences may yield at least a self consistent quantitative picture of the early VMS evolution at solar metallicity. Note, however, that all considered processes are sensitive functions of the initial stellar metal content, i.e. VMS evolution in the Magellanic Clouds may already be quite different from that in our Galaxy.

The fortunate fact that VMS, once they are Wolf-Rayet stars, have almost no memory of their previous evolution, results in the strange situation that we understand their later evolutionary phases better than their main sequence phase. Mass dependent WR mass loss makes WR stars to potential supernova progenitors, while the most likely candidates for black hole formation (at Z=2%) may be stars with initial masses of $\sim$35 $M_\odot$.

Our picture of the evolution of very massive stars has changed dramatically over the last decade (cf. de Loore 1980), which indicates that much progress could be achieved in this field. However, we are still left with a large number of basic problems, which requires much work to be done before we can say we understand very massive stars.

5. Acknowledgements

This work has been supported by the Deutsche Forschungsgemeinschaft through grant La 587/8-1.

References

Deacon J.R., Barlow M.J.: 1991, in: *Wolf-Rayet Stars and Interrelations with Other Massive Stars in Galaxies*, Proc. IAU-Symp. 143, K. van der Hucht et al., eds., Kluwer, p. 558
El Eid M.F., Hartmann D.: 1993, *Astrophysical Journal* **404**, 271

Glatzel W., Kiriakidis M., Fricke K.J.: 1993, *Monthly Notices of the RAS* **262**, L7
de Groot M.J.H., Lamers H.J.G.L.M.: 1992, *Nature* **335**, 422
Hamann W.-R., Koesterke L., Wessolowski U., 1993: *Astronomy and Astrophysics* **274**, 397
Iglesias C.A., Rogers F.J., Wilson B.G.: 1992, *Astrophysical Journal* **397**, 717
Kiriakidis M., Fricke K.J., Glatzel W.: 1993, *Monthly Notices of the RAS* **264**, 50
Kudritzki R.P., Hummer D.G.: 1990, *Annual Review of Astronomy and Astrophysics* **28**, 303
Langer N.: 1989, *Astronomy and Astrophysics* **210**, 93
Langer N.: 1989a, *Astronomy and Astrophysics* **220**, 135
Langer N.: 1991, *Astronomy and Astrophysics* **248**, 531
Langer N.: 1994, in: *Circumstellar Media in the Late Stages of Stellar Evolution*, Cambridge
    University Press, R. Clegg et al., eds., in press
de Loore C.: 1980, *Space Sci. Rev.* **26**, 113
Lamers H.J.G.L.M., Leitherer C.: 1993, *Astrophysical Journal*, in press
Maeder A.: 1981, *Astronomy and Astrophysics* **99**, 97
Maeder A.: 1983, *Astronomy and Astrophysics* **120**, 113
Maeder A.: 1983, *Astronomy and Astrophysics* **120**, 113
Maeder A.: 1991, in: *Instabilities in Evolved Super- and Hypergiants*, C. de Jager, ed.
Meynet G., Maeder A., Schaller G., Schaerer D., Charbonnel C.: 1993, *Astronomy and Astrophysics, Supplement Series*, in press
Pauldrach A., Puls J.: 1990, *Astronomy and Astrophysics* **237**, 409
Schaller G., Schaerer D., Meynet G., Maeder A.: 1992, *Astronomy and Astrophysics, Supplement Series* **96**, 269
Smith L.J., Crother P.A., Prinja R.K.: 1993, *Astronomy and Astrophysics*, in press
Wheeler J.C., Harkness R.P.: 1990, *Rep. Prog. Phys.* **53**, 1467
Vanbeveren D., Packet W.: 1979, *A&A*, **80**, 242
Woosley S.E., Langer N., Weaver T.A.: 1993, *Astrophysical Journal* **411**, 823
Woosley S.E., Langer N., Weaver T.A.: 1994, *Astrophysical Journal*, in prep.

PROPERTIES OF MASSIVE STAR EVOLUTION

A. BRESSAN

Astronomical Observatory of Padua
Vicolo dell'Osservatorio 5
35122-Padova Italy

Abstract. New models of massive stars have been calculated as a part of a project aimed at the construction of an extended grid of stellar tracks for several metallicities, covering all the main evolutionary phases of low, intermediate and massive stars, with updated physics. After a brief description of the input physics, we discuss the properties of the stellar content of young clusters as a function of the age and metal content, in the light of the new evolutionary models. Furthermore we review the global properties of the observed HR diagrams of massive stars in the Solar Vicinity and LMC and discuss the major points of disagreement that emerge from their comparison with those artificially generated from our database of evolutionary tracks.

Key words: Stars: Stellar Evolution – Massive Stars

1. Stellar models with new opacities

We computed an extended grid of stellar models from 0.6 $M_\odot$ to 120 $M_\odot$ with the initial chemical composition Z=0.0004 Y=0.230, Z=0.004 Y=0.240, Z=0.008 Y=0.250, Z=0.02 Y=0.280, Z=0.05 Y=0.352, Z=0.05 Y=0.45 (Bressan et al. 1993 and Fagotto et al. 1993 a,b) . We adopted the OPAL opacities by Iglesias et al. (1992). With respect to the LAOL opacities (Huebner et al 1977) they show two significant enhancements which depend primarily on the adopted metal abundance. The first enhancement, a bump-like structure of a factor of 3 for the solar abundance, occurs at temperatures of a few hundred thousand degrees, and the second occurs at temperatures of about one million degrees, but it is of much smaller amplitude (about 20% only). The network of nuclear reactions accounts for sixteen elements, ^{1}H, ^{2}H, ^{3}He, ^{4}He, ^{7}Be, ^{7}Li ^{12}C, ^{13}C, ^{14}N, ^{16}O, ^{17}O, ^{18}O, ^{20}Ne, ^{22}Ne, ^{25}Mg, and ^{26}Mg and the reaction rates are from Caughlan and Fowler (1988), including the low value for the $^{12}C(\alpha,\gamma)^{16}O$ reaction. Convective overshoot was included for the determination of the extension of the central convective region. The physical motivations for the occurrence of this phenomenon and its efficiency have been recently discussed by Roxburgh (1989) and Zahn (1991). In the formalism of Bressan et al. (1981), we adopted a mean free path *across* the border of the unstable region l=0.5×H_P, which corresponds to about the maximum overshoot distance advocated by Stothers (1991), or the distance *above* the border favoured by Schaller et al. (1992) and Schaerer et al. (1993). The mixing length parameter in the outermost super-adiabatic convective region was set to 1.63, after calibration with the solar model. It is well known that, with this value of the mixing length parameter, the models of massive stars exhibit a strong den-

Space Science Reviews **66**: 373–382, 1994.

© 1994 *Kluwer Academic Publishers. Printed in Belgium.*

A. BRESSAN

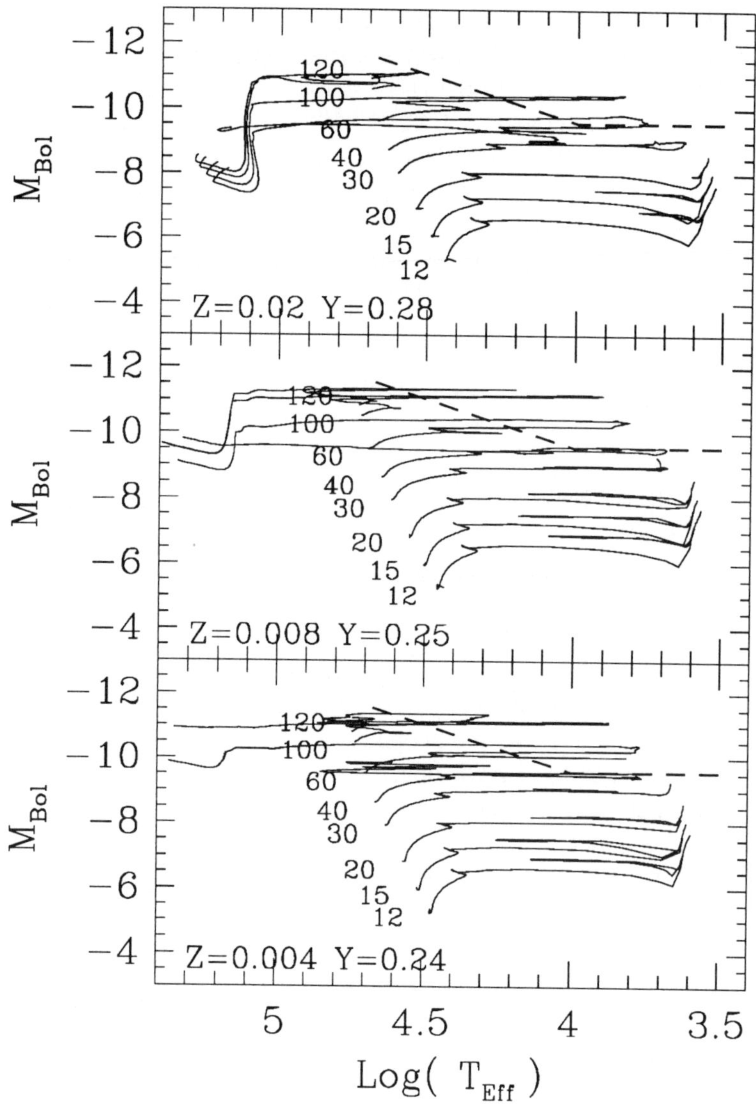

M_{Bol}

$\mathrm{Log}(\, T_{Eff}\,)$

Fig. 1. Evolutionary tracks of massive stars for some selected chemical compositions.

sity inversion at log $T_{Eff} \leq 3.9$ which has been the subject of many theoretical speculations (Maeder 1992 and references therein). In these new models no particular numerical algorithm has been devised to suppress the density inversion. Finally, an overshoot region of size $0.7 \times H_P$ has been considered at the bottom of the convective envelope, according to the suggestion of Alongi et al. (1991).

The evolution of massive stars is dominated by mass loss by stellar wind (see Chiosi & Maeder 1986; Chiosi et al 1992b). In the above calculations we adopted the empirical formulation of the mass loss rate by de Jager et al. (1988) for all evolutionary stages from the main sequence up to the so-called Humphreys-Davidson (1979) limit (HD) in the color magnitude diagram. Beyond this limit, the mass loss rate was increased to 10^{-3} $M_\odot$ yr^{-1} as suggested by the observational data for the Luminous Blue Variables (LBV). If the surface hydrogen fell below 0.3 the star was supposed to begin the Wolf Rayet phase, and the mass-loss rate was modified according to the different types WNL, WNE and WC/WO (Langer 1989). The rates also include a suitable dependence on the metallicity (Kudritzki et al.1989), i.e. $\dot{M}_Z \propto \sqrt{Z}$.

2. The evolution of massive stars

The evolutionary tracks of massive stars for some selected chemical compositions are shown in Fig. 1. The dashed line indicates the HD limit which, in our models, marks the beginning of the LBV phase. Although a more detailed information can be found in Bressan et al. (1993) and Fagotto et al. (1993 a,b), we collected in Table 1 the ages and the durations of several relevant evolutionary phases of massive stars. For each different chemical composition, M indicates the initial mass in solar masses, LBV and Δt the starting age and the duration of the LBV phase in Myr; RSG_i is the age at which the star becomes a Red Supergiant (RSG) for the first time, RSG_f marks the age at which the star leaves the RSG phase and performs the blue loop, and RSG_l indicate the age at which the star enters the final RSG phase. WN_H stems for the age at which the star becomes a WN star (surface-H<0.3) while WN_{He} refers to the time at which the surface hydrogen becomes zero; WC and WO indicate the age at which $^{12}C > ^{14}N$ and $(^{12}C+^{16}O)/^4He > 1$, (surface values by number) respectively. The last three columns indicate the final age of the model and the duration of the hydrogen- and helium- burning phases respectively. Inspection of Fig. 1 and Table 1 shows the role played by mass loss in the evolution of the massive stars of different chemical composition. The properties of advanced evolutionary phases and chemical yields at varying metallicity were thoroughly analysed by Maeder (1991,1993). Here we focuses on the connection between the age and the stellar content of a young cluster and draw, in Fig. 2, the lines which mark the beginning of several relevant evolutionary phases as a function of the initial composition. As can be seen from this figure, a minimum interval of time has to elapse before WR stars appear in a cluster; this minimum age slightly increases from 1.7 Myr to 2.8 Myr when Z decreases from 0.05 to 0.0004. If only the WN type is detected the age is comprised between the two lines labelled WN and WC, while the presence of WC and WO stars indicate a slightly older age and a minimum metal content. WR stars disappear at an age of about 4, 6, 5, 3.3 Myr for Z=0.05, 0.02, 0.008, 0.0004 respectively. A significantly older lower limit to the age of the cluster can be inferred from

TABLE I

Ages and lifetimes (Myr) of relevant evolutionary phases

M	LBV	Δt	RS_i	RS_f	RS_l	WN_H	WN_{He}	WC	WO	t	Δt_H	Δt_{He}
						$Z = 0.05$						
120	-	-	-	-	-	1.75	2.54	2.55	-	3.15	2.53	0.58
100	-	-	-	-	-	1.77	2.55	2.56	-	3.16	2.54	0.58
60	-	-	-	-	-	2.56	-	-	-	3.12	-	0.58
40	3.70	0.0031	-	-	-	3.70	3.72	3.74	-	4.40	3.70	0.65
20	-	-	6.52	7.06	-	-	-	-	-	7.06	6.52	0.52
15	-	-	9.35	10.07	-	-	-	-	-	10.07	9.34	0.68
12	-	-	13.14	14.11	-	-	-	-	-	14.11	13.12	0.91
						$Z = 0.02$						
120	1.90	0.0186	-	-	-	1.91	2.81	2.81	3.24	3.26	2.78	0.46
100	-	-	-	-	-	2.41	3.02	3.02	3.36	3.41	2.97	0.43
60	3.63	0.0035	-	-	-	3.63	3.65	3.66	-	4.15	3.63	0.50
40	4.58	0.0047	-	-	-	4.59	4.62	4.65	-	5.16	4.58	0.55
30	-	-	5.70	5.81	-	6.16	6.16	-	-	6.17	5.69	0.46
20	-	-	8.61	9.23	-	-	-	-	-	9.23	8.60	0.60
15	-	-	12.61	12.99	13.18	-	-	-	-	13.45	12.60	0.80
12	-	-	17.90	18.65	18.70	-	-	-	-	19.08	17.88	1.08
						$Z = 0.008$						
120	3.02	0.0049	-	-	-	2.60	3.14	3.14	3.22	3.34	3.01	0.31
100	3.13	0.0087	-	-	-	2.94	3.26	3.26	3.33	3.46	3.12	0.32
60	3.85	0.0093	-	-	-	3.86	3.98	3.99	4.15	4.25	3.85	0.38
40	4.92	0.0067	-	-	-	5.32	5.33	-	-	5.34	4.91	0.41
30	-	-	6.21	6.35	6.49	-	-	-	-	6.69	6.19	0.47
20	-	-	9.50	9.63	10.06	-	-	-	-	10.16	9.49	0.63
15	-	-	14.10	14.35	14.96	-	-	-	-	15.00	14.09	0.86
12	-	-	19.95	20.50	21.11	-	-	-	-	21.21	19.93	1.17
						$Z = 0.004$						
120	2.99	0.0388	-	-	-	2.82	-	-	-	3.24	2.98	0.25
100	3.13	0.0157	-	-	-	3.14	-	-	-	3.40	3.12	0.27
60	3.81	0.0164	-	-	-	3.89	4.11	4.12	4.12	4.23	3.88	0.34
40	5.05	0.0101	-	-	-	5.47	-	-	-	5.47	5.04	0.42
30	-	-	-	-	6.85	-	-	-	-	6.89	6.39	0.48
20	-	-	9.86	9.98	10.36	-	-	-	-	10.53	9.84	0.65
15	-	-	14.44	14.64	15.33	-	-	-	-	15.34	14.42	0.88
12	-	-	20.32	20.85	21.53	-	-	-	-	21.59	20.29	1.20
						$Z = 0.0004$						
120	2.87	0.0037	-	-	-	2.87	-	-	-	3.12	2.86	0.25
100	2.84	0.0898	-	-	-	3.09	-	-	-	3.34	3.07	0.26
60	4.13	0.1008	-	-	4.23	-	-	-	-	4.23	3.90	0.32
40	5.47	0.0060	-	-	5.48	-	-	-	-	5.48	5.07	0.39
30	-	-	-	-	6.94	-	-	-	-	6.94	6.46	0.46
20	-	-	-	-	10.68	-	-	-	-	10.69	10.02	0.64
15	-	-	14.56	14.62	15.43	-	-	-	-	15.45	14.53	0.88
12	-	-	20.22	20.36	21.46	-	-	-	-	21.51	20.18	1.23

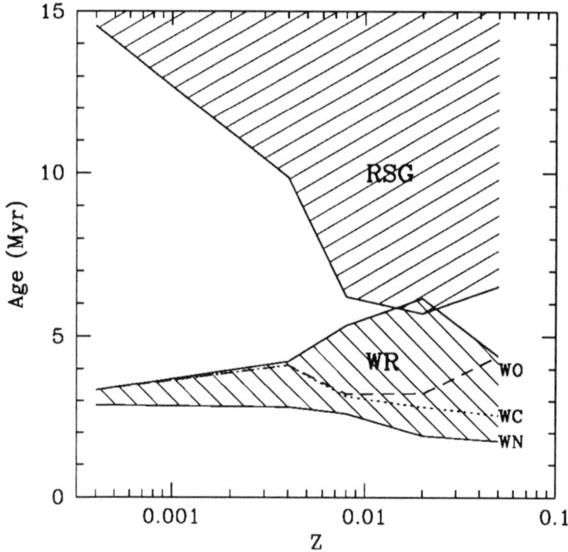

Fig. 2. Ages at which several relevant evolutionary phases appear, as a function of the metallicity. The end of the WR phase is also displayed

the presence of the red supergiants. It is around 6 Myr for Z between 0.05 and 0.008 but rapidly increases to 10 Myr at Z=0.004 and reaches about 15 Myr at still lower metallicities. These limiting ages are clearly less affected by the uncertainties in the mass-loss rates, than those derived from the presence of WR stars, and mainly reflect the transition between the two evolutionary schemes commonly named "case A" and "case B", as the metal content decreases below a certain value. In brief stars of initial mass up to about $30M_\odot$ either perform an extended blue loop in the HRD during central He-burning, as in models with semiconvection evolved with the Ledoux criterion (Chiosi & Summa 1970 case A, Langer et al 1989, Langer 1991) , or ignite helium in the blue and slowly move redward in the HRD diagram during the whole core He-burning phase and, finally, quickly move to the red supergiant region, as in models evolved with the Schwarzschild criterion for the semiconvection (Chiosi & Summa 1970 case B, Brunish & Truran 1982). The intersection of the WR region with that of the RSG in the case of the solar metallicity is due to the behaviour of the model of 30 $M_\odot$ which, during a prolonged stage of red supergiant, looses enough mass to become a WN star shortly before the end of the central helium burning. In our models, this is the only one representative of the evolutionary scheme O-BSG-RSG-WR, if one excludes the transient YSG phase of more massive stars. It is worth pointing out that this evolutionary scheme is very much dependent on the adopted mass-loss rate in the RSG phase. With the adopted parameterization (de Jager et al. 1988), the average rate of mass loss in our solar abundance models of 20 $M_\odot$ and 15 $M_\odot$ during the RSG phase is 2×10^{-6} $M_\odot\,yr^{-1}$ and 1×10^{-6} $M_\odot\,yr^{-1}$ respectively. Feast (1992) showed that the red supergiant variables in LMC follow a well defined period-mass loss relation. Assuming his relation between mass loss and pulsation period and evaluating the period from a period-mass-radius relation (eg. Fox and Wood 1982) we obtained mass loss rates in the RSG phase which were, on the average, 7 and 10 times larger for the models of 20 $M_\odot$ and 15 $M_\odot$, respectively (see also Feast 1992). Looking to the lifetimes of Table 1 it appears that, with mass loss rates of this order and a suitable treatment of internal mixing, it could be possible, at least for the model of 20 $M_\odot$ and solar abundance, to abandon the RSG region and display the features of a WN star (see also Vanbeveren these proceedings). Perhaps a strong support of this idea comes from the observational evidence that some Galactic WN stars, *without* an extended thick atmosphere and thus directly comparable with theoretical models *without* suitable corrections to the effective temperature, fall around the ZAMS of the $20M_\odot$ model (Hamann et al. 1993)

3. The HR diagram of massive stars

The HRDs of the supergiant stars in the solar vicinity (Blaha & Humphreys 1989) and LMC (Fitzpatrick & Garmany 1990) show several common interesting features. The area occupied by the brightest stars appears clearly delimited on its cooler side by a luminosity boundary which declines at decreasing effective temperature, and becomes constant for the yellow and red supergiant stars at $M_{bol} \simeq -9.5$. This suggests that stars brighter than this limit (HD) either possess a very rapid excursion toward the coolest effective temperature or even never experience the red supergiant phase. Furthermore, this boundary is marked by the presence of the luminous blue variables, characterised by recurrent ejections of mass during which they reach about the highest observed mass-loss rates. This fact led Humphreys & Davidson (1979) to first point out that heavy mass-loss, driven by some still unknown interior phenomenon, could prevent the brightest blue supergiants from evolving at cooler temperatures, after central hydrogen exhaustion. Common to both diagrams is the lack in the distribution of the brightest stars near the theoretical zero-age main sequence. This has been attributed to an initial phase during which the cluster remain embedded in the molecular cloud and the stars remain directly undetectable (Garmany et al 1982, Bertelli et al. 1984). The duration of this phase derived from the comparison of the number counts with the theoretical lifetimes agrees with the empirical estimate by Wood & Churchwell (1989) that about $10\div20\%$ of all O stars in the solar vicinity are embedded in molecular clouds and therefore are only indirectly detectable. At luminosities below the HD limit there is a continuous distribution of stars from the zero age main sequence up to the G-spectral type, and a gap between the latter stars and the red supergiants (the Red Hertzsprung Gap -RHG-). The RHG gets narrower at decreasing luminosity, following a diagonal line from $\log T_{eff} \leq 4.2$

to $\log T_{eff} \geq 3.9$, otherwise called the "ledge" (Fitzpatrick & Garmany 1990). The presence of the RHG is usually matched by the evolutionary models of massive stars because they predict that a star crosses this region in a Kelvin-Helmoltz time scale. However the precise shape of the "ledge" can be used to discriminate among the different input physics of stellar models, with particular emphasis to the mixing scheme.

Tuchman & Wheeler (1990), by comparing the models of Brunish & Truran (1982) with the star counts in the HR diagram of LMC, favour the case B evolution of models without core overshoot. On the contrary Fitzpatrick & Garmany (1990) argue that the inclination of the ledge is reproduced only by models which display extended loops (case A). For the chemical composition suitable for the LMC our new models with overshoot and mass-loss follow case A scenario and thus are able to explain the ledge. However the overshoot at the base of the convective envelope is critical for the extension of the blue loop (compare the model of 20 $M_\odot$ Z=0.008 of Fig. 1 with the one of Fig. 3 of Schaerer et al. 1993). Additional support for the case A evolutionary scheme comes from the observations of CNO processed and He-rich material at the surface of some He-rich objects near the main sequence (Kudritzki 1990, Fitzpatrick & Bohannan 1992), suggesting that these stars already underwent the first dredge-up episode and are now in the blue loop of the He-burning phase. It is worth noticing that, after the RSG phase, the surface abundance of nitrogen and helium of the models of Fig. 1 is enhanced by a factor of about 4 and 1.3 respectively with respect to their initial value, while that of carbon and oxygen is reduced to about the 60% and 80% of their initial value, in agreement with the above observations. The continuous distribution of stars from the ZAMS to the ledge constitutes a critical point for the theory, because the latter predicts that the two main phases of nuclear burning are separated by a gap of detectable width in the HR diagram (the Blue Hertzsprung Gap -BHG-). In addition to this, the number counts in the HR diagram of luminous stars within 3 Kpc of the Sun (Blaha and Humphreys 1989) indicate that about 40% of the stars fall outside the region delimited by the theoretical main sequence band, so that these stars should be in the core helium burning phase, while the percentage expected from the lifetimes of Table 1 is about 7%. This number increases to about 11% by accounting for the fact that He-burning stars have initial masses slightly lower than those of H-burning stars of the same luminosity and thus are more favorite by both the initial mass function and the nuclear lifetime. Slightly larger percentages are obtained either adopting a larger value of the $^{12}C(\alpha,\gamma)^{16}O$ reaction rate or neglecting convective overshoot, or both. To cope with these difficulties, Bertelli et al (1984) considered the possibility of a main sequence widening, caused by a suitable increase of the radiative opacity in the zone of the CNO ionization. Although the above discrepancies could be solved at once by invoking a widening of the main sequence band, only the most extreme opacity enhancement adopted by Bertelli et al. (1984) could also explain the anomalous surface abundance observed by Kudritzki and Fitzpatrick

& Bohannan, since it give rise to models that, after ending the H-burning as RSGs, return in the blue. However this strong enhancement is not confirmed by the new opacity calculations.

Alternatively, Tuchman & Wheeler (1990) suggested that a large fraction of the stars inside the BHG are secondaries that have accreted He-rich material from the envelope of a red supergiant primary, and argued that the *peculiar* abundance found from some stars in this region support this idea. Finally, Chiosi et al. (1992a), argued that inhomogeneities in the chemical composition of Galactic as well as LMC supergiants could result into a filling up of the BHG, by affecting both the temperature of the termination of the main sequence band of the metal rich stars and the blue extension of the loop of the metal poor ones. The residual gap could be masked by inadequacies of the conversion from spectral type, color, and apparent magnitudes into $T_{eff}s$ and luminosities, and by uncertainties in the estimate of the reddening and color excess of individual stars. Since the new models allow the study of the effects of a dispersion in metallicity, we performed several simulations of the HR diagram of the LMC supergiants. Fig. 3 shows an extreme case for which we assumed a continuous star formation from 0.2 Myr to 100 Myr with the metallicity randomly distributed between Z=0.001 and Z=0.02, a Salpeter IMF and a limiting visual magnitude of M_V=-4.5. The incompleteness limit estimated by Fitzpatrick and Garmany (1990) for the HR diagram of LMC is also shown in the figure and the number of stars (RSGs excluded) above this limit is about the one contained in their HR diagram. A comparison with the observed diagram of LMC supergiant stars exhibits few distinctive features. First of all, despite the adoption of a limiting visual magnitude compatible with that indicated by Fitzpatrick and Garmany, the predicted number of O type stars above 20 $M_\odot$ is significantly larger than that observed. The argument of the brightest O stars being still embedded in molecular clouds cannot be invoked here, because masses down to 20 $M_\odot$ are involved, with an H-burning lifetime of about 10 Myr. Possible explanations of this discrepancy could be that either the assumption of continuous star formation is wrong, or the real IMF low is significantly steeper than the adopted one (Salpeter), or both. As far as the BHG is concerned, we see that it is still clearly visible even with the adopted spread in metallicity, but now its width is almost comparable to the error bar quoted by Fitzpatrick and Garmany. However the spread in metallicity also affects the morphology of the ledge, which becomes less evident than in the case of a single metal content. As a final remark note that the predicted temperatures of the RGS are slightly hotter (Log $T_{Eff} \simeq 3.6$) than those observed (Log $T_{Eff} \simeq 3.5$).

4. Discussion

The new stellar models computed in Padova, allow the analysis of the effects of the initial composition upon the evolution of massive stars. However, uncertainties in the mass loss rates which are as large as one order of magnitude for

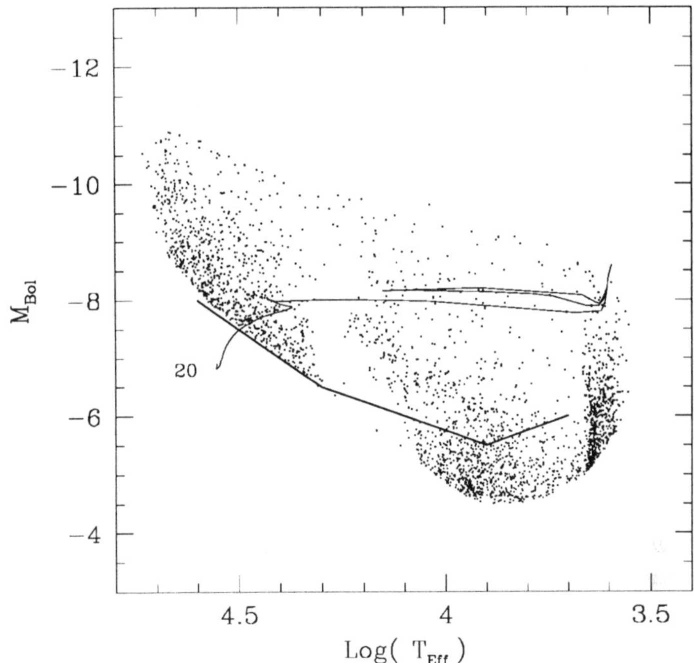

Fig. 3.   The simulated HR diagram of massive stars with metal abundance between 0.001 and 0.02 and brighter than M_V=-4.5. The 20 $M_\odot$ track with Z=0.008 and the limit of incompleteness in the diagram of Fitzpatrick & Garmany are also shown

the RSGs, could deeply modify the evolution of stars between 15 $M_\odot$ and 30 $M_\odot$ and, consequently, our understanding of their most advanced evolutionary phases. Current models reproduce the global properties of the HR diagrams of the brightest stars, but some features require a very detailed knowledge of the internal mixing. Finally the intriguing problem of the excess of the number of evolved stars with respect to that predicted by the theory demands an exhaustive investigation.

References

Alongi, M., Bertelli, G., Bressan, A., Chiosi, C.: 1991 *A&A* **vol. 224**,95
Bertelli, G., Bressan, A., Chiosi, C.: 1984. *A&A* **vol.130**,279
Blaha, C., Humphreys, R. M.: 1989,*AJ* **vol. 98**, 1598
Bressan, A., Bertelli, G., Chiosi, C.: 1981, *A&A* **vol. 102**, 25
Bressan, A., Fagotto, F., Bertelli, G., Chiosi, C.: 1993 *A&A*, **vol. 100**,647
Brunish, W. M., Truran, J. W.: 1982a. *ApJ* **vol. 256**, 247
Caughlan, G. R., Fowler, W. A.: 1988, *Atomic Data Nuc. Data Tables* **vol. 40**, 283
Chiosi, C., Bertelli, G., Bressan, A.: 1992a, In *Instabilities in Evolved Super and Hypergiants*, ed.
    C. de Jager, H. Nieuwenhuijzen. Amsterdam: North Holland, p. 145.
Chiosi, C., Bertelli, G., Bressan, A.: 1992b, ARA&A **vol. 30**, 305

Chiosi, C., Maeder, A.: 1986, ARA&A **vol. 24** , 239
Chiosi, C., Summa, C.: 1970. *Astrophys. Space Sci.* **vol. 8**, 478
de Jager C., Nieuwenhuijzen, H., van der Hucht, K. A.: 1988, *A&AS* **vol. 72**, 259
Fagotto, F., Bressan, A., Bertelli, G., Chiosi, C.: 1993a *A&A*, in press
Fagotto, F., Bressan, A., Bertelli, G., Chiosi, C.: 1993b *A&A*, in press
Feast, M.W.: 1992a, In *Instabilities in Evolved Super and Hypergiants*, ed. C. de Jager, H. Nieuwen-
   huijzen. Amsterdam: North Holland, p. 18.
Fox, M.W., Wood, P.R.: 1982, *ApJ* **vol. 259**,198
Fitzpatrick, E. L., Garmany, C. D.: 1990, *ApJ* **vol. 363**, 119
Fitzpatrick, E. L., Bohannan, B.: 1992, preprint
Garmany, C. D, Conti, P. S., Chiosi, C.: 1982 *ApJ* **vol. 263**, 777
Hamann, W.R., Koesterke, L., Wessolowski, U.: 1993, *A&A* **vol. 274**,397
Huebner, W. F., Merts, A. L., Magee, N. H., Argo, M. F.: 1977, Los Alamos Scientific Laboratory
   Report LA-6760-M
Humphreys, R.M., Davidson, K.: 1979, *ApJ* **vol. 232**, 409
Iglesias, C. A., Rogers, F. J., Wilson, B. G.: 1992, *ApJ* **vol. 397**, 717
Kudritzki, R. P., Pauldrach, A., Puls, J., Abbot, D. C.: 1989, *A&A* **vol. 219**, 205
Kudritzki, R.P.: 1990, In IAU Symp. 148, *The Magellanic Clouds*, eds. R. Haynes, D. Milne Kluver
Langer, N.: 1989, *A&A* **vol. 210**, 93
Langer, N.: 1991, *A&A* **vol. 252**, 669
Maeder, A.: 1991, *A&A* **vol. 242**, 93
Maeder, A.: 1992, In *Instabilities in Evolved Super and Hypergiants*, ed. C. de Jager, H. Nieuwen-
   huijzen. Amsterdam: North Holland, p. 138.
Maeder, A.: 1993, In *New Aspects of Magellanic Cloud Research* Lecture Notes in Physics, eds.
   B. Baschek, G. Klare and J. Lequeux Springer-Verlag **vol. 416**, 284
Roxburgh, I.W.: 1989, *A&A* **vol. 211**, 361
Schaerer, D., Meynet, G., Maeder, A., Schaller, G.: 1993, *A&AS* **vol. 98**, 523
Schaller, G., Schaerer, D., Meynet, G., Maeder, A.: 1992, *A&AS* **vol. 96**,269
Stothers, R.B.: 1991, *ApJ* **vol. 383**, 820
Tuchman, J., Wheeler, J. C.: 1990, *ApJ* **vol. 363**, 255
Wood, D.O.S., Churchwell, E.: 1989. *ApJ* **vol. 340**, 265
Zahn, J.P.: 1991, *A&A* **vol. 252**, 179

NUCLEOSYNTHESIS IN MASSIVE STARS

MOUNIB F. EL EID

Universitats-Sternwarte, Geismarlandstr. 11, D-37083 Gottingen, FRG

Abstract. We discuss three aspects of the nucleosynthesis in massive and intermediate–mass stars during their early evolutionary phases. These are related to the CNO abundances in giant or supergiant stars, to the ^{26}Al yield from massive stars via stellar wind, and to the production of the s–process nuclei in massive stars.

Key words: Stars–Nucleosynthesis

1. Introduction

Nucleosynthesis in stars is a fascinating aspect of their evolution, and a powerful tool to get more insight into their internal structures. The interpretation of the observed abundances in stars offer the possibility to estimate their masses and to find out in which evolutionary phases they are.

In particular, the CNO isotopes observed in giant or supergiant stars are extremely useful for testing our understanding of the physical processes occurring in these important objects. The first part of this paper concentrates on this issue.

The second part focuses on the ^{26}Al production during the early evolution of stars with masses $M \geq 25\,M_\odot$. The yield of this radioactive (mean life $\sim 10^6$ yr) nucleus by stellar wind during the red–supergiant phases or Wolf–Rayet (WR) phases is investigated. The contribution of the WR stars to the ^{26}Al yield is still not established due to many uncertainties in the model computations (mass loss, convection, etc.). The same holds for the other sites which contribute to this yield (supernovae, novae, and AGB stars).

Recent COMPTEL observations (Diehl et al. 1993a) of the 1.8 MeV gamma–ray line, which originates from the decay of ^{26}Al, provided the first detailed map of this γ–line emission of our Galaxy. This map shows that the 1.8 MeV line emission along the galactic plane extends over 60 degrees in longitude, and that the flux distribution is inhomogeneous. The COMPTEL observations seem to exclude a smooth and centrally peaked source population, such as novae (Diehl et al. 1993b)

In the third part of this paper, some comments are given related the production of the heavy nuclei with mass number A=65–90 by the s-process, (the so–called "weak component"). The central He–burning phase of stars with $M \geq 15\,M_\odot$ is considered, and current uncertainties in key nuclear reactions are described.

2. CNO isotopic ratios in giant stars

We have calculated a grid of stellar models in the mass range 1.7 to 15 $M_\odot$ with solar–like initial composition. In this mass range, data of the surface abundances

Space Science Reviews **66**: 383–389, 1994.

© 1994 *Kluwer Academic Publishers. Printed in Belgium.*

in giants or supergiants are available (Harris & Lambert 1984 (HL84); Harris et al. 1988 (HLS88), which can be used to test the stellar models for these objects. This is done by comparing the predicted surface isotopic ratios $^{12}C/^{13}C$, $^{16}O/^{17}O$ and $^{16}O/^{18}O$ after the first or second dredge–up phases with the observed values. A careful coupling between the stellar models and the nucleosynthesis is required to achieve this goal.

Our stellar models have been calculated with the Gottingen dynamical stellar evolution code. A description of the actual version of this code, and more details about these models are given by El Eid (1993). Here, we just mention the main assumptions used in the calculations, and present some results.

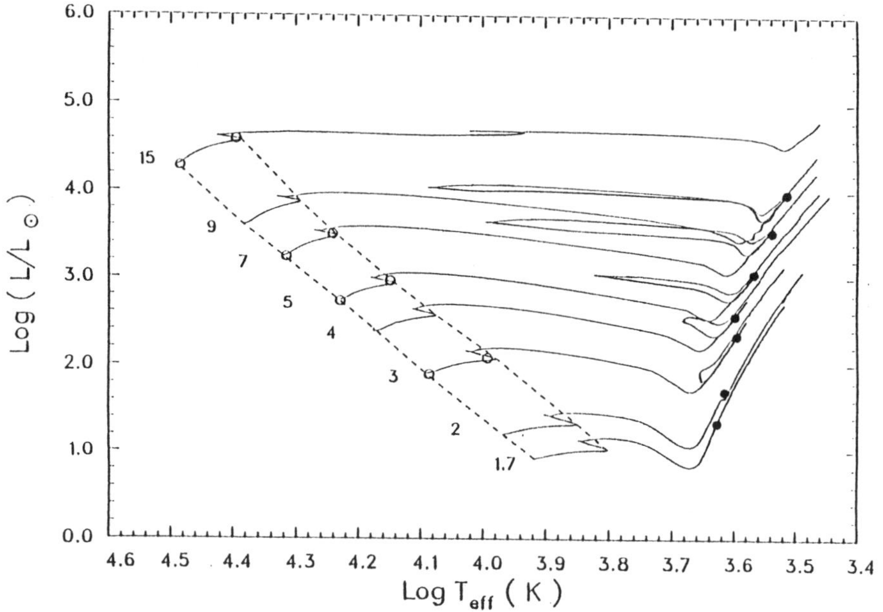

Fig. 1. Theoretical H–R diagram for the stellar masses investigated in this work. The initial composition is (X,Z)=(0.70,0.02) taken from Anders & Grevesse (1989). The superimposed open circles are the results of Stothers & Chin (1993) for the main sequence width. The filled circles indicate the positions beyond which the first dredge–up ceases to change the surface abundances.

We have used the OPAL opacities of Iglesias et al. (1992), and the Schwarzschild criterion for convective instability. In test cases, parameterized core overshooting has been included. The nucleosynthesis has been followed by a detailed network of nuclear reactions. For the H-burning, the pp-chain, CNO tri–cycle, Ne–Na cycle and Mg–Al cycle were included. For He–burning, the relevant reactions which determine the energy production were used. The reaction rates are mostly taken from the compilation of Caughlan & Fowler (1988, hereafter CF88), except of the reactions $^{17}O(p, \alpha)^{14}N$ and $^{17}O(p, \gamma)^{18}F$ which are taken from the revised compilations of Landre et al. (1990). For the $^{12}C(\alpha, \gamma)$ reaction the rate of CF88 is adopted, but multiplied by a factor of 1.7 (see Weaver & Woosley 1993).

Our predictions for the above isotopic ratios are also compared with those from other recent computations (Dearborn 1992 (D92); Schaller et al. 1992 (SCH92); Bressan et al. 1993 (B93)). In this way, we have gained a general overview about the predictions of standard evolutionary models related to the CNO abundances in giant or supergiant stars in the mass range up to 15 $M_\odot$. All these predictions have been compared with observations. Some of our results are summarized in the following.

TABLE I

The surface CNO isotopic ratios obtained in the present model computations. At each mass the first line gives the results after the first dredge–up phase, while the second line the modified values due to the second dredge–up, if any. The quantity [Na/Fe] denotes the enrichment of ^{23}Na relative to iron resulting from the contribution of the Ne–Na cycle. Note that $[X]=\log (X)_{star} - \log (X)_\odot$. The initial mass fraction ratios are from Anders and Grevesse (1989)

$M/M_\odot$	$^{12}C/^{13}C$	$^{16}O/^{17}O$	$^{16}O/^{18}O$	$^{14}N/^{15}N$	$^{12}C/^{14}N$	[Na/Fe]
1.7	19.89	534	551	1018	0.873	0.075
2.0	19.63	162	561	1171	0.762	0.121
3.0	19.14	225	562	1332	0.682	0.152
	18.89	225	562	1350	0.674	0.152
4.0	18.86	297	563	1351	0.684	0.151
	18.50	283	568	1393	0.668	0.155
5.0	18.73	407	564	1344	0.696	0.146
	18.17	378	572	1422	0.662	0.153
7.0	18.27	443	573	1457	0.652	0.162
	17.45	415	584	1595	0.609	0.173
9.0	17.76	518	576	1585	0.608	0.177
	16.75	487	589	1742	0.572	0.183
15.0	15.56	822	564	2194	0.464	0.258
initial	83.00	2465	442	252	2.75	0.0

(a) Fig. 1 shows the evolutionary tracks of our models. The $1.7\,M_\odot$ model was evolved up to He–ignition, the $2\,M_\odot$ model was evolved directly through core He–flash. All other models were evolved beyond core He–burning. The blue loops of the models between 3 and 9 $M_\odot$ occurring during core He–burning are more extended than obtained in models which include overshooting (e.g. SCH92; B93). On the giant branch, our models have lower T_{eff} than found in the computations of SCH92 and B93. This is due to the higher opacities we have used below 6000 K, i.e. outside the range of the OPAL opacity tables. The above features of our models are supported by the data of several giant stars (see El Eid 1993, for details).

(b) Table 1 summarizes the surface CNO isotopic ratios obtained from our stellar models after the first dredge–up and second dredge–up (if present) phases. Note that the enrichment of ^{23}Na relativ to iron is also given.

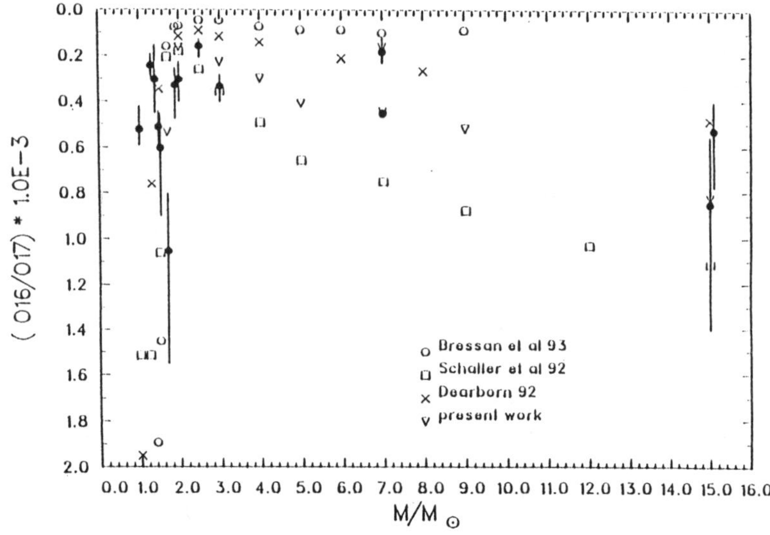

Fig. 2. The ^{16}O/^{17}O ratio versus stellar mass. The observed data (filled circles) and the masses of the observed giants are taken from HLS88. (see text for more details)

(c) In the above computations (including the present ones), the predicted ^{16}O/^{18}O ratio after the first dredge–up phase is found to be close to the initial (solar) value, and insensitive to the stellar mass. This is in good agreement with observations (HL84; HLS88), and implies that ^{18}O survives significant destruction in red giant stars. The weak variation of this isotope with stellar mass also implies that it cannot be used to constrain the maximum depth ot the convective envelope in red giant models.

(d) The predicted ^{12}C/^{13}C ratios after the first (or second dredge–up) phase are found to be close to each other despite of the different assumptions used in the above computations. In particular, overshooting hardly influences these ratios. In the above models, the predicted ^{12}C/^{13}C ratio exhibits a common basic feature: it decreases slowly with increasing stellar mass above $\sim 2\,M_\odot$, while a steep increase is found at lower masses. The former feature is consistent with the observed data on field giants and giants in open galactic clusters (Gilroy 1989). The latter feature is contrary to the observed data. A solution of this problem has not yet been found. The fact that low ^{12}C/^{13}C ratios are observed in giants of masses in the range of core He-flash, doese not seem to be an accident. But it is not yet clear whether the effect of core He–flash can help to resolve the above discrepancy. Eventually non–conventional models, which include rotationally induced mixing are also required (see Gilroy 1989, for a brief discussion).

(d) Fig. 2 shows a comparison between the $^{16}O/^{17}O$ ratios obtained in several computations as well as with the observed data on giant stars according to HLS88. There are several theoretical and observational aspects in this figure.

A difference is visible between the predictions over the whole mass range. At masses $\leq 1.7\,M_\odot$, this deviation is most likely due to the different maximum depth of the convective envelope obtained in these models. At masses above $2\,M_\odot$, the difference is mainly caused by the nuclear reactions destroying ^{17}O. In particular, the still uncertain $^{17}O(p,\alpha)^{14}N$ reaction limits the enhancement of the ^{17}O abundance in these stellar models. This conclusion is based on our comprehensive test computations (El Eid 1993) for the masses above $2\,M_\odot$. In these computations, we have tested the possible range of the above reaction suggested by the compilation of Landre et al. (1991). In Fig. 2, the results of SCH92 are obtained with a relatively high rate for this reaction, while our results correspond roughly to the minimum rate. The values of B93 in Fig. 2 are the lowest, since no destructions of ^{17}O is visible in these computations.

Concerning the observed data in Fig. 2, our investigations show that the masses of many giants are underestimated in the work of HLS88. If these masses are increased, an agreement is achieved with the atmospheric parameters used by HLS88 to derive the surface $^{16}O/^{17}O$ ratios in the giants shown in Fig.2.

We emphasize that not only the recompilation of the $^{17}O(p,\alpha)$ and $^{17}O(P,\gamma)$ reactions are requested, but more observations of massive giants with $M\geq 3\,M_\odot$ are also needed.

3. ^{26}Al production in massive stars

We are currently studying the production of ^{26}Al during the H-burning phases of stars with masses above 25 $M_\odot$. Stars in this mass range experience large mass loss by stellar wind and evolve through the WR phases (cf. SCH92). The yield of ^{26}Al from these stars is still not well determined because of uncertainties in the treatment of mass loss and convective mixing. Clearly, the amount of ejected ^{26}Al increases with initial metallicity, and with increasing stellar mass due to larger mass loss.

Using the same assumptions as described above, we have calculated the aluminum yield for a 25 and 35 $M_\odot$ stars of solar composition. Mass loss rates during the main sequence phase were taken from De Jager et al. (1988). Thereafter, we studied the effects of enhanced mass loss by artificially multiplying the above rate by factors ranging from 2 to 5. We find that the aluminum yield dependes sensitively on the assumed mass loss, in particular during the LBV phase for stars above 40 $M_\odot$. Our calcualtions show that the amount of ^{26}Al ejected in the wind steeply increases with mass from the 25 $M_\odot$ star ($\sim 6\ 10^{-6} M_\odot$) to the 35 $M_\odot$ star ($\sim 4\ 10^{-5} M_\odot$). The latter value is already larger than the yield of the 60 $M_\odot$ star studied by Prantzos (1991). Preliminary results for stars of masses higher than 35 $M_\odot$ suggest that our yield estimates are higher by about a factor of two compared to those obtained by Prantzos (1991). This increase in the yield has also been found by Meynet & Arnould (1993) in the case of a 60 $M_\odot$ star. We note that the actual enhancement is even larger when considering the fact that Prantzos adopted a 50% larger metallicity. A detailed discussion of these issues and the implications for γ–ray astronomy will be the subject of a paper en route.

4. Comments on the s-process nucleosynthesis in massive stars

In the standard scenario, the s–process nucleosynthesis in massive stars with $M \geq 15\,M_\odot$ contributes to the production of the heavy nuclei with mass number A=65–90. The neutrons needed for the synthesis of this so–called "weak component" of the s-process are produced by the $^{22}\mathrm{Ne}(\alpha,\mathrm{n})^{25}\mathrm{Mg}$ reaction towards the end of the central He–burning, at $T \geq 2.5\ 10^8$ K. The efficiency of this single neutron source is limited for to several reasons: (i) the neutrons emitted by the above reaction are partially re–captured by the residual nucleus $^{25}\mathrm{Mg}$, (ii) the abundance of $^{22}\mathrm{Ne}$ is limited, since it is only enriched by the reaction chain $^{14}\mathrm{N}(\alpha,\gamma)^{18}\mathrm{F}(\beta^+)^{18}\mathrm{O}(\alpha,\gamma)^{22}\mathrm{Ne}$, (iii) the production of neutrons is sensitive to the competing reaction $^{22}\mathrm{Ne}(\alpha,\gamma)^{26}\mathrm{Mg}$. Thus, the key reactions which determine the s–process nucleosynthesis in massive stars are $^{18}\mathrm{O}(\alpha,\gamma)^{22}\mathrm{Ne}$, $^{22}\mathrm{Ne}(\alpha,\gamma)^{26}\mathrm{Mg}$ and $^{22}\mathrm{Ne}(\alpha,\mathrm{n})^{25}\mathrm{Mg}$.

In recent experiments (Giesen et al. 1993), α–transfer reactions have been used to calculate the strengths of low energy resonances of the α–capture on $^{18}\mathrm{O}$ and $^{22}\mathrm{Ne}$. The main results of these investigations are as follows: an increase up to a factor 100 is found for the reaction $^{18}\mathrm{O}(\alpha,\gamma)$ in the range T_9=0.10–0.20 due to the resonance at 470 keV (not included previously). The reaction $^{22}\mathrm{Ne}(\alpha,\gamma)$ is strongly influenced by the resonance at 828 keV for $T_9 \leq 0.30$. A possible contribution of a resonance at 633 keV increases significantly the rate of the reaction $^{22}\mathrm{Ne}(\alpha,\mathrm{n})$ in the range $T_9 \leq 0.30$. In addition, an increase of the ratio of the (α,n) and (α,γ) rate is also found to be higher compared to the ratio of CF88.

Preliminary calculations of the s–process in stars of M=15 and 30 $M_\odot$ with these new rates have been performed by Baraffe & El Eid (1993). More extended computations are under way (Wiescher et al. 1993). In the following, we just summarize some basic conclusions obtained from the above calculations. The new higher rates lead to higher values of neutrons per iron seed, hence to a more efficient s–process in massive stars. The production of the s–process nuclei is shifted to higher mass numbers with A=80–90. In particular, the overproduction of the nuclei Sr, Y and Zr is higher than obtained on the basis of the CF88 rates. In the 15 $M_\odot$ star, the neutron production is more efficient, because more

^{22}Ne is destroyed. In the 30 $M_\odot$ star, a complete destruction of ^{22}Ne is achieved due to the higher temperature encountered in this case, and the increased neutron production is due to the higher ratio $(\alpha, n)/(\alpha, \gamma)$. Concerning the s-process yield from massive stars, one should be aware about the possibilty that stars more massive than $30\,M_\odot$ may collpase directly to black holes without contributing to the nucleosynthesis of the heavy elements. This possibilty has recently been emphasized by Brown et al. (1992) and Maeder (1992).

In summary, the s–process in massive stars is far from being established, and needs further theoretical and experimental efforts (see Wiescher et al. 1993 for more details)

References

Anders, E., Grevesse, N.: 1989, *Geochim. Cosmochim. Acta* **53**, 197

Baraffe, I., El Eid, M.F.: 1993, in "Proceeding of the 7th Workshop on Nuclear Astrophys.", eds. W. Hillebrandt & E. Muller, MPA, Garching, in press

Bressan, F., Fagotto, G., Bertelli, G., Chiosi, C.: 1993, *A&AS* 100, 674 (B93)

Brown, G.E., Bruenn, S.W., Wheeler, J.C.: 1992, *Comm. Astrophys.*, **16**, 153

Caughlan, G.R., Fowler, W.A.: 1988 *Atomic Data and Nuclear Data Tables* **40**, 283 (CF88)

Dearborn, D.S.P.: 1992, *Phys. Rep.* **210**, 367

De Jager, C., Nieuwenhuijzen, H., Van der Hucht, K.: 1988, *A&AS* **72**, 259

Diehl, R. et al.: 1993a, *A&AS*, **97**, 181

Diehl, R. et al.: 1993b, *Second Compton Symp.*, Sept. 20, Univ. of Maryland, AIP, in Press

El Eid, M.F.: 1993, CNO Isotopes in Red Giants, submitted to *A&A*, main journal

Giesen, U., Browne, C.P., Gorres, J., Graff, S., Iliadis, C., Trautvetter, H-P, Wiescher, M., Harms, V., Kratz, K.L., Pfeiffer, B., Azuma, R.E., Bucky, M., King, J.D.: 1993, *Nucl. Phys.*, submitted

Gilroy, K.: 1989, *APJ* **347**, 835

Harris, M.J., Lambert, D.L.: 1984 *APJ* **285**, 674 (HL84)

Harris, M.J., Lambert, D.L., Smith, V.V.: 1988, *APJ* **325**, 768 (HLS88)

Iglesias, C.A., Rogers, F.J., Wilson, B.G.: 1992, *APJ* **397**, 717

Landre, V., Prantzos, N., Auger, P., Bogaert, G., Levebre, A., Thibaud, K.P.: 1990, *A&A* **240**, 85

Maeder, A.: 1992, *A&A* **264**, 105

Meynet, G., Arnould, M.: 1993, in *Origin and Evolution of the Elements*, eds. N. Prantzos, E. Vangioni–Flam, and M. Casse, (Cambridge Univ. Press: Cambridge), p. 539

Prantzos, N.: 1993, *APJ* **405**, L55

Stothers, B., Chin, C-W.: 1993, Iron and Molecular Opacities and the Evolution of Pop I Stars, preprint (SC93)

Weaver, T.A., Woosley, S.E.: 1993, Nucleo-synthesis in Massive Stars and the ^{12}C$(\alpha, \gamma)^{16}$O Reaction Rate, *Phys. Rep.*, in press

Wiescher, M., Kappeler, F., Giesen, U., Gorres, J., Baraffe, I., El Eid, M.F., Raiteri, C., Busso, M., Gallino, R., Limongi, M., Chieffi, A.: 1993, in preparation

RESULTS OF EVOLUTIONARY COMPUTATIONS OF PRIMARIES OF CLOSE BINARIES WITH DIFFERENT INITIAL COMPOSITION.

C. DE LOORE AND D. VANBEVEREN

Dept. of physics, Vrije Universiteit Brussel, Brussels, Belgium

Abstract. The present paper summarizes fundamental evolutionary parameters of primaries of close binaries with initial masses between 9 and 60 $M_\odot$ and initial composition appropriate for the Galaxy, the LMC and the SMC. The primary timescales and WR binary timescales are compared with corresponding recent single star predictions.

Key words: Stars:Stellar Evolution – Stars:Massive Binaries

1. Introduction

Massive close binary evolution has been the subject of a large number of papers. For reviews we refer to de Loore (1980) and Vanbeveren (1991, see also these proceedings). The majority of evolutionary computations were performed with Galactic chemical abundances. A limited number of Magellanic Cloud massive close binary evolutionary computations was presented by Hellings and Vanbeveren (1981) and de Loore and Vanbeveren (1992).

Since then new opacity tables became available (OPAL,Iglesias et al.,1992) and we first decided to extend the MC computations mentioned above but with the new OPAL opacities. The simultaneous binary evolutionary code used by de Loore and De Greve (1992) and by De Greve (1993) for Galactic work was updated and adapted for the MC's. While doing so we found a computerbug in the code which made it advisable to recompute the Galactic evolution as well (also these calculations were performed with the OPAL opacities). The detailed results will appear in two papers (de Loore and Vanbeveren, 1993, 1994a). A full analysis will be presented by de Loore and Vanbeveren (1994b). Here we summarize some fundamental parameters for the primaries. Let us recall that these primary results are largely independent from the initial system parameters (period, mass ratio) or from the way the Roche lobe overflow (RLOF) is treated (see also Vanbeveren these proceeding). For the present paper we have also computed the evolution of a 60 $M_\odot$ primary for the Galaxy, the LMC and the SMC.

2. The evolution of primaries

The physical ingredients of our binary code should be entirely compatible with the single star code of Schaller et al. (1993), i.e. the same stellar wind mass loss rate formalism during core hydrogen burning (CHB) and the same treatment of convective core overshooting. A star in a massive binary is considered as a WR star when it is a core helium burning (CHeB) star in the blue part of the HR

Space Science Reviews **66**: 391–394, 1994.

© 1994 *Kluwer Academic Publishers. Printed in Belgium.*

diagram with mass larger than 5 $M_\odot$ (the minimum mass of observed WR binaries). Table 1 gives the core hydrogen burning lifetime (t_{CHB}), the core helium burning lifetime (t_{CHeB}), the mass at the beginning of CHeB (=mass at the end of RLOF or at the end of the LBV phase, see Vanbeveren these proceedings), the WN, WNE, WC lifetime, the mass at the beginning of the WNE or WC phase (remind that the mass at the beginning of the WN phase = mass of the primary at the beginning of the CHeB), the mass at the end of CHeB (=mass of the exploding star).   Figure 1 and 2 compare the binary timescales with massive

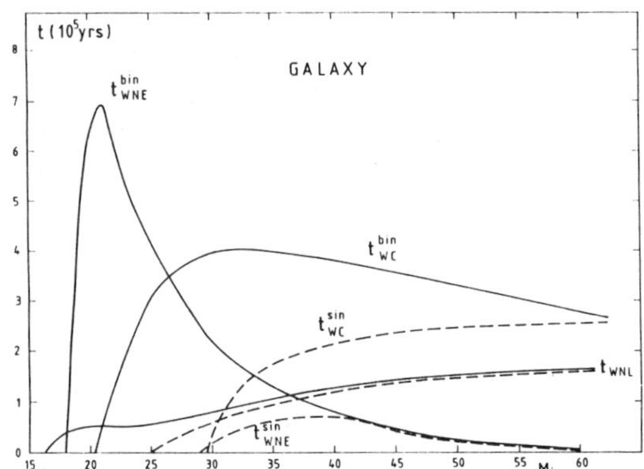

Fig. 1.   Galactic WR lifetimes as a function of initial mass for binary components and for single stars.

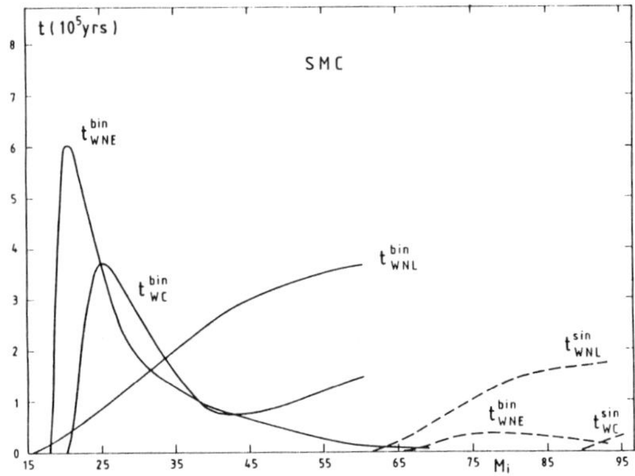

Fig. 2.   SMC WR lifetimes as a function of initial mass for binary components and for single stars.

single star predictions of Schaller et al. (1993) resp. for the Galaxy and for the SMC. Remark the following features:

   a. the overall evolution of primaries of massive close binaries evolving according to case B is very similar in the Magellanic Clouds and in the Galaxy. The

TABLE I

Lifetimes (million ys) of relevant evolutionary phases and mass at the end of RLOF, at the beginning of the WNE phase, at the beginning of the WC phase, just prior to the SN explosion

M_i	t_{CHB}	t_{CHeB}	t_{WN}	t_{WNE}	t_{WC}	M_f	M_{WNE}	M_{WC}	M_{SN}
				$Z = 0.02, X = 0.7$					
60	3.5	0.43	0.12	0.01	0.28	36	23	19.3	6
40	5.5	0.59	0.18	0.06	0.38	18.6	13.9	11.3	5.4
30	6.9	0.71	0.27	0.2	0.4	12.3	9.6	7.1	5
25	8.2	0.8	0.44	0.4	0.32	9.3	7.6	5.2	5
20	10.3	0.83	0.8	0.75	-	6.7	5.4	-	5
15	14.8	1.21	-	-	-	4.5	-	-	4.5
12	20.2	1.59	-	-	-	3.3	-	-	3.3
9	33.23	2.95	-	-	-	2.2	-	-	2.2
				$Z = 0.01, X = 0.74$					
60	4.3	0.44	0.17	0.01	0.26	37	23	19.6	6.2
40	6.1	0.62	0.2	0.06	0.36	19.5	14.1	11.7	5.8
30	7.7	0.73	0.26	0.2	0.41	13.5	9.9	7.2	5
25	9.1	0.74	0.42	0.36	0.36	9.6	7.4	5.3	5
20	11.4	1.03	0.99	0.95	-	6.8	5.3	-	5
15	16.0	1.23	-	-	-	4.5	-	-	4.5
12	22.4	1.68	-	-	-	3.3	-	-	3.3
9	36.3	2.93	-	-	-	2.1	-	-	2.1
				$Z = 0.002, X = 0.76$					
60	4.8	0.39	0.25	0.03	0.14	39	24	20.24	9.9
40	6.3	0.47	0.32	0.06	0.07	25.3	14.7	12.0	9.2
30	7.9	0.64	0.32	0.18	0.26	16.3	9.6	7.4	5
25	9.3	0.81	0.39	0.3	0.36	11.1	7.1	5.3	5
20	11.5	0.72	0.63	0.6	-	7.9	5.1	-	5
15	15.7	1.06	-	-	-	4.5	-	-	4.5
12	22.1	1.58	-	-	-	3.6	-	-	3.6
9	34.2	2.89	-	-	-	2.2	-	-	2.2

mass after RLOF M_f can be expressed as a function of the initial ZAMS mass M_i:

$$\log M_f = -1.054 + 1.459 \log M_i \text{ for Galactic primary components,}$$

$$\log M_f = -1.157 + 1.541 \log M_i \text{ for LMC primary components,}$$

$$\log M_f = -1.160 + 1.578 \log M_i \text{ for SMC primary components.}$$

b. the minimum initial mass of WR binary formation is about 16 $M_\odot$ independent from the metallicity,

c. the mass–luminosity relation of WR binary components is very similar in the MC's and in the Galaxy; combining all evolutionary computations with Galactic and MC chemical abundances, the following relations can be used:

$$\log L/L_\odot = 3.52 + 1.61 \log M/M_\odot \text{ for WNE, WC binary components,}$$

$$\log L/L_\odot = 3.62 + 1.63 \log M/M_\odot \text{ for WNL binary components.}$$

These relations hold for WR binary masses between 5 $M_\odot$ and 30 $M_\odot$.

b. independent from the metallicity, the majority of WNE binary components should originate from initial masses lower than 40 $M_\odot$, above 40 $M_\odot$ the WNE lifetime is very short.

c. the figures illustrate the very large difference between the WR lifetimes predicted by binary evolution and present day single star evolution, i.e. these figures illustrate how wrong one can be if in a theoretical prediction of WR star numbers binaries are omitted.

References

De Greve, J.P.: 1993, *A&AS* **vol. 224**,95
Hellings, P. and Vanbeveren, D: 1981,*A&A* **vol.95**,14
de Loore, C: 1980, *Space Sci. Rev.* **vol.26**,113
de Loore, C. and De Greve, J.P.: 1992,*A&AS* **vol.94**,453
de Loore, C. and Vanbeveren, D: 1992,*A&A* **vol.260**, 273
de Loore, C. and Vanbeveren, D: 1993,*A&AS* **in press**
de Loore, C. and Vanbeveren, D: 1994a,*A&AS* **in press**
de Loore, C. and Vanbeveren, D: 1994b,*A&A* **in press**
Iglesias, C. A., Rogers, F. J., Wilson, B. G.: 1992, *ApJ* **vol. 397**, 717
Schaller, G., Schaerer, D., Meynet, G., Maeder, A.: 1992,*A&A* **vol.96**,269
Vanbeveren, D.: 1991, *Space Sci. Rev.* **vol.56**, 249

THE MASS AND HELIUM DISCREPANCY IN MASSIVE STARS: THE CASE VELA X–1.

D. VANBEVEREN
Dept. of Physics, V.U.B., Pleinlaan 2, 1050 Brussels, Belgium.

A. HERRERO
Instituto de Astrofísica de Canarias, 38200 La Laguna, Tenerife, Spain.

D. KUNZE
Universitats–Sternwarte, Scheinerstr. 1, D–8000 Munchen 80, Germany.

and

M. VAN KERKWIJK
Astrophysical Institute, University of Amsterdam, Kruislaan 403, NL–1098 SJ Amsterdam, The Netherlands.

Abstract. A NLTE–analysis is presented of high S/N spectra of the optical component of the standard massive X–ray binary Vela X–1. In combination with the orbital parameters we conclude that the optical star is highly helium enriched and is significantly overluminous compared to standard evolutionary tracks of massive accretion stars. We then propose a new accretion model able to explain these features.

Key words: Stars:Atmospheres –Stars:Evolution – Stars:Binaries

1. Introduction.

Based on a NLTE–analysis of spectra of a significant number of early type supergiants, Herrero et al. (1992) concluded that the spectroscopic masses (determined from Teff, log g and the distance) were systematically lower than the evolutionary masses. An obvious arbiter for this discrepancy may be an eclipsing binary with an early type supergiant component where the orbital mass is known. One of us (M. v. K.) obtained high S/N spectra of the optical star of the eclipsing standard massive X–ray binary Vela X–1 in the spectral range 4175 – 4525 A. In order to avoid contamination of the X–rays we used the spectra around phase 0. In section 2 a NLTE analysis is presented for this B0.5 I star using the plane–parallel, hydrostatic code of D. Kunze. Section 3 combines the atmospheric results with orbital and X–ray eclipse results in order to determine the most probable location of the supergiant in the HR diagram. In section 4 we present a grid of evolutionary tracks for massive accretion stars in order to compare the evolutionary mass and the spectroscopic mass.

2. Spectroscopic Analysis.

To perform the spectroscopic analysis we use the NLTE model atmospheres described in Herrero et al., 1992. From the final parameters given below it is

Space Science Reviews **66**: 395–400, 1994.

© 1994 *Kluwer Academic Publishers. Printed in Belgium.*

clear that we had to calculate several new models outside the limits of our standard grid. The method of analysis is also described in the reference above.

The spectrum used here covers the range between 4175 and 4525 A. This means that we will use Hγ, HeI 4388 and HeII 4200. The HeII 4541 line, usually included in our analysis, is not in our spectra. A correction of 15 km/s bluewards was applied to the spectrum. A projected rotational velocity of 125 km/s was derived from eleven metal lines of C, N, O, Si and Mg.

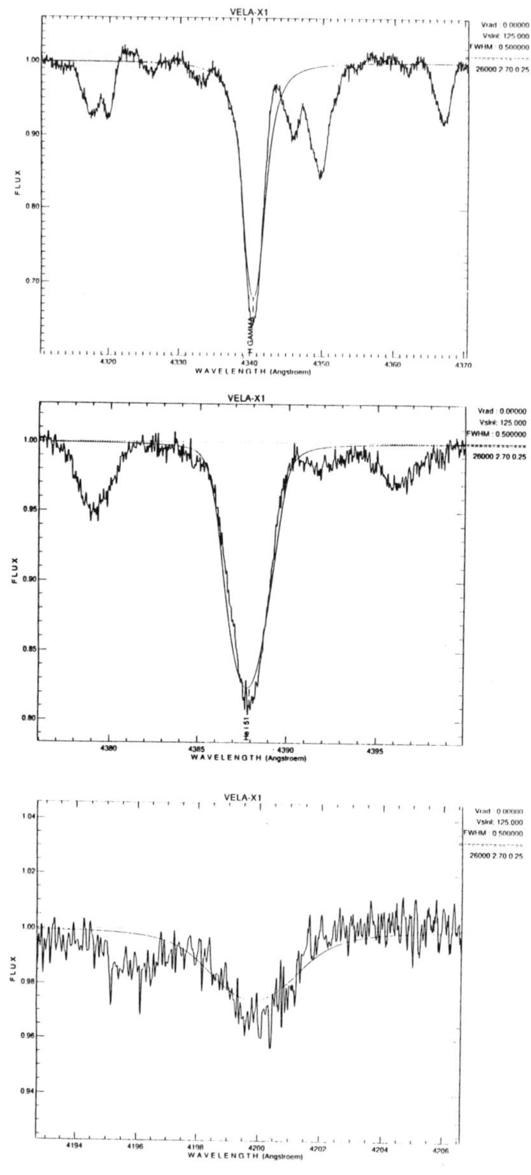

Fig. 1.   Theoretical best fits to the observed lines.

Fig. 1 shows the fits to the observed lines. The best fit is obtained for Teff = 26000 K, log g = 2.70 and $\epsilon = N_{He}/(N_H+N_{He}) = 0.28$. We can see in Fig. 1 that the wind contamination of Hγ in the red wing is large. A strong dilution effect, characteristic of stars of low gravity (see e.g. Herrero et al., 1992), is present in HeI 4471. A slightly lower log g would probably be more adequate. This is indicated not only by the blue wing of Hγ, but also by the HeI 4388 (slightly too broad) and the HeII 4200 (slightly too shallow) lines. Unfortunately, we couldn't bring a model with a lower log g to convergence due to the proximity to the Eddington limit. Also we find from our models that 2.60 would give a worse fit, so that we finally adopt the values given above as the best fit for plane parallel, hydrostatic models.

To correct for the centrifugal force we adopt an inclination angle of 73° (Khruzina and Cherepashchuk, 1986), which means that the rotational velocity is 130 km/s.Remark however tat using 90° would give very similar results.

3. The Most Probable Parameters of The Optical Component of Vela X–1.

From the atmospherical NLTE analysis (section 2): log g = 2.7 uncorrected for centrifugal force, Teff = 26000 K, the number ratio $\epsilon = 0.28$, the rotational velocity = 130 km/s

From the orbit and X–ray eclipse (Rappaport and Joss, 1983; see also Joss and Rappaport, 1984) the following mass and radius are obtained: R = 28 - 35 $R_\odot$, M = 21.5 - 26.5 $M_\odot$ (these values are 95

Combining both sets of results it is possible to determine a fairly narrow range of stellar parameters satisfying the orbital, X–ray eclipse and atmospheric analysis: if M = 21.6 $M_\odot$ – 26.5 $M_\odot$, then R = 31 $R_\odot$ – 35 $R_\odot$, log g = 2.75 – 2.76, log $L/L_\odot$ = 5.5 – 5.7 and the distance d = 1.8 – 2 kpc. Remark that our distance values correpond within the uncertainties to the distance of 2.2 kpc derived by Zuiderwijk and van den Heuvel (1974) from the strenght of the interstellar Ca II lines.

4. Comparison with Evolutionary Results of Massive Mass Gainers.

The evolutionary code uses Rogers-Iglesias opacities, overshooting parameter $\alpha = 0.2$, the stellar wind (SW) mass loss rate during core hydrogen burning is computed using the semi-empirical relation of de Jager et al. (1988), semi-convection is treated according to the SH criterion, Galactic composition i.e. (X,Y,Z) = (0.7, 0.28, 0.02). The method used is as follows: at a certain moment during the core hydrogen burning phase of a massive star, the star is subjected to a mass gain process with a mass gain rate = 0.0005 $M_\odot$/yr. This treatment of the accretion process is more than sufficient for the purpose of the present paper since we are primarily interested in how the mass gainer looks like after the accretion phase and how the star further evolves. The process is stopped when the gainer has a mass between 22 $M_\odot$ and 30 $M_\odot$. The further core hydrogen burning phase is then continued with SW. The values 22 - 30 $M_\odot$ above asure a mass at the end of core hydrogen burning appropriate for the optical counterpart of Vela X–1. It is still a matter of debate how a star react during a (rapid) mass gain process in a binary. We have used two accretion models:

a. the accretion process does not destroy the condition of radiative equilibrium in the outer layers of the mass gainer (the usual accretion model, Benson, 1970a,b).

However accretion of matter from a primary star onto a secondary component (thus also accretion of orbital angular momentum) may cause a (rapid) spinning up of that secondary, thus may cause efficient convection. Furthermore there are situations (late case B or case C) where the real accretion rates are very high. It is therefore not unrealistic to assume that there are situations where the mass gainer becomes completely convective much like in a protostar. This is our second model i.e.

b. the accretion process makes the whole star convective, i.e. the whole star is mixed and becomes homogeneous.

Figure 2 shows evolutionary tracks of accretion stars using accretion model a. The situation where accretion starts close to the ZAMS occurs in binaries with mass ratio smaller than 0.5. The larger the mass ratio the later starts accretion. The figure also shows tracks when the accretion starts near the end of core hydrogen burning (mass ratio close to 1). The numbers along the tracks denote mass values. The most probable position of the optical star of Vela X–1 is given as well.

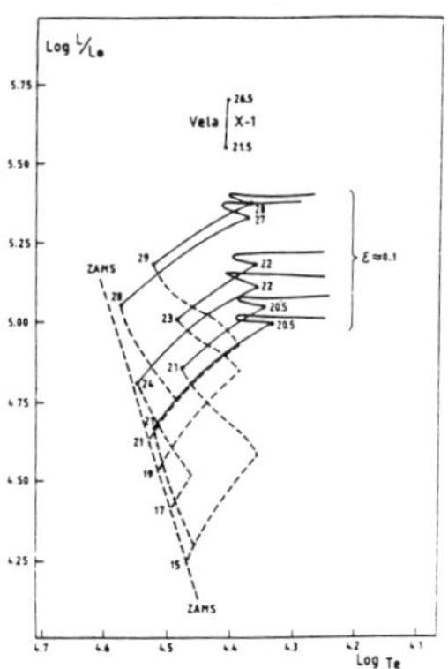

Fig. 2. Evolutionary tracks of accretion stars using accretion model a.

We conclude that the evolutionary prediction does not at all correspond with the observed position, i. e. we note a severe mass discrepancy (or overluminosity). Furthermore all evolutionary sequences predict a ϵ value = 0.09 (the observed value = 0.28), thus we also have a factor 3-4 helium discrepancy. The foregoing conclusion does not depend on the treatment of semi–convevtion.

Figures 3 and 4 show evolutionary tracks of accretion stars using model b described above. In figure 3 (resp. fig. 4) accretion starts when the central hydrogen abundance by weight $X_C = 0.3$ (resp. 0.1)

We notice the following features:

a. when accretion starts later during core hydrogen burning, the resulting hydrogen abundance after homogenization is smaller and thus the overluminosity of the evolutionary tracks after the accretion phase is larger,

b. in figure 4, the accretion process starts near the end of core hydrogen burning (appropriate for binaries with mass ratio close to 1); after the accretion process the stars have a ϵ value = 0.23 (close to the observed value in Vela X–1) and the tracks cross consistently the observed position in the HR diagram of the B0.5Ib component of Vela X–1.

c. obviously semi–convection or the way it is treated is completely unimportant here.

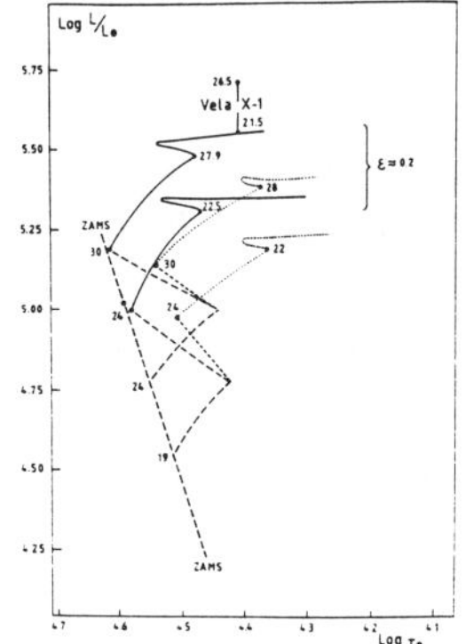

Fig. 3.   Evolutionary tracks of accretion stars using accretion model b; accretion starts when X_C = 0.3.

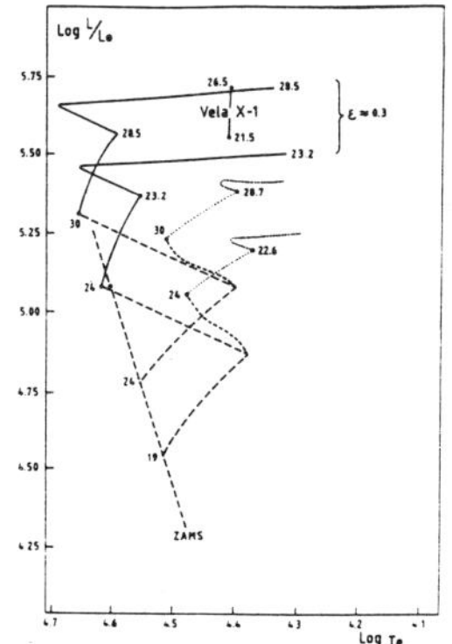

Fig. 4.   Evolutionary tracks of accretion stars using accretion model b; accretion starts when X_C = 0.1.

5. Conclusion

Combining orbital and X-ray eclips observations, and the NLTE spectral analysis of the optical counterpart of the standard massive X-ray binary Vela X–1, it is posible to define a restricted region in the HR diagram for this star and to deduce a value for the helium to hydrogen number ratio. The star is highly overluminous when compared to standard tracks of massive accretion stars and it is significantly helium enriched. Comparison with evolutionary tracks of massive accretion stars we are able to reproduce these 'observations' only if it is assumed that during the accretion phase the gainer (= the progenitor of Vela X–1) becomes completely convective; after the accretion phase the gainer has a considerably reduced hydrogen abundance (X=0.45-0.5 by weight); we expect the star to be overabundant in nitrogen and underabundant in carbon and oxygen.

References

Benson, R.S.: 1970a, *Ph. D. Thesis*, Univ. of California: Berkeley.
Benson, R.S.: 1970b, *Bull. Am. Astron. Soc.* **2**, 295.
de Jager, C., Nieuwenhuyzen, H.and van der Hucht, K.: 1988, *Astron. Astrophys. Suppl.* **72**, 259
Herrero, A., Kudritzki, R.P., Vilchez, J.M., Kunze, D., Butler, K. and Haser, S.: 1992, *Astron. Astrophys.* **261**, 209.
Hiltner, W.A., Werner, J. and Osmer, P.S.: 1972, *Ap. J. Letters* **175**, L19.
Khruzina, T.S. and Cherepashchuk, A.M.: 1986, *Sov. Astron.* **30**, 422.
Maeder, A.: 1990, *Astron. Astrophys. Suppl.* **84**, 139.
Rappaport, S. and Joss, P.C.: 1983, '' in W.H. G. Lewin, E.P.J. van den Heuvel, ed(s)., *Accretion Driven Stellar X-ray Sources*, Cambridge Univ. Press: Cambridge,
Joss, P.C. and Rappaport, S.: 1984, *Ann. Rev. Astron. Astrophys.* **22**, 537.
Zuiderwijk, E.J. and van den Heuvel, E.P.J.: 1974, *Astron. Astrophys.* **35**, 353.

ON THE EVOLUTION OF SECONDARY COMPONENTS
IN MASSIVE CLOSE BINARY SYSTEMS

H. BRAUN and N. LANGER

MPI fur Astrophysik, Karl-Schwarzschild-Str. 1, 85740 Garching, Germany

Abstract. For the evolution of the secondary component of a massive close binary system, it is generally assumed that the mass accretion during core H-burning simply leads to its rejuvenation, i.e. that it evolves like a normal main sequence star with a mass corresponding to its mass after the accretion ceased. We reinvestigate this problem in the framework of a time-dependent semiconvection theory. We find that the process of adaptation of the convective core size to the new (larger) stellar mass may not be completed until core hydrogen depletion, i.e. no rejuvenation occurs. The resulting secondaries show strong differences compared to single stars of same mass.

Key words: massive stars – binaries – accretion

1. Introduction and Input Physics

We have investigated the evolution of accreting massive main sequence stars by utilizing a code which is based on a hydrodynamic stellar evolution code for single stars and includes mass loss according to Kudritzki et al. (1987) and de Jager et al. (1988), OPAL opacities (Rogers and Iglesias 1992) and gravitational energy release due to mass accretion according to Neo et al. (1977). We have calculated models from $12 M_\odot$ to $20 M_\odot$ at a metallicity of $Z = Z_\odot/4$, which accretes various amounts of matter with constant rates of 10^{-4} and $10^{-3} M_\odot/yr$ during core hydrogen burning. We have assumed that the accreted material has the same chemical composition as the surface material of the accreting star. Convection occurs if the Ledoux criterion is fulfilled and semiconvection if $\nabla_{ad} < \nabla < \nabla_{Led}$. Semiconvective mixing is treated as a time-dependent diffusion process with a diffusion coefficient according to Langer et al. (1983). A semiconvective efficiency parameter of $\alpha_{sc} = 0.04$ has been used (cf. Langer et al. 1989), except in one model where $\alpha_{sc} = 0.01$.

The main differences of ours to previous calculations is that we have considered the time scale of semiconvective mixing for the penetration of the mean molecular weight (μ) barrier on top of the convective core. In contrast, in the standard rejuvenation picture (Hellings 1983) the convective core which due to the increased luminosity after the accretion tends to grow in mass is allowed to penetrate the μ-barrier instantaneously.

2. Results

Fig. 1 shows the regions of convection and semiconvection in a $20 M_\odot$ star which is accreted by $3 M_\odot$ at a central helium mass fraction of $Y_c = 0.706$. After accretion has ceased, the convective core remains constant in size and even

Space Science Reviews **66**: 401–404, 1994.

© 1994 *Kluwer Academic Publishers. Printed in Belgium.*

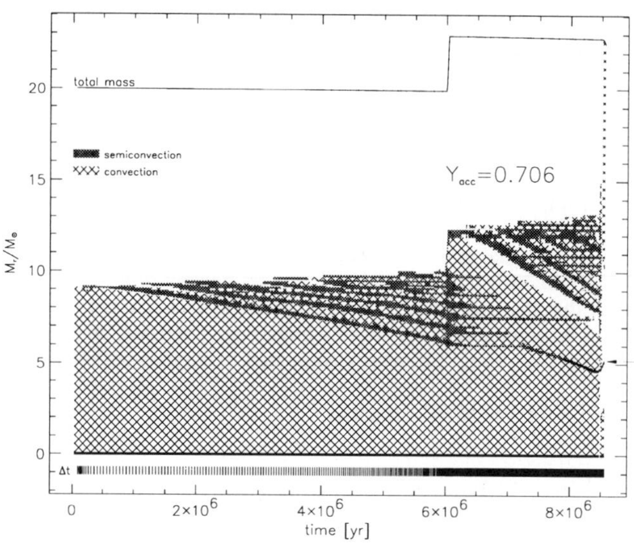

Fig. 1. The convective structure of a $20M_\odot$ star, which accretes $\Delta M = 3M_\odot$ at a central helium mass fraction of $Y_c = 0.706$ during core hydrogen burning. The arrow on the right side indicates the initial size of the helium core in a $23M_\odot$ single star (cf. Fig. 2).

starts to retreat again after a while: due to the increased mass, the luminosity and thus ∇_{rad} are increased all over the star. However, the convective core cannot grow because of the μ barrier ($\nabla_{rad} < \nabla_{Led}$). During the further evolution ∇_{rad} decreases again until $\nabla_{rad} = \nabla_{ad}$ at the convective core boundary, and the core starts to retreat. Note that the retreating core is smaller than that of a single star with the same total mass.

The consequences of this behaviour for the evolutionary track in the HR diagram can be seen in Fig. 2a, which shows the evolutionary tracks of $20M_\odot$ stars accreting $3M_\odot$ at different times. Due to the smaller convective cores, the core hydrogen burning luminosities after accretion are the smaller the later accretion occurs. During the further evolution, hydrogen is mixed by semiconvection from the layers above into the core (but its hydrogen content is still decreasing). Due to this process, the μ-gradient diminishes and, in the model with the earliest accretion (Y_c at the onset of accretion=: $Y_{acc} = 0.432$), vanishes before the core starts to retreat: the star has adopted the same core size and chemical structure as a $23M_\odot$ star, and its further evolution in the HR diagram cannot be distinguished from that of a single star. The models with later accretion do not adapt their convective core to the single star core size, not even until core hydrogen depletion. Consequently, for later accretion the He-core is smaller and the shell hydrogen content higher (cf. Fig. 2b).

Beside the parameter Y_{acc}, there are two other parameters which control whether a star is rejuvenated due to accretion. One of them is the amount of

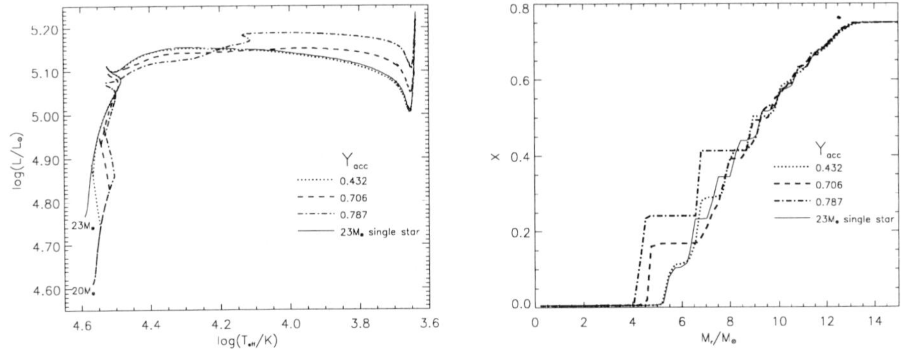

Fig. 2.   a) Evolutionary tracks of three $20M_\odot$ models, which accrete $\Delta M = 3M_\odot$ at different times during core hydrogen burning as denoted by the central helium mass fraction at the occurrence of accretion. In addition the track of a single $23M_\odot$ star is shown. The star with $Y_{\rm acc} = 0.787$ remains blue during core helium burning. $(T_{eff} \approx 4.2)$.
b) Hydrogen profiles at core hydrogen exhaustion for the models in a). The hydrogen burning shell is located on top of the helium core (first "step" of the profile).

accreted material ΔM. The increase in luminosity due to a larger ΔM increases the diffusion coefficient $D_{\rm sc}$ and thus, semiconvective mixing may become efficient enough to increase the hydrogen content of the convective core with time. This happens e.g. in a $20M_\odot$ star accreting $8M_\odot$, where the star rejuvenates even if the accretion occurs as late as $Y_{\rm acc} = 0.787$.

The third important parameter is the initial mass $M_{\rm i}$ of the star. For smaller initial masses, rejuvenation occurs for a wider range of $Y_{\rm acc}$ and ΔM.

The three parameters $Y_{\rm acc}$, ΔM and $M_{\rm i}$ are in principle specified for any given binary system. Only the semiconvection parameter $\alpha_{\rm sc}$, which contains the uncertainty in our semiconvection model, is slightly restricted. Therefore, one may hope to determine it from the study of massive close binary systems. As an example, Fig. 3 shows the evolutionary track of an accreting $20M_\odot$ model with $\alpha_{\rm sc} = 0.01$. This model remains in the blue part of the HRD during core helium burning and even performs its SN explosion as a blue supergiant. Note that in previous calculations of binary models which obtain a blue colour for the pre-supernova stage, the mass transfer starts only when the secondary has already terminated core hydrogen burning, which restricts the initial mass ratio to values very close to 1 (cf. e.g. Podsiadlowski and Joss 1989, de Loore and Vanbeveren 1992).

3. Conclusion

The consideration of semiconvection as a time-dependent mixing process strongly affects stars which accrete matter during core hydrogen burning. The standard rejuvenation picture only applies for large amounts of accreted matter, small

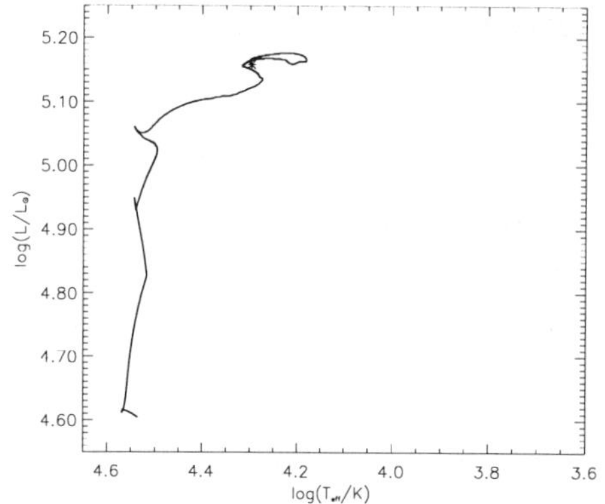

Fig. 3. Evolutionary track of a $20M_\odot$ model with $\Delta M = 3M_\odot$, $Y_{acc} = 0.708$ and $\alpha_{sc} = 0.01$ (cf. Fig. 2a, dashed line). The asterisk indicates the position of SN explosion.

initial masses, early accretion and large α_{sc}. Otherwise, the stars develop smaller helium cores, stay more preferably in the blue part of the HRD during core helium burning and may even explode as a blue supergiant, like the progenitor of SN 1987A.

Acknowledgements

This work is supported in part by the Deutsche Forschungsgemeinschaft through grant La 587/8-1.

References

Hellings P.: 1983, *ApSS*, **96**, 37
de Jager C., Nieuwenhuijzen H., van der Hucht K.A.: 1988, *A&AS*, **72**, 259
Kudritzki A.P., Pauldrach A., Puls J.: 1987, *A&A* , **173**, 293
Langer N., El Eid M.F., Baraffe I.: 1989, *A&AL*, **224**, 217
Langer N., Sugimoto D., Fricke K.J.: 1983, *A&A*, **126**, 207
de Loore, C. and Vanbeveren, D.: 1992, *A&A*, **260**, 273
Neo S., Shigeki M., Nomoto K., Sugimoto D.: 1977, *PASJ*, **29**, 249
Podsiadlowski, Ph. and Joss. P.C.: 1989, *Nature*, **338**, 401
Rogers F.J., Iglesias C.A.: 1992, *ApJS*,**79**, 507

DOUBLE-ZONE MODEL WITH DIFFUSIVE MIXING AND THE MASS AND HELIUM DISCREPANCIES IN OB-STARS

P. DENISSENKOV*
Max-Planck-Institut fur Astrophysik,
Karl-Schwarzschild-Strasse 1, 85748 Garching, Germany

Abstract. A model for massive main sequence (MS) stars is proposed that quantitatively accounts for the mass and helium discrepancies in luminous OB stars. The radiative envelope of the model consists of two zones being mixed by rotationally induced turbulent diffusion during the star's evolution on the MS. The rate of the mixing in the outer zone is assumed to be substantially lower than that in the inner zone. Both, the mass and helium discrepancy, are shown to be due to helium enrichment in the envelope produced by turbulent diffusion. Some arguments to support this double-zone stellar model are given.

Key words: Early-Type Stars – Abundances – Evolution – Rotation – Mixing

1. Introduction

One of the most unexpected contradictions between theory of evolution and observations of massive main sequence stars revealed during the last few years seems to be the so-called "mass discrepancy". It has been found that masses of luminous OB-stars determined from spectroscopic analysis $M_{\rm sp}$ are systimaticaly higher than those obtained by comparing the stars' location in the Hertzsprung-Russel diagram (HRD) with theoretical evolutionary tracks $M_{\rm ev}$. Besides, in some of these stars large atmospheric helium abundances $\varepsilon_{\rm He} \geq 0.16$ have been detected, what in addition gave rise to the "helium discrepancy" problem (see Herrero et al. 1992, hereafter H92, and references therein).

Langer (1992) and Weiss (1993) were the first who supposed that both discrepancies mentioned above might result from internal mixing of the star caused by its rotation. Due to the mixing the star evolves into the region of high luminosities in the HRD which would be occupied by the more massive stars if the latter did not have such mixing.

Let us assume that the observed He enrichment of the OB-stars' atmospheres is produced by rotationally induced turbulent diffusion (Zahn 1983), and that the time $\tau_{\rm mix}$ during which He is transported from the convective core to the stellar surface is proportional to the mass discrepancy $M_{\rm ev} - M_{\rm sp}$. One can expect the atmospheric helium abundance $\varepsilon_{\rm He}$ to grow with the product $D_t \tau_{\rm mix}$, where D_t is the turbulent diffusion coefficient which can be roughly estimated as $D_t \sim (V \sin i)^2 / Rg$ (Zahn 1983), where $V \sin i$ is the projected rotational velocity, R the radius and g the surface gravity of the star. In Fig. 1 (points) we have plotted $\varepsilon_{\rm He}$ as a function of the parameter $P = \log[(M_{\rm ev} - M_{\rm sp})(V \sin i)^2 / Rg]$ by data

* On leave from the Astronomical Institute of the St. Petersburg University (Russia), as an Alexander von Humboldt Fellow

© 1994 *Kluwer Academic Publishers. Printed in Belgium.*

P. DENISSENKOV

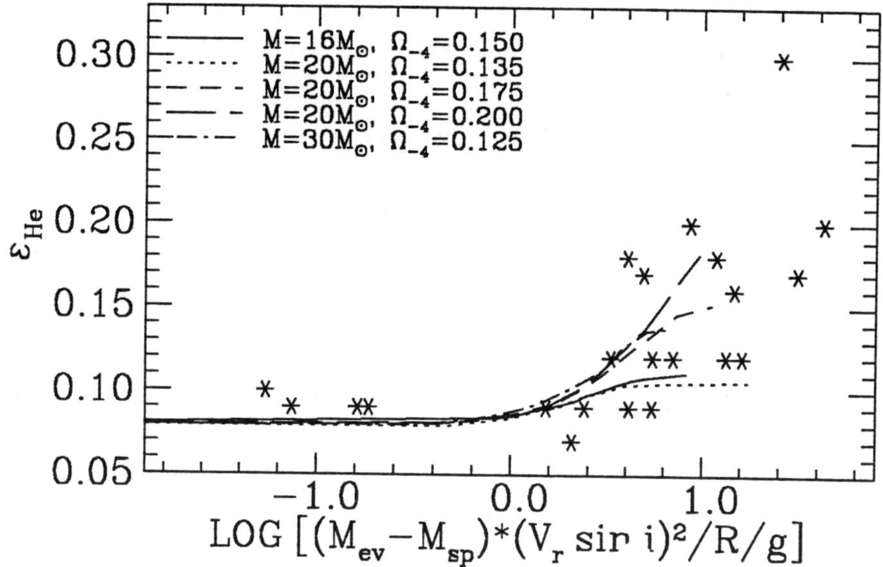

Fig. 1.   Comparison of the observational (points) and theoretical (curves) dependences of the atmospheric helium abundance in OB-stars from the parameter characterising the efficiency of turbulent diffusive mixing.

of H92. We see that ε_{He} does correlate with $P \sim \log(D_t \tau_{mix})$, which can be considered as an observational evidence in favour of the hypothesis about the presence of turbulent diffusion in OB-stars. It should be noted that there are no good correlations of ε_{He} with $\log(M_{ev} - M_{sp})$ and $\log[(V \sin i)^2 / Rg]$ separately.

2. Theoretical models and comparison with observations

In order to quantitatively account for the mass discrepancy and the correlation presented in Fig. 1 (points) we propose the following model.

The radiative envelope of the star is assumed to be mixed by rotationally induced turbulent diffusion in accordance with the description given by Zahn (1983). A special feature of our model is that turbulent diffusion is allowed to mix the stellar matter with the full rate only in the region confined between the convective core boundary M_{cc} and a level with the mass coordinate M_r^{mix} located near the stellar surface (the "inner zone"). In the outer zone (above the point M_r^{mix}) we reduce the rate of mixing dividing it by a factor $RED \gg$ 1. Thus, we consider *a double-zone model with diffusive mixing* which can be specified with the following three parameters: M_r^{mix}, RED, and the rotation parameter $\Omega_{-4} \equiv \Omega/10^{-4}$, where Ω is the surface angular velocity of the star.

The corresponding evolutionary calculations have been performed with the program described by Denissenkov (1993). An important point to mention is that the Schwarzschild criterion for convection has been used to determine the

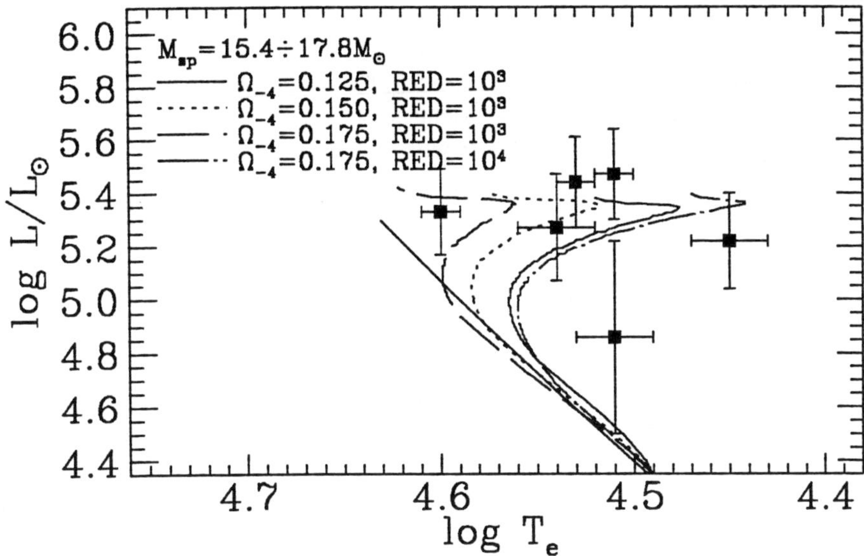

Fig. 2. Comparison of locations in the HRD of OB-stars with $M_{\mathrm{sp}} \approx 16 M_\odot$ and evolutionary tracks with turbulent diffusion calculated under assumption that $M = 16 M_\odot$ and $M_{\mathrm{r}}^{\mathrm{mix}} = 15.5 M_\odot$. The diagonal solid line is the zero-age main sequence.

location of the convective core boundary. As it has been pointed out by Langer (1992), in this particular case the convective core of a rapidly rotating star grows with time due to the helium enrichment in the envelope and the consequently larger luminosity, caused by turbulent diffusive mixing. That did occur in our calculations.

In Fig. 2 evolutionary tracks for the 16 $M_\odot$ double-zone model are compared in the HRD with the location of a group of OB-stars with $M_{\mathrm{sp}} \approx 16 M_\odot$ from H92. The tracks have been calculated under assumption that $M_{\mathrm{r}}^{\mathrm{mix}} = 15.5 M_\odot$, for different values of Ω_{-4} and RED. We see that the calculations reproduce quite well the observed displacement of the OB-stars into the region of higher luminosities. Analysing Fig. 2 we also infer that the above defined quantity τ_{mix} corresponds most probably either to the time required for the star to reach the turning point in the HRD, after that its effective temperature begins to decrease, or simply to the age of the star if it has not yet reached the turning point.

Now we can find theoretical dependences of M_{ev} and $\varepsilon_{\mathrm{He}}$ from the age of the star t by comparing successive locations of the double-zone model in the HRD with the standard evolutionary tracks.

It has turned out that the normalized theoretical mass discrepancy $(M_{\mathrm{ev}} - M)/M$ increases monotonically with the growth of the quantity t/t_{MS} where t_{MS} is the MS life time of the star. Moreover, the dependences of $(M_{\mathrm{ev}} - M)/M$ from t/t_{MS} for different values of M and Ω_{-4} are very similar. This feature, firstly, confirms our assumption that τ_{mix} is proportional to the mass discrepancy and, secondly, explains the strange fact that all the OB-stars independent of their mass and rotational velocity follow approximately the same trend in Fig. 1 (points). Regarding theoretical dependences of $\varepsilon_{\mathrm{He}}$ from $\log P$ (Fig. 1, curves), their shapes look practically identical with that of the observational correlation, but they turn out to be shifted by $\Delta \log P \approx 0.6$ to the left of it. If we decreased four times the diffusion coefficient in our calculations the shift would be compensated. We ascribe this quantitative disagreement to uncertainties in Zahn's equations.

In the conclusion we give some arguments supporting our model.

According to Zahn (1992), in the radiative envelope of the star, where the gradient of μ is usually very small, the velocity of meridional circulations which determines the rate of turbulent diffusive mixing is proportional to a factor E_Ω. For the particular case of uniform rotation $E_\Omega = (1 - \Omega^2/2\pi G\rho)$ and we obtain the well-known classical result that meridional circulations consist of two loops separated by "a quiet zone" located at the density $\rho_* = \Omega^2/2\pi G$.

In the classical case the mass of the layer above the quiet zone, which can be considered as the "outer zone" in our double-zone model, is only about $0.03\ M_\odot$ for the $16\ M_\odot$ star. Let us now suppose that rotation is not uniform, which may correspond better to reality. For $\frac{d\ln\Omega}{d\ln r} = 0.15$ and $\Omega_{-4} = 0.4$ we find the mass of the outer zone to be about $0.7\ M_\odot$, which is not too different from our estimate.

It remains to add that, in order to explain Li abundance peculiarities in F dwarfs and in subgiants, Charbonnel & Vauclair (1992) have also assumed that the rate of turbulent diffusion in the outer loop of meridional circulation is reduced by the factor 10^3.

References

Charbonnel, C., Vauclair, S.: 1992, *Astronomy and Astrophysics*, **265**, 55

Denissenkov, P.A.: 1993, *Preprint MPA*, **733**

Herrero, A., Kudritzki, R.P., Vilchez, J.M., Kunze, D., Butler, K., Haser, S.: 1992, *Astronomy and Astrophysics*, **261**, 209

Langer, N.: 1992, *Astronomy and Astrophysics*, **265**, L17

Weiss, A.: 1993, submitted to *Astronomy and Astrophysics*

Zahn, J.P.: 1983, *Astrophysical Processes in Upper Main Sequence Stars*, 13th Saas-Fee Course, eds. B. Hauck and A. Maeder, 253

Zahn, J.P.: 1992, *Astronomy and Astrophysics*, **265**, 115

THE IMPACT OF SEMI-CONVECTION AND OVERSHOOTING
ON THE SURFACE ABUNDANCES IN MASSIVE STARS

N. MOWLAVI,* M. FORESTINI** and A. JORISSEN***

Institut d'Astronomie et d'Astrophysique,
Universite Libre de Bruxelles, C.P. 165;
Av. F-D. Roosevelt, 50, B-1050 Bruxelles, Belgium

Abstract. We analyse the effects of semi-convection and overshooting on the predicted surface abundances after the first and second dredge-ups in 15 and $20\,M_\odot$ Pop. I stars. Overshooting is applied either to the core boundary or to the boundaries of all convective zones. It is shown that the surface abundances are sensitive to the mixing scheme adopted in the interior. The models including semi-convection lead to lower $^{12}C/^{13}C$ ratios than the other mixing schemes, while models with overshooting predict higher enhancements of sodium at the surface.

1. Introduction

The treatment of convective zones which develop in stars still present short-comings which hinder the proper modelling of their structural evolution. This is especially true in massive stars. In these objects, instabilities against convection indeed develop in several ways. Firstly, the star is characterized by convective cores during the central burning phases. The size of this core can be quite large, amounting to 30% and 15% of the mass of the star during the main sequence (MS) and the core helium burning (CHeB) phases, respectively, in a $15\,M_\odot$ star. The exact location of its boundary is still a matter of debate. A penetration of the bubbles in the stable layers beyond the limit defined by the Schwarzschild criterion is expected, but the extent of this overshooting is still unknown (see for example Zahn, 1991). Secondly, instabilities can develop in the intermediate layers in and above the H-burning shell during the core contraction phase after the central hydrogen exhaustion. These instabilities give rise to intermediate fully convective zones (IFCZ) and semi-convective zones (SCZ), with important energetic consequences for the structural evolution of the star. Thirdly, an IFCZ can, in some cases, develop above the H-burning shell during the core helium burning (CHeB) phase. Finally, we have to mention the convective envelope which plays a unique role in bringing to the surface, at specific phases of evolution, the ashes of nuclear reactions from the deep interior where they are produced (dredge-up phases).

In a previous paper (Mowlavi and Forestini, 1993, MF), we have analysed the effects of semi-convection and of a mild overshooting of 0.20 Hp (where Hp is the pressure scale-height evaluated at the boundaries of the convective

* Boursier I.R.S.I.A.

** Present Adress: Laboratoire d'Astrophysique, Observatoire de Grenoble, BP53X; Rue de la Piscine, 414, F-38041 Grenoble Cedex, France

*** Chercheur qualifie F.N.R.S.

Space Science Reviews **66**: 409–412, 1994.

© 1994 *Kluwer Academic Publishers. Printed in Belgium.*

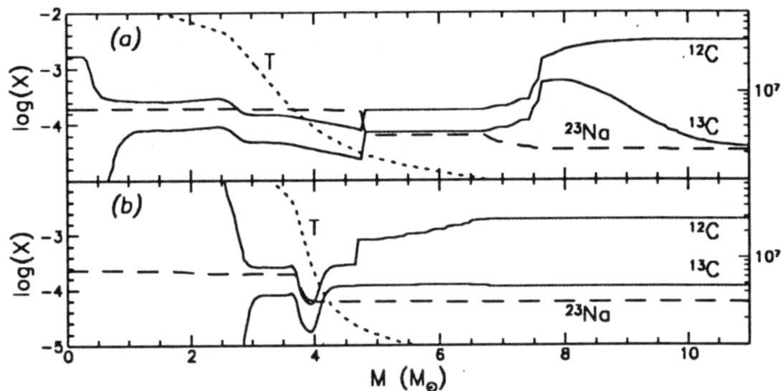

Fig. 1. ^{12}C, ^{13}C and ^{23}Na abundance profiles (in mass fractions)in the $15\,M_\odot$ star calculated with semi-convection *(a)* before central helium ignitium and *(b)* after the core helium exhaustion. The dotted line represent the temperature profile (right scale).

zones), on the structural evolution of Pop. I intermediate-mass and massive stars. Overshooting is applied either to the core boundary or to the boundaries of all convective zones. They are compared to models adopting the standard mixing scheme, i.e. without mixing beyond the Schwarzschild's limit. We refer to MF for a discussion of the structural behavior under these different mixing schemes.

It is the purpose of this paper to analyse the effects of these mixing schemes on the predicted surface abundances for the 15 and $20\,M_\odot$ Pop. I stars. We restrict our study to carbon, whose isotopic ratio is altered in the coolest layers within the H-burning zones, and to sodium, whose abundance is modified at higher temperatures. The surface abundances of these two elements, together with the ^{12}C/^{13}C, ratio are followed until core carbon ignition and compared to values observed in some massive stars.

2. Surface abundances

In the absence of diffusive processes, surface abundances can only be altered by the penetration of the convective envelope in the interior where nuclear reactions have taken place. Such a dredge-up episode happens when the star becomes a red supergiant. For stars less massive than $12\,M_\odot$, a first dredge-up is expected after central hydrogen exhaustion. The situation for massive stars is, however, very sensitive to the development of IFCZs during the core contraction phase. The resulting mixing of matter into the hydrogen burning shell increases its energy production and can in some cases prevent the further penetration of the convective envelope. This is the case in the standard models and in the ones including overshooting. These stars develop large IFCZs and do not experience any dredge-up until the end of their CHeB phase. The $15\,M_\odot$ models with overshooting applied to the core boundary only make exception, however.

Models including semi-convection do experience a first dredge-up before heli-

TABLE I

Surface C and Na abundances (in dex relative to the solar system abundances) and carbon isotopic ratio (by numbers) after the first (first lines) and second (second lines) dredge-ups. The labels m, s, o and c refer to the models calculated with the standard mixing scheme, semi-convection, overshooting applied to all convective zones and core overshooting, respectively.

| | 15 $M_\odot$ | | | | 20 $M_\odot$ | | | |
	m	s	o	c	m	s	o	c
^{12}C	-	-0.22	-	-0.20	-	-0.21	-	-
	-0.21	-0.30	-0.22	-0.21	-0.21	-0.38	-0.24	-0.23
^{23}Na	-	0.23	-	0.25	-	0.24	-	-
	0.25	0.24	0.29	0.25	0.31	0.25	0.34	0.32
^{12}C/^{13}C	-	18.2	-	18.3	-	17.2	-	-
	17.9	15.5	21.0	18.0	19.2	14.5	24.6	24.6

um ignition, and a second one after CHeB. The composition profiles before these two dredge-ups are shown in Fig. 1a and 1b for the 15 $M_\odot$ star. The non-solar abundances in the outer layers at the end of CHeB (Fig. 1b) are the ones resulting from the first dredge-up. The H-burning shell is then located around 3.8 $M_\odot$. A very interesting feature can be seen between 5 and 6.5 $M_\odot$, where the ^{12}C abundance is depleted with respect to its surface abundance. This can be understood as the result of the development of an IFCZ *during* the CHeB, and located above the H-burning shell. This IFCZ mixes the ^{12}C-rich layers left over by retreat of the convective envelope with the ^{12}C-depleted layers in the H-burning zone. Because the ^{13}C abundance is almost constant in these layers, this mixing has basically no effect on this isotope. As a result, the surface ^{12}C/^{13}C ratio after the second dredge-up is lowered. Table 1 shows that ^{12}C/^{13}C reaches 15.5 after the second dredge-up, to be compared with a value of 17.9 predicted by the standard models. Models including semi-convection predict the highest ^{12}C depletion after the second dredge-up, with [^{12}C]=-0.38 dex in the 20 $M_\odot$ star, in contrast with values between -0.20 and -0.24 in the standard models and in the ones with overshooting. As for ^{23}Na, the models predict surface enhancements after the dredge-ups between 0.23 and 0.34 dex. The highest values are found in models with overshooting. We further note that, after the first dredge-up, the surface abundance is no more altered by the second one.

It is well known that the predicted abundances from models with neither diffusion nor rotation cannot explain the observed low carbon isotopic ratios (Smith 1990). For example, the two massive stars analysed by Harris and Lambert (1984), α Sco and α Ori, have ^{12}C/^{13}C ratios as low as 12 ± 3 and 6 ± 1, respectively. Our predictions do also fail to explain such low ratios. However, we note that models with semi-convection can predict a $^{12}C/^{13}C$ ratio close to the one observed in α Sco, within the observational error bars. As for sodium, Boyarchuk et al. (1988) report [Na/Fe] ratios ranging from 0.4 to 0.7 dex in several F supergiants. These values are also difficult to explain by standard models (Denisenkov and Denisenkova, 1990; Prantzos et al., 1991). Our calculations show a sensitivity of the predicted surface ^{23}Na abundance to the adopted mixing scheme. Although models with overshooting present the highest surface sodium alterations, they still cannot explain the largest observed values. Furthermore, the stars observed by Boyarchuk et al. (1988) are yellow supergiants, which means that the star must have experienced a first dredge-up before helium ignition and must perform a blue loop in the H-R diagram during CHeB. Our models including semi-convection are the only ones able to reproduce such features.

3. Conclusion

We have analysed the effects of overshooting and semi-convection on the predicted surface ^{12}C, ^{13}C and ^{23}Na abundances. It is shown that these predictions are sensitive to the adopted mixing schemes. $^{12}C/^{13}C$ varies from 14.5 to 24.6 after the second dredge-up in the $20\,M_\odot$ stars, while the sodium enhancement ranges from 0.25 to 0.34 dex for the same stars. The lowest surface $^{12}C/^{13}C$ ratios after the second dredge-up are predicted by the models including semi-convection, which develop an IFCZ above the H-burning shell during CHeB. In contrast, sodium is more enhanced at the surface of models including overshooting. However, only models performing a blue loop in the H-R diagram would predict surface sodium alterations in yellow supergiants as observed. This feature is only found in models including semi-convection.

References

Boyarchuk, A.A., Gubeny, I., Kubat, I., Lyunbimkov, L.S. and Sakhibullin, N.A.: 1988, *Astrophysika* **28**, 197 and 202

Harris, M.J. and Lambert, D.L.: 1984, *Astrophysical Journal* **281**, 739

Denisenkov, P. and Denisenkova, S.: 1990, *Soviet Astronomy Let.* **16**, 275

Mowlavi, N. and Forestini, M.: 1993, *Astronomy and Astrophysics* , in press

Prantzos, N., Coc, A. and Thibaut, J.P.: 1991, *Astrophysical Journal* **379**, 729

Smith, V.V.: 1990, *Mem. Soc. Astron. Ital.* **61**, 787

Zahn, J.-P.: 1991, *Astronomy and Astrophysics* **252**, 179

IMPROVED BOLOMETRIC CORRECTIONS FOR WR STARS

LINDSEY F. SMITH
School of Physics, University of Sydney, NSW 2006
mail address: 6B/26 Etham Ave, Darling Point, NSW 2027

G. MEYNET
Geneva Observatory, ch. des Maillettes 51, CH-1290 Sauverny, Switzerland

and

J.-C. MERMILLIOD
Institut d'Astronomie de l'Universite de Lausanne, ch. des Maillettes 51, CH-1290
Chavannes-des-Bois, Switzerland

Abstract. Evolutionary models allow an assignment of both a mass and a luminosity to a Wolf-Rayet (WR) star in a cluster, and hence allow a determination of the Bolometric Correction (B.C.). The B.C.'s derived for WN stars range from -4.0 to -6.0 with the expected trend of larger values (in absolute values) for stars with higher excitation spectra. For WC stars, there is little evidence for a similar trend; most observations presented here are consistent with B.C. = -4.5, as found by Smith and Maeder (1989). The convergence of B.C. values derived from evolutionary and atmospheric models is extremely satisfactory, giving increased confidence in both methods.

Key words: Stars: Wolf-Rayet–atmospheres

1. The Method

Given a set of theoretical evolutionary models and a sample of WR stars belonging to open clusters or associations, it is possible to deduce the bolometric corrections (B.C.) for these stars in the following way : 1) The absolute visual magnitude M_v is determined on the base of the observed visual magnitude of the WR star, the distance modulus of the cluster and the reddening in the vicinity of the WR star. Most of the M_v of the WR stars are taken from van der Hucht et al. (1988, hereafter vdH88). 2) We used the stellar models by Meynet et al. (1993) to compute theoretical isochrones in order to assign an age, t, to the cluster and therefore to the WR star. 3) If the age and stage (WNL, WNE or WC) of the WR are known, the theoretical models predict that the mass, $\mathcal{M}$, must fall within a certain range. 4) the luminosity, $\mathcal{L}$, follows from the $\mathcal{M} - \mathcal{L}$ relation predicted by the theoretical models. Thus, if the WR star is in a cluster, its M_v, age, mass and luminosity may be determined and the B.C. follows.

2. The Results

Following the method just described above, we obtained for more than 40 WR stars information on their absolute visual magnitude and on their actual mass. We can plot these data in a M_v vs $\log \mathcal{M}/\mathcal{M}_\odot$ plane. Figure 1 presents the results for the WR in clusters (1a) and in associations (1b). Figure 1a) includes also all

Space Science Reviews **66**: 413–416, 1994.
© 1994 *Kluwer Academic Publishers. Printed in Belgium.*

L. F. SMITH ET AL.

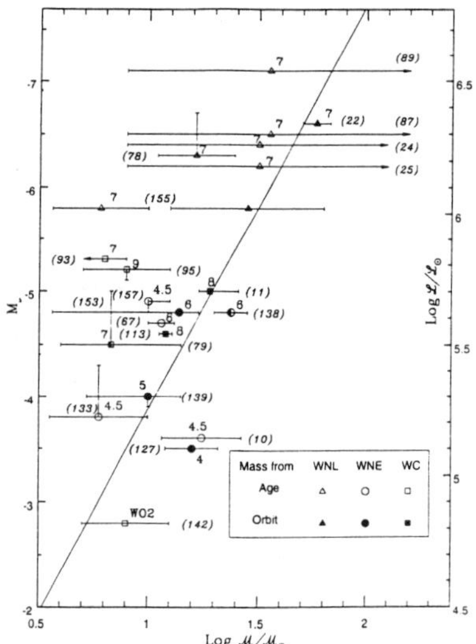

Fig. 1. **a)** M_v versus $\log \mathcal{M}/\mathcal{M}_\odot$ for single stars in clusters and binary stars with mass determinations in both clusters and associations. The diagonal line is for B.C. = -4.5 for WNE, WC and WNL (the latter from lower mass progenitors only, see text). A point falling below the line indicates a larger B.C. (B.C. < -4.5) and vice versa. The subtype, the number of the WR star (in parenthesis) are indicated. The half filled symbols indicate a lower limit from the orbit and an upper limit from the association age. Error bars on M_v are given only when the value we use differs from vdH88 and extends to their value. Error bars in mass correspond to ± 0.1 dex in the age.**b)** Same as in a) for single WR stars in associations.

the spectroscopic binaries, although many of these are in associations rather than clusters. For these WR stars more tight constraints on the mass can be obtained from the study of their orbit. The right axis of Fig. 1 corresponds to B.C. = -4.5, the B.C. derived by Smith & Maeder (1989, hereafter SM89) on the basis of spectroscopic binaries only. The line drawn is:

$$\log \mathcal{L}/\mathcal{L}_\odot = 1.55 + 3.70 \log \mathcal{M}/\mathcal{M}_\odot \tag{1}$$

corresponding to WNE and WC stars from the WNE and WC models by Meynet et al. (1993, for WNL, see below). It is slightly steeper than the $\mathcal{M} - \mathcal{L}$ relation from previous models used by SM89. Points falling below the line indicate a larger B.C. (B.C. < -4.5); points falling above the line indicate a smaller B.C. (B.C. > -4.5).

Considering Fig. 1a), and checking results obtained from WR in associations for consistency, we suggest the B.C.'s summarised in Table 1. Table 1 also gives $T(2/3)$, the temperature at an optical depth equal to 2/3, predicted by the evolutionary models and from model atmosphere fitting of galactic WN stars by Schmutz et al (1989) and of one WC5 star (WR111) by Hillier (1989) and by

Koesterke et al (1991). The number of binaries, of single stars in clusters and of single stars in associations, used to obtain the B.C.'s, are also indicated.

TABLE I

Model and observed T(2/3) in 10^3 K and observed B.C. The number of stars in the different categories, binaries (Bin.), single stars in clusters (Clu.), single stars in associations (Ass.) are indicated.

Subclass	Models	Schmutz et al	B.C.	Number of stars		
				Bin.	Clu.	Ass.
WN7-8	20 - 30	30 - 35	-4.0	2	5	2
WN4.5-5-6	20 - 35	30 - 50	-4.5	3	4	6
WN4	20 - 35	30 - 50	-5.5	1	0	1
WN2-3			-6.0	0	0	2
WC7-9			-3.0 ?	1	2	2
WC8	25 - 40	-	-4.5	2	0	1
WC4-6	40 - 60	60	-4.5	0	0	6
WO	60?	-	-4.5	0	1	0

WN7 stars occupy the high luminosity end of Fig. 1a). One WN8 star belonging to an association falls in the same domain. Most of the WN7 stars are in very young clusters ($\log t < 6.5$, t in years); the exceptions are WR78 and WR155. The mass limits for the "young" WN7 are very wide for a good physical reason. They have evolved from initial masses superior to about 120 $\mathcal{M}_\odot$ which remain WNL for a large portion of their lifetime during which the mass decreases dramatically. Hence the mass is very poorly defined, even in a cluster with well defined age. Error bars for all the "young" WN7 & 8 stars overlap the line. Schmutz (1990) has pointed out that WNL stars, with hydrogen still present, do not strictly follow the WNE-WC $\mathcal{M} - \mathcal{L}$ relationship. However, in general, the WNL fall close to the same relation so long as there is a pure helium (or higher density) core. The reason is that the small hydrogen atmosphere has a large effect on the temperature but a very small effect on the luminosity and the mass. However, for the 120 $\mathcal{M}_\odot$ model with double mass loss (Meynet et al. 1993), the $\mathcal{M} - \mathcal{L}$ relation falls about 0.5 mag. fainter than (1); the reason is that the star has a hydrogen mixture through most of the mass. Thus, for the WNL stars in Fig. 1a), which have a mass of the progenitor equal or superior to $\sim 120\mathcal{M}_\odot$ (all the WN7 plotted in Fig. 1a) except WR78 and WR155), the "B.C. = -4.5" line is a "B.C. = -4.0" line. For WR78 and WR155, the line still corresponds to B.C. = -4.5; however WR78 falls somewhat above the line and also has a B.C. = -4.0 or less. (The position of WR155 is uncertain, there is disagreement between the mass from the age and from the spectroscopic orbit. Both determinations are plotted in Fig.1a).)

In Fig. 1a), three WN6 and one WN5 star, with well defined masses, lie near to the B.C.= -4.5 line, indicating that -4.5 is about right for these subclasses. Three WN4.5 stars have less well defined mass, but straddle the line, making B.C.= -4.5 the best average value. One WN4 star in Fig. 1a) and one each of WN4, WN3 and WN2 in Fig. 1b) fall increasingly far below the line, indicating progressively larger B.C. required. We suggest -5.5 for WN4 and -6.0 for WN2 and WN3.

In Fig. 1a), one WC9 and two WC7 stars fall significantly above the line, indicating B.C. of -3.5 or -3.0. However, two WC8 stars with well defined masses fall near the line. Some representatives of WC5-6 are in associations. The mass limits are too wide to be definitive; however, the midpoints fall near the line. A WO2 star occurs in Fig. 1a) below, but with error bars including the line. Thus for WC8 to WO stars, we suggest B.C.=-4.5 in agreement with previous results based on WR in binaries (Vanbeveren 1982, Lundstrom & Stenholm 1984, SM89).

This method has the big advantage that one views all available data at the same time, with clear error bars, something which is lacking from values derived for individual stars. B.C.'s derived by various methods (binaries, single stars, stellar atmospheres) which were previously divergent are now beginning to converge. This gives increasing confidence in validity of both evolutionary and atmospheric models.

References

Hillier, D.J.: 1989, *ApJ* **347**, 392
Koesterke L., Hamann, W.R., Schmutz, W., Wessolowski, U.: 1991,*A&A* **248**, 166
Lundstrom, I., Stenholm, B.: 1984, *A&AS* **58**, 163
Meynet, G., Maeder, A., Schaller, G., Schaerer, D., Charbonnel, C.: 1993, *A&AS*, in press
Schmutz, W., Hamann, W.R., Wessolowski, U.: 1989, *A&A* **210**, 236
Schmutz, W.: 1990, ASP Conf Ser 7, "Properties of Hot Luminous Stars", Ed. C.D.Garmany, 117
Smith, L.F., Maeder, A.: 1989, *A&A* **211**, 71 (SM89)
Vanbeveren, D.: 1982, *IAU Symp. 99*, eds. C. de Loore & A.J. Willis (Reidel, Dordrecht), p. 117
van der Hucht, K.A., Hidayat, B., Admiranto, A.G., Supelli, K.R., Doom, C.: 1988, *A&A* **199**, 217
    (vdH88)

WOLF-RAYET STARS IN STARBURSTS

Effects of age and metallicity

G. MEYNET

Geneva Observatory, CH-1290 Sauverny, Switzerland

Abstract. Let us suppose that it is possible observationally to determine the number ratio of WR to O stars in a starburst galaxy (cf. e.g. Vacca & Conti 1992) and that one can also have some information on the way the different WR subtypes are distributed (number ratios as WN/WR, WNL/WR etc ...), the question is, what can we deduce from these values on the burst of star formation which gave birth to these WR stars ? Is it possible for instance to constrain the age of the burst (*i.e.* the time elapsed since the beginning of the burst of star formation), its intensity (*i.e.* the ratio of the star formation rate during the burst to that before the burst) or the metallicity of the cloud from which the stars formed ? We present here models of starbursts based on the most recent models for single stars computed by the Geneva group and show that the study of the WR population in a starburst provides very useful insights on the age of the burst and on the metallicity of the star forming zone.

Key words: Stars: Wolf-Rayet – Galaxies: stellar content of

1. The Physical Ingredients of the Starbursts' Models

In this short report we shall study the effects of the age and of the metallicity of the burst on the WR population (by metallicity of the burst, we mean here the initial metallicity of the new born stars). We consider the case of an instantaneous burst of star formation, *i.e.* we suppose that 20 000 stars are born at the same time, and distributed according to an initial mass function of the Salpeter's type ($\frac{dN}{dM} = AM^{-2}$) in the mass range from 8 to 120 $M_\odot$ (it is not useful to consider the lower mass range since we are interested here in number ratios of high mass stars). As stellar models we used the recent grids computed by Schaller et al. (1992) and Schaerer et al. (1993) for solar metallicity and for Z=0.008 and 0.001. The criterion for a star to be considered as a WR, WNL, WNE and WC star are taken as in the papers just quoted. We defined the O stars as all the H-burning stars having an effective temperature superior to 4.477 (30 000 K).

In this first approach, we use models of single stars. As a consequence, our predictions apply to very young starbursts for which only the most massive stars have had time to evolve into the WR stage. For these massive stars, it is likely that the presence of a companion does not much change the evolution, since these stars become WR stars before or just at the end of the hydrogen burning phase. In the case of constant star formation rate, the influence of close binary evolution on the predicted number distribution of WR stars may be quite important, especially at low metallicity as suggested by Maeder (1991) and quantitatively estimated by Vanbeveren & de Loore (1993, see also the review by Vanbeveren in the present volume) and Maeder & Meynet (1994).

Space Science Reviews **66**: 417–420, 1994.
© 1994 *Kluwer Academic Publishers. Printed in Belgium.*

2. Effects of the Age of the Burst

To illustrate the effects of the age of the burst, let us consider the case of the instantaneous burst (IB) represented in Fig. 1a). Six important periods can be distinguished and are commented below.

1) **The period before the WR rich phase of the burst :** For $t < 2.3 \ 10^6 y$. the relative number of WR to O stars is quite low. The massive stars born in the burst have not yet had time to evolve as a WR star.

2) **The WNL phase :** The WR rich phase of the burst begins with a period $(2.3 \ 10^6 < t < 2.9 \ 10^6)$ during which nearly all the WR are WNL stars. This is quite natural since, the WNL stage is the entrance stage in the WR phase.

3) **The "WC & WNL/WNE > 1" phase :** From now on, the WC stars become the dominant type among the WR stars. This would be also the case in a zone of constant star formation rate (at Z=0.020 and for most of the different initial mass stars, this phase is the longest one). Since the most massive stars present very reduced WNE stages (they evolve directly from the WNL phase to the WC one), very few WNE stars are observed at this stage $(2.9 \ 10^6 < t < 4.5 \ 10^6)$.

4) **The "WC & WNL/WNE < 1" phase :** As time goes on, the smaller initial mass stars which become WR stars contribute more and more to the WR population. These stars may present important WNE phases, thus this period $(4.5 \ 10^6 < t < 6.7 \ 10^6)$ is characterised by an important enhancement of the WNE population which becomes as numerous, and even in this case more numerous than the WNL population. Let us just emphasize here that the situation might be quite different in the case most of the WR at this stage owe their existence to roche lobe overflow in close binary systems.

5) **Last period of the WR rich phase :** Our model, based on single star models predict that during this last period, among the WN stars, the WNL stars are again dominant. Indeed the least massive stars which become WR just enter the WNL phase at the end of their life and have no time to proceed further along the WNE stage. Due to the disappearence of the O stars, the number ratios presented in Fig. 1a) become very large.

6) **End of the WR rich phase of the burst :** After about $7 \ 10^6$ years, all the WR, whose progenitors have been born in the burst, have ended their stellar life. It is interesting to note that during the whole WR rich phase, which lasts about $4 \ 10^6$ years all the supernovae have WR as progenitors.

3. Effects of a Change of the Initial Metallicity

A change of initial metallicity, Z, has huge consequences for what concerns the WR population produced by a starburst. Indeed when Z decreases (see Fig. 1), one has for an instantaneous burst that:

1) The duration of the WR rich phase decreases, passing from about $4 \ 10^6$ years at Z=0.020 to $\sim 0.5 \ 10^6$ at Z=0.001. At low Z, due to weaker stellar winds,

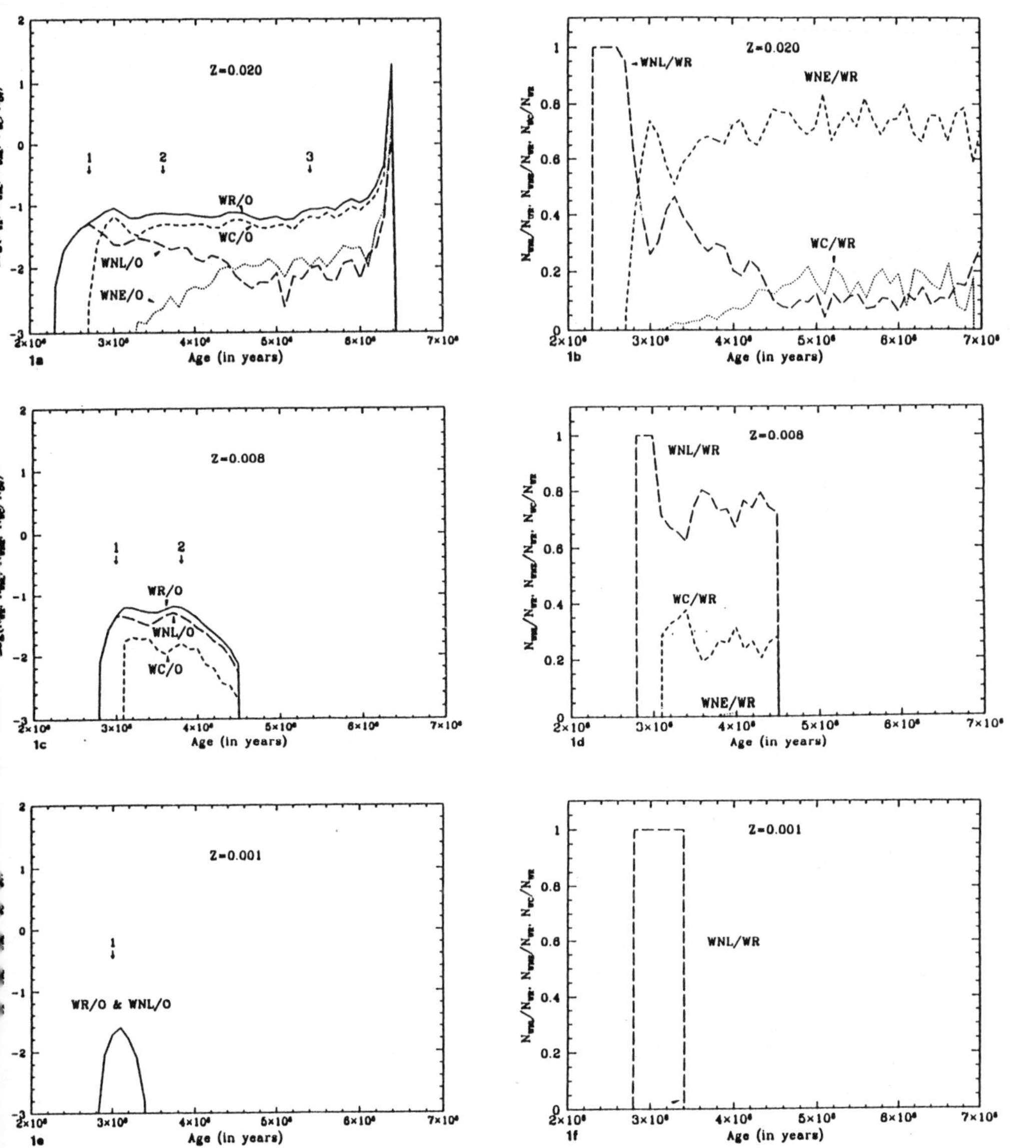

Fig. 1. **a)** Evolution with time of the number ratios (in logarithm) WR/O, WNL/O, WNE/O, WC/O, obtained after an instantaneous burst of star formation which took place at the time $t = 0$. The slope of the IMF was taken equal to -2 ($\frac{dN}{dM} = AM^{-2}$). Solar metallicity models have been used. **b)** Evolution with time of the number ratios WNL/WR, WNE/WR, WC/WR, for the same model of starburst as in a); **c), d)** same as a), b) for models with Z = 0.008; **e), f)** same as a), b) for models with Z = 0.001.

only a small range of very massive stars become WR. This means that lower is Z, narrower in time will be the burst of WR formation which follows the one of star formation. Let us emphasize again that the inclusion in the model of WR stars born in close binary systems could change this situation, since lower initial mass stars might become WR stars thanks to mass transfer mechanisms.

2) At low Z, the WNL type is the dominant one during the WR rich phase of the burst. The distribution of the WR subtypes present less variation with time than at high metallicity. At high metallicity, the changes observed with time of these number ratios, result from important differences between the evolution of massive stars compared to that of low mass stars. At low Z, the range of initial mass stars becoming WR stars is much narrower and thus, nearly all the WR have the same evolution.

3) In contrast with the case at Z=0.020, the low metallicity cases do not long present a huge increase of the WR/O number ratio at the end the WR rich phase. This means that the maximum of the age of a WR at low metallicity is inferior to the maximum of the age of an O star.

4. Conclusions

In this work, we have studied the WR population resulting from an instantaneous burst of star formation. Whatever be the metallicity, starbursts produce high values of the WR/O number ratio. Morover a starburst occuring in a relatively high metallicity zone (solar or higher) may present a WR population which becomes as important or even more important than the O star population. Only WNL stars are predicted to be present in a young low metallicity starburst.

References

Maeder, A.: 1991, *A&A* **242**, 93
Maeder, A., Meynet, G.: 1994, *A&A*, in preparation
Schaerer, D., Meynet, G., Maeder, A., Schaller, G.: 1993, *A&AS*, **98**, 523
Schaller, G., Schaerer, D., Meynet, G., Maeder, A.: 1992, *A&AS*, **96**, 269
Vacca, W.D., Conti, P.S.: 1992, *ApJ*, **401**, 543
Vanbeveren, D., de Loore, C.: 1993, in *Massive Stars: Their Lives in the Interstellar Medium, ASP Conf. Ser.*, **35**, eds. J. P. Cassinelli & E. B. Churchwell, p. 257

A STUDY OF THE SMC CLUSTER NGC 330

C. CHIOSI , A. VALLENARI and F. FAGOTTO
Department of Astronomy, Vicolo Osservatorio 5, 35122 Padua, Italy

G. BERTELLI
Fellow of the National Council of Research, Italy

and

A. BRESSAN and E. NASI
Astronomical Observatory, Vicolo Osservatorio 5, 35122 Padua, Italy

Abstract. We analyze the properties of the SMC cluster NGC 330 on the basis of a new CMD and of published effective temperatures and bolometric magnitudes for a smaller sample of stars (Caloi et al 1993, Stothers & Chin 1992a,b) to check for the kind of mixing taking place in massive stars. The existence of a few stars in the so-called HR gap led Stothers & Chin (1993) to claim that only models with semiconvective mixing and OPAL could match the HR diagram, whereas induced Caloi et al (1993) to conclude that neither semiconvective nor overshoot models are suited. The careful analysis of this cluster with new stellar models indicates that 1) models with convective overshoot and models with semiconvection can both explain the gross features of the HR diagram, 2) models with overshoot ought to be preferred.

1. Introduction

The blue, luminous, and hence young, cluster NGC 330 of the Small Magellanic Cloud has recently drawn much attention because its color-magnitude diagram (CMD) shows two well distinct clumps of blue and red supergiants eluding simple theoretical interpretations.

Stothers & Chin (1992a,b) analyzed the CMD of this cluster in order to assess the kind of mixing taking place in massive stars, i.e. either semiconvection (both the Schwarzschild and the Ledoux criterion) or convective overshoot. The main conclusions were that 1) only stellar models with little or no overshoot at all can reproduce the most salient features of the CMD, namely the maximum effective temperature reached by the hot evolved stars and the difference between the average luminosity of the hot and cool supergiants stars; 2) semiconvective models, in particular those with the Ledoux criterion, are able to give the more convincing representation of the data.

Caloi et al (1993), comparing the location of ten blue supergiants of NGC 330 in the theoretical HRD with the evolutionary tracks by Bencivenni et al (1991) calculated using the Ledoux criterion, found poor agreement between the models and the data. In fact, only four stars of the sample could be assigned to a slow evolutionary phase. More precisely, two stars were located at the top of the main sequence and two at the tip of the blue loop. The remaining six stars were located in regions of the HRD where models predict fast evolutionary rate and hence low probability of detection. They concluded that neither models with semiconvection

Space Science Reviews **66**: 421–424, 1994.
© 1994 *Kluwer Academic Publishers. Printed in Belgium.*

(the Ledoux criterion in their study) nor those with overshoot are able to match the CMD of this clusters.

In this paper, we re-analyze the whole question with the aid of homogeneous grids of stellar models calculated with modern physical input, two sources of opacity (OPAL and LAOL see below), and both the overshoot and semiconvective scheme of mixing. We find that semiconvective models with the Schwarzschild criterion and models with overshoot from the core and the envelope both yield HRDs compatible with the observations. However, details of the HRD seem to favour the case of models with overshoot.

2. Properties of NGC 330

The color excess is taken from Caloi et al (1993) who estimate E_{B-V}=0.06 $\pm$ 0.02, while the extinction law for the SMC, A_V=2.72E_{B-V}, is from Bouchet et al (1985). Following Caloi et al (1993) we adopt the true distance modulus $(m-M)_o$=18.85. In the recent years, many efforts have been devoted to study the metal abundance of NGC 330. Various studies based on different methods suggest a metal content [M/H] ranging from -1.3 to -0.8 or Z=0.001 - 0.003 (Spite et al 1986, Reitermann et al 1990, Bessel 1991, Grebel & Richtler 1992). However, higher metallicity cannot be ruled out, since field stars in the same region of NGC 330 seem to have [M/H]=-0.7 (Spite & Spite 1990). In the following, we will consider stellar models with metallicities in the range Z=0.001 to Z=0.008.

3. Stellar Models

Two different schemes of mixing in the convective regions are considered. In the classical scheme, the Schwarzschild criterion is used to define the boundary of the convective regions and the algorithm of semiconvection is adopted at the border of the convective He-burning cores (Bressan et al 1993, Alongi et al 1993, Fagotto et al 1993). This type of models are shortly referred to as semiconvective. In the overshoot scheme, this follows the formalism by Bressan et al (1981) for the central cores and the revision made by Alongi et al (1991) for the outer envelope. The radiative and conductive opacities are taken either from Huebner et al (1977) (thereinafter LAOL) or Iglesias & Rogers (1992) (thereinafter OPAL). Mass loss by stellar wind is taken into account according to de Jager et al (1988) for all evolutionary stages from the main sequence up to the so-called De Jager limit in the HRD, with the dependence on the metallicity as given by Kudritzski et al (1989). To transform luminosities and effective temperatures into magnitudes and colours of the Johnson UBV system we adopt new release of the library of stellar spectra by Kurucz (1993, private communication).

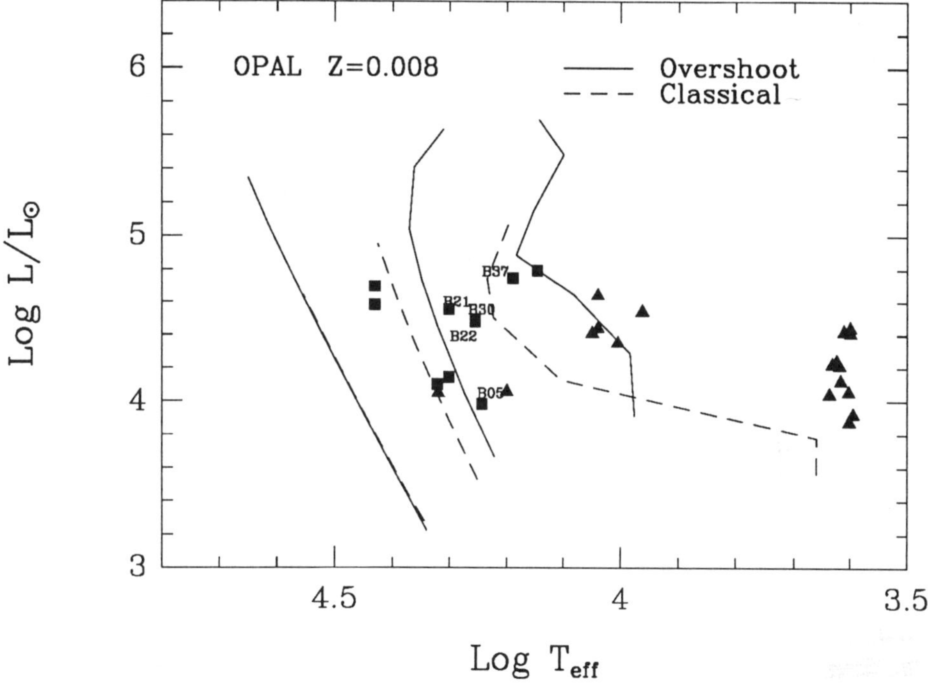

Fig. 1. Supergiants stars from Caloi *et al* (1993) (squares) and from Stothers and Chin (1992) (triangles). The lines indicate the Zero-Age Main Sequence, the termination point of the main sequence and the tip of the loop in the He-burning phase, for overshoot models and classical models using OPAL for Z=0.008.

4. Location of Blue Supergiants on HR Diagram

We begin our analysis by comparing the position in the HRD of the 10 stars by Caloi et al (1993) and a few more by Stothers & Chin (1992a,b), for which independent determinations of the effective temperatures are available. These stars are indicated by filled squares (Caloi et al 1993) and triangles (Stothers & Chin 1992) in the figure below. The comparison is made by drawing in the HRD three loci, namely the zero age main sequence (ZAMS), the termination point of the main sequence (TAMS), and the hottest model during the stationary core He-burning phase therein after referred to as HHE. We assume the metallicity Z=0.008. Figure 1 compares the three loci for semiconvective and overshoot models limited to the case of OPAL. An inspection of the Figure reveals that:

1) Semiconvective models define too a narrow main sequence band so that almost 50% of the blue stars fall in the HR gap, even if this is partially compensated by the bluer extension of the HHE.

2) For metallicity of Z=0.008, OPAL, and overshoot models, only two stars (B22 and B30) are left in the HR gap, all the others fall in regions of stationary

burning. This is basically due to OPAL and the inclusion of envelope overshoot.

Furthermore, the comparison of our models OPAL and overshoot with those by Bencivenni et al (1991) and Stothers & Chin (1992) reveals that their main sequence bands and HHEs were much narrower and cooler. This explains their negative conclusions about models with overshoot.

5. Conclusions

The main conclusions of this study are as follows:

1) Contrary to what claimed in previous studies, models including both core and envelope overshoot and OPAL well reproduce the main features of the CMD and the location on the HRD of most of the reference stars.

2) Paradoxically, semiconvective models with the Schwarzschild criterion although able to fit the overall morphology of the CMD and HRD somewhat fail to hit the main target as they leave outside the regions of stationary burning a larger number of stars.

References

Alongi, M., Bertelli, G., Bressan, A., Chiosi, C.: 1991, *AA* **244**, 95

Alongi, M., Bertelli, G., Bressan, A., Chiosi, C., Fagotto, F., Greggio L., Nasi E.: 1993, *AAS* **97**, 851

Bouchet, P., Lequeux, J., Maurice, E., Prevot, L., Prevot-Bournichon, M. L.: 1985, *AA* **149**, 330

Bressan, A., Bertelli, G., Chiosi, C.: 1981, *AA* **102**, 25

Bressan, A., Fagotto F., Bertelli, G., Chiosi, C.: 1993, *AAS* **100**, 647

Caloi, V., Cassatella, A., Castellani, V., and Walker, A.: 1993, *AA* **271**, 109

Carney, B. W., Janes, K. A., Flower, P. J.: 1985, *AJ* **90**, 1196

de Jager C., Nieuwenhuijzen H., van der Hucht K.A.: 1988, *AAS* **72**, 259

Fagotto, F., Bressan, A., Bertelli G., Chiosi, C.: 1993, *submitted* ,

Feast, M. W., Black, C.: 1980, *MNRAS* **191**, 285

Grebel, E.K., Richtler, T.: 1992, *AA* **253**, 365

Huebner, W. F., Merts, A. L., Magee, N. H., Argo, M. F.: 1977, *Los Alamos Scientific Laboratory Report*, LA-6760-M

Iglesias, C. A., Rogers, F. J.: 1992, *ApJ* **379**, 507

Kudritzski R.P., Pauldrach A., Puls J., Abbot D.C.: 1989, *AA* **219**, 205

Stothers, R. B., and Chin, C. W.: 1992a, *ApJ* **390**, L35

Stothers, R. B., and Chin, C. W.: 1992b, *ApJ* **390**, 136

Reitermann, A., Baschek, B., Stahl, O., Wolf, B.: 1990, *AA* **234**, 109

Spite, M., Cayrel, R., Francois, P., Richtler, T., Spite, F. : 1986, *AA* **168**, 197

Spite, F., Richtler, T., Spite, M.: 1991, *AA* **252**, 557

Spite, M, Spite F.: 1990, *AA* **234**, 67

THE EVOLUTION OF MASSIVE STARS TO EXPLOSION

J. C. WHEELER
Department of Astronomy
University of Texas at Austin

and

D. A. SWARTZ*
X-Ray Astronomy Branch
NASA Marshall Space Flight Center

Abstract. We review the possible evolutionary paths from massive stars to explosive endpoints as various types of supernovae associated with Population I and hence with massive stars: Type II-P, Type II-L, Type Ib, Type Ic, and the hybrid events SN 1987K and SN 1993J. We identify SN 1954A as another hybrid event from the evidence for both H and He in its spectrum with velocities nearly the same as SN 1983J. Evidence for ejected ^{56}Ni mass of ~ 0.07 M$_\odot$ suggests that SN II-P underwent standard iron core collapse, not collapse of an O-Ne-Mg core nor thermonuclear explosion of a C-O core. Most SN II-P presumably arise in single stars or wide binaries of ~ 10 - 20 M$_\odot$. There may be indirect evidence for duplicity in some cases in the form of strong Ba II lines, such as characterized SN 1987A. SN II-L are recognizably distinct from typical SN II-P and must undergo a significantly different evolution. Despite indications that SN II-L have small envelopes that may be helium enriched, they are also distinct from events like SN 1993J that must have yet again a different evolution. The SN II-L that share a common luminosity seem to have ejected a small nickel mass and hence may come from stars with O-Ne-Mg cores. The amount of nickel ejected by the exceptionally bright events, SN 1980K and SN 1979C, remains controversial. SN Ib require the complete loss of the H envelope, either to a binary companion or to a wind. The few identified have relatively large ejecta masses. It is not clear what evolutionary processes distinguish SN Ib's evolving in binary systems from hybrid events that retain some H in the envelope. SN Ic events are both H and He deficient. Binary models that can account for transfer of an extended helium envelope from low mass helium cores, ~ 2 to 4 M$_\odot$, imply C-O core masses that are roughly consistent with that deduced from the ejecta mass plus a neutron star, ~ 2 to 3 M$_\odot$. It is possible that the hybrid events are the result of Roche lobe overflow and that the "pure" events, SN Ib or SN Ic, result from common envelope evolution.

Key words: Stars – Supernovae

1. Introduction

There is a general understanding that massive stars explode as supernovae, at least those that do not collapse directly to make black holes. The task is to make the connection between supernovae and the stars observed through our telescopes which have a range of masses and metallicities and the possibility of a binary companion to alter their basic evolution. There may be other physical processes, the dreaded rotation or magnetic fields, or something yet invented, that break the degeneracy of the evolutionary path at a single mass. There is a growing demand and need to replace the empirical supernova classification system with one based on physics. The evidence for this is the proliferation of "classes" and

* NAS/NRC Resident Research Associate

Space Science Reviews **66**: 425–437, 1994.
© 1994 *Kluwer Academic Publishers. Printed in Belgium.*

the advent of "hybrid" events. The problem is that we are not there yet. We do not know enough to abandon a purely empirical classification system and adopt one based on physics - core collapse or thermonuclear explosion, binary versus single evolution. There are hints but no unambiguous evidence in the vast majority of the observed events. In this review we will muse on the current state of this situation which has been altered somewhat by recent research and thrown in new relief by the discovery of the unambiguous hybrid event, SN 1993J.

We will not attempt to reproduce here the spectral data that define the current supernova classification scheme. Discussion of this can be found in, e.g., Harkness and Wheeler (1990); Wheeler (1990); and Filippenko (1991). Related discussions of the synthesis of evolution and explosion can be found in Branch, Nomoto, & Filippenko (1991) and Wheeler, Swartz, and Harkness (1993).

2. Type II

Type II supernovae (SN II) exhibit a wide range of spectral and photometric properties but, by definition, all display hydrogen Balmer lines in their spectra. Two distinct subtypes can be identified based on the observed B light curve morphology: Those with a phase of nearly constant luminosity following maximum light (SN II-P for 'plateau') and those showing a linear decline (SN II-L). All events within a given subtype are not identical. The subtype distinction may be illusory (Barbon et al. 1979) and, as the duration of the plateau phase is shortened, SN II-P's may become SN II-L's. Young & Branch (1989) confirmed that intrinsic differences in the absolute brightness and the light curve morphology in SN II-P events exist, but also showed that a lack of this diversity occurs for SN II-L. Miller & Branch (1990) and Gaskell (1992) reached similar conclusions for SN II-L, showing that members of this subtype have similar intrinsic magnitudes and light curve shapes. Exceptions to this have been observed, notably the extra-luminous SN 1979C and SN 1980K, but most SN II-L's have the same maximum brightness of $M_B \sim -16.5 \pm 0.13$ (Miller & Branch 1990). In addition to the homogeneity of the SN II-L light curves, all that have been observed in the radio have been detected (Sramek & Weiler 1990) indicating the presence of a strong pre-supernova wind. Several Type II SNe have had radio detections but have not been subtyped by optical photometry. A very weak radio luminosity has been confirmed for the SN II-P SN 1987A. The peculiar Type II SN 1993J is a strong radio emitter.

2.1. TYPE II-P

The generic SN II-P light curve (Doggett & Branch 1985, Kirshner 1990) is well understood (Falk & Arnett 1977, Litvinova & Nadyozhin 1983) as the result of the explosion of a moderately massive red (spectral type M) supergiant. The luminosity during the plateau phase is determined in large part by the progenitor radius and the duration of the plateau by the hydrogen envelope mass, assuming reasonable explosion energies. The plateau phase lasts weeks to months implying a progenitor mass of 10 to 20 $M_\odot$ (and larger if moderate mass loss precedes the explosion). Stars in this mass range grow Fe cores which collapse to form neutron stars or, perhaps, black holes. At the low end of this mass range degenerate O-Ne-Mg cores may form. These probably collapse to form neutron stars as well (Hashimoto et al. 1993 and references therein). The evidence favoring Fe core over O-Ne-Mg core collapse for SN II-P is the amount of radioactive ^{56}Ni ejected. There is evidence (Schmidt et al. 1993, Patat et al. 1993) that most SN II-P eject $\sim$ 0.04 to 0.10 $M_\odot$ of radioactive ^{56}Ni to power the late-time tail (Weaver & Woosley 1980). This is a reasonable value for the ^{56}Ni ejecta mass for Fe core collapse, but is very likely too large to expect from O-Ne-Mg core collapse events. The ^{56}Ni ejecta mass inferred from observations also eliminates the possibility that SN II-P result from thermonuclear explosions of degenerate, low-mass, cores (Woosley 1990).

The fact that a moderately massive hydrogen envelope must remain at the time of the explosion tends to rule out excessively massive stars and stars in interacting binary systems as progenitors of SN II-P. Depending on the assumed mass loss rates and efficiency of convection, stars more massive than $\sim$ 30 $M_\odot$ will lose nearly all of their original hydrogen envelope (Woosley, Langer, & Weaver 1993) whether they evolve as a single star or in a binary system. Stars in the range 10 to 20 $M_\odot$ evolving as the primary star in binary systems also lose a large fraction of their envelopes to mass transfer prior to exploding for a wide range of binary system parameters. Podsiadlowski, Joss, & Hsu (1992) find at most a few solar masses remain of the original 8 $M_\odot$ or larger enveiopes if the mass transfer is roughly conservative. There is certainly no direct observable constraint favoring binary star evolution over single star evolution for progenitors of the majority of SN II-P. They are most likely stars with original main sequence masses between $\sim$ 10 and 20 $M_\odot$ that follow classic single star evolution. A hint that some SN II-P do arise in binary systems comes from the strong Ba II lines that characterize a few of them, including SN 1987A, but also one or two classical SN II-P, SN 1985H (Wheeler, Harkness, and Barkat 1988), and SN 1988A (Turatto et al. 1993). Classical barium stars are thought to be the product of binary evolution in lower mass stars (McClure 1983) and that may also be the case here. The events with strong Ba may be the secondaries in binary systems that have accreted from a companion, perhaps an AGB star.

2.2. TYPE II-L

The empirical evidence for homogeneity among most SN II-L is probably the most stringent constraint on the progenitor mass, radius, and previous evolution. Light curve calculations (Swartz, Wheeler, & Harkness 1991; Blinnikov & Bartunov 1993) require a relatively low hydrogen mass at the time of explosion to prevent the plateau phase characteristic of the SN II-P events. The total ejecta mass is probably between $\sim$ 2 and $\sim$ 5 $M_\odot$. There is again no evidence for binary evolution.

For typical SN II-L, the light curve can be fit with an O-Ne-Mg core collapse ejecting a helium-enhanced envelope of 3 to 5 $M_\odot$ (hydrogen mass $\sim 1\ M_\odot$) and of initial radius $R \sim 400 R_\odot$ (Swartz, Wheeler, & Harkness 1991). Note that such a radius is comparable to that deduced for SN 1993J (Wheeler and Filippenko 1994), and implies that the progenitors may have been G or K, rather than M, supergiants. A low ^{56}Ni mass ($\sim 0.01\ M_\odot$) is needed to produce the contrast in brightness between the light curve peak and the onset of the radioactive tail and the amplitude of that tail. This value is consistent with the highly uncertain production of ^{56}Ni by the O-Ne-Mg core collapse mechanism (Hashimoto et al. 1993). There is too little data to pinpoint the source of energy powering this portion of the light curve. The observed late-time tail is defined by only a few events, two of which are the anomalously bright SN 1979C and SN 1980K. The predicted ^{56}Ni mass for the majority of the SN II-L that seem to have a common peak luminosity is therefore speculative and may in fact be spurious. The relative dimness of the light curves at the end of the linear phase is, however, difficult to reconcile with a nickel mass much in excess of $\sim 0.01\ M_\odot$.

The progenitor envisioned by this scenario would evolve as a single star from an initial mass of ~ 8 to 10 $M_\odot$, experience some mass loss (consistent with the radio, UV, and X-ray observations which imply the ejecta interacts with a circumstellar nebula) and explode at the tip of the asymptotic giant branch after its surface layers have been helium-enhanced. The lower mass limit is set by the fact that most of the SN II-L are not bright enough on the tail to have undergone a thermonuclear explosion, so that the mass must be large enough to lead to non-degenerate carbon ignition. Another constraint on the lower limit is the fact that all stars below about 7 $M_\odot$ lose their envelopes to planetary nebulae. The upper mass limit is, currently, only crudely constrained by the need to have homogeneity among events and thus to exclude the formation of an Fe core. In this scheme, a significant fraction of the envelope must be retained and not lost in the formation of a planetary nebula. Why the planetary nebula ejection mechanism should shut off just below this mass range is not clear.

The progenitors may be identified with the s-processed and lithium-enhanced AGB stars in the LMC and the SMC with $M_{bol} = $ -6 to -7 (Smith and Lambert 1989, 1990; Wheeler, 1990). The stars with O-Ne-Mg cores would occupy the upper end of the AGB at $M_{bol} = -7.1$. It is not clear whether such stars are sufficiently helium enriched, though standard evolution suggests that convective core contraction on the main sequence and dredge-up by the convective envelope will increase the helium abundance while hydrogen will be depleted by mass loss. How the mass loss is achieved has not been resolved. Lithium enrichment follows the formation of a "hot-bottom" envelope in which the burning shell is within the convective envelope (in the most extreme case) so that the fresh lithium is rapidly mixed outward to prevent its destruction (Scalo and Ulrich 1973, Sackmann and Boothroyd 1992). The hot-bottom burning may increase the helium abundance. The conditions that give hot-bottom envelopes tend to also violate the core mass-luminosity relation (Blocker & Schoenberner 1991). If the core mass is less than the Chandrasekhar mass for stars with a luminosity corresponding to the tip of the AGB, then it is not clear that any AGB stars can explode. If this were the case, then the interpretation of SN II-L as being from the range $\sim$ 8 - 10 $M_\odot$ would need reconsideration. The question of envelope structure of stars with O-Ne-Mg cores, hot bottom or not, and hence how bright they would be, requires more investigation.

An alternative single, massive star scenario for SN II-L combines a standard Fe core collapse with a low mass hydrogen envelope and strong interaction with the pre-supernova circumstellar nebula to produce the observed light curve (Blinnikov & Bartunov 1993; see also Woosley, Langer, & Weaver 1993). It is not clear how this model can be reconciled with the observed homogeneity of most SN II-L events nor how mass loss in single star evolution can reduce the quantity of hydrogen to $\sim$ 1 $M_\odot$.

Podsiadlowski, Joss, & Hsu (1992) have shown that stars of original main sequence mass 10 to 15 $M_\odot$, evolving as the primary star in binary systems, tend to lose all but $\sim$ 1 to 3 $M_\odot$ of their hydrogen envelope and explode with radii ranging from $\sim$ 400 to 750 $R_\odot$ and masses ranging from 4.5 to 7.3 $M_\odot$. This may be a sufficiently narrow range of pre-supernova conditions to provide the observed homogeneity. A much broader range of mass and radius results if mass and angular momentum are lost from the system during the mass loss phase. The amount of ^{56}Ni produced over a broader range of initial mass would be essentially the same as for the SN II-P models and would be excessive if, indeed, the observed late-time tails of most SN II-L are radioactively powered and sub-luminous. Note that this sort of binary evolution cannot be easily invoked for both SN II-L and for hybrid events like SN 1993J (see § 4), since the two classes are observationally distinct.

The best studied examples of SN II-L, SN 1979C and SN 1980K are, as often happens, special cases. Blinnikov & Bartunov (1993) have shown that the light curve of the exceptionally bright SN 1979C can be reproduced if the pre-supernova mass loss rate is large, consistent with recent empirical estimates (Lundqvist & Fransson 1988), and if the progenitor radius is $\sim$ 6000 $R_\odot$. Alternatively, Swartz, Wheeler, & Harkness (1991) showed that SN 1979C could arise from the thermonuclear explosion of a C-O core near the Chandrasekhar mass within a 1 $M_\odot$ helium-enriched extended envelope though they neglected the contribution from the wind interaction to the light curve. Swartz et al. assigned somewhat smaller initial radii to SN 1979C and SN 1980K than did Blinnikov & Bartunov, about 1000 $R_\odot$, but still larger than the radius they adopted for the more routine SN II-L. SN 1979C and SN 1980K may thus have M supergiant progenitors whereas other SN II-L had K supergiant progenitors.

There is some evidence on the late-time tail to suggest that SN 1979C flattened to have a tail with the same slope and amplitude as normal SN II-P events

(Turatto, et al. 1990; Patat et al. 1993). If this were the case, then the thermonuclear scenario which requires a large nickel ejecta, ~ 0.5 $M_\odot$, becomes untenable. Unfortunately, the data is insufficient to firmly establish the slope of the tail and the amplitude depends on an uncertain phase since the maximum was not well observed. The other bright SN II-L, SN 1980K, does seem to have a relatively shallow tail, but again the amplitude, and hence the implied nickel mass depends on the choice of phase. Swartz et al. adopted a nickel mass of 0.03 $M_\odot$ for SN 1980K, only a factor of 2 less than the value for SN 1987A. A recent SN II-L, SN 1990K, may have had an intermediate peak luminosity and a tail comparable to SN 1987A (Cappellaro 1993).

Spectroscopically, there is little evidence to differentiate between the various scenarios posed above. For SN 1979C, Branch et al. (1981) found that the spectra near maximum light are roughly consistent with solar metallicity. They could fit neither the shape nor strength of the prominent H-α feature which was nearly pure emission, not a classical P-Cygni profile. Chugai (1991) showed how the observed feature could arise from a P-Cygni profile combined with a broad pure emission component. The emission component comes from the outer fast-moving layers that are heated by the shock produced by the interaction of the ejecta with the pre-existing wind.

Swartz, Wheeler, & Harkness (1991) predict substantial helium enrichment of the envelope. This does not necessarily lead to a helium-dominated optical spectrum, however, since helium requires a source of non-thermal excitation to emit strong lines. Radioactive heating provides this non-thermal excitation in other SNe types (Lucy 1991, Swartz et al. 1993a,b), but the small ejected ^{56}Ni mass in SN II-L may be insufficient to power the helium emission especially if the radioactive material is not mixed outward. Hydrogen will supply a source of free electrons which tend to further suppress the helium lines (Harkness, et al. 1987, Swartz, et al. 1993a). For example, no optical helium lines appear in model spectra of SN 1987A (Swartz, Harkness, & Wheeler 1989, Swartz 1990) although there is ~ 6 $M_\odot$ of helium in the ~ 13 $M_\odot$ ejected. The available SN II-L observations (Branch et al. 1981, Uomoto & Kirshner 1986) show no evidence for helium.

Thermonuclear explosions produce upwards of 0.5 $M_\odot$ of radioactive ^{56}Ni and similar amounts of heavy and intermediate mass elements. The spectra of a thermonuclear-born SN 1979C model (Swartz, Wheeler, & Harkness 1991) may be rich in lines of these elements as well as helium, but spectral calculations have yet to be carried out. The role of the wind interaction on the spectrum formation (Chugai 1991) is uncertain. This wind component comes from the outer, thermally emitting, hydrogen-rich regions and would be common to both the deflagration and core collapse explosion models. It would be difficult to distinguish between these models if the wind component dominates the spectrum.

3. Type I

Two subclasses of hydrogen deficient (Type I) supernovae are commonly believed to arise from core collapse explosions in massive stars. They are spectroscopically characterized near maximum light by the presence (SN Ib) or absence (SN Ic) of strong optical helium lines while they have similar late-time ($t \gtrsim 100$ days) optical spectra (Wheeler & Harkness 1990, Harkness & Wheeler 1990, Filippenko 1991). Their late-time light curve slopes are close to the slope of the ^{56}Co decay rate, or perhaps a little steeper in the case of SN 1983N (Clocchiatti and Wheeler, 1994), indicating a range of progenitor masses from ~ 5 to 10 $M_\odot$ (Ensman and Woosley 1988; Swartz and Wheeler 1991). The SN Ic light curves observed to date seem to be rather homogeneous and are consistent with ejecta masses of ~ 1 to 2 $M_\odot$. SN Ib are detectable in the radio (Sramek & Weiler 1990) and at least one SN Ic, SN 1990B, has also been observed in the radio (Van Dyk et al. 1993). SN Ib and SN Ic are associated with star forming regions (Van Dyk 1992). There are very few clear examples of SN Ib.

3.1. TYPE IB

The presence of helium and the lack of hydrogen implies SN Ib progenitors are helium cores of massive stars that have lost their *entire* hydrogen envelopes. Allowing for a neutron star remnant, the SN Ib light curves suggest the core masses range from ~ 5 to ~ 10 $M_\odot$ corresponding to main sequence masses of ~ 15 to 30 $M_\odot$. Such stars would be expected to explode following Fe core collapse to form a neutron star (and, for the most massive progenitors, possibly a black hole). The amount of radioactive material produced is uncertain (Schlegel and Kirshner 1989; Swartz and Wheeler 1991) but would probably be similar to the amount produced in the Fe core collapse SN II-P events, ~ 0.07 $M_\odot$.

The hydrogen envelope can conceivably be shed by a single star of large initial mass, but lower mass SN Ib's could have evolved in a binary system. In either case, the observable properties of the progenitor are uncertain and no likely progenitor systems have been identified, although Wolf-Rayet stars have been broadly discussed. In the binary star model, it is probable that a fraction of the hydrogen envelope would remain at the time of explosion (Podsiadlowski, Joss, & Hsu 1992) and that a range of hybrid SN Ib's, of low mass, could be observed with a range of hydrogen line strengths. This is not obviously the case but see § 4. SN 1984L ejected a relatively large mass (Swartz & Wheeler 1991), produced very strong helium lines (Harkness et al. 1987), and was spectroscopically and photometrically well monitored. No hydrogen was seen in its spectrum.

Alternatives to the core collapse model include off-center detonations in accreting white dwarf stars (e.g Branch 1988) and slow deflagrations in C-O white dwarfs (Woosley 1990). Neither model can easily explain the broader light curve events, the spectra at early or late time, the association with star forming regions, nor the observed radio emission. No theoretical spectra corresponding to these models have been published.

3.2. TYPE IC

Based on the narrow peak and rapid post-maximum decline of the SN Ic light curves, Shigeyama et al. (1990) argued SN Ic's are low mass versions of the helium star SN Ib scenario. The lack of helium in the observed spectrum is not consistent with this picture. Swartz et al. (1993b) and Swartz (1993) showed that SN Ic's must be helium deficient based on model spectral calculations. The late-time spectra of SN Ic's (and SN Ib's) are dominated by lines of oxygen and calcium. SN Ic explosions are therefore low-mass metal-rich events.

The helium layers of a massive star can be lost in a wind in single star evolution, but the initial star must be extremely massive and the resulting oxygen core may be too massive to reproduce the SN Ic light curve. Woosley, Langer, & Weaver (1993) found that all stars that lose their hydrogen envelopes early in helium burning may converge to a relatively narrow final mass range (~ 2.5 to 5 $M_\odot$) assuming a reasonable estimate of the mass loss rate. They found that a 60 $M_\odot$ star could evolve to a ~ 4.3 $M_\odot$ oxygen core while an 85 $M_\odot$ star begins to collapse while still containing more than 8 $M_\odot$. The light curve of the 60 $M_\odot$ model was too broad to reproduce the SN Ic light curve.

For moderate initial masses, the hydrogen and helium layers can only be removed by interaction with a binary companion. As with hydrogen transfer in binary systems in SN Ib formation, it is probable that a fraction of the helium would remain at the time of explosion and that a range of hybrid SN Ic's could be observed with a range of helium line strengths. No such event seems to have yet been identified.

To account for the deduced ejecta mass and a neutron star remnant, the immediate pre-supernova progenitor mass must be ~ 2 to 3 $M_\odot$. For the progenitors stars to have lost their helium as well as hydrogen layers to a binary companion, the helium core mass must fall in a restricted range where extended helium envelopes can form. The mass of the helium core should thus have been ~ 2 to 4 $M_\odot$ (Battacharya & Van den Heuvel 1991). This corresponds to a total main sequence mass of 10 to 15 $M_\odot$. The estimated mass of the immediate progenitor, presumed to be a bare C-O core of ~ 2 to $3 M_\odot$, corresponds to main sequence masses of again about 10 to 15 $M_\odot$, or perhaps a little larger. Thus the main sequence mass of SN Ic progenitors could be comparable to those for SN Ib. Though the statistics are poor, many more SN Ic than SN Ib have been detected (Muller et al. 1992), suggesting that SN Ic should come from less massive stars than do SN Ib.

The process of loss of the helium envelope by mass transfer through Roche lobe overflow could, in principle, occur in either the primary or the secondary. In this case, SN Ic events could have either an unevolved or a compact companion. The small separation and small mass ejection required to leave two neutron stars bound following the secondary's explosion may require evolution that leads to mass transfer from the helium core of the secondary (Battacharya & Van den Heuvel 1991). This is also the qualitative condition to make a progenitor of a SN Ic and so some SN Ic may be involved in the production of binary pulsars. Yamaoka, Shigeyama, & Nomoto (1993) note that non-degenerate C-O core companions to neutron stars may be more common than He star companions because of the tendency to lose the helium envelope.

Alternatively, tidal instability, instead of helium envelope Roche lobe overflow, may remove helium from the core. This occurs through a common envelope phase (Romani 1993) followed by mass and angular momentum loss from the system. To account for SN Ic, the common envelope phase must naturally turn off after considerable mass loss, but before a complete merger. For low-mass helium cores, the tendency of the radius to expand (Yamaoka, Shigeyama, & Nomoto 1993) may guarantee that the common envelope process continues until the oxygen core is exposed. Thus a common envelope process may be more thorough in removing an extended envelope than the Roche lobe overflow process. It is necessarily the secondary that explodes in such a case because it is necessary to have a compact companion to generate the common envelope when Roche lobe transfer commences. If this is the primary route to SN Ic, then the progenitors should all have neutron star companions and two neutron stars, either bound or unbound, should be left after the explosion.

Again, as for SN Ib, the direct observed progenitor systems, presumably bare low-mass C-O cores with main sequence or possibly neutron star companions, remain unidentified.

4. Hybrid Events: SN 1954A, SN 1987K and SN 1993J

There appears to be a range of conditions over which supernovae can occur but few 'hybrid' objects that bridge the gaps between the limited number of well-defined supernova types. The prototype for this behavior is SN 1987K (Filippenko 1988), and the best observed is SN 1993J, the recent event in M81 (Wheeler and Filippenko 1994). The early spectra of SN 1993J contained hydrogen features typical of SN II. Optical helium lines appeared approximately 25 to 30 days after the explosion (Filippenko & Matheson 1993, Swartz et al. 1993a). Roughly 2 $M_\odot$ was ejected (Nomoto et al. 1993; Podsiadlowski et al. 1993) with a helium-rich ($X \sim 0.2$) envelope of ~ 0.3 to 0.5 $M_\odot$ (Swartz et al. 1993a). This corresponds to a helium core mass of ~ 3 to 4 $M_\odot$ at the time of explosion and an initial main sequence mass of ~ 15 $M_\odot$. Based on the deduced initial mass, the explosion followed Fe core collapse. The amount of ^{56}Ni produced is uncertain at present, but estimated to be $\sim 0.05 - 0.1$ $M_\odot$. The star must have lost most of its hydrogen envelope to a binary companion. It appears to be a hybrid object between SN II and SN Ib (Swartz et al. 1993a, Filippenko, Matheson, & Ho 1993).

If the AGB star scenario of Swartz, Wheeler, & Harkness (1991) for SN II-L is correct then SN 1993J is *not* related to these objects but to the more massive

SN II-P's. The fact that the best examples of SN Ib (SN 1984L and SN 1983N) never showed any evidence for hydrogen implies that SN 1993J may not be a simple extension of SN Ib's but that it requires some significantly different evolutionary process related to its mass or binary configuration. It may be that systems like SN 1993J result from Roche lobe overflow that automatically leaves a finite envelope behind as it becomes helium rich and retreats from the Roche lobe (Podsiadlowski, et al. 1993), but that SN Ib result when common envelope evolution ensues and the envelope is removed much more efficiently. If SN Ic's require binary interaction to remove the helium layers then it is interesting to note that SN 1993J did not lose its helium envelope even though it, too, very likely evolved in a binary system and otherwise seems to have the right core mass to undergo such an effect. It may be that retention of the small H envelope prevents the He envelope from expanding to a giant configuration.

We would argue, aided by hindsight based on SN 1993J, that the first hybrid event on record is SN 1954A which has previously been tentatively classified as a SN Ib because of the strong He I lines in the old photographic spectra (Branch 1972). Branch discussed the possible identification of H-α in the spectrum obtained 48 days after the maximum, but was uncomfortable because that implied a velocity, $\sim$ 11,800 km s^{-1}, which was substantially larger than the velocity implied for the helium lines, $\sim$ 7200 km s^{-1}. Allowing for the distortion of the plate response, one can now see that the spectrum clearly showed the P-Cygni profile of H-α indented by He I λ6678 as displayed so clearly in SN 1993J at about the same epoch. Furthermore, SN 1993J showed very similar velocities for the H-α and He I lines at about the same epoch. The H-β and H-γ lines are not prominent, but they may also be distorted by the photographic response.

This identification of SN 1954A as a hybrid event like SN 1993J removes one of the awkward discrepancies associated with it. SN 1954A had a very steep late time light curve, reminiscent of a SN Ia or SN Ic (Wheeler and Harkness 1990), and considerably steeper than the other SN Ib with which it has been tentatively classified. The light curve of SN 1954A seems to be somewhat steeper than that of SN 1993J, a point that requires further investigation, but otherwise fits comfortably in this class with low ejecta mass.

5. Summary

We would like to be able to delineate a continuous transition from one supernova type to another that correlates with well-defined parameters of stellar evolution theory such as the initial mass of the star and the binary separation. We can make tentative suggestions for the progenitor masses of the various supernova types and at least point at situations where duplicity seems to be important.

SN II-L appear to correspond to single stars of mass $\sim$ 8 to 10 M$_\odot$. Alternatively, ignoring the constraint that SN II-L form a homogeneous class, SN II-L may originally be as massive as SN II-P but lose a portion of their envelopes before exploding. SN 1979C may have been just at the lower end of this range, $\sim$ 7 to 8 $M_\odot$, and exploded following carbon deflagration, or it could have been at the upper end and have undergone a substantial interaction with a circumstellar nebula. SN II-P correspond to single stars of somewhat higher mass, $\sim$ 10 to 20 $M_\odot$. Stars of even higher mass are statistically rare and may often collapse directly to make black holes with no optical display.

Stars of mass similar to SN II-P that lose their envelopes in binary systems become SN Ib of moderate ejecta mass. If the entire envelope is not lost, hybrid objects like SN 1993J, SN 1987K, and SN 1954A result. Single stars of even higher mass lose their hydrogen and most of their helium in a wind and may correspond to the more massive SN Ib's. In a binary system, stars that are stripped to bare helium cores may, under some circumstances, undergo further mass loss and become SN Ic's. If the progenitor is the secondary in the binary system, it may become a SN Ic by virtue of tidal mass loss through a common envelope to a compact companion. The initial mass of such a secondary is then moderate.

The fact that SN II-L envelopes may be helium enriched should not be confused with, and is probably unrelated to, the fact that SN 1993J (and SN Ib's) have helium in their spectra. In the SN II-L's the helium probably is dredged from the core but is difficult to observe in the spectra because of the veiling effects of hydrogen. The helium observed in SN Ib and hybrid events arises from the nearly pure helium content of their cores.

The evolutionary properties distinguishing SN Ib, SN 1993J, and SN II events is difficult to explain. Is there a mass above which stars lose all their hydrogen and explode as SN Ib's or is the distinguishing characteristic the nature of the envelope loss process, Roche lobe overflow versus a common envelope process? Are there really only a few well-defined supernova types or are the poor statistics obscuring a broader diversity of supernovae corresponding to the wide variety of favorable conditions for supernova production? More data and further study will help to fill in the still unclear picture.

We thank Philip Podsiadlowski and Stefano Benetti for useful conversations. This work was supported in part by NSF Grant 9218035, NASA Grant NAGW-2905, the R. A. Welch Foundation, and by NASA through grant number GO-2563.01-87A from the Space Telescope Science Institute, which is operated by the Association of Universities for Research in Astronomy, Inc., under NASA contract NAS5-26555.

References

Barbon, R., Ciatti, F., & Rosino, L.: 1979, *A&A* **72**, 287
Battacharya, D., & Van den Heuvel, E. P. J. : 1991, *Phys. Rep.* **203**, 1
Blinnikov, S. I., & Bartunov, O. S.: 1993, *A&A* **273**, 106
Blocker, T., & Schonberner, D. : 1991, *A&A* **244**, L43
Branch D.: 1972, *A&A* **16**, 247

Branch, D.: 1988, in K. Nomoto, ed., *Lecture Notes in Physics, vol. 305*, Springer:New York, 281
Branch, D. et al. : 1981, *ApJ* **244**, 780
Branch, D., Nomoto, K., & Filippenko, A. V. : 1991, *Comments Astrophys.* **15**, 221
Cappellaro, E.: 1993, preprint.
Clocchiatti, A., & Wheeler, J. C.: 1994, in preparation.
Chugai, N. N. : 1991, *MNRAS* **250**, 513
Doggett, J. B., & Branch, D.: 1985, *AJ* **90**, 2303
Ensman, L. M., & Woosley, S. E.: 1988, *ApJ* **333**, 754
Falk, S. W., & Arnett, W. D. : 1977, *ApJS* **33**, 515
Filippenko, A. V. : 1988, *AJ* **96**, 1941
Filippenko, A. V.: 1991, in I. J. Danziger and K. Kjaer, eds., *SN 1987A and Other Supernovae*,
    ESO:Garching, 343
Filippenko, A. V., & Matheson, T.: 1993, *IAU Circ.*, No. 5777
Filippenko, A. V., Matheson, T., & Ho, L. C.: 1993, *ApJ* **415**, L103
Gaskell, C. M.: 1992, *ApJ* **389**, L17
Harkness, et al. : 1987, *ApJ* **317**, 355
Harkness, R. P., & Wheeler, J. C.: 1990, in A. G. Petschek, ed., *Supernovae*, Springer:New York,
    1
Hashimoto, M., Iwamoto, K., & Nomoto, K.: 1993, *ApJ* **414**, L105
Kirshner, R. P.: 1990, in A. G. Petschek, ed., *Supernovae*, Springer:New York, 59
Litvinova, I. Yu., & Nadyozhin, D. K.: 1983, *Ap. Sp. Sci.* **89**, 89
Lucy, L. B.: 1991, *ApJ* **383**, 308
Lundqvist, P. & Fransson, C.: 1988, *A&A* **192**, 221
McClure, R. D.: 1983, *ApJ* **268**, 264
Miller, D. L., & Branch, D.: 1990, *AJ* **100**, 530
Muller, R. A., et al.: 1992, *ApJ* **384**, L9
Nomoto, K., Suzuki, T., Shigeyama, T., Kumagai, S., Yamaoka, H., & Saio, H.: 1993, *Nature* **364**,
    507
Patat, F., Barbon, R., Cappellaro, E., & Turatto, M.: 1993, *A&A*, submitted
Podsiadlowski, Ph., Joss, P. C., & Hsu, J. L. L.: 1992, *ApJ* **391**, 246
Podsiadlowski, Ph., Hsu, J. L. L., Joss, P. C., & Ross, R. R.: 1993, *Nature* **364**, 509
Romani, R. W.: 1993, preprint
Sackmann, I.-J. & Boothroyd, A. I.: 1992, *Apj* **392**, L71
Scalo, J. M. & Ulrich, R. K.: 1973, *ApJ* **183**, 151
Schlegel, E. M., & Kirshner, R. P.: 1989, *AJ* **98**, 577
Schmidt, B. P. et al.: 1993, *AJ* **105**, 2236
Shigeyama, T., Nomoto, K., Tsujimoto, T., & Hashimoto, M. -A.: 1990, *ApJ* **361**, L23
Smith, V. V., & Lambert, D. L.: 1989, *ApJ* **345**, L75
Smith, V. V., & Lambert, D. L.: 1990, *ApJ* **361**, L69
Sramek, R. A., & Weiler, K. W.: 1990, in A. G. Petschek, ed., *Supernovae*, Springer:New York,
    76
Swartz, D. A.: 1993, *ApJ*, submitted
Swartz, D. A., Clocchiatti, A., Benjamin, R., Lester, D. F., & Wheeler, J. C.: 1993a, *Nature* **365**,
    232
Swartz, D. A., Filippenko, A. V., Nomoto, K., & Wheeler, J. C.: 1993b, *ApJ* **411**, 313
Swartz, D. A., Harkness, R. P., & Wheeler, J. C.: 1989, *Nature* **337**, 439
Swartz, D. A., & Wheeler, J. C.: 1991, *ApJ* **379**, L13
Swartz, D. A., Wheeler, J. C., & Harkness, R. P.: 1991, *ApJ* **374**, 266
Turatto, M., Cappellaro, E., Barbon, R., Della Valle, M., Ortolani, S., & Rosino, L.: 1990, *AJ* **100**,
    771
Turatto, M., Cappellaro, E., Benneti, S., & Danziger, I. J.: 1993, M.N.R.A.S., in press.
Uomoto, A., & Kirshner, R. P.: 1986, *ApJ* **308**, 685
Van Dyk, S.: 1992, *AJ* **103**, 1788
Van Dyk, S. D., Sramek, R. A., Weiler, K. W., & Panagia, N.: 1993, *ApJ* **409**, 162
Weaver, T. A., & Woosley, S. E.: 1980, in R. E. Meyerott and G. H. Gillespie, eds., *Supernova*

Spectra, AIP:New York, 15

Wheeler, J. C.: 1990, in J. C. Wheeler, T. Piran, and S. Weinberg, eds., *Supernovae*, World Scientific: Singapore, 1

Wheeler, J. C., & Filippenko, A. V.: 1994 in R. A. McCray and Z. Wang, eds., *Supernovae and Supernova Remnants*, Cambridge University Press: Cambridge, in press

Wheeler, J. C., Harkness, R. P., & Barkat, Z: 1988, *Lecture Notes in Phys.* **305**, 305

Wheeler, J. C., & Harkness, R. P.: 1990, *Rep. Prog. Phys.* **53**, 1467

Wheeler, J. C., Swartz, D. A., & Harkness, R. P.: 1993, *Phys. Rep.* **227**, 113

Woosley, S. E.: 1990, in A. G. Petschek, ed., *Supernovae*, Springer:New York, 182

Woosley, S. E., Langer, N., & Weaver, T. A.: 1993, *ApJ* **411**, 823

Yamaoka, H., Shigeyama, T., & Nomoto, K.: 1992, *A&A*,in press

Young, T. R., & Branch, D.: 1989, *ApJ* **342**, L79

SN 1987A AND SN 1993J: TESTING STELLAR EVOLUTION THEORY?

PHILIPP PODSIADLOWSKI

Institute of Astronomy, Cambridge CB3 0HA, UK

Abstract. Nearby supernovae like SN 1987A and SN 1993J provide valuable constraints on the late evolution of massive stars. For this purpose, we review evolutionary models for the progenitor of SN 1987A and confront them with five observational/theoretical tests we devised. We show that single-star models (with the possible exception of rapid-rotation models) fail at least two of these tests, while two binary models (accretion and merger models) are consistent with all available constraints. We conclude that it is most likely that the progenitor of SN 1987A had a binary companion, either at the time of the explosion or at least in the not-too-distant past, and that SN 1987A should therefore not be used to calibrate single stellar evolution theory. For SN 1993J, we infer from the presupernova photometry and the early light curve that its progenitor was a $\sim 15\,M_\odot$ star that lost almost all of its hydrogen-rich envelope prior to the supernova. This seems to require that the progenitor underwent stable case C mass transfer. We discuss future observational tests of binary models for both supernovae.

Key words: SN 1987A – SN 1993J – Massive Stars – Binaries

1. Introduction

The recent, relatively nearby supernovae SN 1987A in the Large Magellanic Cloud (LMC; distance $\sim 50\,\mathrm{kpc}$) and SN 1993J in M81 (distance $\sim 3.3\,\mathrm{Mpc}$) have become the best studied supernovae in history and may therefore provide important constraints on the late evolution of massive stars. However, both supernovae are very different from ordinary type II supernovae. It is also not clear whether they should be best discussed within the framework of single or binary stellar evolution theory. In Sections 2 – 5 of this review, we confront the various models for SN 1987A with the observational constraints (for a more detailed, earlier review, see Podsiadlowski 1992). In Section 2, we devise five tests a successful model has to fulfill. In Sections 3 and 4, we review the various single and binary models for the progenitor, respectively, and rigorously apply these tests. In Section 5, we summarize our conclusions and present further future tests. In Section 6, we present a model for SN 1993J and discuss its implications for stellar evolution theory.

2. Constraints on the Progenitor of SN 1987A

The problem of devising tests for models of the progenitor of SN 1987A consists of finding constraints that, on one hand, are well established and, on the other hand, are as restrictive as possible. As our main tests we have chosen the three major anomalies of the progenitor (its compactness, the ring of ejected material surrounding it, and its chemical anomalies), and the supernova explosion itself. In addition, a successful model has not only to be able to explain the main

Space Science Reviews **66**: 439–453, 1994.

© 1994 *Kluwer Academic Publishers. Printed in Belgium.*

features of this supernova, but it also has to be consistent with all other, general observational and theoretical constraints for massive stars.

2.1. THE COMPACTNESS OF THE PROGENITOR

One of the major surprises of this supernova event was that the progenitor (Sk $-69°202$) was a blue supergiant of spectral type B3 I (Rousseau et al. 1978) rather than a red supergiant, the generally expected precursor for Type II supernovae (e.g., Falk and Arnett 1977; Woosley and Weaver 1985). While there have been theoretical calculations in the past which produced red as well as blue presupernova models (see, e.g., Lamb et al. 1976; Brunish and Truran 1982), these models were not consistent with the observed distribution of massive stars in the Hertzsprung-Russell (H-R) diagram (Humphreys 1984; Fitzpatrick and Garmany 1990) which show that, even in the LMC, stars as massive as $\sim 40\,M_\odot$ pass through an extended red-supergiant phase.

2.2. THE RING AROUND THE PROGENITOR

Observations of the circumstellar nebula around SN 1987A with the *NTT* by Wampler et al. (1990) and the *HST* by Jacobson et al. (1991) have revealed that the circumstellar material around the supernova, first discovered with the *IUE* satellite (Fransson et al. 1989), has the morphology of a narrow ring rather than that of a spherical shell. The overabundance of this material in nitrogen relative to carbon and oxygen (Fransson et al. 1989) suggests that it is mainly composed of material which has been processed by nuclear reactions deep inside the progenitor and which has subsequently been ejected. The ring-like geometry of these ejecta implies an axisymmetric, but highly non-spherical structure of the envelope of the progenitor and/or its winds. Hydrodynamical calculations of the ring, in the framework of an interacting wind model (e.g., Blondin & Lundqvist 1993), require a density enhancement in the equatorial plane of ~ 20 relative to the polar direction in the red-supergiant wind from the progenitor. The origin of this non-sphericity provides a severe constraint for models of the progenitor. A plausible mechanism for generating the required asymmetry is the flattening of the progenitor's envelope caused by rapid rotation (Chevalier and Soker 1989). However, Chevalier and Soker showed, using straightforward angular momentum considerations, that a single star which was rapidly rotating on the main sequence would be a slow rotator in any subsequent supergiant phase and could not be significantly flattened at the time of the supernova explosion.

2.3. CHEMICAL ANOMALIES

There is substantial, observational evidence for anomalous chemical abundances in the hydrogen-rich envelope of the progenitor and in the presupernova ejecta (for an earlier summary see Truran and Weiss 1990). A few weeks after the supernova explosion, the spectra revealed an overabundance of barium and other s-process elements (Williams 1987; Danziger et al. 1988). These observations imply an apparent enhancement of these elements in the progenitor's hydrogen-rich envelope (note that these elements are only produced during helium burning and should therefore not be found in the hydrogen-rich envelope). Detailed atmosphere calculations by Hoflich (1988) and Mazzali, Lucy and Butler (1992) have confirmed the overabundances of s-process elements and possibly helium in the outer envelope. Evidence for an enhanced helium abundance (He/H $\simeq$ 0.2) is also found in the circumstellar material (Allen, Meikle and Spyromilio 1989; Wang 1991). Hoflich finds that only helium and s-process elements are enriched, while other elements are consistent with scaled solar abundances. This strongly suggests that these apparent enhancements are caused by real abundance effects and not by deficiencies in the atmosphere calculations (non-LTE effects, etc.), since it would be a coincidence, if only those elements that are related by the nuclear process by which they were produced (s processing during helium burning), but have unrelated atomic and spectroscopic properties, showed these anomalies. All of these anomalies can be reconciled, if the progenitor's envelope had been mixed thoroughly with a significant fraction of material that underwent nuclear burning (hydrogen burning and at least partial helium burning) before the supernova explosion.

2.4. The Supernova Explosion

The supernova explosion itself provides further important constraints for possible supernova progenitors. Comparisons of hydrodynamical calculations of the explosion with the observed supernova light curve and detailed atmosphere studies are all consistent with an immediate supernova progenitor of radius $R \sim 3 \times 10^{12}$ cm, core mass $M_c \sim 6\, M_\odot$, and envelope mass $M_{\mathrm{env}} \gtrsim 5\, M_\odot$ (see, e.g., Shigeyama, Nomoto and Hashimoto 1988; Hoflich 1988; Woosley, Pinto and Ensman 1988; Arnett and Fu 1989).

2.5. Independent Observational/Theoretical Constraints

A successful model for the progenitor of this supernova has to be consistent with the general theory of massive stars (or, at least, explain why the general theory may not be applicable in this particular case). For example, a generic model for SN 1993J should also be able to account for the detailed distribution of massive stars in the H-R diagram (the location and extent of the red-supergiant branch, the ratio of blue to red supergiants, the main-sequence width, etc.).

3. Single-Star Models for the Progenitor

3.1. The Low-Metallicity Model

In many ways the simplest model to explain a blue progenitor is the model in which the progenitor never becomes a red supergiant and ends its life as a blue supergiant because of an unusually low metallicity ($\sim 1/3\, Z_\odot$; Arnett 1987; Hillebrandt et al. 1987; Truran and Weiss 1987). The model is attractive, because it requires metallicity as the only free parameter. However, this model cannot explain SN 1987A, since it predicts that massive stars in the LMC should never reach the Hayashi line, whereas the red-supergiant branches of both the LMC and the Small Magellanic Clouds are populated up to a mass of $\sim 40\, M_\odot$ (Humphreys 1984; Fitzpatrick and Garmany 1990). In addition, a low-metallicity model has no mechanism for producing the ring around the progenitor and the chemical anomalies.

3.2. Extreme-Mass-Loss Models

Maeder (1987) and Wood and Faulkner (1988) proposed mass loss as the primary cause for a blue-supergiant progenitor. In their models, the progenitor would have first become a red supergiant, but returned to the blue-supergiant region, after it had lost most of its hydrogen-rich envelope. These models require not only very high mass-loss rates (typically 5 to 10 times larger than empirical rates; Chiosi and Maeder 1986; Dupree 1986), but also extreme fine-tuning of those rates. For slightly different rates, the star would end its life either as a red supergiant or as a Wolf-Rayet star.

While this model can explain the observational characteristics and perhaps some of the chemical anomalies (because processed material may have been exposed to the surface, it cannot be appropriate for SN 1987A, since it cannot reproduce the supernova light curve, which requires a much larger envelope mass (see, e.g., Woosley 1988). In addition, the model also provides no mechanism for producing the ring around the supernova.

3.3. The Restricted-Convection Model

A very different evolutionary scenario was first suggested by Woosley, Pinto and Ensman (1988; also see Langer, El Eid and Baraffe 1989; Weiss 1989), who showed that — under certain, limited circumstances — it is possible that a star first becomes a red supergiant, but returns to the blue-supergiant region in the H-R diagram near the end of its evolution (see Fig. 1). One of the most important assumptions in this model, which gives it its name, is the use of the Ledoux criterion for convective instability (Ledoux 1947) instead of the usually preferred Schwarzschild criterion (Schwarzschild 1906). This suppresses semiconvection above the convective hydrogen-burning core and produces smaller hydrogen-exhausted cores with shallower H-He composition gradients in the hydrogen-rich envelope above the core. As a result, the progenitor spends most or all of its helium-core-burning phase as a red supergiant, but becomes a blue supergiant shortly before the supernova explosion. Whether the star returns to the blue also depends on several other conditions; in the models by Woosley et al. (1988), the metallicity and the helium abundance both have to be low ($Z \lesssim 1/3\, Z_\odot$ and $Y \lesssim 0.23$), and mass loss relatively unimportant. Convective overshooting from the core must also be unimportant, since even a moderate amount of convective

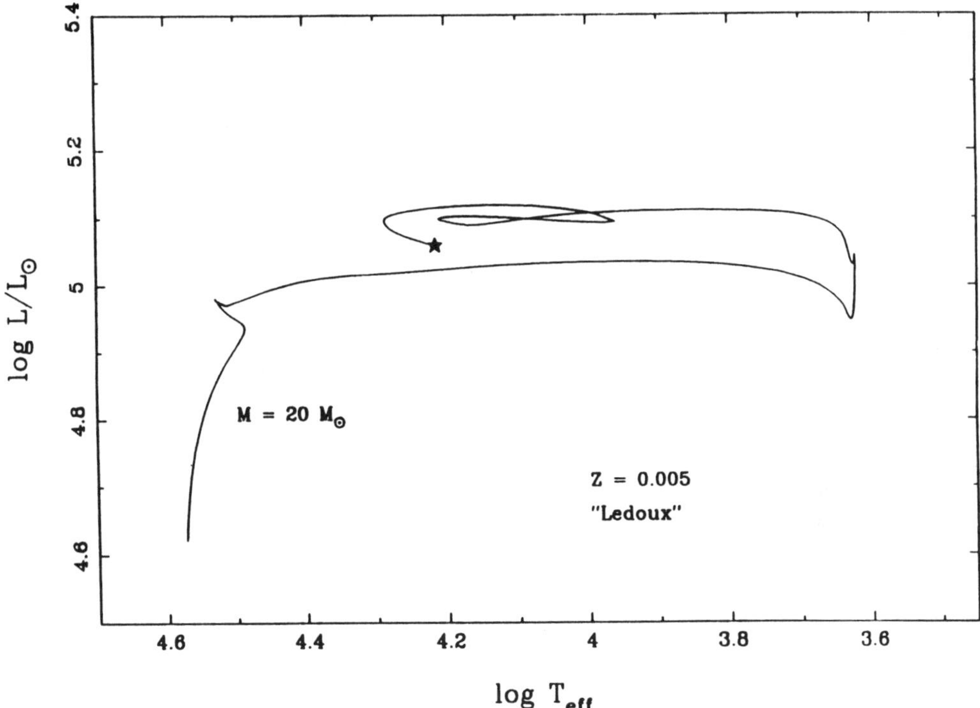

Fig. 1. The evolutionary track in the H-R diagram for a star of mass $M = 20\,M_\odot$ in a restricted-convection model ($Z = 0.005$, no mass loss, no convective overshooting; from Weiss 1989).

overshooting would prevent the final transition to the blue (see, e.g., Langer 1991b).

This model can explain a blue progenitor and the characteristics of the supernova explosion. It may also be able to account for the ratio of blue to red supergiants observed in the LMC. The main problems of this model are that it cannot explain most of the chemical anomalies and that it provides no plausible mechanism for the formation of the ring around the progenitor. In addition, it may also have difficulties explaining some of the details of the distribution of stars in the H-R diagram. Most importantly, the progenitors never reach the true Hayashi line in their red-supergiant phase (Tuchman and Wheeler 1989); the red extreme in these models is between 4000 and 6000 K. This effective temperature range is almost completely empty in the observational H-R diagram. While it is possible to produce models with cooler red giant branches (by decreasing the mixing length), these models would not return to the blue after core helium burning (El Eid 1993).

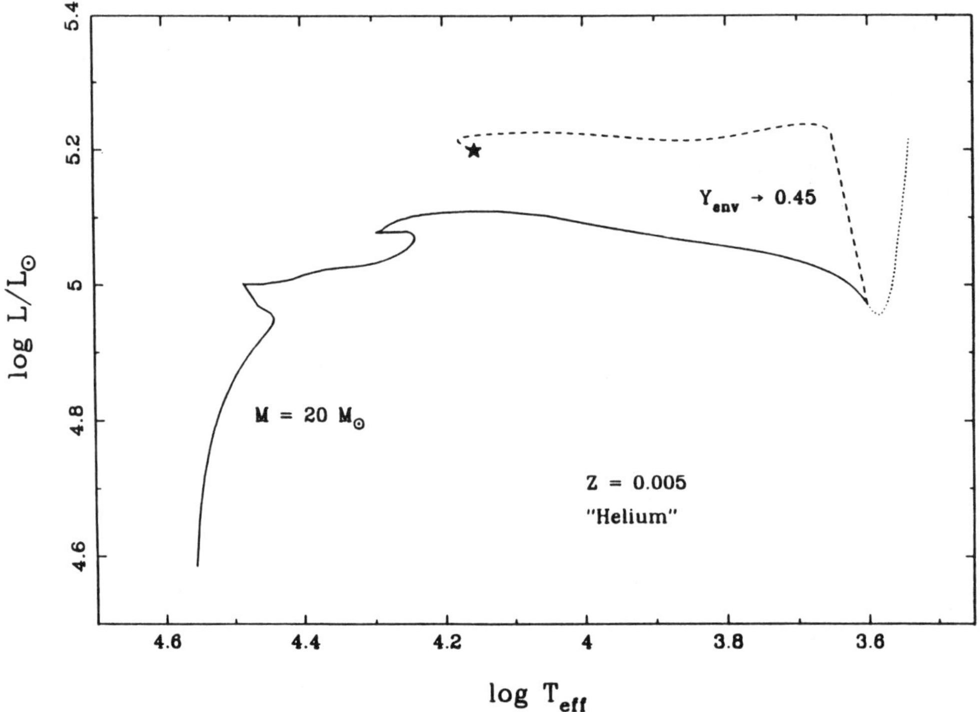

Fig. 2. The evolutionary track in the H-R diagram for a star of initial mass $M = 20\,M_\odot$ in the helium-enrichment model ($Z = 0.005$, mass loss according to De Jager et al. 1988; the helium abundance in the envelope is (artificially) increased from $Y = 0.24$ to $Y = 0.45$ where indicated).

3.4. THE HELIUM-ENRICHMENT MODEL

Saio, Kato and Nomoto (1988) proposed an evolutionary scheme in which the progenitor also undergoes a blue-red-blue evolution (see Fig. 2). They adopted the standard Schwarzschild criterion for convection and use relatively large mass loss rates (~ 5 times larger than the empirical rates). At the end of helium core burning (which mainly takes place in the blue-supergiant region), they artificially mix the hydrogen-rich envelope with a large fraction of the helium-rich interior. This produces a decrease of the opacity in the radiative envelope and triggers a rapid contraction of the envelope. The star then ends its life as a blue supergiant.

This model can explain a blue progenitor, the supernova event itself and perhaps (by design) the nitrogen and helium anomalies (Saio, Nomoto and Kato 1988). However, the model also has problems accounting for the distribution of stars in the H-R diagram. Since almost all of the helium-core-burning phase is spent in the blue supergiant region, the model does not produce a sufficient number of red supergiants to be consistent with the observed H-R diagram. However, the main problem of this scenario lies in the identification of a mixing mechanism that (spontaneously?) dredges up a substantial amount of helium-rich material.

Saio et al. propose meridional circulation as a mixing mechanism. However, it is well known from the theory of meridional circulation (see, e.g., Mestel 1965, Tassoul and Tassoul 1984) that even a small chemical gradient prevents the mixing of material across chemical gradients (via so-called μ-currents), as long as the star is not rotating close to break-up velocity. On the other hand, a star that is not breaking up on the main sequence, will not be close to break-up in a subsequent red-supergiant phase.

3.5. ROTATION MODELS

Rotational mixing (caused by meridional circulation or baroclinic instabilities) has been invoked by various authors (Weiss, Hillebrandt and Truran 1988; Ramadurai and Wiita 1989; Langer 1991a) in order to explain the large overabundance of nitrogen in the circumstellar material (Fransson et al. 1989). In these models, unlike the model by Saio et al. (1988), rotational mixing occurs in the pre-main-sequence or early main-sequence phase, in which the star has not yet built up a significant chemical gradient that would suppress rotational mixing. The calculations of Weiss et al. (1988) and Langer (1991a), which differ somewhat in their assumptions, are able to reproduce the observed abundance ratios of the CNO elements, but they produce only a moderate enhancement in helium.

In a more extreme version of this scenario Langer (1992) showed that, for very rapidly rotating stars (near break-up?), helium can diffuse to the surface and produce a significant increase in the helium abundance in the envelope. The evolution in the H-R diagram is similar to the restricted-convection model, except that the star spends very little time as a red supergiant (hence this model cannot be a generic model for massive stars in the LMC).

A rapid-rotation model could potentially avoid many of the problems most other single-star models are confronted with. Just as in those models, a blue progenitor could explain the observed supernova explosion. In addition, such a model may be able to reproduce many of the chemical anomalies (although probably not the barium anomaly) and could perhaps provide a sufficient rotational asymmetry to produce the supernova ring (in particular, if the star avoids a red-supergiant phase in which magnetic braking would provide an efficient slow-down mechanism). The model has some of the same problems as the restricted-convection model in accounting for the detailed distribution of stars in the H-R diagram (e.g., the location of the red-supergiant branch). However, on the whole, a rapid-rotation model appears to be the most promising of the single-star models, although there are still a number of open questions concerning its details.

4. Binary Models for the Progenitor

All of the models discussed in the previous section assume single-star evolution. However, most stars actually occur in binary or multiple systems (e.g., Abt and

Levy 1976, 1978; Kraicheva et al. 1978, 1979; Duquennoy and Mayor 1991). Indeed, it is more likely that any exploding, massive star has a companion at the time of the explosion than that it is single. In many cases, the companion will be relatively distant and not affect the presupernova evolution of the progenitor (although it may still affect the explosion and the evolution of the remnant). However, in a substantial fraction of cases (20 – 40 % of stellar systems; Podsiadlowski, Joss and Hsu 1992; Tutukov, Yungel'son and Iben 1992), the companion is close enough to interact with the progenitor and to completely alter its presupernova evolution and the resulting supernova explosion.

4.1. COMPANION MODELS

One class of binary models for the progenitor of SN 1987A was based on the idea that the star which exploded was not Sk $-69°202$, but instead a previously undetected companion star, either a helium star (Fabian et al. 1987) or a stripped supergiant (Joss et al. 1988).

While both of these models are consistent with the constraints imposed by stellar and binary evolution, they are inconsistent with the observed features of the SN 1987A explosion (see, e.g., Hsu et al. 1991). Furthermore, the central prediction of these models, the survival of Sk $-69°202$, is now ruled out by the low luminosity of the observed bolometric light curve (Suntzeff et al. 1991).

4.2. ACCRETION MODELS

Accretion of mass from a companion can drastically alter a star's further evolution. Two types of accretion models have been proposed. In the models by Podsiadlowski and Joss (1989) and De Loore and Vanbeveren (1992), the progenitor ends its evolution as a blue supergiant, because it has accreted a substantial amount of mass from its companion after its hydrogen-core-burning phase. (If accretion takes place, when the progenitor is still on the main sequence, it would be rejuvenated as a main-sequence star and subsequently mimic the evolution of a more massive, single star; see, e.g., Hellings 1983; but also see Braun and Langer 1994). On the other hand, Barkat and Wheeler (1989) and Tuchman and Wheeler (1990) emphasize the possible increase of the helium abundance in the envelope (either because of accretion-induced dredge-up or the accretion of helium-rich material) which favors blue-supergiant progenitors.

The first type of evolution is illustrated in Figure 3, which shows the evolutionary track in the H-R diagram of a star of 15 $M_\odot$ that accretes 5 $M_\odot$ during the helium-core-burning phase (solid and dashed curves). Because of the added mass, the immediate supernova progenitor is a blue supergiant instead of a red supergiant, as it would have been without the accretion phase (dotted curve). The primary (the mass donor), on the other hand, is most likely to have already completed its own evolution and to have become a neutron star (or black hole). Thus, this model predicts that, at the time of the explosion, the blue-supergiant

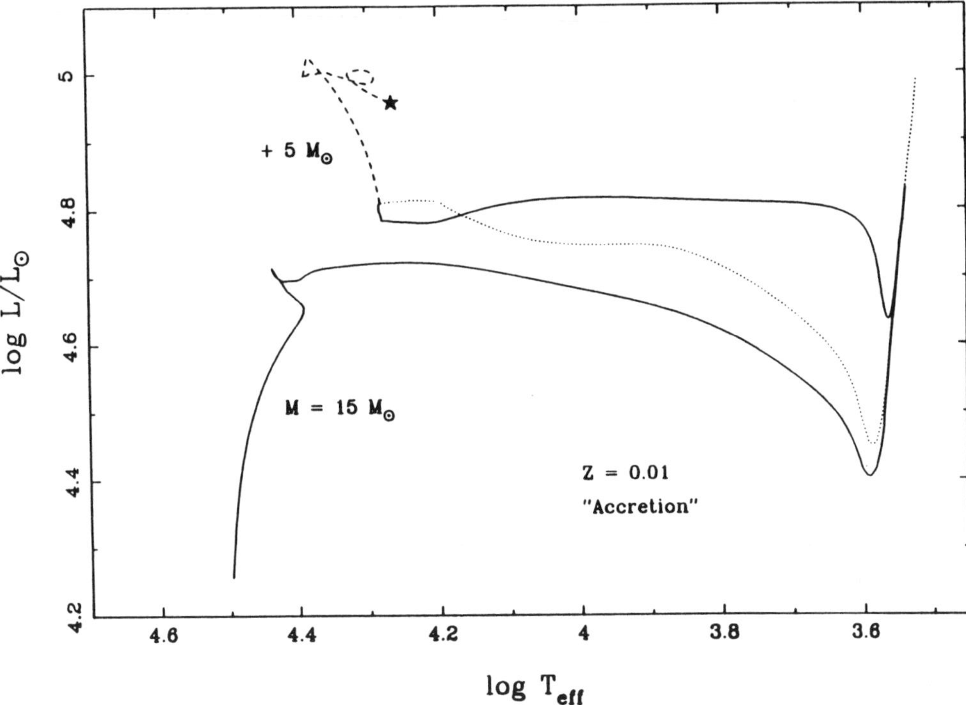

Fig. 3.   The evolutionary track in the H-R diagram for a star of mass $M = 15\,M_\odot$ with and without accretion in the accretion model ($Z = 0.01$, mass loss according to De Jager et al. 1988, moderate amount of convective overshooting).

progenitor still has a companion, which is either a compact star or, less likely, an ordinary star (Podsiadlowski and Joss 1989).

This model can explain a blue progenitor and appears to be consistent with all stellar and binary constraints (for a discussion of the relevant aspects of binary evolution, see Podsiadlowski et al. 1992). It is worth emphasizing that this model provides a generic mechanism to produce blue-supergiant progenitors and does not require fine-tuning of the input parameters. In particular, it is insensitive to the assumed metallicity, the treatment of convection, the details of the mass-transfer process, or the amount of mass that may be lost by the supernova progenitor in a stellar wind. It requires only that at least several solar masses more be accreted by the progenitor than is subsequently lost via a stellar wind, and that the accretion phase takes place after the termination of the main-sequence phase of the progenitor. The final color of the progenitor (which reflects its compactness) depends on the amount of matter that has been accreted. Podsiadlowski et al. (1992) estimated that $\sim 5\,\%$ of all massive stars should end their evolution as blue supergiants because they accreted matteror because they merged with a companion (see Section 4.3).

In an accretion model, the progenitor is expected to be rapidly rotating immediately after the accretion phase and possibly at the time of the supernova explosion. This can provide plausible mechanisms for the origin of the circumstellar ring (Podsiadlowski 1991). For example, one possibility is that the ring is the product of an excretion-disk-like outflow (similar to the slowly expanding disks in some models for Be stars; see, e.g., Poeckert 1982) that is swept up by a subsequent fast stellar wind (in the Be-star phase, the system would be very similar to the binary radio pulsar PSR 1259-63; Johnston et al. 1992).

The model may also be able to explain many of the chemical anomalies of the progenitor (in particular the nitrogen and the helium anomaly), since the accreted material will have experienced partial CNO processing and will be helium-enriched. In addition, it is possible that the core of the progenitor will be spun up to nearly break-up velocity (in particular, if the mass-gaining star has a convective envelope). If core spin-up takes place, meridional currents may be able to overcome the stabilizing effects of chemical gradients (Mestel 1965; see, however, Tassoul and Tassoul 1984) and produce large-scale mixing of the star, including the dredge-up of helium and perhaps s-processed material.

4.3. MERGER MODELS

The merger models for SN 1987A (Hillebrandt and Meyer 1989; Podsiadlowski, Joss and Rappaport 1990; also see Chevalier and Soker 1989) are closely related to the accretion models. Instead of accreting part of the companion, the companion merges with the progenitor and is completely dissolved in the latter's envelope. As a consequence, the progenitor becomes a blue supergiant, either because of the increased mass in the envelope (Podsiadlowski et al. 1990 and Fig. 4) or because of the dredge-up of helium (Hillebrandt and Meyer 1989), or both.

The merger of two stars is generally believed to be one of the possible, relatively frequently occurring outcomes of dynamically unstable mass transfer (Paczynski 1970; also see the discussion in Podsiadlowski et al. 1992). It is likely that, during the merger process, helium and possibly even some of the products of helium burning from the primary's core (Meyer and Meyer-Hofmeister 1979; Iben and Tutukov 1985), provided that the primary is an evolved star. In addition, three-dimensional simulations of the common-envelope phase (Livio and Soker 1988) suggest that the merger process is accompanied by substantial mass loss and that most of the mass is lost in the equatorial plane. This will automatically lead to ring- or disk-like structures surrounding the merged system. Thus, the final outcome of the merger of two stars is a rapidly rotating single blue supergiant, which may be thoroughly mixed and surrounded by a ring of ejected material. Such a model could explain all of the constraints of this supernova event.

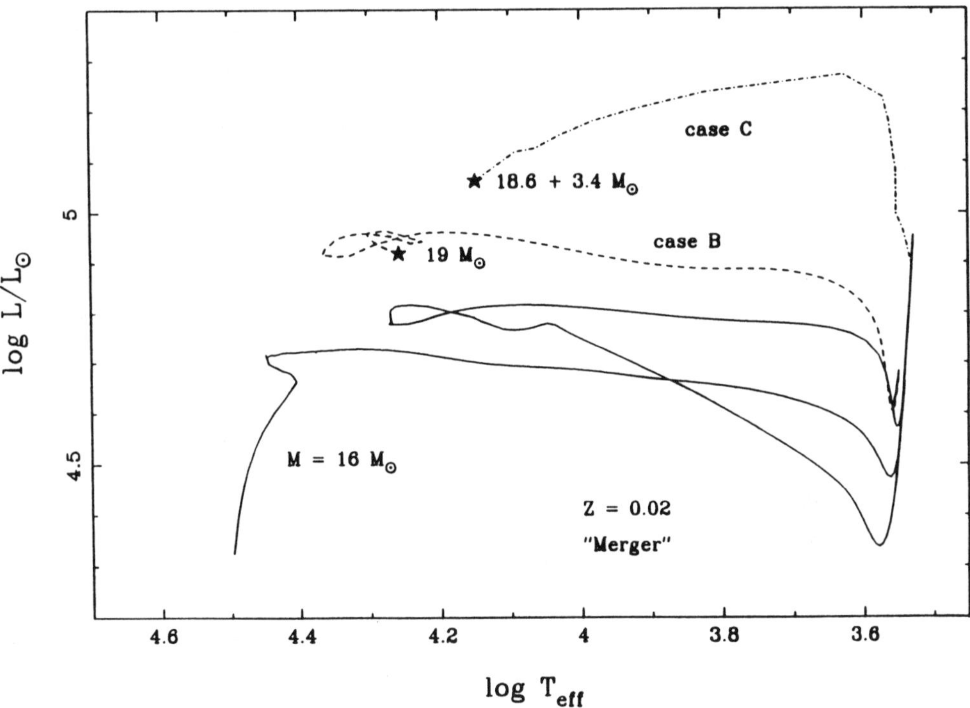

Fig. 4.   The evolutionary tracks in the H-R diagram of a $16\,M_\odot$ star (solid curve) and two illustrative merger calculations (dashed and dot-dashed curves; $Z = 0.02$, no mass loss, moderate amount of convective overshooting; from Podsiadlowski et al. 1990).

5. Discussion and Tests

In Table 1, we summarize the results of applying the tests developed in Section 2 to the various single and binary models discussed in the previous two sections. On face value, all single-star models (except for the rapid-rotation model) and the binary companion models fail at least two of the five tests. Thus, the main conclusion of our analysis is that SN 1987A cannot be used directly to test or calibrate standard stellar evolution theory (i.e., the theory of single, slowly rotating stars), since it appears that the progenitor was either a member of an interacting binary or a very rapid rotator. This conclusion can only be avoided, if some of the strongest constraints we used can be argued away (e.g., the ring and the chemical anomalies). This, however, appears very unlikely. On the other hand, SN 1987A may provide important constraints for rapid-rotation or binary models. Before this can be done with some confidence, some further tests of the presently viable models will be necessary.

The most straightforward confirmation of any model would be the detection of a compact companion star, as predicted in the binary accretion model (note that,

	blue progenitor	circumstellar ring	chemical anomalies	supernova explosion	evolution of massive stars
low-metallicity models	yes	no	no	yes	no
extreme-mass-loss models	yes	no	?	no	no
restricted-convection models	yes	no	no	yes	?
helium-enrichment model	yes	no	?	yes	no
rapid-rotation model	yes	?	?	yes	?
companion models	no	?	?	no	yes
accretion models	yes	yes	?	yes	yes
merger models	yes	yes	?	yes	yes

TABLE 1: Summary of the tests devised in Section 2 and applied to the various single and binary models for the progenitor of SN 1987A (see Sections 3 and 4).

in the binary merger model, no companion is expected). An important parameter in many single-star models (including the rapid-rotation model) is the actual metallicity of the progenitor. These models seem to require a substantially smaller metallicity than the value found in the most detailed analyses of young LMC stars (Russell and Bessell 1989; Russell and Dopita 1990). Fortunately, it should be possible to resolve this issue with future observations and calculations. High-resolution observations of stars 2 and 3 in the Sk $-69°202$ system, combined with modern atmosphere calculations (e.g., as presented here by Kudritzki 1994), should allow not only a determination of the overall metallicity of stars in the association to which the progenitor of SN 1987A belonged, but also of individual element abundances, which can then be used to construct custom-made opacity tables (note that the most recent, increased opacities [Rogers and Iglesias 1992] may already pose severe problems for single-star models).

Finally, in the accretion and the merger models, it is possible that, after the accretion or merger phase, rotational mixing dredges up helium and the products of helium burning. If material from the layer, in which the reaction $^{13}C(\alpha, n)\,^{16}O$, one of the major neutron sources for s-processing, is dredged up and ^{13}C, produced in the hydrogen-burning shell, is dredged down, this would provide an efficient environment for continued s-processing. This could not only explain the barium anomaly, but also provide further chemical tests for these models (for more details and further tests, see Podsiadlowski 1992).

6. SN 1993J: Evidence for stable case C mass transfer?

Supernova 1993J in the spiral galaxy M81 is the brightest supernova since SN 1987A and, like the latter, appears to be another peculiar type II supernova. Its early light curve is characterized by a very sharp initial peak (lasting for less than ten days) followed by a less rapid secondary brightening, which was qualitatively similar to the secondary brightening observed in SN 1987A.

Photometric constraints on the progenitor (Aldering et al. 1993) suggest that the progenitor probably was a K supergiant with a mass of $\sim 15\ M_\odot$ (although the progenitor cannot yet be identified unambiguously). In addition, the early light curve of the supernova, in particular its initial very sharp peak, imply that the progenitor had lost all but a few tenths of a solar mass of its hydrogen-rich envelope (e.g., Nomoto et al. 1993; Podsiadlowski et al. 1993; Woosley et al. 1993), i.e., was a stripped supergiant.

It is unlikely that a $15\ M_\odot$ star would lose almost all of its envelope in a stellar wind, since this would require a much larger initial main-sequence mass than is consistent with the photometric constraints on the progenitor, as well as significant fine-tuning of the mass-loss rate (otherwise the progenitor would be either a normal red supergiant or a Wolf-Rayet star). A more probable scenario is that the progenitor was a member of a close binary system and underwent stable case C mass transfer (i.e., as a red supergiant after helium core-burning; see Podsiadlowski et al. 1992). As a result of Roche-lobe overflow, the progenitor will lose most of its hydrogen-rich envelope, of which a significant fraction will be accreted by the companion. However, when the mass in the hydrogen-rich envelope has decreased below a certain critical mass ($\sim 0.3\ M_\odot$ for a progenitor of $\sim 15\ M_\odot$), the envelope of the progenitor will shrink significantly so that the star will no longer fill its Roche lobe, and mass transfer will cease. The primary will initially still have the appearance of a red supergiant (although of somewhat earlier type), but will have only a residual hydrogen-rich envelope with a mass of at most a few tenths of a solar mass. This is precisely the type of progenitor required to explain the early light curve.

Hsu et al. (1994) have modeled this type of evolution in the context of SN 1993J. They find that mass transfer can be stable and a common-envelope phase be avoided provided that the initial mass ratio of the system is not too far from 1 and that the primary loses several solar masses before it fills its Roche lobe, possibly in an enhanced stellar wind. After the supernova, the system will remain bound and may resemble PSR 1259-63 (Johnston et al. 1992). One of the attractive features of this model is that it can be rigorously tested, since the companion star, which is presently hidden below the photosphere of the supernova, should ultimately reappear. If the supernova lead to the formation of a pulsar, periodic modulation of its pulse period may allow the direction confirmation of such a binary scenario.

If the progenitor of SN 1993J was a member of a binary system, it would also not provide a good case study for the calibration of the theory of single, massive stars. However, it might provide conclusive proof for the existence of stable (as opposed to dynamical) case C mass transfer, for which there has only been indirect evidence in the past (see the discussion in Podsiadlowski et al. 1992).

References

Abt, H. A. and Levy, S. G.: 1976, *Astrophysical Journal, Supplement Series* **30**, 273

Abt, H. A. and Levy, S. G.: 1978, *Astrophysical Journal, Supplement Series* **36**, 241

Aldering, G., Humphreys, R. M. and Richmond, M.: 1993, preprint

Allen, D. A., Meikle, W. P. S. and Spyromilio, J.: 1989, *Nature* **342**, 403

Arnett, W. D.: 1987, *Astrophysical Journal* **319**, 136

Arnett, W. D. and Fu, A.: 1989, *Astrophysical Journal* **340**, 396

Barkat, Z. and Wheeler, J. C.: 1989, *Astrophysical Journal* **342**, 940

Blondin, J. M. and Lundqvist, P.: 1993, *Astrophysical Journal* **405**, 337

Braun, H. and Langer, N.: 1994 *this workshop*

Brunish, W. M. and Truran, J. W.: 1982, *Astrophysical Journal* **49**, 447

Chevalier, R. A. and Soker, N.: 1989, *Astrophysical Journal* **341**, 867

Chiosi, C. and Maeder, A.: 1986, *Annual Review of Astronomy and Astrophysics* **24**, 329

Danziger, I. J. et al.: 1988, '' in M. C. Kafatos and A. G. Michalitsianos, ed(s)., *Supernova 1987A in the Large Magellanic Cloud*, Cambridge Unversity Press: Cambridge, 37

De Jager, C., Nieuwenhiujzen, H. and van der Hucht, K.: 1988, *Astronomy and Astrophysics, Supplement Series* **72**, 259

De Loore, C. and Vanbeveren, D.: 1992, *Astronomy and Astrophysics* **260**, 273

Dupree, A. K.: 1986, *Annual Review of Astronomy and Astrophysics* **24**, 377

Duquennoy, A. and Mayor, M.: 1991, *Astronomy and Astrophysics* **248**, 485

El Eid, M. F.: 1993, private communication

Fabian, A. C., Rees, M. J., van den Heuvel, E. P. J. and van Paradijs, J.: 1987, *Nature* **328**, 323

Falk, S. W. and Arnett, W. D.: 1977, *Astrophysical Journal, Supplement Series* **33**, 515

Fitzpatrick, E. L. and Garmany, C. D.: 1990, *Astrophysical Journal* **363**, 119

Fransson, C. et al.: 1989, *Astrophysical Journal* **336**, 429

Hellings, P.: 1983, *Astrophysics and Space Science* **96**, 37

Hillebrandt, W., Hoflich, P, Truran, J. W. and Weiss, A.: 1987, *Nature* **327**, 597

Hillebrandt, W. and Meyer, F.: 1989, *Astronomy and Astrophysics* **219**, L3

Hoflich, P.: 1988, *Proc. Astron. Soc. Australia* **7**, 434

Hsu, J. J. L., Podsiadlowski, Ph., Joss, P. C. and Ross, R. R.: 1994, in preparation

Hsu, J. J. L, Ross, R. R., Podsiadlowski, Ph. and Joss, P. C.: 1991, '' in I. J. Danziger and K. Kjar, ed(s)., *ESO/EIPC Workshop, Supernova 1987A and Other Supernovae*, ESO: Garching, 37

Humphreys, R. M.: 1984, '' in A. Maeder and A. Renzini, ed(s)., *Observational Tests of the Stellar Evolution Theory*, Reidel: Dordrecht, 279

Iben, I., Jr. and Tutukov, A. V.: 1985, *Astrophysical Journal, Supplement Series* **58**, 661

Jacobson, P. et al.: 1991, *Astrophysical Journal* **369**, L63

Johnston, S., Manchester, R. N., Lyne, A. G., Bailes, M., Kaspi, V. M., Guojun, Q. and D'Amico, N.: 1992, *Astrophysical Journal* **387**, L37

Joss, P. C., Podsiadlowski, Ph., Hsu, J. J. L. and Rappaport, S.: 1988, *Nature* **331**, 237

Kraicheva, Z. T., Popova, E. I., Tutukov, A. V. and Yungel'son, L. R.: 1978, *Sov. Astr.* **22**, 670

Kraicheva, Z. T., Popova, E. I., Tutukov, A. V. and Yungel'son, L. R.: 1979, *Sov. Astr.* **23**, 290

Kudritzki, R.: 1994 this workshop

Lamb, S. A., Iben, I., Jr. and Howard, W. M.: 1976, *Astrophysical Journal* **207**, 209

Langer, N.: 1991a, *Astronomy and Astrophysics* **243**, 155

Langer, N.: 1991b, *Astronomy and Astrophysics* **252**, 669

Langer, N.: 1992, *Astronomy and Astrophysics* **265**, L17

Langer, N., El Eid, M. F. and Baraffe, I.: 1989, *Astronomy and Astrophysics* **224**, L17

Ledoux, P.: 1947, *Astronomy and Astrophysics* **105**, 305

Livio, M. and Soker, N.: 1988, *Astronomy and Astrophysics* **329**, 764

Maeder, A.: 1987, '' in J. I. Danziger, ed(s)., *Proc. ESO Workshop on SN 1987A*, ESO, Garching, 251

Mazzali, P. A., Lucy, L. B. and Butler, K.: 1992, *Astronomy and Astrophysics* **258**, 399

Mestel, L.: 1965, '' in L. H. Aller and D. B. Mclaughlin, ed(s)., *Stellar Structure Vol. 8*, Chicago University Press: Chicago, 465

Meyer, F. and Meyer-Hofmeister, E.: 1979, *Astronomy and Astrophysics* **78**, 167

Nomoto, K. et al.: 1993, *Nature* **364**, 507

Pazcynski, B.: 1970, '' in K. Gyldenkerne and R. M. West, ed(s)., *Mass Loss and Evolution in Close Binaries*, Copenhagen University Press: Copenhagen, 139

Podsiadlowski, Ph.: 1991, *Nature* **350**, 654

Podsiadlowski, Ph.: 1992, *Publications of the ASP* **104**, 717

Podsiadlowski, Ph., Hsu, J. J. L., Joss, P. C. and Ross, R. R.: 1993, *Nature* **364**, 509

Podsiadlowski, Ph. and Joss, P. C.: 1989, *Nature* **338**, 401

Podsiadlowski, Ph., Joss, P. C. and Rappaport, S.: 1990, *Astronomy and Astrophysics* **227**, L9

Podsiadlowski Ph., Joss, P. C. and Hsu, J. J. L.: 1992, *Astrophysical Journal* **391**, 246

Poeckert, R.: 1982, '' in M. Jaschek and H.-G. Groth, ed(s)., *Be Stars*, Reidel: Dordrecht, 453

Ramadurai, S. and Wiita, P. J.: 1989, *Comm. Astrophysics* **13**, 107

Rogers, F. J. and Iglesias, C. A.: 1992, *Astrophysical Journal, Supplement Series* **79**, 507

Rousseau, J., Martin, N., Prevot, L., Rebeirot, E., Robin, A. and Brunet, J. P.: 1978, *Astronomy and Astrophysics, Supplement Series* **31**, 243

Russell, S. C. and Bessell, M. S.: 1989, *Astronomy and Astrophysics, Supplement Series* **70**, 865

Russell, S. C. and Dopita, M. A.: 1990, *Astronomy and Astrophysics, Supplement Series* **74**, 93

Saio, H., Kato, M. and Nomoto, K.: 1988, *Astrophysical Journal* **331**, 388

Saio, H., Nomoto, K. and Kato, M.: 1988, *Nature* **334**, 508

Schwarzschild, K.: 1906, *Gott. Nachr.* **41**,

Shigeyama, T., Nomoto, K. and Hashimoto, M.: 1988, *Astronomy and Astrophysics* **196**, 141

Suntzeff, N. B., Phillips, M. M., Depoy, D. L., Elias, J. H. and Walker, A.R.: 1991, *Astronomical Journal* **102**, 1118

Tassoul, M. and Tassoul, J.-L.: 1984, *Astrophysical Journal* **279**, 384

Truran, J. W. and Weiss, A.: 1987, '' in I. J. Danziger, ed(s)., *Proc. ESO Workshop on Supernova 1987A*, ESO, Garching, 271

Truran, J. W. and Weiss, A.: 1990, *Comm. Astrophys.* **14**, 195

Tuchman, Y. and Wheeler, J. C.: 1989, *Astrophysical Journal* **344**, 835

Tuchman, Y. and Wheeler, J. C.: 1990, *Astrophysical Journal* **363**, 255

Tutukov, A. V., Yungel'son, L. R. and Iben, I., Jr.: 1992, *Astrophysical Journal* **386**, 197

Wampler, E. J. et al.: 1990, *Astrophysical Journal* **362**, L13

Wang, L.: 1991, *Astronomy and Astrophysics* **246**, L69

Weiss, A.: 1989, *Astrophysical Journal* **339**, 365

Weiss, A., Hillebrandt, W. and Truran, J. W.: 1988, *Astronomy and Astrophysics* **197**, L11

Williams, R. E.: 1987, *Astrophysical Journal* **320**, L117

Wood, P. R. and Faulkner, D. J.: 1988, *Proc. Astr. Soc. Australia* **7**, 75

Woosley, S. E.: 1988, *Astrophysical Journal* **330**, 218

Woosley, S. E., Eastman, R. G., Weaver, T. A., Pinto, P. A.: 1993, *Astrophysical Journal,* in press

Woosley, S. E., Pinto, P. A. and Ensman, L.: 1988, *Astrophysical Journal* **324**, 466

Woosley, S. E. and Weaver, T. A.: 1985, '' in J. Audouze and N. Mathieu, ed(s)., *Nucleosynthesis and its Implications on Nuclear and Particle Physics*, Reidel: Dordrecht, 145

CHEMICAL AND PHOTOMETRIC EVOLUTION OF ELLIPTICAL GALAXIES

C. CHIOSI

Department of Astronomy, Vicolo Osservatorio 5, 35122 Padua, Italy

Abstract. In this paper we present the new chemical-spectro-photometric models of population synthesis by Bressan, Chiosi & Fagotto (1993). The models are specifically designed for elliptical galaxies. They include the presence of dark matter and galactic winds triggered by the energy deposit from supernovae and winds of massive stars. The models are aimed at studying the UV-excess and its dependence on the metallicity, the color-magnitude relation, and the color evolution as a function of the redshift. It is shown that in order to explain the color-magnitude relation as a result of galactic winds, the energy input from massive stars is required. Supernovae alone cannot provide sufficient energy to start galactic wind before the metallicity and hence colors have got saturated. We show that the main source of the UV-excess are the old, hot HB and AGB manque stars of high metallicity present in varying percentages in the stellar content of a galaxy. Since in our model the mean and maximum metallicity are ultimately driven by the mass of the galaxy, this provides a natural explanation for the observed correlation between UV-excess and metallicity. Finally, looking at the color evolution as function of the redshift, we suggest that a sudden change occurring in the color (1550-V) at the onset of the old, hot HB and AGB manque stars of high metallicity, is a good age indicator. The detection of this feature at a certain redshift would impose firm constraints on the underlying cosmological model of the universe.

1. Introduction

Accurate models of spectro-photometric evolution of galaxies of different morphological type are basic to many key questions of galaxy formation and cosmology. For nearby galaxies the observed integrated spectral energy distribution (ISED) is used to understand the properties of the their stellar populations and to infer the past history of star formation. For more and more distant galaxies the observed magnitude-redshift relation, the color-redshift relation, the galaxy number counts, and the redshift distribution of galaxies are compared to the theoretical predictions from spectro-photometric models to cast light on the Hubble constant H_0, the deceleration parameter q_o, and the redshift of galaxy formation z_{for}.

Recently, Bressan, Chiosi & Fagotto (1993), thereinafter BCF, have presented new chemical, spectro-photometric models of population synthesis particularly tailored for elliptical galaxies with the aim of understanding the causes of the UV excess and the color-magnitude relation. In short, the UV region of the ISED of elliptical galaxies shows an excess that correlates with the Mg_2 index or equivalently the metallicity, while the colors get redder at decreasing magnitudes. Several competing hypotheses have been advanced to explain the UV light, each invoking different types of stars going from young massive stars to a variety of old, low mass stars, and finally some unusual stages of stellar evolution, such as accretion disks around White Dwarfs (WD) in binary systems or hot components

Space Science Reviews **66**: 455–470, 1994.
© 1994 *Kluwer Academic Publishers. Printed in Belgium.*

of binary stars (see Greggio & Renzini 1990 for details). Most of the existing models attribute the UV light to the so-called P-AGB stars evolving toward the WD cooling sequence (e.g. Bruzual & Charlot 1993, Magris & Bruzual 1993).

The color-magnitude relation is commonly interpreted as due to increasing average metallicity with total galaxy mass (Faber 1977, Dressler 1984, Vader 1986), and the effect of galactic winds driven by supernova explosions (Matteucci & Tornambe' 1987, Arimoto & Yoshii 1987, Yoshii & Arimoto 1987).

Recent observations both of the UV spectra (Ferguson & Davidsen 1993, Ferguson et al. 1991) and colors of elliptical galaxies (Bower et al. 1992, Marsiaj 1992), together with the advancements in the theory of stellar models and population synthesis impose to look for models of elliptical galaxies accounting for the above observational facts in a self consistent fashion (see BCF for details).

In this paper we report on the analysis made by BCF limited to a selected number of topics, namely, the UV excess, the color-magnitude relation, and potential probes of galaxy ages.

2. Library of Stellar Tracks and Spectra

The library of stellar models includes all the evolutionary sequences calculated by Alongi et al. (1993), Bressan et al. (1993), Fagotto et al. (1993a,b) for several combinations of the chemical parameters Y and Z. The mass range goes from 0.6 $M_\odot$ to 120 $M_\odot$, and the evolutionary phases go from the main sequence to either the start of the thermally pulsing regime of the asymptotic giant branch phase (TP-AGB) or central C-ignition as appropriate for the initial mass of the stars. Each set contains 52 evolutionary sequences with suitable mass spacing. The parameters Y and Z obey the relation $\Delta Y/\Delta Z = 2.5$, which is a lower limit to the estimates obtained by Pagel et al. (1992). The following combinations of abundances are considered [Y=0.230, Z=0.0004], [Y=0.230, Z=0.001], [Y=0.250, Z=0.008], [Y=0.280, Z=0.020], [Y=0.352, Z=0.05], and [Y=0.475, Z=0.100].

Since the detailed discussion of the properties of these models is beyond the scope of the paper, we limit ourselves to highlight a few relevant points. All the models but those for the composition [Y=0.475, Z=0.1], are calculated with the the radiative opacities by Iglesias et al. (1992) based on the abundance scale by Grevesse (1991). The models with [Y=0.475, Z=0.1] are obtained with the classical opacities of Hubner et al. (1977). The contribution from molecular opacities is according to Bessell et al. (1989, 1991). The extension of the convective regions, both central cores and external envelopes, is calculated with overshoot according to the formalism by Bressan et al. (1981) and Alongi et al. (1991) for the core and the envelope, respectively. The mixing length parameter of the outermost super-adiabatic convection is 1.63 the local H_p. This choice constrains the models with the chemical composition and age of the Sun to fit its luminosity and effective temperature. Therefore, the evolutionary sequences and accompanying isochrones constitute the kind of **calibrated data base** required by the population synthesis models. The most interesting result is the behaviour of low mass, high metallicity stars during the core He-burning and AGB phase. As already found by Brocato et al. (1990), Castellani & Tornambe' (1991), Horch et al. (1992), and Dorman et a. (1993), these stars depart from the standard location in the CMD. In brief at high metallicity the evolution does not proceed toward and along the AGB, but toward a slow phase taking place at high effective temperatures without going back toward the AGB. For solar and twice solar metallicity, the blue phase begins during the helium shell burning. For three times solar metallicities it begins during the late stages of core He-burning. For still higher metallicities a large fraction of the core He-burning lifetime is spent at very high effective temperatures and high luminosities.

The library of stellar spectra in use is from Kurucz (1992). It contains spectral energy distributions expressed in $erg\ s^{-1}\ cm^{-2}\ A\ steradian^{-1}$ over a wide range of gravities, $T_{eff}s$, and metallicities. Each spectrum goes from $90A$ to $1.6 \times 10^6 A$, with a mean resolution of 20 A. Because sometimes the isochrones in the CMD happen to reach T_{eff} and gravities not included in the spectral library, this has been extended both to high and low effective temperatures. The extension to high temperatures is less of problem, whereas that to low temperatures is a cumbersome affair. This has been accomplished with the aid of the observed spectra in the catalogs of Lancon & Rocca-Volmerange (1992), Terndrup et al. (1990, 1991) and Straizys & Sviderskiene (1972), and the corresponding colors (V-K) and (J-K). See BCF for all details.

3. Fundamentals of Population Synthesis

The integrated monochromatic flux generated by the stellar content of a galaxy of age T is defined as

$$F_\lambda(T) = \int_0^T \int_{M_L}^{M_U} \Psi(t, Z)\ \phi(M) f_\lambda(M, \tau', Z)\ dM\ dt \qquad (1)$$

where $\Psi(t, Z)$ is the stellar birth-rate (SFR), $\phi(M)$ is the initial mass function (IMF), and $f_\lambda(M, \tau', Z)$ is the monochromatic flux of a star of mass M, metallicity $Z(t)$, and age $\tau' = T - t$. The IMF is according to the Salpeter law $dN/dM = \phi(M) = A \times M^{-x}$, where $x = 2.35$ and A is a normalization constant to be fixed by the following condition

$$\int_{M_*}^{M_U} \phi(M) \times M \times dM = \zeta \qquad (2)$$

where ζ is the fraction of the total mass in form of stars stored in the IMF above M_*. Finally, the total mass of a SSP is normalized to $1 M_\odot$.

The upper limit of integration is $M_U = 120 M_\odot$, while the lower limit M_* is the minimum mass contributing to the chemical enrichment of the interstellar medium over a time scale comparable to the total lifetime of a galaxy. This mass is approximately equal to $1M_\odot$.

The integrated flux of a galaxy is considered as the convolution of many single stellar populations (SSP), i.e. coeval, chemically homogeneous assembles of stars with age τ' and metallicity $Z(t)$. The flux from individual SSPs is derived from accurate isochrone calculations according to the method by Bertelli et al. (1990).

4. Broad Band Colors of SSPs

The most interesting features of broad band colors like (U-B), (B-V), (V-J), and (V-K) of SSPs to look at, are the variations caused by appearance of AGB and RGB stars in a SSP.

In brief, as a SSP ages two characteristic groups of stars appear and, in virtue of their intrinsic properties, dominate the light and the colors of the SSP. They are the AGB and RGB stars. According to the elementary theory of stellar evolution, as a SSP ages first AGB and later both RGB and AGB stars appear, their onset occurring at two precise values of age, t_{AGB} and t_{RGB}, respectively. Following Renzini (1981), the relative contribution of the AGB and RGB stars to the integrated bolometric luminosity of the SSP is fixed by the so-called Fuel Consumption Theorem or in other words it is proportional to the amount of fuel burned in these stages (see also Renzini & Buzzoni 1986). This of course depends on details of stellar structure that are not discussed here for the sake of brevity. Because of the precise values of t_{AGB} and t_{RGB}, the appearance of AGB and RGB stars are expected to engender a sort of a *rapid* variation in the spectrophotometric properties of the SSP, that Renzini & Buzzoni (1986) have named *Phase Transitions*. Finally, because both types of stars are confined within quite narrow ranges of late spectral types, one may expect that their appearance should be accompanied by prominent increases in the near-infrared (JHKL) luminosity, and hence in the colors like (V-J), (V-H), (V-K), (V-L), etc. The detailed study of the evolution of SSP colors by BCF shows that a rapid variation in coincidence with the onset of the AGB stars is visible only in colors like (V-J) and (V-K), whereas no variation is seen when RGB stars come in. Indeed, their effect on the colors is always masked by the already existing AGB stars. This is in full agreement with the precepts of the Fuel Consumption Theorem.

5. Models for the Chemical Evolution of Elliptical Galaxies

A galaxy is considered as made of two components, namely the luminous material of mass M_L embedded in a massive halo of dark matter of mass M_D, whose existence does not influence the system but for the gravitational effect. Furthermore, both components are supposed to be constant with time, i.e. the closed box description is adopted. Indeed since our models allow for the effect of galactic winds, M_L will suddenly decrease when galactic winds occur. However, till this particular stage the evolution (chemical) of the system obeys the simple laws of the closed box model (Tinsley 1980). Galactic winds are supposed to occur when the energy input from supernova explosions and ejection of stellar winds from massive stars equates the binding energy of the gas mass $M_g(t)$ (variable with time because of star formation). In the following, the initial value of M_L, labeled M_L^0, is used to rank our models as a function of the mass.

Binding Energies. According to the dynamical models by Bertin et al. (1992) and Saglia et al. (1992), the binding energy of the gas under the influence of dark matter is expressed as

$$\Omega_g(t) = -G\frac{M_g(t)M_L}{R_L}\left[\alpha_L + \frac{M_D}{M_L}\frac{1}{2\pi}(\frac{R_L}{R_D})[1 + 1.37(\frac{R_L}{R_D})]\right] \qquad (3)$$

Typical values of α_L and R_L/R_D to be used in the above equations are $\alpha_L = 0.5$, $R_L/R_D = 0.2$, which in turn lead to $M_L/M_D = 0.2$ (see Bertin et al. 1992 for more details). The radius R_L is estimated with the aid of the results by Saito (1979a,b) and Arimoto & Yoshii (1987), who obtain

$$R_L = 26.1 \times (M_L/10^{12}M_\odot)^{0.45} \quad kpc \qquad (4)$$

Energy Input from Supernovae and Stellar Winds. The thermal energy of the gas is determined by the energy input from Type I and Type II supernovae and stellar winds ejected by massive stars (see also Theis et al. 1992). The energy input from the three components are

$$E_{th}(t)_i = \int_0^t \epsilon_{SN}(t - t')R_i(t')M_L dt' \quad ergs \qquad (5)$$

where i stands for SNI, SNII, and W. Accordingly, R_{SNI} and R_{SNII} are the rates of supernova production, while R_W is the rate of gas ejection by massive stars in form of stellar winds. The time t' is either the explosion time or the time at which stellar winds occur.

The decrease of the thermal energy content of a supernova remnant as a function of time strictly follows the Cox (1972) law, with a characteristic cooling time $t_{cs} = 5.7 \times 10^4 \epsilon_0^{4/17} n_0^{-9/17}$ yr, where $n_0(t) = \rho_g(t)/m_H$ is the number density of the interstellar medium. The peak energy is 7.52×10^{50} ergs when the remnant is younger than t_{cs} and 2.2×10^{50} ergs when the remnant is older than t_{cs}.

In the case of stellar the winds, the thermal content is estimated in the following way: a typical massive star (say in the range $20M_\odot$ to $120M_\odot$) in the course of evolution ejects about half of its mass in form of a wind with terminal velocity of about $2,000\,Km\,s^{-1}$ (see Chiosi & Maeder 1986) and, therefore, injects into the interstellar medium an energy of about $\epsilon_{W0} = 1/2(M/2)(Z/Z_*)^{0.75}V^2$ ergs, where the term $(Z/Z_*)^{0.75}$ takes into account that both the mass-loss rates and terminal velocities depend on the metallicity (Kudritzki et al. 1987, Theis et al.

1992). The reference metallicity is $Z_* = 0.06$. Part of this energy will be radiated away and part will contribute to heat up the gas. In analogy with supernova remnants, we assume that this energy will cool down with the same law as for the energy injected by supernovae but with a different cooling time scale t_{cw}. In the models below we adopt $t_{cw} = 15 \times 10^6$ yr. This assumption stands on the notion that the energy content in the wind is continuously refueled by the radiative energy output from the emitting star.

It is easy to figure out that stellar winds can contribute as much as classical supernova explosions and, therefore, they cannot be neglected in evaluating the total energy required to power galactic winds.

The rate of Type I supernovae $R_{SNI}(t)$ is taken from Greggio & Renzini (1983). This stands on the evolutionary scenario, in which the progenitors of Type I supernovae are in close binary systems containing a degenerate WD, produced by a low or intermediate mass primary component, which accretes material from the companion. When the mass of the WD (either He or CO) becomes larger than a critical value, the ignition of nuclear burning in highly degenerate conditions causes the explosion of the star. See Greggio & Renzini (1983), Matteucci & Greggio (1986), and Matteucci & Tornambe' (1987) for more details and referencing. The rate is calculated assuming for the total mass M_B of the binary system the range 3 to $8M_\odot$, and adopting a suitable distribution function for the mass fraction $\mu = M_2/M_B$ which extends up to $\mu = 0.5$ (see the discussion by Matteucci & Greggio 1986 for all details). The rate of Type II supernovae is simply calculated by adding the contribution from stars in the mass range 8 to $16M_\odot$ exploding as single objects to the that from the mass range 16 to $120M_\odot$. Finally, the rate of mass injection by stellar winds is simply calculated for stars in the range 30 to $120M_\odot$.

The total thermal energy stored into the interstellar medium is therefore $E_{th}(t) = E_{th}(t)_{SNI} + E_{th}(t)_{SNII} + E_{th}(t)_W$, and finally a galactic wind is supposed to occur when $E_{th}(t) \geq \Omega_g(t)$.

Basic Equations of Chemical Evolution. The equation governing the chemical evolution of the models are not given here for the sake of brevity. They can be found in BCF, and Chiosi (1986), together with all details concerning the calculations of the stellar yields of various elements (see also Matteucci 1991). The IMF is the same as in calculation of the ISED of SSPs.

Star Formation Rate. The rate of star formation is assumed to depend on the gas mass according to

$$\Psi(t) = \nu M_g(t)^k = \nu(M_L^0)^{k-1} G(t)^k \tag{6}$$

where ν and k are adjustable parameters, and the star formation rate is normalized to M_L^0. Linear and quadratic dependencies of the star formation rate on the gas content, were first proposed by Schmidt (1959) and have been adopted ever since because of their simplicity (see Larson 1991 for a recent review). The parameter ν fixes the time scale of star formation. According to Arimoto &

Yoshii (1987) a typical time scale of star formation in elliptical galaxies is of the order of the free-fall time scale or as short as the collisional time scale between collapsing clouds of gas, i.e. in the range 10^7 to 10^8 yr. Expressing the age in units of 10^9 yr, this implies $\nu \simeq 10 \div 20$.

Model Parameters. Each model is characterized by a number of parameters, namely M_L^0 (varied from 3 to $0.05 \times 10^{12} M_\odot$), the ratios R_L/R_D and M_L/M_D in the gravitational potential (both assumed constant and equal to 0.2), the star formation rate, initial mass function (k, ν, and ζ), and cooling time for stellar wind t_{cw} (all these calibrated on the observational properties of elliptical galaxies), and finally the age of the galaxy T_{GAL} (in the range $12 \div 17 \times 10^9$ yr).

6. Results for the Closed Models

In the following we summarize the main properties of the models limited to those that according to the thorough analysis by BCF best reproduce the UV excess and the color-magnitude relation of elliptical galaxies.

Powering the Galactic Winds. Galactic winds are basically powered by stellar winds from massive stars and at a less extent by supernova explosions. This is mainly due to the different cooling time associated to the two mechanisms: the cooling time of stellar winds is significantly longer than that of supernova remnants (15×10^6 yr versus $\simeq 10^5$ yr).

Varying the Gas Dependence of the Star Formation. The gas dependence of the star formation rate bears very much on the relative epoch at which galactic winds start in galaxies of different mass. In general, going going from $k = 1$ to $k = 2$, the delay in the age of gas ejection between massive and low mass galaxies increases, and for certain combinations of the parameters the galactic wind never starts.

Varying the Time Scale of Star Formation. As expected the parameter ν governing the time scale of star formation concurs to determine the age at which galactic winds take place. Keeping fixed all other parameters, the higher ν, the earlier the galactic winds occur.

Varying ζ. The fraction of the IMF stored above $1 M_\odot$ affects the metallicity and the rate of energy input by supernova explosions and stellar winds. As expected, at increasing ζ, the gas metallicity goes up. However, at the same time the galactic wind tends to occur earlier so that the two effects tend to balance each other.

Evolution of $G(t)$ and $Z(t)$. The gas fraction $G(t)$ and metallicity $Z(t)$ obey the laws of the closed model, i.e. both are independent of the mass of the galaxy, which is simply a scale factor. However, depending on the galaxy mass and of course on the choice for k, ν, and ζ, the age at which galactic winds start and hence the gas falls abruptly to zero and the metallicity gets frozen, is different. At later epochs galaxies will be refueled in gas by dying stars, but since this is supposed not to give rise to further star formation, its presence is totally irrelevant to the chemical model.

Metallicity Distribution. The stellar content of the models span a wide range of metallicities, which if gas is retained for sufficient time can reach high values. This is shown by the fractionary cumulative mass distribution of living stars at the present time as function of the metallicity. For models with $k = 1$, $\nu = 20$, and $\zeta = 0.4$, even in the extreme case of a galaxy with $M_L^0 = 3 \times 10^{12} M_\odot$, a large fraction of the stars (about 35 to 40%) have metallicities lower than solar, and only a small fraction amounting to about 5% is able to reach metallicities in the range as high as Z=0.08.

General Remarks. Summarizing the results of this analysis, it is clear that the main properties of the models, namely, occurrence of galactic winds, maximum and mean metallicities, are determined by the time scale of cooling associated to stellar winds, and the dependence of the star formation rate on the gas content.

7. Color-Magnitude Relation for Elliptical Galaxies

The color-magnitude relation adopted in this paper refers to the sample of early type galaxies in the Coma and Virgo clusters studied by Bower et al. (1992). The mean redshift of Virgo and Coma are $z = 0.003$ and $z = 0.023$, respectively, which makes the K-corrections to colors and magnitudes quite small. To compare models with observations we have adopted the distance modulus $(m - M)_o = 31.54$ for Virgo (Branch & Tammann 1992) and applied to Coma the shift $\delta(m-M)_o = 3.65$ (Bower et al. 1992).

The colors of an elliptical galaxy are determined by the blend of many stellar populations with different metallicities and ages in proportions that reflect the past history of star formation and chemical enrichment. Higher metallicities and/or older ages yield redder colors. The main problem with the interpretation of the color-magnitude relation in terms of galactic winds, is that galaxies of different mass must reach a suitable degree of mean metal enrichment before the onset of galactic winds. Furthermore, too high metallicities must be avoided otherwise too red colors would result. The degree of metal enrichment in a closed model is determined by the gas fraction $G(t)$, which in turn is regulated by the efficiency of star formation through the key parameters k, ν and ζ.

The thorough analysis by BCF indicates that the color-magnitude is reproduced by models with $t_{cw} = 15 \times 10^6$ yr, $k = 1$, $\nu = 20$, $\zeta = 0.40$. and ages in the range $12 \div 15 \times 10^9$ yr. This is shown in Fig.1.

8. On the Nature of the UV-Excess in Elliptical Galaxies

Ultraviolet observations of elliptical galaxies have shown that these apparently old stellar systems emit a significant fraction of their luminosity in the far UV (short-ward of about 2,000 A), otherwise known as the *ultraviolet excess*. Although the UV flux may vary from galaxy to galaxy, the remaining spectrum down to the

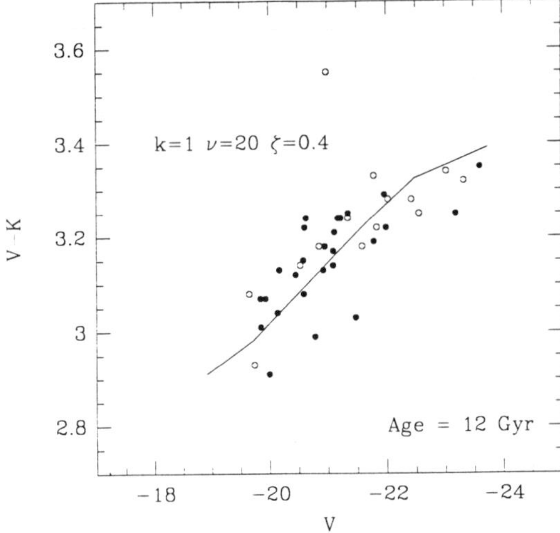

Fig. 1. The color-magnitude relation for models with $t_{cw} = 15 \times 10^6$yr, $k = 1$, $\nu = 20$, $\zeta = 0.40$ and age 12×10^9 yr. Open and filled circles are the data by Bower et al. (1992) for Virgo and Coma galaxies, respectively (see the text for details).

infrared is virtually identical in all galaxies, apart from some effects short-ward of about 5,000 A. This UV emission has a minimum level in M32 and reaches intensities which are almost an order of magnitude higher in other galaxies such as NGC4486 and NGC4649 (Code & Welch 1979; Bertola et al. 1980, 1982; Burstein et al. 1988). The source of this emission seems to be distributed across the galaxies in the same way as the normal, cool component. The study of Burstein et al. (1988) has shown that the ultraviolet-optical color $(1550 - V)$, defined as the difference between the average flux in the range 1250 and 1850 A and in the V-band, correlates with the Mg_2 index, which is customarily taken as a measure of the galaxy metallicity. In other words, the higher the ultraviolet excess, the higher is the metallicity of the galaxy. The observations by Ferguson et al. (1991) and Ferguson & Davidsen (1993) with the Hopkins Ultraviolet Telescope (HUT) add further information on the UV emission. The spectra (see the case of NGC1399, a giant elliptical galaxy with strong UV emission) show that the UV-excess falls off short-ward of about 1,000 A, indicating that the light is dominated by stars with temperatures less than 25,000 K. This result and the (1550-V) versus Mg_2 relation put strong constraints on the models for the UV-excess not yet fully appreciated.

8.1. Candidates to the UV Emission

Several competing hypotheses have been advanced to account for the UV light of ellipticals, each invoking different types of hot stars going from young massive stars to a variety of old, low mass stars, and finally to some unusual hot stage of stellar evolution that is overabundant in the old population of ellipticals, such as accretion disks around WDs in binary systems or, in general, hot component of binary stars. The three options have been critically examined by Greggio & Renzini (1990) to whom the reader should refer for details. The situation is as follows:

Young Stars. A recent episode of star formation is often invoked to account for the UV upturn (e.g. Bruzual 1983; Burstein et al. 1988; Bica & Alloin 1987, 1988; Rocca-Volmerange 1989). Since all ellipticals observed so far show evidence of UV emission and the lifetime of a typical massive star ($M \geq 2M_\odot$) is short, it follows that only continuing star formation is likely to be involved. The main flaw of the model is the lack of a simple explanation of the relation between $(1550 - V)$ and Mg_2. Furthermore, those few galaxies which clearly show evidence of ongoing star formation (e.g. NGC205) do not follow this correlation.

Old, Low-Mass Stars. This option includes all kinds of evolved low mass stars, which for whatever conditions are able to spend a significant fraction of their life in the suitable range of luminosity and effective temperature to become a powerful source of UV radiation, and at the same time to match the constraints imposed by the $(1550 - V) - Mg_2$ relation and the HUT observations. In short, if for whatever reason (either mass loss or growth of the inner core or both) the mass of the outer envelope of an evolved, low mass star falls below a certain critical value, the object rapidly shifts to high effective temperatures either at constant or increasing luminosity.

According to the scrutiny made by Greggio & Renzini (1990), there are several potential candidates:

i) The post red giant stars (P-RGB) that can be generated by unusually high rate of mass loss during the RGB phase and high values of $\Delta Y / \Delta Z$. Their existence is a matter of speculation.

ii) The hot horizontal branch stars (H-HB) and their progeny named AGB-manque. This option is typical of low mass stars with very high metallicity which can spend a significant fraction of their He-burning phase at high effective temperatures and avoid the classical AGB phase. This can be accomplished either with a suitable values of $\Delta Y / \Delta Z$ and strong dependences of the mass loss rates on the metallicity (Greggio & Renzini 1990) or even with canonical mass loss rates and suitable values of $\Delta Y / \Delta Z$ (Castellani & Tornambe 1991; Horch et al. 1992; Fagotto et al. 1993a,b; Dorman et al. 1993).

iii) The post early AGB stars (P-EAGB). These stars have normal HB phase but depart from AGB before the start of the thermally pulsing regime of the He-burning shell and eventually join the scheme of the AGB-manque.

iv) The H- and He-Burning P-AGB Stars. This is the standard channel for producing sources of UV radiation. These stars leaving the AGB during either a thermal pulse (He-burning P-AGB) or between two pulses (H-burning P-AGB). Since the body of evolutionary tracks for the He-burning scheme is very limited, most of the studies of the UV-excess are based on the H-burning P-AGB stars.

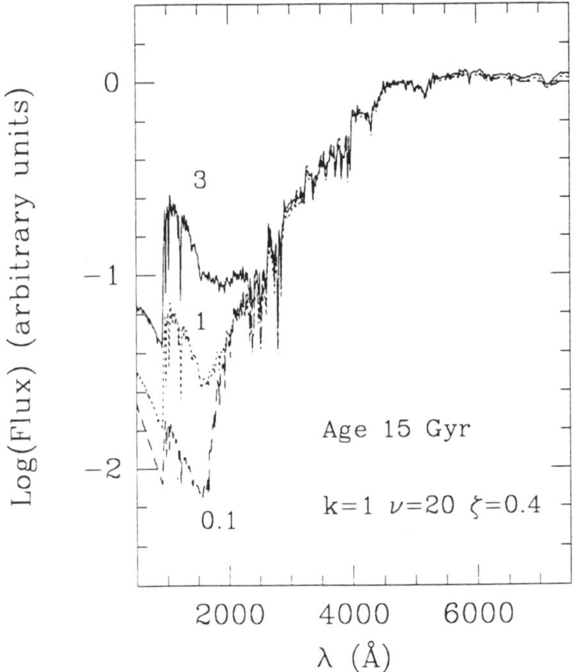

Fig. 2. The spectra of model galaxies with $t_{cw} = 15 \times 10^6$ yr, $k = 1$, $\nu = 20$, $\zeta = 0.40$, and age of 15×10^9 yr. The mass M_L^0 annotated along each curve is in units of $10^{12} M_\odot$. All the ISEDs have been normalized to coincide in the visual region of the spectrum. See the text for details.

However there are several drawbacks in the H-burning P-AGB scheme. In short, on the side of stellar models it depends entirely on the adopted relation between the mass of the core and the mass of the residual envelope (see Greggio & Renzini 1990 for details). Furthermore, the H-burning shell at the surface does not release enough energy to account for the strongest UV emissions (see Greggio & Renzini 1990). Finally, there is the embarrassing result that the characteristic temperature of the H-burning P-AGB stars (similar remark applies also to the He-burning P-AGB mode) is too hot to explain the pronounced turn-over near the Lyman limit in observed spectra (Ferguson et al. 1991).

Binary Stars. Owing to the complexity of this scheme, and the large variety of possible situations, the effect of binary stars has not yet been considered. This topic has been discussed by Renzini & Buzzoni (1986) to whom we refer for more details.

8.2. UV-Excess of Elliptical Galaxies

The ISED of three model galaxies with different mass and same age (15×10^9 yr) is shown in Fig.2. The superposition is made imposing coincidence of the flux in the V region of the spectrum. The ISED in the UV region markedly depends

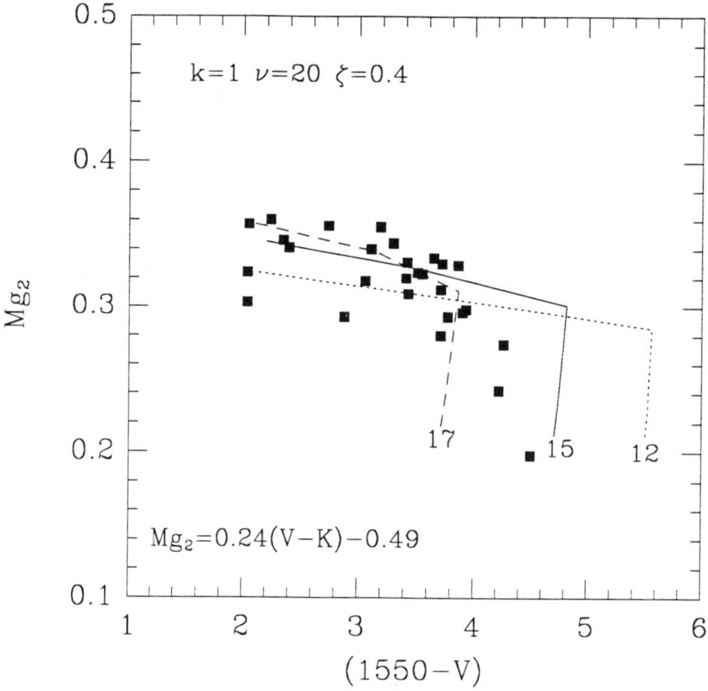

Fig. 3. Comparison of the theoretical relation Mg_2 versus (1550-V) with the data by Burstein et al. (1988) (filled squares). Mg_2 is obtained from the (V-K) color according to the analytical fit due to Buzzoni et al. (1992) at the bottom of the panel. Each line corresponds to a different value of the age (in units of 10^9 yr. Along each curve, the galaxy mass and maximum metallicity increase from left to right.

on the mass and hence the maximum metallicity of the galaxy.

In the $3 \times 10^{12} M_\odot$ galaxy, in which about 5% the star mass reaches metallicities as high as 0.07 to 0.08, the UV-excess peaks at about $1,000A$ and falls off at shorter wavelength. The peak, whose intensity is only a factor of four smaller than the flux in the visual region as indicated by the observational data (see Ferguson et al. 1991 and Ferguson & Davidsen 1993), is generated by the H-HB and AGB-manque stars. The component short-ward of $1,000A$ is due to the fraction of old stars evolving on a short time scale towards and along the WD cooling sequence (classical PNs and PN-like phase of the metal-rich component).

Conversely, in the case of the $1 \times 10^{11} M_\odot$ galaxy, whose maximum metallicity is only $Z \simeq 0.05$, the UV flux is generated by the classical P-AGB and HB stars because the maximum metallicity is not high enough to activate the H-HB and AGB-manque source. The flux gets its maximum intensity at about $400A$, and it is about a factor of thirty lower than in the visual region.

The slope of the ISED in the wavelength range $1,000A$ to $2,000A$ does not significantly differ passing from one source to another. This explains why a formal reproduction of the UV-excess in terms of the sole P-AGB stars can be found if one neglects the information from the spectral range below $1,000A$ indicating cooler temperatures of the source (Ferguson et al. 1991, Ferguson & Davidsen 1993), the arguments provided by the Fuel Consumption Theorem on the lack of sufficient energy to account for the observed flux (Greggio & Renzini 1990), and finally the lack of a robust correlation between the UV-excess and the metallicity of the parent galaxy (Bertelli et al. 1989).

There is another aspect worth being mentioned, i.e. the different age at which the UV-excess by evolved stars begins. As expected, the age varies with the kind of source at work. For the $3 \times 10^{12} M_\odot$ galaxy the UV-excess sets in and grows to full amplitude when the age gets older than about 7.6×10^9 yr, whereas for $1 \times 10^{11} M_\odot$ galaxy this occurs when the age is about 12.5×10^9 yr.

Finally, the theoretical ISEDs presented by BCF remarkably agree with the observational data by Burstein et al. (1988) for typical galaxies with strong (NGC4649) and intermediate (NG1404), and low (M32) UV-excess, respectively.

8.3. Theoretical $(1550 - V) - Mg_2$ Relation

The above trend is also confirmed by the relation between the color $(1550 - V)$ and Mg_2 shown in Fig.3 together with the observational data by Burstein et al. (1988). Since the library of stellar spectra in usage does not possess the resolution in wavelength required to calculate the Mg_2 index according to the definition by Faber et al. (1977), an empirical calibration of this index as a function of the (V-K) color is used to check the relation between Mg_2 and (1550-V).

The Mg_2 index is derived from the (V-K) color with the aid of the calibration by Buzzoni et al. (1992), which fits the data for elliptical galaxies by Davies et al. (1987) and Frogel et al. (1978).

The results are shown in Fig.3 for three different values of the age as indicated. Along each line, the metallicity and the galaxy mass decrease from left to right. As expected, the relation Mg_2 versus (1550-V) now shows a marked dependence both on metallicity and age, which perhaps could explain part of the scatter in the observational data.

Considering all the uncertainties affecting the above procedures, we dare to conclude that our models successfully reproduce the correlation between UV-excess and Mg_2 index.

9. Cosmological Evolution of Elliptical Galaxies: Age Probes

In this section, instead of discussing the evolution of the colors as a function of the redshift z under the effect of the K- and E-corrections (see Guiderdoni & Rocca-Volmerange 1987 for a definition) we focus the attention on the possibility of dating galaxies by means of the UV-excess. For purposes of illustration, we limit the discussion below to cases with $H_0 = 50 \ Km \ s^{-1} \ Mpc^{-1}$ and $z_{for} = 5$. The corresponding galaxy ages are 16×10^9 yr and 12.26×10^9 yr, respectively. The value adopted for H_0 agrees with the recent determination by Sandage &

Tammann (1990) who find $H_0 = 52 \pm 2$ Km s^{-1} Mpc^{-1}. A confident estimate of z_{for} is still a matter of debate, with discordant conclusions by different authors. It goes from the small value of $z_{for} \leq 3$ suggested by Wyse (1985), to values in the range $3 \leq z_{for} \leq 5$ according to Yoshii & Takahara (1988) and Eisenhard & Lebofsky (1987), to the high values quoted by Dunlop et al. (1989), who examining a sample of radio galaxies conclude that z_{for} should be larger than 10. More recently, Buzzoni et al. (1993) looking at the colors in distant clusters of galaxies and assuming the Sandage & Tammann (1990) determination of the Hubble constant, conclude that galaxies have formed more likely beyond $z_{for} = 5$. Finally, there are also observational evidences: for instance galaxies are observed up to $z = 3.4$ (Lilly 1988) and $z = 3.8$ (Chambers et al. 1990), quasars are seen up to $z = 4.9$ (Schneider et al. 1991, 1992), which implies that the oldest galaxies should have formed at redshifts not smaller than $z_{for} = 5$.

BCF have shown that signatures in the color evolution (like those caused by the appearance of AGB stars) manifesting themselves at early epochs cannot be used as age probes in cosmological context. On the contrary, the onset of the UV-excess caused by the old, high metallicity stars is a very promising candidate (see Renzini & Buzzoni 1986; Greggio & Renzini 1990). The most favourable object to look at is a massive (luminous) object with strong UV-excess, such as our model with $M_L^0 = 3 \times 10^{12} M_\odot$, $k = 1$, $\nu = 20$, and $\zeta = 0.40$. In the rest frame, following the interruption of star formation, the color $(1550 - V)$ suffers from a marked drop-off from $(1550 - V) < 2$ to $(1550 - V) > 6$ till it eventually gets blue again at much older ages in coincidence with the appearance of the hot luminous stars sources of UV radiation. The onset of the UV-excess occurs at the age of 7.6×10^9 yr, to which different values of z_{uv} correspond depending on the assumed H_0, q_0, and z_{for}. Limiting the discussion to the case of $H_0 = 50$ Km s^{-1} Mpc^{-1} and $z_{for} = 5$, the expected values for z_{uv} are $z_{uv} = 0.81$ for $q_0 = 0$, and $z_{uv} = 0.34$ for $q_0 = 0.5$. Other values of z_{uv} are possible at varying H_0, q_0, z_{for}. Detection of the UV upturn can be made by looking at the variation as a function of the redshift either of the spectra or of suitable colors in the UV pass-bands of the Hubble Space Telescope.

It goes without saying that the age in question depends the adopted $\Delta Y / \Delta Z$ and efficiency of mass loss during the RGB and AGB phases. The explanation of the UV-excess in elliptical galaxies imposes some constraints both on $\Delta Y / \Delta Z$ and mass loss. Our choices for both constitute a sort of lower limit, and therefore the corresponding age of the UV-upturn is a sort of upper limit. However, much higher $\Delta Y / \Delta Z$ and mass loss rates are not very likely.

It follows from this that not only the UV-excess is a powerful probe of galaxy ages but detection of its onset at a certain redshift together with the measurements of its fall off as a function of the redshift would allow us to establish useful relations between H_0, q_0 and z_{for}. Future observations will certainly make feasible the detection of this important evolutionary effect.

References

Alongi M., Bertelli G., Bressan A., & Chiosi, C.: 1991, *A&A* **244**, 95.

Alongi, M., Bertelli, G., Bressan, A., Chiosi, C., Fagotto, F., Greggio L., & Nasi, E.: 1993, *A&AS* **97**, 851.

Arimoto, N., & Yoshii, Y.: 1987, *A&A* **173**, 23.

Bertelli, G., Betto, R., Bressan, A., Chiosi, C., Nasi, E., & Vallenari, A.: 1990, *A&AS* **85**, 845.

Bertelli, G., Chiosi C., & Bertola, F.: 1989, *ApJ* **392**, 522.

Bertin, G., Saglia, R. P., & Stiavelli, M.: 1992, *ApJ* **384**, 423.

Bertola, F., Capaccioli, M., Holm, A. V., & Oke, J. B.: 1980, *ApJ* **298**, L37.

Bertola, F., Capaccioli, M., & Oke, J. B.: 1982, *ApJ* **254**, 494.

Bessell, M. S., Brett, J. M., Scholz, M., & Wood, P. R.: 1989, *A&AS* **77**, 1.

Bessell, M. S., Brett, J. M., Scholz, M., & Wood, P. R. 1991, *A&AS* **87**, Erratum, 621.

Bica, E., & Alloin, D.: 1987, *A&A* **186**, 49.

Bica, E., & Alloin, D.: 1988, *A&A* **192**, 98.

Bower, R. G., Lucey, J. R., & Ellis, R. S.: 1992, *MNRAS* **254**, 589.

Branch, D., & Tammann, G. A.: 1992, *ARA&A* **30**, 359.

Bressan, A., Bertelli G., & Chiosi, C.: 1981, *A&A* **102**, 25.

Bressan, A., Bertelli G., Fagotto, F., & Chiosi, C.: 1993b, *A&AS* **100**, 647.

Bressan, A., Chiosi, C. & Fagotto, F.: 1993, *A&AS*, submitted, (BCF).

Brocato, E., Matteucci, F., Mazzitelli, I., & Tornambe', A.: 1990, *ApJ* **349**, 458.

Bruzual, G. A.: 1983, *ApJ* **273**, 205.

Bruzual A., G. & Charlot, S.: 1993, *ApJ* **405**, 538.

Burstein, D., Bertola, F., Buson, L. M., Faber, S. M., & Lauer, T. R.: 1988a, *ApJ* **328**, 440.

Buzzoni, A., Chincarini, G., & Molinari, E.: 1993, *ApJ* in press.

Castellani, M., & Tornambe', A.: 1991, *ApJ* **381**, 393.

Chambers, K., Miley, G., & van Breugel, W.: 1990, *ApJ* **363**, 21.

Chiosi, C.: 1986, in *Nucleosynthesis and Stellar Evolution: 16th Saas-Fee Course*, ed. B. Hauck, A. Maeder, & G. Meynet, (Geneva: Geneva Obs.), 199.

Chiosi, C., & Maeder, A.: 1986, *ARA&A* **24**, 329.

Code, A. D., & Welch, G. A.: 1979, *ApJ* **228**, 95.

Cox, D. P.: 1972, *ApJ* **178**, 159.

Davies, R. L., Burstein, D., Dressler, A., Faber, S. M., Lynden-Bell, D., Terlevich, R. J., & Wegner, G.: 1987, *ApJS*, **64**, 581.

Dorman, B., Rood, R. T., O'Connell, R. W.: 1993, *ApJ*, in press.

Dressler, A.: 1984, *ApJ* **181**, 512.

Dunlop, J. S., Guiderdoni, B., Rocca-Volmerange, B, Peacock, J. A., & Longair, M. S.: 1989b, *MNRAS* **240**, 257.

Eisenhard, P. R., & Lebofsky, M. J.: 1987, *ApJ* **316**, 70.

Faber, S. M.: 1977, eds. B. M. Tinsley & R. B. Larson, *The Evolution of Galaxies and Stellar Populations*, (Yale University Observatory: New Heaven), p. 157.

Faber, S. M., Burstein, D., & Dressler, A.: 1977, *AJ*, **82**, 941.

Fagotto, F., Bressan, A., Bertelli, G., & Chiosi, C.: 1993a, *A&AS*, in press.

Fagotto, F., Bressan, A., Bertelli, G., & Chiosi, C. 1993b, *A&AS*, in press.

Ferguson, H. C., & Davidsen, A. F.: 1993, *ApJ* **408**, 92.

Ferguson, H. C., Davidsen, A. F., Kriss, G. A., Blair, W. P., et al.: 1991, *ApJ* 382, L69.

Frogel, J. A., Persson, S. E., Aaronson, M., Matthews, K.: 1978, *ApJ* **220**, 75.

Greggio, L., & Renzini, A.: 1983, *A&A* **118**, 217.

Greggio, L., & Renzini, A.: 1990, *ApJ* **364**, 35.

Grevesse, N.: 1991, *A&A* **242**, 488.

Guiderdoni, B., & Rocca-Volmerange, B.: 1987, *A&A* **186**, 1.

Horch, E., Demarque, P., & Pinsonneault, M.: 1992, *ApJ*, **388**, L53.

Hubner, W. F., Merts, A. L., Magee, N. H., & Argo, M. F.: 1977, Los Alamos Scientific Laboratory Report LA-6760-M.

Iglesias, C. A., Rogers, F. J., & Wilson, B. G.: 1992, *ApJ* **397**, 717.

Kudritzki, K. P., Pauldrach, A., & Puls, J.: 1987, *A&A*, **173**, 293.

Kurucz, R. L.: 1992, eds. B. Barbuy & A. Renzini, *The Stellar Populations of Galaxies*, (Kluwer Academic Publishers: Dordrecht), p. 225.

Lancon, A., & Rocca- Volmerange, B.: 1992, *A&AS* **96**, 593.

Larson, R. B.: 1991, ed. D. L. Lambert, *Frontiers of Stellar Evolution*, ASP Conference Series, Vol. 20, p. 571.

Lilly, S.: 1988, *ApJ* **333**, 161.

Magris, G. C., & Bruzual, G. A.: 1993, *ApJ*, in press.

Marsiaj, P.: 1992, Thesis, University of Padua, Italy.

Matteucci, F.: 1991, ed. D. L. Lambert, *Frontiers of Stellar Evolution*, ASP Conf. Ser. vol. 20, p. 539.

Matteucci, F., & Greggio, L.: 1986, *A&A* **154**, 279.

Matteucci, F., & Tornambe', A.: 1987, *A&A* **185**, 51.

Pagel, B. E. J., Simonson, E.A., Terlevich, R. J., & Edmunds, M.,G.: 1992, *MNRAS* **255**, 325.

Renzini, A.: 1981, *Ann. Phys. Fr.* **6**, 87.

Renzini, A., & Buzzoni, A.: 1986, eds. C. Chiosi & A. Renzini, *Spectral Evolution of Galaxies*, (Reidel:Dordecht), p. 195.

Rocca-Volmerange, B.: 1989, *MNRAS* **236**, 47.

Saglia, R. P., Bertin, G., & Stiavelli, M.: 1992, *ApJ* **384**, 433.

Saito, M.: 1979a, *PASJ* **31**, 181.

Saito, M.: 1979b, *PASJ* **31**, 193.

Sandage, A., & Tammann, G. A.: 1990, *ApJ* **365**, 1.

Schmidt, M.: 1959, *ApJ* **129**, 243.

Schneider, D.P., Schmidt, M., & Gunn, J.E.: 1991, *AJ* **102**, 837.

Schneider, D.P., Van Gorkom, J.H., Schmidt, M., & Gunn, J. E.: 1992, *AJ* **103**, 1451.

Straizys, V., & Sviderskiene, Z.: 1972, *Bull. Vilnius Obs.* **35**, 3.

Terndrup, D. M., Frogel, J. A., & Withford, A. E.: 1990, *ApJ* **357**, 453.

Terndrup, D. M., Frogel, J. A., & Withford, A. E.: 1991, *ApJ* **378**, 742.

Theis, Ch., Burkert, A., & Hensler, G.: 1992, *A&A* **265**, 465.

Tinsley, B. M.: 1980, *Fundamentals of Cosmic Physics* **5**, 287.

Vader, J. P.: 1986, *ApJ* **306**, 390.

Wyse, R. F. G.: 1985, *ApJ* **299**, 593.

Yoshii, Y., & Arimoto, N.: 1987, *A&A* **188**, 13.

Yoshii, Y., & Takahara, F.: 1988, *ApJ* **299**, 593.

A MODEL FOR THE CHEMICAL EVOLUTION OF THE SOLAR NEIGHBORHOOD USING METALLICITY DEPENDENT YIELDS

LETICIA CARIGI

Centro de Investigaciones de Astronoma (CIDA), A.P. 264, Merida, Venezuela

and

GUSTAVO BRUZUAL A.[*]

Landessternwarte Heidelberg-Konigstuhl, 69117 Heidelberg, Germany

Abstract. We discuss the results from a chemical evolution model of the local galactic disk which takes into account stellar yields, lifetimes, remnants, and supernova progenitor masses which depend on the initial metallicity of the collapsing clouds. The detailed evolution of H, He, C, O, Fe, and of the heavy elements (Z) is followed dropping the instantaneous recycling approximation. Our results reproduce the majority of the observational constraints.

Key words: Chemical Evolution - Chemical Abundance - Stellar Evolution

1. Model

The stellar yields (SY) are fundamental input quantities in a chemical evolution model. Maeder (1992) and Schaller et al. (1992) have computed SY, main sequence lifetimes (τ_{MS}), mass of remnants, and mass limits for the formation of type II SN (SNII), for stars in the mass range from 1 to 120 $M_\odot$ of two different initial metallicities (Z = 0.001 and 0.020). Here we present results from an infall type chemical evolution model for the solar neighborhood computed under the following assumptions:

1) We adopt the Scalo (1986) Initial Mass Function (IMF), constant in time and defined in the mass range from 0.1 to 120 $M_\odot$.

2) The star formation rate (SFR) is proportional to the gas surface mass density, $\Psi = \nu \sigma_{gas}$, where the efficiency $\nu = 0.60$ Gyr^{-1} is constant in time.

3) We include two types of SN: (a) Type Ia (SNIa), which originate from C-deflagration in a C-O white dwarf (Greggio & Renzini 1983). Aa denotes the fraction of binary systems which become SNIa ($Aa = 0.05$). (b) Type II (SNII), which result from the runaway core collapse of massive single stars of mass between 7.5 $M_\odot$ and the lower mass limit for WR star progenitors (Maeder 1992).

4) We use the following nucleosynthesis results: (a) Stars with $m < 0.8$ $M_\odot$ do not enrich the interstellar medium (ISM). (b) For low and intermediate mass stars (LIMS), $0.8 \leq m \leq 7.5$ $M_\odot$, SY_{He} and the remnant masses are taken from Maeder (1992) and SY_C , SY_O , and SY_Z from Renzini & Voli (1981; for Z = 0.004 and 0.020, with $\alpha = 1.5$ and $\eta = 1/3$). (c) For high mass stars (HMS), 7.5

---

[*] On leave from Centro de Investigaciones de Astronoma, Merida, Venezuela

Space Science Reviews **66**: 471–475, 1994.

© 1994 *Kluwer Academic Publishers. Printed in Belgium.*

TABLE I

Model Predictions at 13 Gyr

Model	σ_{gas}	$\dot{\sigma}_a(t)$	σ_T	$\Psi/\bar{\Psi}$	RSNIa	RSNII
Best	4.59	0.50	70.0	0.43	$2.49\ 10^{-3}$	$9.34\ 10^{-3}$
C	10.03	0.32	45.0	0.71	$4.34\ 10^{-3}$	$1.58\ 10^{-2}$
Obs.	6.6 ± 2.5	0.28-0.71	46-71	0.4-2.0	$8h^2\ 10^{-3}$	$3h^2\ 10^{-2}$
ref.	RB92	M89	KG91	M92	EBM89	EBM89

$< m \leq 120$ M$_\odot$, SY and the remnant masses are taken from Maeder (1992) for the case of high mass loss rate. The Fe mass expelled by a SNII was chosen to be 0.075 M$_\odot$ (Nomoto et al. 1990). d) The chemical products of C-deflagration and remnant properties are those prescribed by model W7 of Nomoto et al. (1984). The stellar properties are interpolated linearly in Z and in mass.

5) To compare with the observed rates of gas infall $\dot{\sigma}_f(t)$ and SN (RSN), we assume that these rates are constant in space. A galactic radius equal to 15 kpc is used. The comparison is presented at a model age of 13 Gyr.

2. Results and Discussion

Tables I and II summarize the main results from our model. The RSN in Table I are expressed in units of $pc^{-2}Gyr^{-1}$. The observed values correspond to the averages of Evans et al. (1989) for a uniform disk galaxy of radius 15 kpc and $L_B = 2 \times 10^{10} L_{B\odot}$ (or $\approx 28.3 L_{B\odot} pc^{-2}$). $h = H_o/100$. For comparison we include the predictions for the *classical* model (C). In this model we use the stellar yields and other stellar properties computed for Z=Z$_\odot$: SY for HMS are taken from Maeder (1985) and Arnett (1991); and τ_{MS} from Gusten & Mezger (1983). This model is computed for the Salpeter (1955) IMF, and we assume $\nu = 0.23$, and $Aa = 0.09$. In our best model, the Scalo IMF predicts a lower number of HMS than the Salpeter IMF used in model C. Hence the former needs a faster SFR to reach the observed abundances and requires a higher gas consumption. Our best model does not succeed in reproducing $\Delta Y/\Delta Z$, although we have improved previous estimates. The opposite behavior of SY_{He} and SY_Z with the initial Z and the number of LIMS causes an increment of this ratio, although not enough to reach the observed value. The model also fails to reproduce the observed C and O abundances (see Table II).

The curvature of the predicted evolution of [Fe/H] at early times differs from observations (Figure 1a). We think that this problem, as well as the high consumption of interstellar gas, can be solved with a different SFR, which is higher toward the beginning of the evolution and lower at recent epochs. The predicted

TABLE II

Solar System Age and Abundances by Mass

Model	Age	H	He	C	O	Fe	Z	$\frac{\Delta Y}{\Delta Z}$
	Gyr	10^{-1}	10^{-1}	10^{-3}	10^{-3}	10^{-3}	10^{-2}	[a]
Our	4.6	7.110	2.705	4.877	6.166	1.187	1.843	2.5
C	5.3	7.237	2.565	3.080	9.345	1.207	1.982	0.9
Obs. [b]	4.5	7.068	2.743	3.057	9.557	1.181	1.886	3.5 [c]
$\pm\sigma$		0.177	0.165	0.282	0.771	0.027	0.160	0.7

behavior of [C/Fe] versus [Fe/H] is in agreement with observations (Figure 1b). At early epochs the predicted [C/Fe] is high, due to the different percentages of ejected elements from HMS; then, the following generations of HMS and the LIMS enrich the ISM more efficiently in C than the SNIa enrich it in Fe. In Table II: (a) $\frac{\Delta Y}{\Delta Z}$ is given at 13 Gyr. The observations are from (b) Anders & Grevesse (1989) and (c) Peimbert (1986). Due to the the large amount of O ejected by the first stellar generations, the model can reproduce [O/Fe] > 0.5 (Figure 1c). At [Fe/H] $= -0.7$, the [O/Fe] ratio shows a change of slope due to the death of the bulk of the LIMS and new HMS. Moreover the Fe abundance increases due to the large number of SNIa.

Since the detailed dependence of SY on Z is not known, we tried a $\sqrt{Z}$ approximation for HMS, which produced better agreement with the observed C and O abundances, without modifying other model results (Figures 1b and 1c, short-dashed lines).

Possible inconsistencies between the stellar models of Maeder (1992) and Renzini & Voli (1981) can affect the abundances values. We experimented with a model in which we assumed that the LIMS do not enrich the ISM with C. The result is shown in Figure 1b with long-dashed lines.

The G dwarf-Fe distribution is consistent with the observational constraints when we use an infall e-folding constant of 3.5 Gyr.

3. Conclusions

Our main conclusions can be summarized as follows:

1) The inclusion of the dependence of SY on the initial Z affects greatly the evolution of the chemical elements.

2) The predicted C and O abundance and the [C/Fe] ratios for $t > 4$ Gyr do not agree well with the observational constraints. This could be due to a non-linear dependence of the SY evolution with initial Z or/and to inconsistencies among the SY of HMS and LIMS. Therefore, it is essential to know how the SY vary

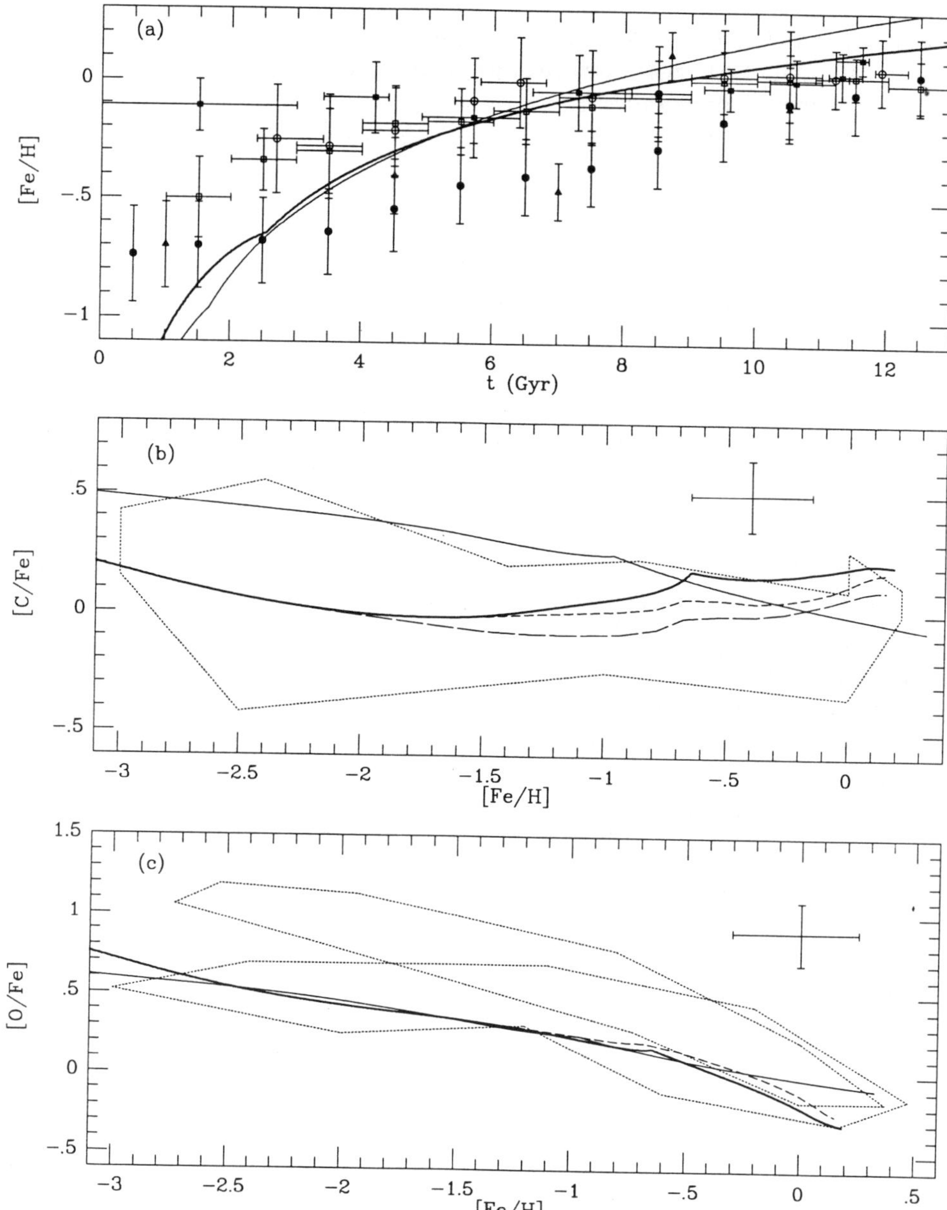

Fig. 1. (a) Predicted Age-[Fe/H] relation. Observations are taken from: open squares, Twarog (1980); filled squares, Carlberg et al. (1985); filled triangles, Nissen et al. (1985); open circles, Meusinger et al. (1991); and filled circles, Rana & Basu (1992). (b) Predicted evolution of the [C/Fe] *vs.* [Fe/H] relation. The observational data collected by Abia et al. (1991) lie in the area encircled by dotted lines. (c) Predicted evolution of the [O/Fe] *vs.* [Fe/H] relation. Polygonals enclose the areas containing corrected data by Bessell et al. (1991). The crosses show typical errors. In all frames: Thick solid line: all basic assumptions are considered. Short-dashed line: the dependence of SY on Z for HMS stars is approximated as $\sqrt{Z}$. Long-dashed line: $SY_C = 0$ for LIMS. Thin solid line: *classical* model.

with the initial stellar Z, for stars in a wide mass range.

3) We predict $\Delta Y/\Delta Z = 2.5$, improving previous estimates, but still not reaching the highest reported observational values.

4) The predicted solar system age and abundance, and the age-Fe relation are well reproduced.

References

Abia, C., Canal, R., & Isern, J.: 1991, *ApJ* **366**, 198

Anders, E., & Grevesse, N.: 1989, *Geochim. Cosmochim. Acta* **53**, 197

Arnett, W. D.: 1991, '' in D. L. Lambert, ed(s)., *Frontiers of Stellar Evolution, ASP Conf. Ser. 20*, , 389

Bessell, M. S., Sutherland, R. S., & Ruan K.: 1991, *ApJ* **383**, L71

Carlberg, R. G., Dawson, P. C., Hsu, T., & Vanderberg, D. A.: 1985, *ApJ* **294**, 674

Evans, R., van den Bergh, S., & McClude, R. D.: 1989, *ApJ* **345**, 752 (EBM89)

Greggio, L., & Renzini, A.: 1983, *A&A* **118**, 217

Gusten, R. & Mezger, P. G.: 1983, *Vistas Astron* **26**, 159

Kuijken, K., & Gilmore, G.: 1991, *ApJ* **367**, L9 (KG91)

Maeder, A.: 1985, '' in J. Audouze & N. Mathieu, ed(s)., *Nucleosynthesis and Its Implications on Nuclear and Particle Physics, NATO A SI Series, Vol. 163*, , 207

Maeder, A.: 1992, *A&A* **264**, 105

Meusinger, H., Reimann, H.-G, & Stecklum, B.: 1991, *A&A* **245**, 57 (M92)

Meusinger, H.: 1992, *A&A* **266**, 190

Mirabel, I. F.: 1989, '' in Tenorio-Tagle, G., Moles, M., & Melnick J., ed(s)., *IAU Colloq. 120, Structure and Dynamics of the Interstellar Medium*, , 396 (M89)

Nissen, P. E., Edvardsson, B., & Gustafsson, B.: 1985, '' in I. J. Danziger, F. Matteucci, & K. Kja, ed(s)., *Production and Distribution of C, N, O Elements*, , 131

Nomoto, K., Thielemann, F. K., & Yokoi, K.: 1984, *ApJ* **286**, 644

Nomoto, K., Shigeyama, T., & Tsujimoto, T.: 1990, '' in G. Michaud & A. Tutukov, ed(s)., *IAU Symp. 145, Evolution of Stars: The Photospheric Abundance Connection*, , 21

Peimbert, M.: 1986, *PASP* **98**, 1057

Rana, N. C., & Basu, S.: 1992, *A&A* **265**, 499 (RB92)

Renzini, A., & Voli, M.: 1981, *A&A* **94**, 175

Salpeter, E. E.: 1955, *ApJ* **121**, 161

Scalo, J. M.: 1986, *Fund. Cosmic Phys* **2**, 1

Schaller, G., Schaerer, D., Meynet, G. & Maeder, A.: 1992, *A&ASS* **96**, 269

Twarog, B. A.: 1980, *ApJ* **242**, 242

POSTER LIST

1. S. Heap, J. Holbrook, E. Malumuth, W. Waller: HST IMAGARY OF THE STARBURST NUCLEUS OF M83=NGC5236.

2. S. Heap, E. Malumuth: PLANETARY CAMERA OBSERVATIONS OF R136.

3. S. Heap, I. Hubeny, T. Lanz, I. Reid, J. Morland: NON-LTE ANALYSIS OF THE HOT SUBDWARF BD75325.

4. T. Stecher, UIT Team: ULTRAVIOLET IMAGING TELESCOPE OBSERVATIONS OF MASSIVE STARS IN M31, M33, THE LMC AND THE SMC.

5. R. Ruelas- Mayorga: LUMINOSITY FUNCTION OF THE STARS IN THE GALACTIC BULGE.

6. A. Nota, G. de Marchi, C. Leitherer THE POPULATIONS OF MASSIVE STARS IN R136 FROM FOC ULTRAVIOLET OBSERVATIONS.

7. P. Grison, J-P. Beaulieu: THE EROS PROJECT.

8. D. Bomans, E. Grebel: BLUE AND RED SUPERGIANTS IN NGC330: OBSERVATIONS AND THEORY.

9. E. Grebel, W. Roberts, J-M Will, K. de Boer: Be STARS IN THE MAGELLANIC CLOUD CLUSTERS.

10. A. Brown, E. de Geus, P. de Zeeuw: THE STELLAR CONTENT OF THE OBI ASSOCIATION.

11. W. Roberts, E. Grebel: YOUNG METAL DEPENDENT ISOCHRONES IN FILTER SYSTEMS.

12. C. Esteban: THE FUNDAMENTAL RAMETERS OF WR STARS SURROUNDED BY RING NEBULA M33.

13. J. Parker: A DISSECTION OF 30 DORADUS AND A DISCUSSION OF OTHER MAGELLANIC CLOUD OB ASSOCIATIONS.

14. W. Bagnuolo, D. Gies: THE PHYSICAL PROPERTIES AND EVOLU-
TION OF FOUR O-TYPE CLOSE BINARIES.

15. A. Zavagno, L. Deharveng, J. Caplan: A NEW YOUNG STELLAR
OBJECT IN THE S187 COMPLEX.

16. L. Penny, W. Bagnuolo, D. Gies: DOPPLER TOMOGRAPHY OF O-
TYPE BINARY SYSTEMS.

17. D. Gies, A. Fullerton: EXPANDING ORBITAL PARAMETER SPACE
FOR O-TYPE BINARIES: 15MON AND HD53975.

18. K. Annuk: PHOTOMETRY OF THE MULTIPLE WOLF-RAYET SYS-
TEM HD211853 (GP Cep).

19. K. Annuk: THE WOLF-RAYET STAR HD192641: A BINARY OR A
SINGLE?

20. V. Niemela, G. de Castro: CLOSE COMPANIONS OF WR STARS.

21. A. Underhill, G. Hill: HD190918: ARBITER BETWEEN MODELS FOR
WOLF-RAYET STARS.

22. U. Oberlack, R. Diel, T. Montmerle, N. Prantzos, P. von Ballmoos: IMPLI-
CATIONS OF 26Al EMISSION AT 1. 8MeV FROM THE VELA REGION.

23. P. Morris, H. Lamers, P. Conti, G. Kningsberger: DEPENDENCE OF
THE STELLAR/ATMOSPHERIC PARAMETERS AND THE CONTINUUM
SHAPES ON FREE-FREE SCATTERING IN WOLF-RAYET STARS.

24. C. Jordi, F. Figueras, J. Torra, E. Masana, R. Asiain: EFFECTIVE TEM-
PERATURES AND SURFACE GRAVITIES IN EARLY-TYPE STARS.

25. W. Rolleston: SPECTROSCOPIC ANALYSIS OF B-TYPE STARS IN
THE SMALL AND LARGE MAGELLANIC CLOUDS.

26. P. Crowther, L. Smith, D. J. Hillier: TAILORED ANALYSIS OF 23
GALACTIC WN STARS.

27. A. Neto, R. Viotty, A. Cassatella, G. Baratta: ONE CENTURY OF SPEC-
TROSCOPIC OBSERVATIONS OF ETA CARINAE: A TOOL FOR THE LBV
PHENOMENA.

28. L. Smith, P. Crowther, R. Prinja: He3-519: A NEW GALACTIC LUMINOUS BLUE VARIABLE.

29. A. van Genderen, M. de Groot, P. Th: THE EVOLUTIONARY STATUS OF ETA CARINAE BASED ON HISTORICAL AND MODERN OPTICAL AND INFRARED PHOTOMETRY.

30. R. Viotti, G. Baratta, M. Barylak, A. Cassatella, A. Neto, V. Polcaro, C. Rossi: THE NATURE OF THE LUMINOUS BLUE VARIABLE AG CARINAE.

31. K. Brownsberger, W. Vacca: INRINSIC STELLAR PARAMETERS AND THE LYMAN CONTINUUM FLUXES FOR WC STARS IN THE GALAXY AND THE LMC.

32. R. Kingsburgh, M. Barlow, P. Storey: OXYGEN AND CARBON ABUNDANCES DOR THE WO STARS.

33. R. Kingsburgh, M. Barlow: DR1: AN EXTRAGALACTIC WO STAR AND ITS SURROUNDING HII REGION.

34. K. Venn: LTE AND NLTE-ABUNDANCES IN SUPERGIANTS.

35. N. Netzer: THE GRADUAL ACCELERATION OF THE OUTFLOW OF VX-SAGITARII: A STELLAR EVOLUTIONARY EFFECT.

36. D. Schaerer, W. Schmutz: HYDRODYNAMIC ATMOSPHERE MODELS FOR HOT LUMINOUS STARS.

37. M. Runacres, R. Blomme: SMOOTH WINDS AND CLUMPED WINDS: INFLUENCE ON THE MASS-LOSS OF ZETA Pup.

38. K. Vyverman, W. Van Rensbergen, D. Vanbeveren: ON THE RELATION BETWEEN THE MASS LOSS RATES AND THE STELLAR PARAMETERS OF OB-STARS.

39. L. Auer, G. Koeningsberger: CONSTRAINTS ON THE WIND STRUCTURE IN WR STARS.

40. I. Stevens: COLLIDING STELLAR WINDS IN EARLY-TYPE BINARY SYSTEMS.

41. L. Bianchi, J. Hutchings: THE STELLAR WINDS OF MASSIVE STARS

IN M 31.

42. K. Simon, E. Sturm: NLTE ANALYSIS OF HOT STARS.

43. M. Corocan, J. Swank, S. Heap, S. Stone: CHARACTERISCTICS OF X-RAY EMISSION FROM HOT STARS.

44. E. Rey, J. Puls, A. Herrero: A NEW LINE FORMATION CODE FOR UNIFIED MODEL ATMOSPHERES.

45. U. Wessolowski, W. Hamman,I. Koesterke: NON-LTE SPECTRAL ANALYSES OF WOLF-RAYET STARS: WR2, THE ONLY GALACTIC REP-RESENTATIVE OF SUBTYPE WN2.

46. S. Haser, D. J. Lennon, R. Kudritzki, J. Puls: THE STELLAR WIND OF AN O8. 5I(f) STAR IN M31 REVISITED.

47. S. Haser, N. Walborn, R. Kudritzki, D. J. Lennon: HUBBLE SPACE TELESCOPE SPECTROSCOPY OF MASSIVE HOT STARS IN THE MAG-ELLANIC CLOUDS.

48. S. Haser, J. Puls, R. Kudritzki: O-STAR WINDS IN THE MAGELLANIC CLOUDS AND THE MILKY WAY: UV P CYGNI PROFILES AND A MOD-IFIED SEI-METHOD.

49. N. Rons: TURBULENCE AND P CYGNI PROFILE FEATURES IN THE WIND OF O-TYPE STARS.

50. A. Feldmeier, J. Puls: TIME-DEPENDENT WINDS IN O-STARS.

51. R. Schulte-Ladbeck, G. Clayton, C. Leitherer, L. Drissen, C. Robert, A. Nota, J. Parker: WIND ASYMMETRIES IN MASSIVE STARS.

52. R. Schulte-Ladbeck, D. J. Hillier, K. Nordsieck: A SPECTRPOLARIMET-RIC SURVEY OF WOLF-RAYET STARS.

53. D. Lennon, D. Wobig, R. Kudritzki, O. Stahl: THE ATMOSPHERIC COMPOSITION, EXTINCTION AND LUMINOSITY OF THE LBV R71.

54. D. Lennon, B. Seufert, R. Kudritzki, A. Herrero: SPECTROSCOPY OF MASSIVE O-TYPE STARS IN THE MILKY WAY AND LMC CLUSTERS.

55. D. Lennon, P. Mazzali, F. Pasian, P. Bonifacio, V. Castellani: A SPEC-

TROSCOPIC ANALYSIS OF B STARS IN THE SMC CLUSTER NGC330.

56. R. Steinitz, U. Goldstein: He-PECULIAR STARS.

57. T. Gang, C. Gummersbach, A. Kaufer, J. Kovacs, H. Mandel, O. Stahl, B. Wolf: SPECTRAL VARIATIONS OF HD160529.

58. O. Stahl, T. Gang, C. Gummersbach, A. Kaufer, J. Kovacs, H. Mandel, T. Szeifert, B. Wolf: PERIODIC SPECTRAL VARIATIONS OF THETA 1 ORIONIS C.

59. L. Kaper, G. Hammerschlag-Hensberge, J. van Loon: A MULTIWAVELENGTH STUDY OF STELLAR WIND STRUCTURE IN HIGH-MASS X-RAY BINARIES.

60. P. Denissenkov: DOUBLE-ZONE MODEL WITH DIFFUSE MIXING AND THE MASS AND He DISCREPANCY.

61. A. Claret, A. Gimenez: APSIDAL MOTION IN MASSIVE BINARY STARS.

62. J. Figueiredo, J. -P. De Greve, R. Hilditch: OB-TYPE BINARIES: MODELS VERSUS OBSERVATIONS.

63. R. Hilditch, G. Hill: COMPARISON IN BINARY EVOLUTION MODELS WITH THREE OB-TYPE SYSTEMS.

64. M. Shara, A. Moffat, L. Smith, M. Potter, V. Niemela: TEN MORE NEW GALACTIC WOLF-RAYET STARS.

65. A. Chiosi, A. Vallenari, F. Fagotto, G. Bertelli, A. Bressan, E. Nasi: A STUDY OF THE SMC CLUSTER NGC330.

66. L. Smith, G. Meynet, J. -C. Mermilliod: IMPROVED BOLOMETRIC CORRECTIONS OF WR STARS FROM EVOLUTIONARY MODELS.

67. D. Schaerer: UPDATED WOLF-RAYET MODELS: C/O OPACITIES.

68. E. Kemnade, W. Glatzel: THE INFLUENCE OF MASS LOSS ON STELLAR STABILITY.

69. M. Kiriakidis, M. El Eid, K. Fricke, W. Glatzel: THE STABILITY OF WOLF RAYET STARS.

70. W. Glatzel, M. Kiriakidis, K. Fricke: THE STABILITY OF MASSIVE STARS.

71. H. Braun, N. Langer: ON THE EVOLUTION OF SECONDARY COMPONENTS IN MASSIVE CLOSE BINARY SYSTEMS.

72. C. Ritossa: INFLATIONS AND DEFLATIONS OF MASSIVE STARS.

73. C. de Loore, D. Vanbeveren: EVOLUTIONARY SEQUENCES FOR CLOSE BINARY STARS FOR VARIOUS CHEMICAL COMPOSITIONS.

74. N. Mowlavi, M. Forestini, A. Jorrison: SEMI-CONVECTION AND OVERSHOOTING IN MASSIVE STARS.

75. D. Vanbeveren, A. Herrero, D. Kunze, M. van Kerkwijk: THE MASS AND He-DISCREPANCY: THE CASE OF VELA X-1.

76. A. Bressan, C. Chiosi, F. Fagotto: SPECTRO-PHOTOMETRIC EVOLUTION OF ELLIPTICAL GALAXIES.

77. M. Hafner, B. Baschek, R. Wehrse: SUPERNOVA PHOTOSPHERES: PARAMETERS AND INFORMATION CONTENT OF THEIR SPECTRA.

78. K. Cananzi, R. Augarde, J. Lequeux: A SPECTRAL SYNTHESIS OF STELLAR POPULATIONS USING BALMER LINES.

79. A. Bruzual, G. Magris: POPULATION SYNTHESIS AND EMISSION LINES FROM GALAXIES: THE UPPER END OF THE IMF.

80. A. Bruzual, L. Carigi: A MODEL FOR THE CHEMICAL EVOLUTION OF THE SOLAR NEIGHBOURHOUD USING METALLIC DEPENDENT YIELD.

81. L. Deng: THE EVOLUTION WITH TURBULENT DIFFUSION.

82. G. Meynet: WOLF- RAYET STARS IN STARBURSTS.

83. J. Smolinski, J. Climenhaga, Y. Huang, S. Jiang, M. Schmidt, O. Stahl: THE COMPLEXITY OF THE H α PROFILE NOF THE HYPERGIANT HR8752 DURING 1970-1992.

84. I. Voloskina, V. Lyuty: THE ADDITIONAL RADIATION OF Cyg X-1

AT PRIMARY MINIMUM.

85. P. Whitelok: VARIABILITY OF ETA CARINAE AND AG CARINAE.

86. M. Beech, R. Mitalas: MASSIVE STAR FORMATION: THE ACCRE-TION PARADIGM.

87. F. Fagotto, A. Bressan, G. Bertelli, C. Chiosi: NEW GRIDS OF EVOLU-TIONARY MODELS.

88. G. Bertelli, A. Bressan, C. Chiosi, F. Fagotto, E. Nasi: NEW ISOCHRONES FROM LOW (Z=0. 0004) TO HIGH (Z=0. 05) METALLICI-TY.

89. R. Voors, H. Lamers, O. Stahl: VARIATIONS IN THE $H\alpha$ PROFILE OF P Cygni.

90. S. Heap, I. Hubeny, W. Schmutz: MONITORING THE WIND OF χ Per WITH GHRS: VARIABILITY OF Si IV 1393 , 1403 AT LOW VELIOCITIES.

SUMMARY

THE EDITORS

The Meeting 'Evolution of Massive Stars' considered of course 'Structure and Evolution of Massive Stars' but focussed also on 'Structure and Spectra of the Atmospheres of Massive Stars' since the meeting was as well scheduled explicitly as 'A confrontation between Theory and Observations'.

Extended stellar atmospheres and internal structure have been treated simultaneously in a Massive Star Meeting since they have to be considered as belonging to the same celestial object. It is just one of the challenges of Modern Astrophysics to join the topics of 'Structure and Evolution of the Stars' and 'Stellar atmospheres into one coherent theory. The meeting provided a beautifull survey of recent achievements, the actual state of research on massive stars showing the ways for future research in the various fields.

In the following we give some critical remarks and a subjective summary of striking results.

1. Distribution and numbers of stars

No evolutionary model can properly be tested performing number statistics on the distribution of various types of massive stars if one has not the disposal of a 'complete' set of observations. Long term observations on small telescopes for a large number of stars accompanied by carefull reduction provides very usefull information.

In the Galaxy and in the Magellanic Clouds the number distribution of different types of stars can be obtained by direct number counts; methods are being developped in order to deduce relative star numbers (WR/O) from the spectra of distant galaxies (WR galaxies). Before a comparison between these observations and theoretical evolutionary predictions is meaningful, either from star burst models or assuming continuous star formation, we not only have to be sure about completeness of the observational data but we need an answer on the following questions regarding the direct number count data:

-why do we observe so few massive O type stars close to the theoretically predicted ZAMS. Are we dealing here with a incompleteness phenomenon or with something inherent to the formation process of massive stars?

-why do we observe so many stars outside the theoretically predicted main sequence band? Is it due to incorrect or incomplete stellar structure processes or is it due to wrong Teff determinations of B and early A supergiants. Let us remind that it is the Teff region between 35000 K and 20000 K which looks overpopulated and this region corresponds to a B-V difference which equals 0.04 only. Much more complete NLTE Teff determinations are needed for the

Space Science Reviews **66**: 485–490, 1994.
© 1994 *Kluwer Academic Publishers. Printed in Belgium.*

B supergiants in this domain before definite conclusions can be drawn. As was advocated frequently in the past by different authors and demonstrated again during this meeting, comparison of theoretical prediction with well populated and completely surveyed clusters may be much more conclusive. However it has become clear that a study of theoretically predicted star numbers and the comparison to observations is meaningless if single star evolutionary results are not combined with corresponding close binary evolution. This is the case for star burst models as well as for continous star formation models. The same holds surely for population systhesis models.

2. Observations of individual stars

What kind of observations of individual massive stars are of fundamental importance for stellar evolution?

2.1. BASIC PARAMETERS OF SINGLE STARS

It is quite obvious that the knowledge of the HRD position of a star of any type and its chemical composition is fundamental for stellar evolution. The meeting has shown the remarkable progress of NLTE photospheric analysis of O type stars and B supergiants. The inclusion of large number of lines, especially iron, and thus the inclusion of blanketing effects in NLTE computations makes the determination of stellar parameters very reliable. In stars with large stellar wind mass loss rates, the stellar wind significantly influences the photospheric analysis so that both effects have to be treated simultaneously.

A few years ago the mass and helium discrepancy in supergiants was introduced. In some cases the mass discrepancy can be removed by including sphericity, line–blanketing and the effect of stellar winds; the helium discrepancy however remains. NLTE analysis of CNO elements in a number of these suspected helium enriched stars demonstrates that indeed we are dealing here with CNO processed matter at the stellar surface. It is a task for stellar evolution to find out how CNO processed mass layers can show up at the stellar surface during OB supergiant phases.

There is now overwhelming evidence that WR stars are bare helium cores of massive stars. Although NLTE codes for WR stars are not yet as complete as NLTE codes for O type stars, abundance determinations of H, He, C, N and O elements are very much in agreement with evolutionary prediction (CNO cycle and 3α–process elements).

A great semi–empirical novelty of WR stars also discussed during this meeting is the location of the WR stars in the HR diagram, especially the location of the WN stars of the Galaxy and the LMC. Compared to standard evolution, this location implies that stars with mass = 20 $M_\odot$ (and even lower) must form WR stars, even in the LMC.

2.2. BASIC PARAMETERS OF BINARIES

About 35-40 % of all massive stars in the Galaxy have a companion close enough that during further evolution they will interact. Whether this percentage is metallicity dependent is not known.

Observational results which need our attention as far as evolution of massive binaries is concerned:

-although the number of observations are still scares, it looks as if the WR binary populations in the MC's show very similar characteristics as the Galactic population,

- the X-ray binary Cyg X-3, with a period of a few houres, seems to be a WR star with a compact companion. How such a system can be formed is still very uncertain (spiral-in process ?),

- the optical counterpart of the standard massive X–ray binary Vela X–1 is significantly helium enriched (a factor 4) and is highly overluminous (undermassive or shows a severe mass discrepancy),

- a new and carefull statistical study of a large sample of pulsars reveals that on the average the runaway velocities are much larger than previously thought, i.e. 250–300 km/s. When the observed large runaway velocities of pulsars reflect the asymmetry of the supernova ejecta, it can be concluded that the survival probability of a massive binary during its first SN explosion is very low. Although the predicted number of OB type runaways formed as a consequence of a previous binary SN explosion is not affected, we can conclude now that only a small fraction are expected to have a compact companion.

The larger disruption probability also removes the problem of the possible large number of WR+compact companions expected from theory in the past (when the survival probability was thought to be large) but not observed (especially the expected hard X–rays are not observed).

May be this larger disruption probability also explains why there are so few double pulsars observed.

2.3. STELLAR WIND MASS LOSS RATES

From evolutionary point of view we only need the value of the mass loss rate as a function of basic stellar parameters. Theoretical M_{dot} values have always been lower than empirical values (typically a factor 2). With the new hydrodynamical models including an increasing number of metal lines, this difference decreases systematically, the theoretical values approaching the observed ones. From evolutionary point of view it is thus advisable to study the influence of stellar wind mass loss by means of semi–empirical M_{dot} formulas. Such formulas can be determined with some degree of confidence for Galactic OB type stars; the dependence on metallicity can then be included based on mass loss theory. One advice: the use of sophisticated regression formulas give no better results than simple ones. On the contrary, simple ones have more physical significance.

Using current M_{dot} formalisms it can be concluded that mass loss by stellar wind during the OB phase affects stellar evolution only if $M \geq 40\ M_\odot$ in the Galaxy, $M \geq 55\ M_\odot$ in the LMC and $M \geq 80\ M_\odot$ in the SMC. Even there the effect is not very large but we leave it to the reader concluding whether in general mass loss by stellar wind during the OB phase is important or unimportant for stellar evolutionary computations.

The mass loss rates of red supergiants (RSG) are badly known although their influence on stellar evolution may be magnitudes larger than OB star mass loss. The uncertainty is much larger than the canonical factor two. A NLTE star by star study of a large sample of red supergiants may prove to be very important for massive star evolution.

The mass loss rates of WR stars: from observational point of view we have a reasonable idea of the order of magnitude of M_{dot} values but the dependence on basic stellar parameters remains a matter of faith. When the influence of this mass loss process on massive star evolution is investigated, one has the big advantage that small changes in the M_{dot} formalism during the WR phase of a massive star implies significantly different WN/WC number ratios. Observed WN/WC ratios and their dependence on metallicity are therefore of great help for evolution.

LBV's although few in number play a very important role for the evolution of the very massive stars (VMS), i.e. for initial masses larger than 40-50 $M_\odot$. The mass loss rates observed in these stars are large enough in order to prohibit a star to evolve to the red part of the HR diagram. Binaries with period large enough that the primary first encounters an LBV instability phase before it fills its critical Roche lobe, will avoid a classical mass transfer phase (the LBV scenario for very massive close binaries ,VMCB).

3. Massive star evolution

The opacities needed in order to calculate the stellar interior are in good shape (OPAL opacities) however there are still many basic uncertainties in present massive star evolutionary computations:

- the extend of the convective core (overshooting); by comparing evolutionary predictions with a variety of observations, one gets the impression that overshooting in massive stars should be small. However physics (of course also within assumptions and limitations) predicts large overshooting. It seems save to conclude that physics predicts a maximum value whereas the real value must be obtained by extensive comparison with all observations.

- the role of semi–convection and how to treat this semi–convection; the way semi–convection is treated is essential for single star evolution (the RSG phase) and critically affects the predicted B/R supergiant number ratio. However this number ratio is also critically dependent on the adopted RSG mass loss rates. Since they are still poorly known, it is still premature to promote any kind of treatment based on the B/R number ratio only.

It is interesting to notice that semi–convection or the way it is treated is unimportant for the evolution of primaries in close binaries. Also for the majority of secondaries it is unimportant. It only slightly modifies the gainers evolution if the mass ratio of the initial binary was close to one and only a small amount of mass lost by the loser during Roche lobe overflow is accreted by the gainer.

- the role of rotation and its influence on internal mixing.

- does the star succeed in keeping its outer layers in radiative equilibrium during rapid accretion in binaries or is the star homogenized as a consequence of rotationally induced mixing? The latter assumption has been used succesfully (the results were presented during the meeting) in order to explain the overluminosity (mass discrepancy) and helium enrichment of the optical star of the X–ray binary Vela X–1.

- Present single star evolution is unable to explain the observed HRD location of WN stars, especially LMC models do not fit. The reason may be the very large uncertainty in the mass loss rates of red supergiants.

- Evolution provides a model for any type of supernova explosion. The progenitor of SN 1993J must have been a binary component but SN 1987A is still a controversial case. Single star evolution is able to explain the blue progenitor but with extreme fine tuning of a variety of parameters. Binary evolution however and in particular the behaviour of the mass gainer in a binary with initial mass ratio very close to one gives a natural explanation for the blue progenitor.

The meeting collected 127 people in Brussels, the capital of Europe, and stored up the world's knowledge about one fabulous topic: massive stars. Scientists from all over the world being specialists in a wide range of scientific disciplines contributed jointly to a better understanding of massive stars. It was a very good feeling to see the appreciation of mutual complementarity between theoreticians and observers; between experts in stellar atmospheres and stellar interiors; between predictors of single star and binary star evolution;...

Fundamental knowledge of structure and evolution of massive stars cannot grow steadily without the integrated and tuned contributions of many.

The meeting showed that insight and knowledge has made great progress lately in many disciplines connected with Massive Star Research. Some progress can simply be called decisive. So, the spectrum of an early type star can now be produced by a uniquely defined atmosphere on top of star being defined by the same fundamental parameters. Of course such —although decisive— progress cannot be called final since the last unexplained spectral detail might hide a fundamental ignorance.

The actual state of the art of other aspects was very clearly pictured at the meeting. Even so that the hope for some of them ever making a final progress has to be evaluated differently. In earthly situations the dynamics of clouds and layers of air certainly obey Newtonian mechanics but their local behaviour is nevertheless o so difficult to predict. A needful prediction has thus to, use more than classical mechanics alone in order to be adequately described and precisely parametrised. Also for Early Type Stars the importance of many hydrodynamical effects has clearly been understood and demonstrated mechanically but will be very hard (never?) calculated precisely without adjustment to observational data. Nobody has to be ashamed to have to appeal on good old HR diagram and the modern techniques to determine positions precisely in it to decide to what extend the qualitatively predicted hydrodynamical events really will take place

All researchers on Massive Stars may be really happy to work on a very alive subject. Everybody in this community will certainly be satisfied with her/his own achievements; but even more curious about the progress made by her/his colleagues in various fields of Massive Star research. It will be a joy for many of those who have left Brussels soon after the meeting to see each other again in a next meeting after which they will tell: now we know better how to work further.